LONDON MATHEMATICAL SOCIETY LECTURE NOTE SERI

Managing Editor: Professor Endre Süli, Mathematical Institute, University of (
Woodstock Road, Oxford OX2 6GG, United Kingdom

The titles below are available from booksellers, or from Cambridge University I
www.cambridge.org/mathematics

386 Independence-friendly logic, A.L. MANN, G. SANDU & M. SEVENSTER
387 Groups St Andrews 2009 in Bath I, C.M. CAMPBELL *et al* (eds)
388 Groups St Andrews 2009 in Bath II, C.M. CAMPBELL *et al* (eds)
389 Random fields on the sphere, D. MARINUCCI & G. PECCATI
390 Localization in periodic potentials, D.E. PELINOVSKY
391 Fusion systems in algebra and topology, M. ASCHBACHER, R. KESSAR & B. OLIVER
392 Surveys in combinatorics 2011, R. CHAPMAN (ed)
393 Non-abelian fundamental groups and Iwasawa theory, J. COATES *et al* (eds)
394 Variational problems in differential geometry, R. BIELAWSKI, K. HOUSTON & M. SPEIGHT (eds)
395 How groups grow, A. MANN
396 Arithmetic differential operators over the p-adic integers, C.C. RALPH & S.R. SIMANCA
397 Hyperbolic geometry and applications in quantum chaos and cosmology, J. BOLTE & F. STEINER (eds)
398 Mathematical models in contact mechanics, M. SOFONEA & A. MATEI
399 Circuit double cover of graphs, C.-Q. ZHANG
400 Dense sphere packings: a blueprint for formal proofs, T. HALES
401 A double Hall algebra approach to affine quantum Schur–Weyl theory, B. DENG, J. DU & Q. FU
402 Mathematical aspects of fluid mechanics, J.C. ROBINSON, J.L. RODRIGO & W. SADOWSKI (eds)
403 Foundations of computational mathematics, Budapest 2011, F. CUCKER, T. KRICK, A. PINKUS & A. SZANTO (eds)
404 Operator methods for boundary value problems, S. HASSI, H.S.V. DE SNOO & F.H. SZAFRANIEC (eds)
405 Torsors, étale homotopy and applications to rational points, A.N. SKOROBOGATOV (ed)
406 Appalachian set theory, J. CUMMINGS & E. SCHIMMERLING (eds)
407 The maximal subgroups of the low-dimensional finite classical groups, J.N. BRAY, D.F. HOLT & C.M. RONEY-DOUGAL
408 Complexity science: the Warwick master's course, R. BALL, V. KOLOKOLTSOV & R.S. MACKAY (eds)
409 Surveys in combinatorics 2013, S.R. BLACKBURN, S. GERKE & M. WILDON (eds)
410 Representation theory and harmonic analysis of wreath products of finite groups, T. CECCHERINI-SILBERSTEIN, F. SCARABOTTI & F. TOLLI
411 Moduli spaces, L. BRAMBILA-PAZ, O. GARCÍA-PRADA, P. NEWSTEAD & R.P. THOMAS (eds)
412 Automorphisms and equivalence relations in topological dynamics, D.B. ELLIS & R. ELLIS
413 Optimal transportation, Y. OLLIVIER, H. PAJOT & C. VILLANI (eds)
414 Automorphic forms and Galois representations I, F. DIAMOND, P.L. KASSAEI & M. KIM (eds)
415 Automorphic forms and Galois representations II, F. DIAMOND, P.L. KASSAEI & M. KIM (eds)
416 Reversibility in dynamics and group theory, A.G. O'FARRELL & I. SHORT
417 Recent advances in algebraic geometry, C.D. HACON, M. MUSTAŢĂ & M. POPA (eds)
418 The Bloch–Kato conjecture for the Riemann zeta function, J. COATES, A. RAGHURAM, A. SAIKIA & R. SUJATHA (eds)
419 The Cauchy problem for non-Lipschitz semi-linear parabolic partial differential equations, J.C. MEYER & D.J. NEEDHAM
420 Arithmetic and geometry, L. DIEULEFAIT *et al* (eds)
421 O-minimality and Diophantine geometry, G.O. JONES & A.J. WILKIE (eds)
422 Groups St Andrews 2013, C.M. CAMPBELL *et al* (eds)
423 Inequalities for graph eigenvalues, Z. STANIĆ
424 Surveys in combinatorics 2015, A. CZUMAJ *et al* (eds)
425 Geometry, topology and dynamics in negative curvature, C.S. ARAVINDA, F.T. FARRELL & J.-F. LAFONT (eds)
426 Lectures on the theory of water waves, T. BRIDGES, M. GROVES & D. NICHOLLS (eds)
427 Recent advances in Hodge theory, M. KERR & G. PEARLSTEIN (eds)
428 Geometry in a Fréchet context, C.T.J. DODSON, G. GALANIS & E. VASSILIOU
429 Sheaves and functions modulo p, L. TAELMAN
430 Recent progress in the theory of the Euler and Navier–Stokes equations, J.C. ROBINSON, J.L. RODRIGO, W. SADOWSKI & A. VIDAL-LÓPEZ (eds)
431 Harmonic and subharmonic function theory on the real hyperbolic ball, M. STOLL
432 Topics in graph automorphisms and reconstruction (2nd Edition), J. LAURI & R. SCAPELLATO
433 Regular and irregular holonomic D-modules, M. KASHIWARA & P. SCHAPIRA
434 Analytic semigroups and semilinear initial boundary value problems (2nd Edition), K. TAIRA
435 Graded rings and graded Grothendieck groups, R. HAZRAT

436 Groups, graphs and random walks, T. CECCHERINI-SILBERSTEIN, M. SALVATORI & E. SAVA-HUSS (eds)
437 Dynamics and analytic number theory, D. BADZIAHIN, A. GORODNIK & N. PEYERIMHOFF (eds)
438 Random walks and heat kernels on graphs, M.T. BARLOW
439 Evolution equations, K. AMMARI & S. GERBI (eds)
440 Surveys in combinatorics 2017, A. CLAESSON *et al* (eds)
441 Polynomials and the mod 2 Steenrod algebra I, G. WALKER & R.M.W. WOOD
442 Polynomials and the mod 2 Steenrod algebra II, G. WALKER & R.M.W. WOOD
443 Asymptotic analysis in general relativity, T. DAUDÉ, D. HÄFNER & J.-P. NICOLAS (eds)
444 Geometric and cohomological group theory, P.H. KROPHOLLER, I.J. LEARY, C. MARTÍNEZ-PÉREZ & B.E.A. NUCINKIS (eds)
445 Introduction to hidden semi-Markov models, J. VAN DER HOEK & R.J. ELLIOTT
446 Advances in two-dimensional homotopy and combinatorial group theory, W. METZLER & S. ROSEBROCK (eds)
447 New directions in locally compact groups, P.-E. CAPRACE & N. MONOD (eds)
448 Synthetic differential topology, M.C. BUNGE, F. GAGO & A.M. SAN LUIS
449 Permutation groups and cartesian decompositions, C.E. PRAEGER & C. SCHNEIDER
450 Partial differential equations arising from physics and geometry, M. BEN AYED *et al* (eds)
451 Topological methods in group theory, N. BROADDUS, M. DAVIS, J.-F. LAFONT & I. ORTIZ (eds)
452 Partial differential equations in fluid mechanics, C.L. FEFFERMAN, J.C. ROBINSON & J.L. RODRIGO (eds)
453 Stochastic stability of differential equations in abstract spaces, K. LIU
454 Beyond hyperbolicity, M. HAGEN, R. WEBB & H. WILTON (eds)
455 Groups St Andrews 2017 in Birmingham, C.M. CAMPBELL *et al* (eds)
456 Surveys in combinatorics 2019, A. LO, R. MYCROFT, G. PERARNAU & A. TREGLOWN (eds)
457 Shimura varieties, T. HAINES & M. HARRIS (eds)
458 Integrable systems and algebraic geometry I, R. DONAGI & T. SHASKA (eds)
459 Integrable systems and algebraic geometry II, R. DONAGI & T. SHASKA (eds)
460 Wigner-type theorems for Hilbert Grassmannians, M. PANKOV
461 Analysis and geometry on graphs and manifolds, M. KELLER, D. LENZ & R.K. WOJCIECHOWSKI
462 Zeta and L-functions of varieties and motives, B. KAHN
463 Differential geometry in the large, O. DEARRICOTT *et al* (eds)
464 Lectures on orthogonal polynomials and special functions, H.S. COHL & M.E.H. ISMAIL (eds)
465 Constrained Willmore surfaces, Á.C. QUINTINO
466 Invariance of modules under automorphisms of their envelopes and covers, A.K. SRIVASTAVA, A. TUGANBAEV & P.A. GUIL ASENSIO
467 The genesis of the Langlands program, J. MUELLER & F. SHAHIDI
468 (Co)end calculus, F. LOREGIAN
469 Computational cryptography, J.W. BOS & M. STAM (eds)
470 Surveys in combinatorics 2021, K.K. DABROWSKI *et al* (eds)
471 Matrix analysis and entrywise positivity preservers, A. KHARE
472 Facets of algebraic geometry I, P. ALUFFI *et al* (eds)
473 Facets of algebraic geometry II, P. ALUFFI *et al* (eds)
474 Equivariant topology and derived algebra, S. BALCHIN, D. BARNES, M. KEDZIOREK & M. SZYMIK (eds)
475 Effective results and methods for Diophantine equations over finitely generated domains, J.-H. EVERTSE & K. GYORY
476 An indefinite excursion in operator theory, A. GHEONDEA
477 Elliptic regularity theory by approximation methods, E.A. PIMENTEL
478 Recent developments in algebraic geometry, H. ABBAN, G. BROWN, A. KASPRZYK & S. MORI (eds)
479 Bounded cohomology and simplicial volume, C. CAMPAGNOLO, F. FOURNIER- FACIO, N. HEUER & M. MORASCHINI (eds)
480 Stacks Project Expository Collection (SPEC), P. BELMANS, W. HO & A.J. DE JONG (eds)
481 Surveys in combinatorics 2022, A. NIXON & S. PRENDIVILLE (eds)
482 The logical approach to automatic sequences, J. SHALLIT
483 Rectifiability: a survey, P. MATTILA
484 Discrete quantum walks on graphs and digraphs, C. GODSIL & H. ZHAN
485 The Calabi problem for Fano threefolds, C. ARAUJO *et al*
486 Modern trends in algebra and representation theory, D. JORDAN, N. MAZZA & S. SCHROLL (eds)
487 Algebraic combinatorics and the Monster group, A.A. IVANOV (ed)
488 Maurer–Cartan methods in deformation theory, V. DOTSENKO, S. SHADRIN & B. VALLETTE
489 Higher dimensional algebraic geometry, C. HACON & C. XU (eds)
490 C^∞-algebraic geometry with corners, K. FRANCIS-STAITE & D. JOYCE
491 Groups and graphs, designs and dynamics, R.A. BAILEY, P. J. CAMERON & Y. WU (eds)
492 Homotopy theory of enriched Mackey functors, N. JOHNSON & D. YAU
493 Surveys in combinatorics 2024, F. FISCHER & R. JOHNSON (eds)
494 K-theory and representation theory, R. PLYMEN & M.H. ŞENGÜN (eds)
495 Polygraphs: from rewriting to higher categories, D. ARA *et al*

London Mathematical Society Lecture Note Series: 496

Groups St Andrews 2022 in Newcastle

Edited by

C. M. CAMPBELL
University of St Andrews

M. R. QUICK
University of St Andrews

E. F. ROBERTSON
University of St Andrews

C. M. RONEY-DOUGAL
University of St Andrews

D. I. STEWART
University of Manchester

Shaftesbury Road, Cambridge CB2 8EA, United Kingdom

One Liberty Plaza, 20th Floor, New York, NY 10006, USA

477 Williamstown Road, Port Melbourne, VIC 3207, Australia

314–321, 3rd Floor, Plot 3, Splendor Forum, Jasola District Centre, New Delhi – 110025, India

103 Penang Road, #05–06/07, Visioncrest Commercial, Singapore 238467

Cambridge University Press is part of Cambridge University Press & Assessment, a department of the University of Cambridge.

We share the University's mission to contribute to society through the pursuit of education, learning and research at the highest international levels of excellence.

www.cambridge.org
Information on this title: www.cambridge.org/9781009563222

DOI: 10.1017/9781009563208

When citing this work, please include a reference to the DOI 10.1017/9781009563208

First published 2025

Printed in the United Kingdom by TJ Books Limited, Padstow Cornwall

A catalogue record for this publication is available from the British Library

A Cataloging-in-Publication data record for this book is available from the Library of Congress

ISBN 978-1-009-56322-2 Paperback

Contents

	Introduction	*page* vii
1	Finite group schemes *Michel Brion*	1
2	Algorithms for polycyclic groups *Bettina Eick*	53
3	The spread of finite and infinite groups *Scott Harper*	74
4	Discrete subgroups of semisimple Lie groups, beyond lattices *Fanny Kassel*	118
5	Complete reducibility and subgroups of exceptional algebraic groups *Alastair J. Litterick, David I. Stewart and Adam R. Thomas*	191
6	Axial algebras of Jordan and Monster type *Justin McInroy and Sergey Shpectorov*	246
7	An introduction to the local-to-global behaviour of groups acting on trees and the theory of local action diagrams *Colin D. Reid and Simon M. Smith*	295
8	Finite groups and the class-size prime graph revisited *Víctor Sotomayor*	341
9	Character bounds for finite simple groups and applications *Pham Huu Tiep*	360
10	Generalized Baumslag-Solitar groups: a topological approach *Mathew Timm*	407

Introduction

Groups St Andrews 2022 was the eleventh conference of the Groups St Andrews series. It was originally planned as "Groups St Andrews 2021" and scheduled for August 2021 at Newcastle University in the United Kingdom. In common with everyone on the planet, the organizers had to revise their plans due to the global COVID pandemic and so the conference actually took place in 2022. By this point, some semblance of normality had returned and academics were able to travel to attend conferences again and sit in lecture theatres together learning mathematics. Even so, a number of colleagues unfortunately still dropped out at short notice due to contracting COVID just before they intended to travel to Newcastle. Despite these disruptions, there were 70 mathematicians able to attend the meeting. The members of the Organising Committee of Groups St Andrews 2022 were Colin Campbell, Martyn Quick, Edmund Robertson and Colva Roney-Dougal (all from St Andrews) and David Stewart (Newcastle).

The academic business of the conference ran for seven days from Sunday 31st July to Saturday 6th August. Four main speakers delivered four talks each, surveying areas of contemporary development in group theory and related areas: Michel Brion (Institut Fourier, Université Grenoble Alpes), Fanny Kassel (Institut des Hautes Études Scientifiques), Denis Osin (Vanderbilt University) and Pham Huu Tiep (Rutgers University). There were also invited speakers delivering one-hour plenary talks: Miklos Abert (Alfréd Rényi Institute of Mathematics), Bettina Eick (Technische Universität Braunschweig), Scott Harper (University of Bristol), Julia Pevtsova (University of Washington) and Simon Smith (University of Lincoln). In addition, there were 40 contributed short talks, from which Alexander Hulpke (Colorado State University) and Péter Pál Pálfy

(Alfréd Rényi Institute of Mathematics) were invited to give one-hour versions.

A lively social programme ran through the conference. A wine reception was held on the Monday evening shortly after the final lecture of the day and the conference dinner took place on the Tuesday evening. There was no academic business on the afternoon of Wednesday 3rd August and many delegates took advantage of the conference excursion to Hadrian's Wall. They had the opportunity to walk along the Wall and to explore some of the Roman Forts along it. As with previous Groups St Andrews conferences, **The Daily Group Theorist** ran throughout, publishing each day's activities together with additional contributions from the delegates. We thank the various editors of this fine, and by now traditional, publication.

The Organisers are grateful to a number of organisations whose support ensured the success of the conference. Groups St Andrews 2022 was organised in partnership with the Clay Mathematics Institute. It was supported by the Heilbronn Institute for Mathematical Research (HIMR), the UKRI/EPSRC Additional Funding Programme for Mathematical Sciences, and by the London Mathematical Society and the Edinburgh Mathematical Society. As well as supporting the expenses of the invited speakers, the grants obtained were used to support postgraduate research students and participants from Scheme 5 countries.

These proceedings contain substantial surveys written by some of the invited speakers together with some other papers, also of a survey nature, contributed by delegates.

Finally, the St Andrews-based members of the Organising Committee would like to thank the local organiser, David Stewart, for his herculean efforts to ensure that the conference ran smoothly. As the first paragraph of this introduction indicates, the COVID pandemic required us to be flexible and make considerable adjustments from the original plans. Without David's efforts, this would not have been possible.

Colin M. Campbell
Martyn R. Quick
Edmund F. Robertson
Colva M. Roney-Dougal
David I. Stewart

1
Finite group schemes

Michel Brion[a]

Abstract

These extended notes give an introduction to the theory of finite group schemes over an algebraically closed field, with minimal prerequisites. They conclude with a brief survey of the inverse Galois problem for automorphism group schemes.

1.1 Introduction

Finite group schemes are broad generalizations of finite groups; they occur in algebraic geometry, number theory, and the structure and representations of algebraic groups in positive characteristics. Unlike finite groups which exist on their own, finite group schemes depend on an additional data: a base, for example a commutative ring.

This text is an introduction to finite group schemes over an algebraically closed field. In characteristic 0, these may be identified with finite groups, as follows from Cartier's theorem (see Theorem 1.4.13 for a direct proof). But these form a much wider class in characteristic $p > 0$, as it includes the finite-dimensional restricted Lie algebras (also called p-Lie algebras). In fact, such Lie algebras form the building blocks of finite group schemes, together with finite groups; see Corollary 1.5.14 for a precise statement.

Many notions and results of group theory extend to the setting of finite group schemes, sometimes with more involved proofs; for example,

[a] Université Grenoble Alpes, Institut Fourier, 100 rue des Mathématiques, 38610 Gières, France.
Michel.Brion@univ-grenoble-alpes.fr

Lagrange's theorem, which requires substantial developments on quotients (see Corollary 1.5.13). Still, the topic leaves much room for developments, e.g., the notion of conjugacy class is unsettled (several approaches are discussed in the appendix of the recent preprint [17]).

The theory of finite group schemes over a field k is often presented as part of that of algebraic groups (in the sense of group schemes of finite type), see [7, 8, 25]. This yields a broader view of the topic and many natural examples, but also requires quite a few results from commutative algebra and algebraic geometry.

This text aims at presenting some fundamental structure results for finite group schemes, with minimal prerequisites: basic notions of algebra and familiarity with linear algebra. For this, we deal mainly with finite schemes (rather than algebraic schemes). These can be viewed in three ways:

- algebraically, via finite-dimensional algebras (i.e., k-algebras of finite dimension as k-vector spaces),
- geometrically, via finite sets equipped with a finite-dimensional local algebra at each point,
- functorially, via points with values in finite-dimensional algebras.

We will start with the first viewpoint, where finite group schemes are identified with finite-dimensional Hopf algebras, and mainly work with the second and third ones.

The structure of this text is as follows. Section 1.2 begins with three motivating examples which will be reconsidered at later stages. We then describe a classical correspondence between finite sets and their rings of k-valued functions, where k is an algebraically closed field. These rings are exactly the reduced finite-dimensional (commutative, associative) k-algebras. Next, we define finite schemes via finite-dimensional algebras, and obtain structure results for these; in particular, Theorem 1.2.13. We then turn to the functor of points, which yields simple formulations of basic operations such as the sum and product of finite schemes. This section ends with a brief overview of notions and results on more general schemes.

In Section 1.3, we introduce finite group schemes, and generalize basic notions of group theory to this setting: (normal) subgroups, group actions, semi-direct products. Then we define infinitesimal group schemes (also known as connected, or local), and obtain a first structure result: every finite group scheme is the semi-direct product of an infinitesimal group scheme and a finite group (Theorem 1.3.13).

Section 1.4 develops Lie algebra methods for studying infinitesimal group schemes; these present some analogies with connected Lie groups. We begin with the Lie algebra of derivations of an algebra; in characteristic $p > 0$, this is a restricted Lie algebra via the pth power of derivations. We then give overviews of restricted Lie algebras, and infinitesimal calculus on affine schemes. Next, we introduce the Lie algebra of an affine group scheme, and use it to show that finite group schemes are reduced in characteristic 0 (Theorem 1.4.13). Returning to positive characteristics, we define Frobenius kernels, present a structure result for these (Theorem 1.4.23), and some applications, e.g. to finite group schemes of prime order.

Section 1.5 deals with quotients of affine schemes by actions of finite group schemes. The intuitive notion of quotient as an orbit space does not extend readily to this setting, e.g. for infinitesimal group schemes as they have a unique k-point. A substitute is the categorical quotient, for which we obtain a key finiteness property (Theorem 1.5.4). Next, we discuss quotients by free actions and applications to the structure of finite group schemes (Corollaries 1.5.13 and 1.5.14).

The final Section 1.6 is a brief survey of some recent developments on automorphism group schemes in projective algebraic geometry. It focuses on a version of the inverse Galois problem in this setting, which asks whether a given group scheme can be realized as the full automorphism group scheme of a projective variety. The answer is positive for finite groups by a classical result (see [13, 20, 19]), but negative for many abelian varieties as recently shown in [18, 1] (see Theorem 1.6.7 for a precise statement). Also, the answer is positive in the setting of connected algebraic groups (in particular, infinitesimal group schemes) and connected automorphism group schemes; see Theorem 1.6.5, based on [4].

The exposition is essentially self-contained in Sections 1.2 and 1.3, which consider almost exclusively finite (group) schemes. Sections 1.4 and 1.5 also deal with affine (group) schemes, and rely on a few results for which we could find no direct proof; most notably, basic properties of quotients by free actions (Theorem 1.5.12). In these sections, we also use some fundamental results of commutative algebra, for which an excellent reference is [11]. Section 1.6 is more advanced, and involves notions and results of algebraic geometry which can be found in [15].

This text presents only the first steps in the theory of finite group schemes. Here are some suggestions for further reading: [26, Chap. III] for more on this topic, [37] for affine group schemes, [25] for algebraic groups (both over an arbitrary field), [29] for finite commutative group schemes over a perfect field, and [34] over an arbitrary base.

Notation and conventions. We fix an algebraically closed field k of characteristic $p \geq 0$. By an **algebra**, we mean a commutative associative k-algebra A with identity element, unless otherwise mentioned. The **dimension** of A is its dimension as a k-vector space. Given $a_1, \ldots, a_m \in A$, we denote by $(a_1, \ldots, a_m)$ the ideal of A that they generate. The polynomial algebra in n indeterminates over k is denoted by $k[T_1, \ldots, T_n]$.

1.2 Finite schemes

1.2.1 Motivating examples

Example 1.2.1 Let n be a positive integer and consider the nth power map

$$k^* \longrightarrow k^*, \quad x \longmapsto x^n.$$

This is a group homomorphism with kernel the group $\mu_n(k)$ of nth roots of unity in k. If $p = 0$ or n is prime to p, then $\mu_n(k)$ is a cyclic group of order n. Also, if $p > 0$ then $\mu_p(k)$ is trivial, since $x^p - 1 = (x-1)^p$. This still holds when k is replaced with any field extension. But if k is replaced with an algebra R having nonzero nilpotent elements, then the group of pth roots of unity $\mu_p(R)$ is nontrivial.

For any algebra R, we may view $\mu_p(R)$ as the set of algebra homomorphisms $f : A \to R$, where $A = k[T]/(T^p - 1)$. Indeed, such a homomorphism f is uniquely determined by $f(t)$, where t denotes the image of T in A.

More generally, we have for any n and any algebra R

$$\mu_n(R) = \mathrm{Hom}_{\mathrm{alg}}(k[T]/(T^n - 1), R),$$

where the right-hand side denotes the set of algebra homomorphisms. This suggests a way to encode the nth roots of unity by the algebra $k[T]/(T^n - 1)$, of dimension n (regardless of the characteristic).

Example 1.2.2 Assume that $p > 0$ and consider the pth power map, also called the Frobenius map

$$F : k \longrightarrow k, \quad x \longmapsto x^p.$$

This is a ring homomorphism with trivial kernel. But again, if k is replaced with an algebra R having nonzero nilpotents, then the pth power map has a nontrivial kernel,

$$\alpha_p(R) = \{x \in R \mid x^p = 0\} = \mathrm{Hom}_{\mathrm{alg}}(k[T]/(T^p), R).$$

This kernel is encoded by the p-dimensional algebra $k[T]/(T^p)$, equipped with additional structures which will be introduced in §1.3.1.

In the next, more advanced example, we will freely use some results on elliptic curves which can be found in [32].

Example 1.2.3 Let E be an elliptic curve with origin 0. Then E is a commutative group with neutral element 0. Thus, for any positive integer n, we have the multiplication map

$$n_E : E \longrightarrow E, \quad x \longmapsto nx.$$

If $k = \mathbb{C}$ then $E \simeq \mathbb{C}/\Lambda$ as a group, where Λ is a lattice in $\mathbb{C}$; as a consequence, $\Lambda \simeq \mathbb{Z}^2$ as a group. Thus, the kernel of n_E (the n-torsion subgroup of E) satisfies

$$\mathrm{Ker}(n_E) \simeq \left(\frac{1}{n}\Lambda\right)/\Lambda \simeq \Lambda/n\Lambda \simeq (\mathbb{Z}/n\mathbb{Z})^2.$$

In particular, $\mathrm{Ker}(n_E)$ has order n^2.

This still holds over an arbitrary (algebraically closed) field k of characteristic p, if $p = 0$ or if n is prime to p. Also, the endomorphism n_E of E has degree n^2 for any $n > 0$. But the structure of its kernel depends on the curve E if $p > 0$ divides n. For instance, $\mathrm{Ker}(p_E)$ has order p if E is ordinary, and is trivial if E is supersingular. The supersingular elliptic curves form only finitely many isomorphism classes.

To get a more uniform description of n-torsion subgroups, one considers the schematic kernel $E[n]$. This is a finite group scheme of order n^2 regardless of the characteristic, as we will see in Remark 1.5.11.

1.2.2 Algebras of functions on finite sets

Given a finite set E, we denote by $\mathcal{O}(E)$ the set of maps $f : E \to k$. Then $\mathcal{O}(E)$ is an algebra for the operations of pointwise addition and multiplication; we have an isomorphism of algebras $\mathcal{O}(E) \simeq k^n$, where $n = |E|$. We will investigate the assignment $E \mapsto \mathcal{O}(E)$ in a series of observations and lemmas.

For any $x \in E$, we denote by

$$\mathrm{ev}_x : \mathcal{O}(E) \longrightarrow k, \quad f \longmapsto f(x)$$

the evaluation at x. Then ev_x is an algebra homomorphism, and hence its kernel $\mathfrak{m}_x$ is a maximal ideal of $\mathcal{O}(E)$. Also, we define $\delta_x \in \mathcal{O}(E)$ by

$$\delta_x(y) = \begin{cases} 1 & \text{if } y = x \\ 0 & \text{otherwise.} \end{cases}$$

Then $(\delta_x)_{x \in E}$ is a basis of the k-vector space $\mathcal{O}(E)$, which satisfies

$$\delta_x^2 = \delta_x \quad (x \in E), \qquad \delta_x \delta_y = 0 \quad (x, y \in E, y \neq x), \qquad \sum_{x \in E} \delta_x = 1.$$

The idempotents of the ring $\mathcal{O}(E)$ (i.e. those $f \in \mathcal{O}(E)$ such that $f^2 = f$) are exactly the partial sums $\delta_F = \sum_{x \in F} \delta_x$, where $F \subset E$.

Lemma 1.2.4 *Every algebra homomorphism $u : \mathcal{O}(E) \to k$ is of the form ev_x for a unique $x \in E$.*

Proof Since $\sum_{x \in E} \delta_x = 1$, there exists $x \in E$ such that $u(\delta_x) \neq 0$. Then $u(\delta_x) = 1$ as $\delta_x^2 = \delta_x$. Let $y \in E \setminus \{x\}$, then $\delta_x \delta_y = 0$ and hence $u(\delta_y) = 0$. Thus, $u = \mathrm{ev}_x$. □

Next, consider another finite set F. Then every map $\varphi : E \to F$ yields a map

$$\varphi^* : \mathcal{O}(F) \longrightarrow \mathcal{O}(E), \quad g \longmapsto g \circ \varphi$$

which is clearly an algebra homomorphism.

Lemma 1.2.5 *Every algebra homomorphism $u : \mathcal{O}(F) \to \mathcal{O}(E)$ is of the form φ^* for a unique $\varphi : E \to F$.*

Proof Let $x \in E$, then the composition $\mathrm{ev}_x \circ u : \mathcal{O}(F) \to k$ is an algebra homomorphism. By Lemma 1.2.4, there exists a unique $y \in F$ such that $\mathrm{ev}_x \circ u = \mathrm{ev}_y$, that is, $u(g)(x) = g(y)$ for all $g \in \mathcal{O}(F)$. So the statement holds for the map $\varphi : E \to F$, $x \mapsto y$ and for no other map. □

We now consider the product $E \times F$ with projections $\mathrm{pr}_E : E \times F \to E$, $\mathrm{pr}_F : E \times F \to F$. Then one may readily check that the map

$$\mathrm{pr}_E^* \otimes \mathrm{pr}_F^* : \mathcal{O}(E) \otimes \mathcal{O}(F) \longrightarrow \mathcal{O}(E \times F), \delta_x \otimes \delta_y \longmapsto \delta_{(x,y)} \tag{1.1}$$

is an isomorphism of algebras. Likewise, consider the sum $E \sqcup F$ with inclusion maps $i_E : E \to E \sqcup F$, $i_F : F \to E \sqcup F$, then the map

$$(i_E^*, i_F^*) : \mathcal{O}(E \sqcup F) \longrightarrow \mathcal{O}(E) \times \mathcal{O}(F) \tag{1.2}$$

is an isomorphism of algebras.

Remark 1.2.6 Let A be an algebra, and $f : A \to k$ an algebra homomorphism. Then the kernel $\mathfrak{m}$ of f is a maximal ideal, and $A = k \oplus \mathfrak{m}$ where k is the line spanned by the identity element; this identifies f with the projection $A \to k$. In particular, f is uniquely determined by $\mathfrak{m}$.

If A is finite-dimensional (as a k-vector space), then every maximal ideal $\mathfrak{m}$ is the kernel of a unique algebra homomorphism to k. Indeed, the quotient $A/\mathfrak{m}$ is a field extension of k of finite degree, and hence equals k as the latter is algebraically closed. This yields a bijection between algebra homomorphisms from A to k and maximal ideals of A.

Clearly, every algebra $\mathcal{O}(E)$ is **reduced**, i.e., it has no nonzero nilpotent element. We will now obtain a converse:

Lemma 1.2.7 *Let A be a reduced finite-dimensional algebra, and denote by E the set of algebra homomorphisms $f : A \to k$.*

(i) The set E is finite and the assignment

$$A \longrightarrow \mathcal{O}(E), \quad a \longmapsto (f \mapsto f(a)) \tag{1.3}$$

is an isomorphism of algebras.

(ii) Every quotient algebra of A is reduced.

Proof (i) In view of Lemma 1.2.5, it suffices to show that there exists an algebra isomorphism $A \simeq \mathcal{O}(F)$ for some finite set F.

Assume that there exist nonzero ideals B, C of A such that $A = B \oplus C$. Let $1 = e + f$ be the corresponding decomposition of the identity element of A; then we easily obtain $e^2 = e$, $f^2 = f$ and $ef = 0$, and hence B (resp. C) is a subalgebra of A with identity element e (resp. f). Since A is reduced, so are B and C. Using the isomorphism (1.2) and induction on the dimension of A, we may thus assume that A admits no such decomposition.

Let $a \in A \setminus \{0\}$ and consider the multiplication map

$$a_A : A \longrightarrow A, \quad b \longmapsto ab \tag{1.4}$$

(so that the assignment $a \mapsto a_A$ is the regular representation of A). Then a_A is an endomorphism of the finite-dimensional vector space A, and hence satisfies

$$A = \mathrm{Ker}(a_A^n) \oplus \mathrm{Im}(a_A^n) \qquad (n \gg 0). \tag{1.5}$$

Moreover, $\mathrm{Ker}(a_A^n)$ and $\mathrm{Im}(a_A^n)$ are ideals of A, and $\mathrm{Im}(a_A^n) \neq 0$ as A is reduced. By our assumption, it follows that $A = \mathrm{Im}(a_A^n)$ for $n \gg 0$.

In particular, a_A is injective, and hence a is invertible. So A is a field; arguing as in Remark 1.2.6, it follows that $A = k$.

(ii) Let I be an ideal of $\mathcal{O}(E)$. Since $\sum_{x \in E} \delta_x = 1$, we have $I = \sum_{x \in E} I\delta_x$, where $I\delta_x \subset \mathcal{O}(E)\delta_x = k\delta_x$. As a consequence, $I = \oplus_{x \in F} k\delta_x$ for a unique subset $F \subset E$. Then $\mathcal{O}(E)/I \simeq \mathcal{O}(E \setminus F)$ is indeed reduced. □

Combining Lemmas 1.2.4, 1.2.5 and 1.2.7, we obtain:

Proposition 1.2.8 *The assignment $E \mapsto \mathcal{O}(E)$ yields a bijective correspondence from finite sets (and maps between such sets) to reduced finite-dimensional algebras (and homomorphisms between such algebras).*

The inverse correspondence is denoted by $A \mapsto \operatorname{Spec}(A)$ (the **spectrum** of the algebra A).

For any maps of finite sets $E \xrightarrow{\varphi} F \xrightarrow{\psi} G$, we have $(\psi \circ \varphi)^* = \varphi^* \circ \psi^*$. Thus, the category of finite sets is equivalent to the opposite of the category of reduced finite-dimensional algebras.

1.2.3 Finite schemes and finite-dimensional algebras

Definition 1.2.9 The **category of finite schemes** is the opposite category to that of finite-dimensional algebras.

In more concrete terms, finite schemes are finite-dimensional algebras, with morphisms going the other way round.

A basic example of a nonreduced algebra is the **algebra of dual numbers** $k[T]/(T^2) = k[\varepsilon] = k \oplus k\varepsilon$, where $\varepsilon^2 = 0$.

Remark 1.2.10 With the above definition, some properties of finite schemes follow readily from the dual properties of algebras. For example, any two finite schemes X, Y admit a **product**, i.e., a finite scheme Z equipped with morphisms $\mathrm{pr}_X : Z \to X$, $\mathrm{pr}_Y : Z \to Y$ (the projections) which satisfy the following universal property: for any finite scheme W equipped with morphisms $f : W \to X$, $g : W \to Y$, there exists a unique morphism $h : W \to Z$ such that $f = \mathrm{pr}_X \circ h$ and $g = \mathrm{pr}_Y \circ h$.

Indeed, any two algebras A and B admit a "coproduct", namely, the tensor product $A \otimes B$ equipped with the homomorphisms $A \to A \otimes B$, $a \mapsto a \otimes 1_A$ and $B \to A \otimes B$, $b \mapsto 1_A \otimes b$.

In view of the universal property, the above scheme Z is unique up to isomorphism; we will use the standard notation $Z = X \times Y$.

We will obtain a structure result for finite-dimensional algebras (Theorem 1.2.13). For this, we recall some notions from commutative algebra.

Let R be a commutative ring. The set of nilpotent elements of R is an ideal, called the **nilradical**; we denote it by $\mathfrak{n} = \mathfrak{n}(R)$. The quotient ring $A/\mathfrak{n} = A_{\mathrm{red}}$ is reduced, and $\mathfrak{n}$ is the smallest ideal with this property. Clearly, $\mathfrak{n} \subset \mathfrak{m}$ for any maximal ideal $\mathfrak{m}$ of R.

The ring R is **indecomposable** if it has no nontrivial decomposition into a direct product of rings. Equivalently, R has no idempotent $e \neq 0, 1$ (this notion appeared implicitly in the proof of Lemma 1.2.7).

Also, R is **local** if it has a unique maximal ideal $\mathfrak{m}$. Equivalently, $\mathfrak{m}$ is an ideal of R and every $x \in R \setminus \mathfrak{m}$ is invertible. The quotient ring $R/\mathfrak{m}$ is then a field, called the **residue field** of R.

We now record two auxiliary results:

Lemma 1.2.11 *Let A be a finite-dimensional algebra. Then A is indecomposable if and only if it is local. Under this assumption, the maximal ideal $\mathfrak{m}$ is the nilradical of A, with residue field k. Moreover, we have $\mathfrak{m}^n = 0$ for $n \gg 0$.*

Proof Assume that A is local with maximal ideal $\mathfrak{m}$. If $e \in A$ is indecomposable, then $e(1-e) = 0$. Thus, we have either $e \in \mathfrak{m}$ or $1 - e \in \mathfrak{m}$. In the former case, $1 - e$ is invertible, hence $e = 0$. In the latter case, we obtain similarly $e = 1$. Thus A is indecomposable.

Conversely, assume that A is indecomposable. To show that A is local, we argue as in the proof of Lemma 1.2.7. Let $a \in A$, then for $n \gg 0$, we have $\mathrm{Ker}(a_A^n) = 0$ or $\mathrm{Im}(a_A^n) = 0$ in view of the decomposition (1.5). Thus, a is nilpotent or invertible. As a consequence, A is local and its maximal ideal $\mathfrak{m}$ is the nilradical. We have $A/\mathfrak{m} = k$ by Remark 1.2.6.

It remains to show that $\mathfrak{m}^n = 0$ for $n \gg 0$. Since the powers $\mathfrak{m}^n$ form a decreasing sequence of subspaces of A, we have $\mathfrak{m}^n = \mathfrak{m}^{n+1}$ for $n \gg 0$. This yields a finite-dimensional vector space $V = \mathfrak{m}^n$ equipped with commuting nilpotent endomorphisms $u_1, \ldots, u_N$ (the multiplication maps by elements of a basis of $\mathfrak{m}$) such that $V = u_1(V) + \cdots + u_N(V)$. So the dual vector space V^* comes with commuting nilpotent endomorphisms, the transposes $u_1^T, \cdots, u_N^T$. If $V \neq 0$ then these endomorphisms have a common nonzero kernel, i.e., there exists a nonzero $f \in V^*$ such that $f \circ u_i = 0$ for $i = 1, \ldots, n$. But then $f(V) = 0$, a contradiction. □

Lemma 1.2.12 *Let A be a local finite-dimensional algebra, $\mathfrak{m}$ its maximal ideal, and $a_1, \ldots, a_m \in \mathfrak{m}$. Then the following conditions are equivalent:*

(i) *The algebra A is generated by $a_1, \dots, a_m$.*
(ii) *The ideal $\mathfrak{m}$ is generated by $a_1, \dots, a_m$.*
(iii) *The vector space $\mathfrak{m}/\mathfrak{m}^2$ is generated by the images of $a_1, \dots, a_m$.*

Proof (i)⇒(ii) Let $a \in \mathfrak{m}$. There exists $P \in k[T_1, \dots, T_m]$ such that $a = P(a_1, \dots, a_m)$. Then the constant term of P must be 0, and hence $a \in (a_1, \dots, a_m)$.

Since the implication (ii)⇒(iii) is obvious, it remains to prove that (iii)⇒(i). For this, we use the decreasing filtration of A by the powers $\mathfrak{m}^n$, where $n \geq 0$, and the associated graded $\mathrm{gr}(A) = \bigoplus_{n \geq 0} \mathfrak{m}^n/\mathfrak{m}^{n+1}$. Then $\mathrm{gr}(A)$ is a graded algebra generated by $\mathfrak{m}/\mathfrak{m}^2$, and hence by the images $\bar{a}_1, \dots, \bar{a}_m$ of $a_1, \dots, a_m$. Given a nonzero $a \in A$, there exists a unique integer $n \geq 0$ such that $a \in \mathfrak{m}^n \setminus \mathfrak{m}^{n+1}$ (since $\mathfrak{m}^r = 0$ for $r \gg 0$). Then there exists a polynomial P as above such that $\bar{a} = P(\bar{a}_1, \dots, \bar{a}_m)$, where $\bar{a}$ denotes the image of a in $\mathfrak{m}^n/\mathfrak{m}^{n+1}$, and $\bar{a}_i$, the image of a_i in $\mathfrak{m}/\mathfrak{m}^2$ for $i = 1, \dots, m$. This means that $a - P(a_1, \dots, a_m) \in \mathfrak{m}^{n+1}$. We now conclude by decreasing induction on n, using again the vanishing of $\mathfrak{m}^n$ for $n \gg 0$. □

Theorem 1.2.13 *Let A be a finite-dimensional algebra, and denote by E the (finite) set of algebra homomorphisms $f : A \to k$.*

(i) *The assignment $A \to \mathcal{O}(E)$, $a \mapsto (f \mapsto f(a))$ (1.3) induces an isomorphism of algebras $A_{\mathrm{red}} \xrightarrow{\sim} \mathcal{O}(E)$.*
(ii) *For any $x \in E$, the idempotent $\delta_x \in \mathcal{O}(E)$ lifts to a unique idempotent $e_x \in A$. Moreover, $e_x e_y = 0$ for all distinct $x, y \in E$, and $\sum_{x \in E} e_x = 1$. The idempotents of A are exactly the partial sums $\sum_{x \in F} e_x$, where $F \subset E$.*
(iii) *We have $A = \prod_{x \in E} A e_x$ and each $A e_x$ is a local algebra.*

Proof (i) Let $\pi : A \to A_{\mathrm{red}} = A/\mathfrak{n}$ denote the projection. Since every homomorphism of algebras $f : A \to k$ sends $\mathfrak{n}$ to 0, the composition with π yields a bijection from $\mathrm{Hom}_{\mathrm{alg}}(A_{\mathrm{red}}, k)$ to $\mathrm{Hom}_{\mathrm{alg}}(A, k)$. So the assertion follows from Lemma 1.2.7.

We now show (ii) and (iii) simultaneously. Since the algebra A is finite-dimensional, it admits a decomposition $A = A_1 \times \cdots \times A_n$ where each A_i is indecomposable, and hence local (Lemma 1.2.11). Thus, $A_i = k e_i \oplus \mathfrak{m}_i$, where e_i is the identity element of A_i, and $\mathfrak{m}_i$ the nilradical. So $\mathfrak{m}_1 \times \cdots \times \mathfrak{m}_n$ is an ideal of A contained in $\mathfrak{n}$, and the quotient $A/\mathfrak{m}_1 \times \cdots \times \mathfrak{m}_n \simeq k^n$ is reduced. It follows that $\mathfrak{m}_1 \times \cdots \times \mathfrak{m}_n = \mathfrak{n}$; moreover, we may identify E with $\{1, \dots, n\}$. Via this identification, each $\delta_i \in \mathcal{O}(E)$ lifts to the idempotent e_i of A. Moreover, $e_i e_j = 0$ for all distinct i, j, and $\sum_{i=1}^n e_i = 1$.

Next, we show that every idempotent $e \in A$ can be written as $\sum_{i \in F} e_i$ for some $F \subset \{1, \ldots, n\}$. We have $e = (t_1 e_1 + x_1, \ldots, t_n e_n + x_n)$ where $t_i \in k$ and $x_i \in \mathfrak{m}_i$ for $i = 1, \ldots, n$. Since $e^2 = e$, we obtain $t_i^2 + 2t_i x_i = t_i$ and $x_i^2 = x_i$. Thus, $x_i = 0$ as x_i is nilpotent. This implies the assertion.

This assertion implies in turn that each e_i is the unique lift of δ_i, completing the proof. □

In view of Theorem 1.2.13, we may reformulate the definition of finite schemes in more geometric terms:

Definition 1.2.14 A finite scheme X consists of a finite set E together with local finite-dimensional algebras $\mathcal{O}_{X,x}$ for each $x \in E$. We then say that $\mathcal{O}_{X,x}$ is the **local ring of X at x.**

With this definition, the algebra associated with X is

$$A = \mathcal{O}(X) = \prod_{x \in E} \mathcal{O}_{X,x}$$

and we still write $X = \operatorname{Spec}(A)$. We say that X is **local** if A is local.

Given another finite scheme Y with data $(F, (\mathcal{O}_{Y,y})_{y \in F})$, a morphism of finite schemes $f : X \to Y$ consists of a map $\varphi : E \to F$ together with algebra homomorphisms $\mathcal{O}_{Y,\varphi(x)} \to \mathcal{O}_{X,x}$ for all $x \in E$ (as follows by considering the dual homomorphism $f^* : \mathcal{O}(Y) \to \mathcal{O}(X)$ and using Theorem 1.2.13).

Remark 1.2.15 A basic invariant of a local scheme X is the dimension of $\mathcal{O}(X)$, also known as the **length** of X. There is a unique local algebra of dimension 1 (resp. 2) up to isomorphism, namely, k (resp. $k[\varepsilon]$). But this fails in dimension 3, since $k[T]/(T^3)$ and $k[T_1, T_2]/(T_1^2, T_1 T_2, T_2^2)$ are nonisomorphic (as follows e.g. from Lemma 1.2.12). In higher dimensions, there may be infinitely many nonisomorphic local algebras, for example

$$k[T_1, T_2]/(P, T_1^n, T_1^{n-1} T_2, \ldots, T_1 T_2^{n-1}, T_2^n),$$

where P is a homogeneous polynomial of degree $n - 1$ in T_1, T_2, and $n \geq 5$ (exercise).

Definition 1.2.16 Let X, Y be finite schemes. We say that Y is a **subscheme** of X if the algebra $\mathcal{O}(Y)$ is a quotient of $\mathcal{O}(X)$. Then the morphism $i : Y \to X$ corresponding to the projection $\mathcal{O}(X) \to \mathcal{O}(Y)$ is called an **immersion**.

With the above notation, this is equivalent to F being a subset of E, and $\mathcal{O}_{Y,y}$ being a quotient of $\mathcal{O}_{X,i(y)}$ for any $y \in F$. Also, note that the subschemes of X correspond bijectively to the ideals of $\mathcal{O}(X)$, by assigning to Y the kernel of the projection $\mathcal{O}(X) \to \mathcal{O}(Y)$.

Definition 1.2.17 Let $X_1, \ldots, X_n$ be finite schemes, and $A_1, \ldots, A_n$ the corresponding algebras. The **sum** $X = X_1 \sqcup \cdots \sqcup X_n$ is the finite scheme $\mathrm{Spec}(A_1 \times \cdots \times A_n)$.

With an obvious notation, we then have $E = E_1 \sqcup \cdots \sqcup E_n$ and $\mathcal{O}_{X,x} = \mathcal{O}_{X_i,x}$ for all $x \in X_i$ $(i = 1, \ldots, n)$. The projections $A_1 \times \cdots \times A_n \to A_i$ correspond to immersions $X_i \to X$. Also, Theorem 1.2.13 may be reformulated as follows: every finite scheme has a unique decomposition into a sum of local finite schemes.

1.2.4 The reduced subscheme

We first obtain a useful addition to the structure theorem for finite-dimensional algebras (Theorem 1.2.13):

Proposition 1.2.18 *(i) Every finite-dimensional algebra A admits a largest reduced subalgebra A^{red}. Moreover, the composition $A^{\mathrm{red}} \to A \to A/\mathfrak{n} = A_{\mathrm{red}}$ is an isomorphism.*

(ii) Every homomorphism of finite-dimensional algebras $f : A \to B$ induces homomorphisms $f_{\mathrm{red}} : A_{\mathrm{red}} \to B_{\mathrm{red}}$, $f^{\mathrm{red}} : A^{\mathrm{red}} \to B^{\mathrm{red}}$ such that the diagram

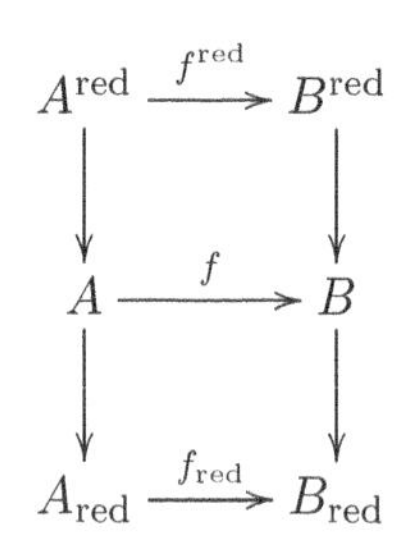

commutes.

(iii) For any finite-dimensional algebras A, B, we have natural isomorphisms of algebras

$$A^{\mathrm{red}} \otimes B^{\mathrm{red}} \xrightarrow{\sim} (A \otimes B)^{\mathrm{red}}, \quad A_{\mathrm{red}} \otimes B_{\mathrm{red}} \xrightarrow{\sim} (A \otimes B)_{\mathrm{red}}.$$

Proof (i) With the notation of Theorem 1.2.13, the subalgebra $B = \prod_{x \in E} k e_x$ is reduced, the composition $B \to A \to A_{\mathrm{red}}$ is an isomorphism, and B contains every idempotent of A. It follows that B contains every reduced subalgebra C of A, since C is spanned by its idempotents in view of Lemma 1.2.7. This yields the assertion.

(ii) The commutativity of the top square follows from the fact that every quotient of a reduced algebra is reduced (Lemma 1.2.7 again). The commutativity of the bottom square is readily checked from the definitions.

(iii) By Lemma 1.2.7 once more and the isomorphism (1.1) (or a direct argument), the tensor product of any two reduced finite-dimensional algebras is reduced. This easily implies the assertion. □

Definition 1.2.19 A finite scheme X is **reduced** if the algebra $\mathcal{O}(X)$ is reduced.

In view of Proposition 1.2.8, the category of reduced finite schemes is equivalent to that of finite sets via the assignments $X \mapsto X(k)$ and $E \mapsto \mathcal{O}(E)$. Moreover, Proposition 1.2.18 translates as follows in the language of schemes:

Corollary 1.2.20 *Every finite scheme X has a largest reduced subscheme X_{red}. Moreover, there exists a unique morphism $r : X \to X_{\mathrm{red}}$ such that $r \circ i = \mathrm{id}_{X_{\mathrm{red}}}$, where i denotes the immersion $X_{\mathrm{red}} \to X$. The formations of X_{red} and r are functorial and commute with products.*

1.2.5 The functor of points

Definition 1.2.21 Let $X = \mathrm{Spec}(A)$ be a finite scheme, and R a finite-dimensional algebra. An R-**valued point** of X is a homomorphism of algebras $u : A \to R$. The set of R-valued points of X is denoted by $X(R) = \mathrm{Hom}_{\mathrm{alg}}(\mathcal{O}(X), R) = \mathrm{Hom}(\mathrm{Spec}(R), X)$.

With the above notation, $X(k)$ is the finite set E of Definition 1.2.14; its points are also known as the k-**points** or k-**rational points** of X. Also, every morphism of finite schemes $f : X \to Y$ induces a map $f(R) : X(R) \to Y(R)$ given by precomposition with the dual homomorphism $f^* : \mathcal{O}(Y) \to \mathcal{O}(X)$. The map $f(k) : X(k) \to Y(k)$ is the map $\varphi : E \to F$ of the above definition.

Note that $\mathrm{Spec}(k)(R) = \mathrm{Hom}_{\mathrm{alg}}(k, R)$ is a unique point for any k-algebra R. As a consequence, for any $x \in X(k)$ viewed as a morphism $x : \mathrm{Spec}(k) \to X$, we obtain a point $x(R) \in X(R)$.

Every algebra homomorphism $u : R \to S$ induces a map $X(u) : X(R) \to X(S)$ via postcomposition. Moreover, $X(\mathrm{id}_R) = \mathrm{id}_{X(R)}$ and

$X(v \circ u) = X(v) \circ X(u)$ for any algebra homomorphisms $R \xrightarrow{u} S \xrightarrow{v} T$. This yields a functor from the category of finite-dimensional algebras to that of sets: the **functor of points** of X, that we denote by h_X.

Given a functor F from finite-dimensional algebras to sets, there exists a natural isomorphism

$$\mathrm{Hom}(h_X, F) \xrightarrow{\sim} F(X), \quad u \longmapsto u(X)(\mathrm{id}_X),$$

where the left-hand side denotes the set of morphisms of functors, also known as natural transformations (Yoneda's lemma, see [10, Lem. VI-1]). In particular, we obtain natural isomorphisms

$$\mathrm{Hom}(h_X, h_Y) \simeq h_Y(X) = \mathrm{Hom}(X, Y) \simeq \mathrm{Hom}_{\mathrm{alg}}(\mathcal{O}(Y), \mathcal{O}(X)).$$

It follows that every finite scheme is uniquely determined by its functor of points. So we may view finite schemes as **representable functors** from finite-dimensional algebras to sets, i.e., functors of the form h_X.

Some operations on finite schemes have a simple formulation in terms of their functors of points. For example, given finite schemes $X_1, \ldots, X_n$, their sum satisfies

$$(X_1 \sqcup \cdots \sqcup X_n)(R) = X_1(R) \sqcup \cdots \sqcup X_n(R)$$

for any finite-dimensional algebra R. Also, given two finite schemes X, Y, we have a functorial bijection

$$X(R) \times Y(R) \xrightarrow{\sim} (X \times Y)(R)$$

via the tensor product of algebra homomorphisms. In other words, the functor $R \mapsto X(R) \times Y(R)$ is represented by the product $X \times Y$.

More generally, given two morphisms of finite schemes $f : X \to Z$, $g : Y \to Z$, we may consider the functor

$$R \longmapsto X(R) \times_{Z(R)} Y(R) = \{(u, v) \in X(R) \times Y(R) \mid f(R)(u) = g(R)(v)\}.$$

Then this functor is represented by the finite scheme

$$W = \mathrm{Spec}\left(\mathcal{O}(X) \otimes_{\mathcal{O}(Z)} \mathcal{O}(Y)\right)$$

where $\mathcal{O}(X)$ (resp. $\mathcal{O}(Y)$) is a $\mathcal{O}(Z)$-algebra via f^* (resp. g^*). Indeed, this follows easily from the universal property of the tensor product of algebras.

Definition 1.2.22 With the above notation and assumptions, the finite scheme W is called the **fibered product** of X and Y above Z, and denoted by $X \times_Z Y$.

In particular, consider a morphism of finite schemes $f : X \to Y$, and a k-point of Y viewed as a morphism $y : \mathrm{Spec}(k) \to Y$. Then the fibered product $X \times_Y y$ is called the (schematic) **fiber of f at y** and denoted by X_y.

The functor of points of X_y satisfies

$$X_y(R) = \{u \in X(R) \mid f(R)(u) = y(R)\} \tag{1.6}$$

for any finite-dimensional algebra R. In particular, $X_y(k)$ is the set-theoretic fiber $f(k)^{-1}(y)$.

Also, note that $X \times_Z Y$ is the subscheme of $X \times Y$ with ideal generated by the $f^*(h) \otimes 1 - 1 \otimes g^*(h)$, where $h \in \mathcal{O}(Z)$. As a consequence, X_y is the subscheme of X with ideal $f^*(\mathfrak{m}_y)\mathcal{O}(X)$, where $\mathfrak{m}_y$ denotes the maximal ideal of y in $\mathcal{O}(Y)$.

Remark 1.2.23 Some notions and results of this section extend to the setting of **affine schemes**; these form the opposite category to that of algebras (without finiteness condition). More specifically, the constructions of sums and fibered products extend unchanged; also, affine schemes may be viewed as representable functors from algebras to sets, via their functor of points. We will still use the notations $X = \mathrm{Spec}(A)$ and $A = \mathcal{O}(X)$ in the setting of affine schemes.

A basic example of affine scheme is the **affine n-space**

$$\mathbb{A}^n = \mathrm{Spec}(k[T_1, \ldots, T_n]),$$

where n is a positive integer. We have $\mathbb{A}^n \simeq \mathbb{A}^1 \times \cdots \times \mathbb{A}^1$ (n times), since $k[T_1, \ldots, T_n] \simeq k[T] \otimes \cdots \otimes k[T]$. Moreover, $\mathbb{A}^n(R) = R^n$ for any algebra R.

But there are important differences between affine and finite schemes. For example, an affine scheme may well have no k-point (just consider a nontrivial field extension of k, e.g., the field of rational functions $k(T)$). The subclass of (affine) **schemes of finite type**, also known as **algebraic schemes**, is better behaved in this respect; these correspond to the finitely generated algebras, i.e., those isomorphic to a quotient $A = k[T_1, \ldots, T_n]/I$, where I is an ideal of $k[T_1, \ldots, T_n]$. The functor of points of $X = \mathrm{Spec}(A)$ satisfies

$$X(R) = \{(x_1, \ldots, x_n) \in R^n \mid P(x_1, \ldots, x_n) = 0 \text{ for all } P \in I\}.$$

In particular, $X(k)$ is the set of zeros of I in k^n; such a set is known as an (affine) **algebraic set**.

By Hilbert's basis theorem (see [11, Thm. 1.2]), every ideal I as above is finitely generated, and hence every algebraic set is the set of common zeros of finitely many polynomials. Also, by Hilbert's Nullstellensatz (see [11, Thm. 1.6]), the maximal ideals of A are exactly the kernels of elements of $X(k) = \mathrm{Hom}_{\mathrm{alg}}(A, k)$. Moreover, A is finite-dimensional if and only if $X(k)$ is finite.

A new feature of affine schemes is the **Zariski topology**. For algebraic schemes, it can be defined as the topology on $X(k)$ with closed sets being the zeros of ideals of $k[T_1, \dots, T_n]$ containing I. These ideals can be identified with the ideals J of A, and they correspond bijectively with the **closed subschemes** $Y = \mathrm{Spec}(A/J)$ of $X = \mathrm{Spec}(A)$ (then $Y(k)$ is closed in $X(k)$). Thus, every affine algebraic scheme is isomorphic to a closed subscheme of an affine space. Also, for finite schemes, the Zariski topology is just the discrete topology, i.e., every subset is closed.

Every algebraic scheme $X = \mathrm{Spec}(A)$ has a largest closed reduced subscheme X_{red} corresponding to the nilradical of A; moreover, $X_{\mathrm{red}}(k) \xrightarrow{\sim} X(k)$. But A may have no largest reduced subalgebra; for example, the nonreduced algebra $k[T_1, T_2]/(T_2^2)$ is generated by its reduced subalgebras $k[T_1]$ and $k[T_1 + T_2]$. Still, A has a largest reduced finite-dimensional subalgebra: the span of its idempotents (see e.g. [25, Prop. 1.29]).

An affine scheme X is called **connected** if the algebra $\mathcal{O}(X)$ is indecomposable. Every affine algebraic scheme X is the sum of finitely many connected schemes, and these are affine algebraic as well. Moreover, X is connected if and only if $X(k)$ is connected relative to the Zariski topology (see loc. cit.).

Remark 1.2.24 We briefly present some further aspects of scheme theory, which will only be used in Section 1.6; we refer to [10] for a user-friendly introduction to schemes.

There is a notion of (not necessarily affine) schemes. These are obtained by gluing affine schemes, like manifolds in differential geometry; the Hausdorff property is replaced with the property that the diagonal is closed. Schemes are equipped with the Zariski topology; the notion of closed subscheme extends readily to this setting.

A basic example is the **projective n-space** $\mathbb{P}^n$, obtained by gluing appropriately $n + 1$ copies of $\mathbb{A}^n$ (corresponding to the nonvanishing of homogeneous coordinates). The **projective schemes** are those isomorphic to a closed subscheme of some projective space.

A scheme is called **of finite type** (or **algebraic**) if it admits an open covering by finitely many affine schemes of finite type. Every projective scheme is of finite type. Also, every scheme of finite type has only finitely many connected components.

A scheme is **reduced** (resp. **integral**) if the algebra $\mathcal{O}(U)$ is reduced (resp. integral) for any open affine subset U. A **variety** is an integral scheme of finite type. For example, $\mathbb{A}^n$ and $\mathbb{P}^n$ are varieties, as well as elliptic curves.

By taking k-points, every closed reduced subscheme of $\mathbb{A}^n$ is identified with an **algebraic subset** of k^n, i.e., the set of common zeros of polynomials in n variables. The subvarieties of $\mathbb{A}^n$ correspond to **irreducible** algebraic subsets (those that are not the union of proper closed subsets). Likewise, we may define the algebraic subsets of $\mathbb{P}^n(k)$ as the sets of common zeros of homogeneous polynomials in $n+1$ variables; then the closed subvarieties of $\mathbb{P}^n$ can be identified with the irreducible algebraic subsets of $\mathbb{P}^n(k)$.

1.3 Finite group schemes

1.3.1 Basic definitions and examples

A group structure on a set G is given by two maps $\mu : G \times G \to G$, $(x, y) \mapsto xy$ (the multiplication map) and $\iota : G \to G$, $x \mapsto x^{-1}$ (the inverse map), together with an element $e \in G$ (the neutral element) which satisfy the group axioms. These translate into the commutativity of the following diagrams:

$$\begin{array}{ccc} G \times G \times G & \xrightarrow{\mu \times \mathrm{id}} & G \times G \\ \downarrow{\scriptstyle \mathrm{id} \times \mu} & & \downarrow{\scriptstyle \mu} \\ G \times G & \xrightarrow{\mu} & G \end{array} \tag{1.7}$$

(i.e., μ is associative),

$$\begin{array}{ccccc} G & \xrightarrow{(\mathrm{id},\iota)} & G \times G & \xleftarrow{(\iota,\mathrm{id})} & G \\ & {\scriptstyle e}\searrow & \downarrow{\scriptstyle \mu} & \swarrow{\scriptstyle e} & \\ & & G & & \end{array} \tag{1.8}$$

(i.e., ι is the inverse map), and

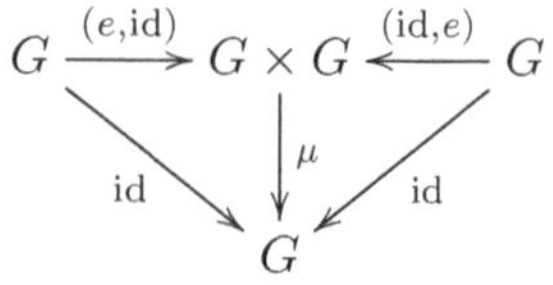

$$\begin{array}{ccccc} G & \xrightarrow{(e,\mathrm{id})} & G \times G & \xleftarrow{(\mathrm{id},e)} & G \\ & {\scriptstyle \mathrm{id}} \searrow & \downarrow {\scriptstyle \mu} & \swarrow {\scriptstyle \mathrm{id}} & \\ & & G & & \end{array} \tag{1.9}$$

(i.e., e is the neutral element). Here $e : G \to G$ denotes the constant map $g \mapsto e$.

Next, let $A = \mathcal{O}(G)$. In view of the isomorphism (1.1), we may identify $\mathcal{O}(G \times G)$ with $A \otimes A$, and $\mathcal{O}(G \times G \times G)$ with $A \otimes A \otimes A$. By Lemma 1.2.5, the data of the multiplication $\mu : G \times G \to G$ is equivalent to that of an algebra homomorphism

$$\Delta = \mu^* : A \longrightarrow A \otimes A.$$

Likewise, ι corresponds to an algebra endomorphism

$$S = \iota^* : A \longrightarrow A,$$

and e to an algebra homomorphism

$$\varepsilon = e^* : A \longrightarrow k.$$

Moreover, the commutative diagrams (1.7), (1.8), (1.9) correspond to commutative diagrams

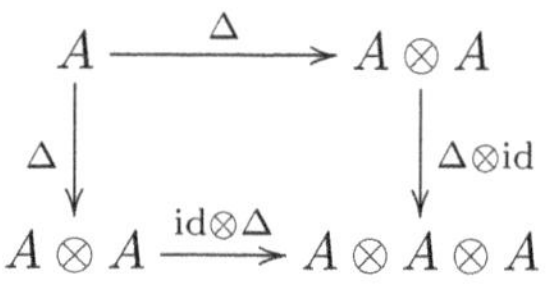

$$\begin{array}{ccc} A & \xrightarrow{\Delta} & A \otimes A \\ {\scriptstyle \Delta} \downarrow & & \downarrow {\scriptstyle \Delta \otimes \mathrm{id}} \\ A \otimes A & \xrightarrow{\mathrm{id} \otimes \Delta} & A \otimes A \otimes A \end{array} \tag{1.10}$$

$$\begin{array}{ccccc} & & A & & \\ & {\scriptstyle \varepsilon} \swarrow & \downarrow {\scriptstyle \Delta} & \searrow {\scriptstyle \varepsilon} & \\ A & \xleftarrow{\mathrm{id} \otimes S} & A \otimes A & \xrightarrow{S \otimes \mathrm{id}} & A \end{array} \tag{1.11}$$

$$\begin{array}{ccccc} & & A & & \\ & {\scriptstyle \mathrm{id}} \swarrow & \downarrow {\scriptstyle \Delta} & \searrow {\scriptstyle \mathrm{id}} & \\ A & \xleftarrow{\mathrm{id} \otimes \varepsilon} & A \otimes A & \xrightarrow{\mathrm{id} \otimes \varepsilon} & A \end{array} \tag{1.12}$$

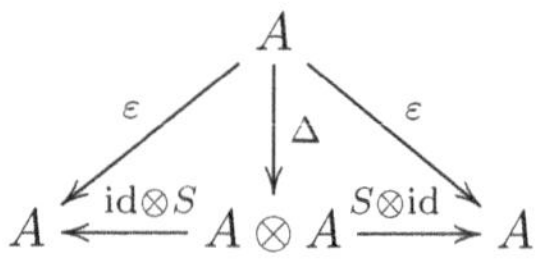

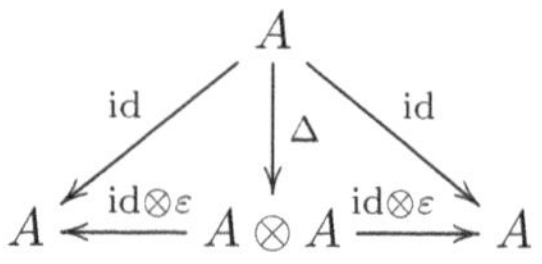

Here we denote again by $\varepsilon : A \to A$ the composition of $\varepsilon : A \to k$ with the inclusion of k into A.

Definition 1.3.1 A **Hopf algebra** is an algebra A equipped with algebra homomorphisms $\Delta = \Delta_A : A \to A \otimes A$ (the **comultiplication**), $S = S_A : A \to A$ (the **antipode**) and $\varepsilon = \varepsilon_A : A \to k$ (the **augmentation**) such that the diagrams (1.10), (1.11) and (1.12) commute.

Given two Hopf algebras A, B, a **homomorphism of Hopf algebras** $u : A \to B$ is an algebra homomorphism such that $\Delta_B \circ u = (u \otimes u) \circ \Delta_A$.

The latter condition corresponds to the equality $f(xy) = f(x)f(y)$ for a group homomorphism $f : G \to H$ and for all $x, y \in G$.

Actually, the notion of Hopf algebra is more general, and does not assume that A is commutative, nor that k is algebraically closed. Also, the data of Δ determines S, ε uniquely, and every homomorphism of Hopf algebras $u : A \to B$ satisfies $S_B \circ u = u \circ S_A$ and $\varepsilon_B \circ u = u \circ \varepsilon_A$ (see [37, §2.1]).

By Proposition 1.2.8 and the isomorphism (1.1), the category of finite groups is equivalent to the opposite category of reduced finite-dimensional Hopf algebras. This motivates the following:

Definition 1.3.2 The **category of finite group schemes** is the opposite to that of finite-dimensional Hopf algebras.

Equivalently, a finite group scheme is a finite scheme G equipped with morphisms $\mu : G \times G \to G$, $\iota : G \to G$ and with a k-point e such that the diagrams (1.7), (1.8) and (1.9) commute. For any finite-dimensional algebra R, we obtain a group structure on the set $G(R)$ with multiplication map $\mu(R)$, inverse map $\iota(R)$, and neutral element $e \in G(k) \subset G(R)$. Given $x, y \in G(R)$, we denote $\mu(x, y)$ by xy, and $\iota(x)$ by x^{-1}.

Alternatively, we may view finite group schemes as **representable group functors**, i.e., representable functors from the category of finite-dimensional algebras to that of groups.

Definition 1.3.3 A **subgroup scheme** of a finite group scheme G is a subscheme H such that $H(R)$ is a subgroup of $G(R)$ for any finite-dimensional algebra R. We then write $H \leq G$.

Definition 1.3.4 Let $f : G \to H$ be a homomorphism of finite group schemes. The **kernel** $\mathrm{Ker}(f)$ is the fiber of f at e_H.

We have $\mathrm{Ker}(f)(R) = \{x \in G(R) \mid f(R)(x) = e_H\} = \mathrm{Ker}(f(R))$ for any finite-dimensional algebra R. As a consequence, $N = \mathrm{Ker}(f)$ is a **normal subgroup scheme** of G, i.e., $xyx^{-1} \in N(R)$ for any such R and any $x \in G(R)$, $y \in N(R)$. We then write $N \trianglelefteq G$.

Definition 1.3.5 The **order** of a finite group scheme G is the dimension of the algebra $\mathcal{O}(G)$.

As for finite schemes, the category of reduced finite group schemes is equivalent to that of finite groups via $G \mapsto G(k)$. A quasi-inverse functor sends every finite group F to the algebra $\mathcal{O}(F)$. (This algebra is canonically isomorphic to the dual of the group algebra $k[F]$, but generally not to $k[F]$ itself. Indeed, $\mathcal{O}(F)$ is always commutative, but $k[F]$ is commutative if and only if F is commutative). Also, the notion of order of a finite group scheme generalizes that for finite groups. We will freely identify finite groups with the associated group schemes.

We now reconsider our first examples from §1.2.1:

Example 1.3.6 Given a positive integer n, let $A = k[T]/(T^n - 1)$. Then A is a k-algebra of dimension n, and one may check that A is a Hopf algebra relative to the homomorphisms

$$\Delta : A \longrightarrow A \otimes A, \quad t \longmapsto t_1 \otimes t_2,$$

$$S : A \longrightarrow A, \quad t \longmapsto t^{n-1},$$

$$\varepsilon : A \longrightarrow k, \quad t \longmapsto 1,$$

where t denotes the image of T in A. The corresponding finite group scheme is the group scheme of nth roots of unity, denoted by μ_n. Indeed, we have for any finite-dimensional algebra R

$$\mu_n(R) = \{r \in R \mid r^n = 1\}.$$

Example 1.3.7 If $p > 0$ then the local p-dimensional algebra $B = k[T]/(T^p)$ is a Hopf algebra relative to the homomorphisms

$$\Delta : B \longrightarrow B \otimes B, \quad t \longmapsto t_1 \otimes 1 + 1 \otimes t_2,$$

$$S : B \longrightarrow B, \quad t \longmapsto -t,$$

$$\varepsilon : B \longrightarrow k, \quad t \longmapsto 0$$

with a similar notation. The corresponding finite group scheme α_p satisfies for any finite-dimensional algebra R

$$\alpha_p(R) = \{r \in R \mid r^p = 0\}.$$

Note that μ_p and α_p are isomorphic as schemes, since their algebras are isomorphic via $A \to B$, $t \mapsto t + 1$. One may show that these algebras are not isomorphic as Hopf algebras; equivalently, μ_p and α_p are not isomorphic as group schemes (see Example 1.4.17 for an alternative proof). So we obtain three distinct group schemes of order p, namely, μ_p, α_p and the cyclic group $\mathbb{Z}/p\mathbb{Z}$. We will see in Corollary 1.4.26 that these yield all group schemes of order p.

Definition 1.3.8 A finite group scheme G is **infinitesimal** if $G(k) = \{e\}$; equivalently, the algebra $\mathcal{O}(G)$ is local.

Infinitesimal group schemes are also called **local** or **connected**. If $p = 0$ then every infinitesimal group scheme is trivial (Theorem 1.4.13 below). This fails if $p > 0$ in view of the above examples of μ_p and α_p.

1.3.2 Actions of finite group schemes, semi-direct products

Definition 1.3.9 An **action** of a finite group scheme G on a finite scheme X is a morphism of schemes

$$\alpha : G \times X \longrightarrow X, \quad (g, x) \longmapsto g \cdot x$$

such that $g \cdot (h \cdot x) = gh \cdot x$ and $e \cdot x = x$ for any algebra R and any $g, h \in G(R)$, $x \in X(R)$.

The former condition is equivalent to the commutativity of the square

$$\begin{array}{ccc} G \times G \times X & \xrightarrow{\mu \times \mathrm{id}} & G \times X \\ {\scriptstyle \mathrm{id} \times \alpha} \downarrow & & \downarrow {\scriptstyle \alpha} \\ G \times X & \xrightarrow{\alpha} & X \end{array}$$

and the latter condition, to the commutativity of the triangle

$$\begin{array}{ccc} X & \xrightarrow{(\mathrm{e},\mathrm{id})} & G \times X \\ & {\scriptstyle \mathrm{id}} \searrow & \downarrow {\scriptstyle \alpha} \\ & & X \end{array}$$

Let $A = \mathcal{O}(X)$. Then the data of a G-action α on X is equivalent to that of a homomorphism of algebras

$$\rho = \alpha^* : A \longrightarrow \mathcal{O}(G) \otimes A$$

such that the following diagrams commute:

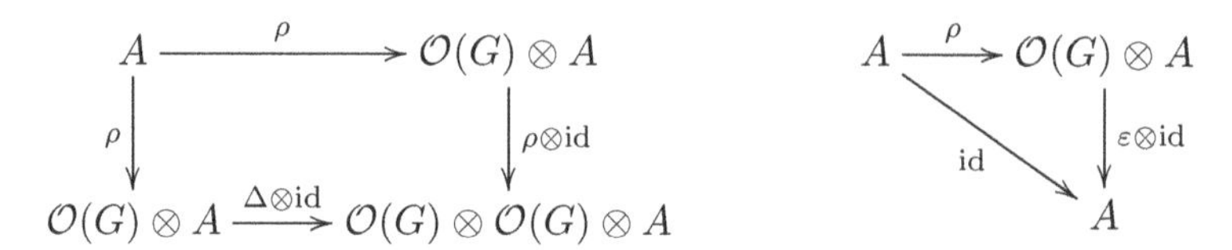

where $\Delta = \mu^* : \mathcal{O}(G) \to \mathcal{O}(G) \otimes \mathcal{O}(G)$ and $\varepsilon = e^* : \mathcal{O}(G) \to k$. We then say that A is a G-**algebra**, and X is a G-**scheme**. The map ρ is called the **co-action**; it equips A with the structure of a (left) comodule over the Hopf algebra $\mathcal{O}(G)$.

From the functorial viewpoint, a G-action on X is an action of the group $G(R)$ on the set $X(R)$ for any finite-dimensional algebra R, which is functorial in R.

For example, the projection $\mathrm{pr}_X : G \times X \to X$ is an action, called the **trivial action**. It corresponds to the trivial action of $G(R)$ on $X(R)$ for any R as above, and to the co-action $1 \otimes \mathrm{id} : A \to \mathcal{O}(G) \otimes A$, $a \mapsto 1 \otimes a$. In the opposite direction, a G-action is called **faithful** (or **effective**) if no proper subgroup scheme acts trivially.

Remark 1.3.10 Consider again a finite G-scheme X with action morphism α. For any $g \in G(k)$, we obtain an automorphism $\alpha(g, -)$ of the scheme X, and hence an algebra automorphism $\alpha(g, -)^*$ of A. This yields in turn an action of $G(k)$ on A by algebra automorphisms, where each g acts via the inverse of $\alpha(g, -)^*$.

If G is reduced, then the data of a G-action on X is equivalent to that of a $G(k)$-action on A by algebra automorphisms. More specifically, the algebra $\mathcal{O}(G) \otimes A$ is identified with the set of maps $G(k) \to A$ equipped with pointwise addition and multiplication. This identifies the co-action $\rho : A \to \mathcal{O}(G) \otimes A$ with the map $a \mapsto (g \mapsto g \cdot a)$.

This construction can be generalized as follows: let R be a finite-dimensional algebra, and $g \in G(R) = \mathrm{Hom}(\mathcal{O}(G), R)$. Composing $\alpha^* : A \to \mathcal{O}(G) \otimes A$ with $g \otimes \mathrm{id} : \mathcal{O}(G) \otimes A \to R \otimes A$, we obtain an algebra homomorphism $A \to R \otimes A$, and hence an R-algebra endomorphism $g^* : R \otimes A \to R \otimes A$. One may check that g^* is an automorphism with inverse $(g^{-1})^*$; moreover, this yields an action of $G(R)$ on $R \otimes A$ by R-algebra automorphisms, where g acts via $(g^{-1})^*$. This action is functorial in R, and determines the G-action on X uniquely. Moreover, G acts faithfully on X if and only if the group $G(R)$ acts faithfully on $R \otimes A$ for any finite-dimensional algebra R.

Definition 1.3.11 Let $u : G \to H$ be a homomorphism of finite group schemes. Let X be a G-scheme, and Y an H-scheme. A morphism of schemes $f : X \to Y$ is **equivariant** if we have $f(g \cdot x) = u(g) \cdot f(x)$ for any algebra R and any $g \in G(R)$, $x \in X(R)$.

Example 1.3.12 Every finite group scheme G acts on itself by left multiplication: $(x, y) \mapsto xy$ for any finite-dimensional algebra R and any $x, y \in G(R)$. Also, G acts on itself by right multiplication ($(x, y) \mapsto yx^{-1}$) and by conjugation ($(x, y) \mapsto xyx^{-1}$). Moreover, every homomorphism of finite group schemes $f : G \to H$ is equivariant relative to either of these actions.

Next, let N, H be finite group schemes and $\alpha : H \times N \to N$ an action by group automorphisms, i.e., $x \cdot yz = (x \cdot y)(x \cdot z)$ for any finite-dimensional algebra R and any $x \in H(R)$, $y, z \in N(R)$. We may then form the semi-direct product $N(R) \rtimes H(R)$ for any such R. This yields a group functor, which is clearly represented by the product scheme $N \times H$ equipped with the appropriate multiplication and inverse morphisms, and with the neutral element (e_N, e_H). The corresponding finite group scheme is the **semi-direct product** $G = N \rtimes H$. We have $N \trianglelefteq G$ and $H \leq G$.

We now come to the main result of this section:

Theorem 1.3.13 *Let G be a finite group scheme. Then the reduced subscheme G_{red} is a subgroup scheme. Moreover, G has a largest infinitesimal subgroup scheme G^0, and G^0 is normal in G. We have $G = G^0 \rtimes G_{\mathrm{red}}$.*

Proof Recall from Corollary 1.2.20 that the formation of the reduced subscheme X_{red} commutes with products. In view of the commutative diagrams (1.7), (1.8) and (1.9), it follows that G_{red} is a subgroup scheme.

Likewise, the formation of $r = r_X : X \to X_{\mathrm{red}}$ commutes with products, and hence $r_G : G \to G_{\mathrm{red}}$ is a homomorphism. Let K be its kernel, then we have $G = K \rtimes G_{\mathrm{red}}$. Moreover, K is infinitesimal, since r is bijective on k-valued points. If I is an infinitesimal subgroup scheme of G, then $r_I : I \to I_{\mathrm{red}}$ is trivial. By functoriality, it follows that I is a subgroup scheme of $\mathrm{Ker}(r_G) = K$. So K is the largest infinitesimal subgroup scheme of G. □

Example 1.3.14 Assume that $p > 0$ and let n be a positive integer. Then the group scheme α_p admits an action of μ_n by group automorphisms, via $x \cdot y = xy$. So we may form the semi-direct product $G = \alpha_p \rtimes \mu_n$.

If n is prime to p, then one may readily check that $G_{\mathrm{red}} = \mu_n$. As a consequence, G_{red} is not a normal subgroup scheme of G.

On the other hand, if $n = p$ then G is a noncommutative infinitesimal group scheme of order p^2.

Remark 1.3.15 As for affine schemes, we may define the category of **affine group schemes** as the opposite to that of Hopf algebras. Alternatively, the affine group schemes are the representable group functors from the category of algebras. The notions of action, equivariant morphism, semi-direct product extend readily to the setting of affine group schemes. Also, we have an obvious notion of **closed** subgroup scheme. The (schematic) kernel of a homomorphism of affine group schemes is a closed normal subgroup scheme.

An affine group scheme G is called **algebraic** if its underlying scheme is algebraic, i.e., the algebra $\mathcal{O}(G)$ is finitely generated. Then G is called an **algebraic group** for simplicity. Basic examples of affine algebraic groups include:

- the **multiplicative group** $\mathbb{G}_m$, corresponding to the Hopf algebra $k[T, T^{-1}] \simeq k[T_1, T_2]/(T_1T_2 - 1)$ with comultiplication, antipode and augmentation given by $T \mapsto T_1 \otimes T_2$, $T \mapsto T^{-1}$ and $T \mapsto 1$ (compare with Example 1.3.6). The corresponding group functor is given by $R \mapsto (R^\times, \times)$, where $R^\times$ denotes the group of invertible elements of the algebra R,
- the **additive group** $\mathbb{G}_a$, corresponding to the Hopf algebra $k[T]$ with comultiplication, antipode and augmentation given by $T \mapsto T_1 \otimes 1 + 1 \otimes T_2$, $T \mapsto -T$ and $T \mapsto 0$ as in Example 1.3.7. The corresponding group functor is given by $R \mapsto (R, +)$,
- the **general linear group** GL_n, which represents the group functor $R \mapsto \mathrm{GL}_n(R)$ (the group of invertible $n \times n$ matrices with coefficients in R). Its Hopf algebra is $k[T_{ij}, 1 \leq i, j \leq n][1/\det(T_{i,j})]$, where the T_{ij} are the matrix coefficients. According to the formula for the product of matrices, the comultiplication satisfies

$$\Delta(T_{ij}) = \sum_{\ell=1}^{n} T_{i\ell}^{(1)} \otimes T_{\ell j}^{(2)}$$

  with an obvious notation. Likewise, the antipode is given by the inverse of matrices, and the augmentation is the evaluation map at the identity matrix.

Note that $\mathbb{G}_m = \mathrm{GL}_1$ and $\mathbb{G}_a$ is the closed subgroup scheme of GL_2 with ideal $(T_{1,1} - 1, T_{2,1}, T_{2,2} - 1)$.

Likewise, given a finite-dimensional vector space V, we may define the general linear group GL_V. A homomorphism of affine group schemes $\rho : G \to \mathrm{GL}_V$ is called a **linear representation of G in V**. We say that ρ is **faithful** if its kernel is trivial; then ρ yields an isomorphism of G onto a closed subgroup scheme of GL_V (see [25, Cor. 3.35]).

If G is a finite group scheme, then its action on itself by left multiplication yields a faithful linear representation in $\mathcal{O}(G)$ (as follows from Remark 1.3.10). Thus, G is isomorphic to a closed subgroup scheme of GL_n, where $n = |G|$.

An algebraic group G is called **linear** if it is isomorphic to a closed subgroup scheme of some general linear group. Then G is clearly affine; conversely, every affine algebraic group is linear (see [25, Cor. 4.10]).

Finally, the structure theorem 1.3.13 extends partially to any affine algebraic group G: the reduced subscheme G_{red} is a subgroup scheme and the connected component of e in G is a normal subgroup scheme, denoted by G^0. Moreover, the connected components of G are exactly the cosets gG^0, where $g \in G(k)$. In particular, there are only finitely many such cosets, and $G = G^0 G_{\mathrm{red}}$. Finally, $G^0_{\mathrm{red}} = G^0 \cap G_{\mathrm{red}}$ is a group variety.

Remark 1.3.16 We will also encounter nonaffine group schemes; these may be defined as schemes G equipped with morphisms $\mu : G \times G \to G$, $\iota : G \to G$ and with $e \in G(k)$ satisfying the group axioms. The notion of **algebraic group** extends readily to this setting. For example, every elliptic curve E equipped with a k-point 0 has a unique structure of an algebraic group with neutral element 0 (see [32, §III.2]).

1.4 Lie algebras and applications

1.4.1 The Lie algebra of derivations of an algebra

Definition 1.4.1 A **derivation** of an algebra A is a k-linear map $D : A \to A$ which satisfies the Leibniz rule:

$$D(ab) = aD(b) + D(a)b \quad (a, b \in A). \tag{1.13}$$

We denote by $\mathrm{Der}(A)$ the set of derivations of A. For any $D \in \mathrm{Der}(A)$ and $a \in A$, the map $aD : A \to A$, $b \mapsto aD(b)$ is a derivation. This yields an A-module structure on $\mathrm{Der}(A)$, which is in particular a k-vector space. If A is finite-dimensional, then $\mathrm{Der}(A)$ is finite-dimensional as well.

Given $D_1, D_2 \in \mathrm{Der}(A)$, the commutator

$$[D_1, D_2] = D_1 \circ D_2 - D_2 \circ D_1$$

is easily seen to be a derivation. Moreover, the commutator map is bilinear, antisymmetric (that is, $[D, D] = 0$ for all $D \in \mathrm{Der}(A)$), and satisfies the Jacobi identity:

$$[D_1, [D_2, D_3]] + [D_2, [D_3, D_1]] + [D_3, [D_1, D_2]] = 0$$

for any $D_1, D_2, D_3 \in \mathrm{Der}(A)$. So $\mathrm{Der}(A)$ is a **Lie algebra**.

Given $D \in \mathrm{Der}(A)$ and a positive integer n, we obtain

$$D^n(ab) = \sum_{m=0}^{n} \binom{n}{m} D^m(a) D^{n-m}(b) \quad (a, b \in A) \tag{1.14}$$

by induction on n, where $D^n = D \circ \cdots \circ D$ (n times). If $p = \mathrm{char}(k) > 0$, then it follows that D^p is a derivation for any $D \in \mathrm{Der}(A)$.

Example 1.4.2 Let A be the polynomial ring $k[T_1, \ldots, T_n]$. Then $\mathrm{Der}(A)$ is a free A-module with basis the partial derivatives $\partial_i : P \to \partial P / \partial T_i$ ($i = 1, \ldots, n$). The Lie algebra structure on $\mathrm{Der}(A)$ is given by

$$[P\partial_i, Q\partial_j] = P(\partial_i Q)\partial_j - (\partial_j P) Q \partial_i$$

for all $P, Q \in A$ and all i, j. If $p > 0$ then $\partial_i^p = 0$ for all i.

More generally, let $P_1, \ldots, P_m \in k[T_1, \ldots, T_n]$ and consider the quotient $A = k[T_1, \ldots, T_n]/(P_1, \ldots, P_m)$. Denote by $t_1, \ldots, t_n$ the images of $T_1, \ldots, T_n$ in A. Then the assignment $D \mapsto (D(t_1), \ldots, D(t_n))$ identifies the A-module $\mathrm{Der}(A)$ with the kernel of the "Jacobian matrix" $(a_{ij})_{1 \le i \le n, 1 \le j \le m}$, where a_{ij} denotes the image of $\partial_i P_j$ in A.

As a consequence, $\mathrm{Der}(A)$ is a finitely generated A-module for any finitely generated algebra A.

Example 1.4.3 Let n be a positive integer, and $A = k[T]/(T^n)$. Then A is a local algebra with basis $1, t, \ldots, t^{n-1}$, where t denotes the image of T, and with maximal ideal $\mathfrak{m} = (t)$. For any $D \in \mathrm{Der}(A)$, we have $nt^{n-1}D(t) = 0$ by the Leibniz rule. Moreover, by the preceding example or a direct argument, the map $D \mapsto D(t)$ identifies the A-module $\mathrm{Der}(A)$ with the ideal $I = \{f \in A \mid nt^{n-1}f = 0\}$. We have $I = \mathfrak{m}$ if $p = 0$ or p does not divide n, and $I = A$ otherwise.

In the former case, there exists a unique $D_1 \in \mathrm{Der}(A)$ such that $D_1(t) = t$ (it arises from the derivation Td/dT of $k[T]$). Moreover, the

$D_i = t^i D_1$, where $0 \le i \le n-2$, form a basis of the vector space $\mathrm{Der}(A)$. We have the commutation relations

$$[D_i, D_j] = \begin{cases} (j-i)D_{i+j-1} & \text{if } i+j \le n, \\ 0 & \text{otherwise.} \end{cases} \tag{1.15}$$

In particular, $\dim(\mathrm{Der}(A)) = n-1$ and $\mathrm{Der}(A)$ stabilizes the powers $\mathfrak{m}^i = (t^i)$ for $i = 1, \ldots, n-1$. As a consequence, the Lie algebra $\mathrm{Der}(A)$ is solvable.

In the latter case (where $p > 0$ divides n), there exists a unique $D_0 \in \mathrm{Der}(A)$ such that $D_0(t) = 1$ (arising from $d/dT \in \mathrm{Der}(k[T])$) and the vector space $\mathrm{Der}(A)$ has basis the $D_i = t^i D_0$, where $0 \le i \le n-1$. These satisfy the relations (1.15). In particular, $\dim(\mathrm{Der}(A)) = n$ and $\mathrm{Der}(A)$ does not stabilize $\mathfrak{m}$. But it stabilizes the powers $\mathfrak{m}^i$, where i is a positive multiple of p. If $n > p$ then $\mathfrak{m}^{n-p}$ is a nonzero subspace of A, stable by $\mathrm{Der}(A)$ and killed by D_0^p but not by D_0. As a consequence, the Lie algebra $\mathrm{Der}(A)$ is not simple. If $n = p$ then $\mathrm{Der}(A) = \mathrm{Der}\,(k[T]/(T^p))$ is the **Witt algebra**; it is solvable when $p = 2$, and simple when $p \ge 3$ (exercise).

Lemma 1.4.4 *Let A be an algebra.*

(i) We have $D(e) = 0$ for any $D \in \mathrm{Der}(A)$ and any idempotent $e \in A$.
(ii) If $A = B \times C$ is decomposable, then the natural map $\mathrm{Der}(B) \times \mathrm{Der}(C) \to \mathrm{Der}(A)$ is an isomorphism of Lie algebras.
(iii) Assume that A is finite-dimensional. Then $\mathrm{Der}(A) = 0$ if and only if A is reduced.

Proof (i) Since $e = e^2$, we have $D(e) = 2eD(e)$ and hence $eD(e) = 2e^2D(e) = 2eD(e) = D(e)$. It follows that $D(e) = 0$ as desired.

(ii) Denote by e the identity element of B; then $B = eA$. By (i), every $D \in \mathrm{Der}(A)$ satisfies $D(e) = 0$, and hence $D(B) \subset B$ in view of the Leibniz rule. Clearly, $D|_B \in \mathrm{Der}(B)$. Likewise, $D|_C \in \mathrm{Der}(C)$; this implies readily the statement.

(iii) If A is reduced, then it is spanned by its idempotents by Lemma 1.2.7. So $\mathrm{Der}(A) = 0$ by (i).

Conversely, assume that $\mathrm{Der}(A) = 0$. Using (ii), we may further assume that A is indecomposable; then A is local by Theorem 1.2.13. If $A \ne k$ then its maximal ideal $\mathfrak{m}$ is nonzero, and hence there exists $n \ge 2$ such that $\mathfrak{m}^{n-1} \ne 0 = \mathfrak{m}^n$ (Lemma 1.2.11). Choose a nonzero linear map $f : \mathfrak{m}/\mathfrak{m}^2 \to \mathfrak{m}^{n-1}$ and define a linear map $D : A = k \oplus \mathfrak{m} \to A$ by

$D(1) = 0$ and $D(a) = f(\bar{a})$ for any $a \in \mathfrak{m}$ with image $\bar{a} \in \mathfrak{m}/\mathfrak{m}^2$. Then $D \neq 0$ and one may readily check that D is a derivation. □

1.4.2 Restricted Lie algebras

Given a vector space V, we denote by $\mathfrak{gl}(V)$ the Lie algebra of endomorphisms of V (relative to the commutator map). If $V = k^n$ then we get the Lie algebra of $n \times n$ matrices, denoted by $\mathfrak{gl}_n$.

For any algebra A, the space of derivations $\mathrm{Der}(A)$ is a Lie subalgebra of $\mathfrak{gl}(A)$. If $p > 0$ then $\mathrm{Der}(A)$ is stable by the pth power map of linear maps $\mathfrak{gl}(A) \to \mathfrak{gl}(A)$, $u \mapsto u^p = u \circ \cdots \circ u$ (p times). We then say that $\mathrm{Der}(A)$ is a **restricted Lie subalgebra** of $\mathfrak{gl}(A)$.

The notion of restricted Lie algebra may be defined intrinsincally, as follows. One may check that the pth power map of $\mathfrak{g} = \mathfrak{gl}(V)$ satisfies the following relations:

(i) $(tx)^p = t^p x^p \qquad (t \in k, x \in \mathfrak{g})$,
(ii) $\mathrm{ad}(x^p) = \mathrm{ad}(x)^p \qquad (x \in \mathfrak{g})$,
(iii) $(x+y)^p = x^p + y^p + \sum_{i=1}^{p-1} s_i(x,y) \qquad (x, y \in \mathfrak{g})$.

Here we denote by

$$\mathrm{ad} : \mathfrak{g} \longrightarrow \mathfrak{gl}(\mathfrak{g}), \quad x \longmapsto (y \mapsto [x,y])$$

the adjoint representation, and we set

$$s_i(x,y) = -\frac{1}{i} \sum_u \mathrm{ad}(x_{u(1)}) \mathrm{ad}(x_{u(2)}) \cdots \mathrm{ad}(x_{u(p-1)})(x_1),$$

where u runs through the maps $\{1, 2, \ldots, p-1\} \to \{0, 1\}$ which take the value 0 exactly i times. For example, $s_1(x,y) = [x,y]$ if $p = 2$. We refer to [7, II.7.3.2] or [25, Prop. 10.38] for details.

Definition 1.4.5 A **restricted Lie algebra** is a Lie algebra $\mathfrak{g}$ equipped with a self-map $x \mapsto x^{[p]}$ (the **pth power map**) satisfying the above relations (i), (ii), (iii).

We will see that both notions of restricted Lie (sub)algebras are equivalent. For this, we will use two universal constructions in Lie theory.

First, one associates with every Lie algebra $\mathfrak{g}$, the **enveloping algebra** $\mathrm{U}(\mathfrak{g})$ defined as the quotient of the tensor algebra $\mathrm{T}(\mathfrak{g})$ by the two-sided ideal generated by the elements

$$x \otimes y - y \otimes x - [x,y], \quad x, y \in \mathfrak{g}.$$

Then $\mathrm{U}(\mathfrak{g})$ is an associative algebra with an identity element; it is commutative if and only if $\mathfrak{g}$ is commutative. The natural map $\alpha : \mathfrak{g} \to \mathrm{U}(\mathfrak{g})$ is the universal homomorphism of Lie algebras from $\mathfrak{g}$ to an associative algebra (viewed as a Lie algebra via the commutator map). If $\mathfrak{g}$ is finite-dimensional with basis $x_1, \ldots, x_n$, then the monomials

$$\alpha(x_1)^{i_1} \cdots \alpha(x_n)^{i_n} \quad (i_1, \ldots, i_n \geq 0)$$

form a basis of $\mathrm{U}(\mathfrak{g})$ by the Poincaré–Birkhoff–Witt theorem (see for example [25, Thm. 10.36]). In particular, α is injective.

Next, assume that $p > 0$ and consider a restricted Lie algebra $\mathfrak{g}$ with pth power map $x \mapsto x^{[p]}$. The **restricted enveloping algebra** $\mathrm{U}^{[p]}(\mathfrak{g})$ is the quotient of $\mathrm{U}(\mathfrak{g})$ by the ideal generated by the elements

$$\alpha(x)^p - \alpha(x^{[p]}), \quad x \in \mathfrak{g}$$

(these are contained in the center of $\mathrm{U}(\mathfrak{g})$ in view of (ii)). As above, $\mathrm{U}^{[p]}(\mathfrak{g})$ is an associative algebra with an identity element, equipped with a map $\beta : \mathfrak{g} \to \mathrm{U}^{[p]}(\mathfrak{g})$ which is the universal homomorphism of restricted Lie algebras from $\mathfrak{g}$ to an associative algebra (viewed as a restricted Lie algebra via the commutator and the pth power maps). If $\mathfrak{g}$ is finite-dimensional with basis $x_1, \ldots, x_n$ as above, then the monomials

$$\beta(x_1)^{i_1} \cdots \beta(x_n)^{i_n} \quad (0 \leq i_1, \ldots, i_n \leq p - 1)$$

form a basis of $\mathrm{U}^{[p]}(\mathfrak{g})$ (see [7, Prop. II.7.3.6] or [25, Thm. 10.40]). In particular, $\mathrm{U}^{[p]}(\mathfrak{g})$ is finite-dimensional; we have $\dim \mathrm{U}^{[p]}(\mathfrak{g}) = p^{\dim(\mathfrak{g})}$. Moreover, $\mathrm{U}^{[p]}(\mathfrak{g})$ is commutative if and only if $\mathfrak{g}$ is commutative.

As a consequence, every finite-dimensional restricted Lie algebra $\mathfrak{g}$ is equipped with a faithful finite-dimensional representation, namely, its representation in $\mathrm{U}^{[p]}(\mathfrak{g})$ by left multiplication. This yields:

Proposition 1.4.6 *Every finite-dimensional restricted Lie algebra is isomorphic to a restricted Lie subalgebra of* $\mathfrak{gl}_n$ *for some* n.

We also record the following easy result, see Corollary 1.4.25 for an application.

Lemma 1.4.7 *Every nonzero finite-dimensional restricted Lie algebra contains a restricted Lie subalgebra of dimension* 1.

Proof By Proposition 1.4.6, it suffices to show the assertion for a Lie subalgebra $\mathfrak{g} \subset \mathfrak{gl}_n$, stable by the pth power map.

We first consider the case where $\mathfrak{g}$ consists of nilpotent endomorphisms.

Then there exists a nonzero $x \in \mathfrak{g}$ such that $x^p = 0$, and hence kx is the desired subalgebra.

So we may assume that $\mathfrak{g}$ contains a non-nilpotent element x. We have $x = y + z$ where y is diagonalizable, z is nilpotent and y, z commute; thus, $y \neq 0$. Then $x^{p^n} = y^{p^n}$ for $n \gg 0$, and hence we may assume that x is diagonalizable.

Replacing x with a scalar multiple, we may further assume that 1 is an eigenvalue of x. If all its eigenvalues are in the prime field $\mathbb{F}_p$, then $x^p = x$ and hence kx is the desired subalgebra. Otherwise, $x^p - x$ is a nonzero diagonalizable element of $\mathfrak{g}$ and $\mathrm{Ker}(x) \subsetneq \mathrm{Ker}(x^p - x)$. Iterating this argument completes the proof. □

1.4.3 Zariski tangent spaces

We begin with a slight generalization of the notion of derivation of an algebra: given an algebra homomorphism $u : A \to B$, we say that a linear map $D : A \to B$ is a **derivation** if it satisfies the Leibniz rule: $D(a_1 a_2) = u(a_1)D(a_2) + D(a_1)u(a_2)$ for all $a_1, a_2 \in A$. The set of such derivations is a B-module, denoted by $\mathrm{Der}(A, B)$.

Derivations may be viewed as "infinitesimal algebra homomorphisms". More specifically, recall the algebra of dual numbers, $k[\varepsilon] = k[T]/(T^2)$. For any algebra A, we set $A[\varepsilon] = A \otimes k[\varepsilon]$. We then have the following result, whose proof is a direct verification.

Lemma 1.4.8 *Let $u : A \to B$ be an algebra homomorphism, and $D : A \to B$ a linear map. Then D is a derivation if and only if $u + \varepsilon D : A \to B[\varepsilon]$ is an algebra homomorphism.*

We will mostly consider the case where $B = k$, and view $\mathrm{Der}(A, k)$ as a subspace of the dual vector space A^*. This subspace can be described as follows:

Lemma 1.4.9 *Let $f : A \to k$ be an algebra homomorphism with kernel $\mathfrak{m}$, so that $A/\mathfrak{m} = k$. Then the assignment $D \in \mathrm{Der}(A, k) \mapsto D|_{\mathfrak{m}} \in \mathfrak{m}^*$ induces an isomorphism of vector spaces $\mathrm{Der}(A, k) \xrightarrow{\sim} (\mathfrak{m}/\mathfrak{m}^2)^*$.*

Proof Since $A = k \oplus \mathfrak{m}$, every $D \in \mathrm{Der}(A, k)$ is uniquely determined by $D|_{\mathfrak{m}}$. Moreover, $D|_{\mathfrak{m}^2} = 0$ by the Leibniz rule, and hence $D|_{\mathfrak{m}}$ factors through a unique linear map $\delta : \mathfrak{m}/\mathfrak{m}^2 \to k$. Conversely, given such a map δ, let $D : A \to k$ be the linear map such that $D(1) = 0$ and $D(a) = \delta(\bar{a})$ for any $a \in \mathfrak{m}$ with image $\bar{a} \in \mathfrak{m}/\mathfrak{m}^2$. Then one may readily check that D is a derivation. □

Definition 1.4.10 A **vector field** on an affine scheme X is a derivation of the algebra $A = \mathcal{O}(X)$.

The **Zariski tangent space** of X at $x \in X(k)$ is the vector space

$$T_x(X) = (\mathfrak{m}/\mathfrak{m}^2)^*,$$

where $\mathfrak{m}$ denotes the kernel of the homomorphism $x : A \to k$.

The composition of the map $\mathrm{Der}(A) \to \mathrm{Der}(A, A/\mathfrak{m}) = \mathrm{Der}(A, k)$, $D \mapsto x \circ D$ with the isomorphism $\mathrm{Der}(A, k) \xrightarrow{\sim} T_x(X)$ (Lemma 1.4.9) is the **evaluation map**

$$\mathrm{ev}_x : \mathrm{Der}(A) \longrightarrow T_x(X).$$

If X is a finite scheme, then Theorem 1.2.13 yields an isomorphism $T_x(X) \simeq (\mathfrak{m}_x/\mathfrak{m}_x^2)^*$, where $\mathfrak{m}_x$ denotes the maximal ideal of the local algebra $\mathcal{O}_{X,x}$. Also, the dimension of $T_x(X)$ is the minimal number of generators of this algebra in view of Lemma 1.2.12.

We will obtain an interpretation of the Zariski tangent space in terms of "infinitesimal calculus" on schemes. We keep the setting of Definition 1.4.10. By Lemma 1.4.8, we may identify $\mathrm{Der}(A, k)$ with the set of algebra homomorphisms $\varphi : A \to k[\varepsilon]$ such that $\pi \circ \varphi = x$, where π denotes the algebra homomorphism $k[\varepsilon] \to k$, $\varepsilon \mapsto 0$. This identifies $T_x(X)$ with the fiber at x of the map $X(\pi) : X(k[\varepsilon]) \to X(k)$. Also, we denote by σ the algebra homomorphism $k \to k[\varepsilon]$. Then $\pi \circ \sigma = \mathrm{id}$, and hence $X(\sigma)$ is a section of $X(\pi)$; it sends x to the origin of $T_x(X)$.

Next, consider a morphism of affine schemes $f : X \to Y$ and let $y = f(x) \in Y(k)$. Then we have a commutative square

$$\begin{array}{ccc} X(k[\varepsilon]) & \xrightarrow{f(k[\varepsilon])} & Y(k[\varepsilon]) \\ {\scriptstyle X(\pi)}\downarrow & & \downarrow{\scriptstyle Y(\pi)} \\ X(k) & \xrightarrow{f(k)} & Y(k), \end{array}$$

and hence a map

$$df_x : T_x(X) \longrightarrow T_y(Y),$$

the **differential** of f at x. The algebra homomorphism $f^* : \mathcal{O}(Y) \to \mathcal{O}(X)$ induces linear maps $\mathfrak{m}_y^n \to \mathfrak{m}_x^n$ for all positive integers n, and hence a linear map $\mathfrak{m}_y/\mathfrak{m}_y^2 \to \mathfrak{m}_x/\mathfrak{m}_x^2$. One may readily check that the transpose of the latter map is the differential df_x. Also, differentials satisfy the **chain rule**: for any morphisms of affine schemes

$$X \xrightarrow{f} Y \xrightarrow{g} Z, \quad x \longmapsto y \longmapsto z,$$

we have $d(g \circ f)_x = dg_y \circ df_x$.

For any two affine schemes X, Y and any $x \in X(k)$, $y \in Y(k)$, we have a natural isomorphism

$$T_{(x,y)}(X \times Y) \xrightarrow{\sim} T_x(X) \times T_y(Y) \tag{1.16}$$

given by $(d(\mathrm{pr}_X)_x, d(\mathrm{pr}_Y)_y)$.

If G is an affine group scheme, then $G(k[\varepsilon])$ is a group equipped with a homomorphism $G(\pi) : G(k[\varepsilon]) \to G(k)$ with kernel $T_e(G)$, and with a homomorphism $G(\sigma) : G(k) \to G(k[\varepsilon])$ such that $G(\pi) \circ G(\sigma) = \mathrm{id}$. As a consequence, we have

$$G(k[\varepsilon]) = T_e(G) \rtimes G(k). \tag{1.17}$$

Moreover, every homomorphism of affine group schemes $f : G \to H$ yields a homomorphism of (abstract) groups $f(k[\varepsilon]) : G(k[\varepsilon]) \to H(k[\varepsilon])$, which restricts to the differential $df_e : T_e(G) \to T_e(H)$. If f is the immersion of a closed subgroup scheme, then df_e is injective. In the next subsection, we will equip $T_e(G)$ with the structure of a (restricted) Lie algebra, and show that df_e is a homomorphism of such algebras.

1.4.4 The Lie algebra of an affine group scheme

Let G be an affine group scheme, and $A = \mathcal{O}(G)$ with comultiplication map Δ.

Definition 1.4.11 A derivation D of A is **left invariant** if the diagram

$$\begin{array}{ccc} A & \xrightarrow{D} & A \\ \downarrow{\scriptstyle\Delta} & & \downarrow{\scriptstyle\Delta} \\ A \otimes A & \xrightarrow{\mathrm{id} \otimes D} & A \otimes A \end{array}$$

commutes.

It is easy to check that the left invariant derivations form a restricted Lie subalgebra $\mathrm{Der}^G(A)$ of $\mathrm{Der}(A)$.

Proposition 1.4.12 *The evaluation map* $\mathrm{ev}_e : \mathrm{Der}(A) \to T_e(G)$ *restricts to an isomorphism*

$$\mathrm{Der}^G(A) \xrightarrow{\sim} T_e(G). \tag{1.18}$$

Proof Let $D \in \mathrm{Der}(A)$. Then D is left invariant if and only if the diagram

$$\begin{array}{ccc} A & \xrightarrow{\varphi} & A[\varepsilon] \\ \Big\downarrow{\scriptstyle \Delta} & & \Big\downarrow{\scriptstyle \Delta[\varepsilon]} \\ A \otimes A & \xrightarrow{\mathrm{id} \otimes \varphi} & A \otimes A[\varepsilon] \end{array}$$

commutes, where $\varphi = \mathrm{id} + \varepsilon D$. This is equivalent to the commutativity of the dual diagram

$$\begin{array}{ccc} G \times G \times \mathrm{Spec}(k[\varepsilon]) & \xrightarrow{\mathrm{id} \times \psi} & G \times G \\ \Big\downarrow{\scriptstyle \mu \times \mathrm{id}} & & \Big\downarrow{\scriptstyle \mu} \\ G \times \mathrm{Spec}(k[\varepsilon]) & \xrightarrow{\psi} & G \end{array}$$

where $\psi : G \times \mathrm{Spec}(k[\varepsilon]) \to G$ is the morphism of affine schemes corresponding to the algebra homomorphism $\varphi : A \to A[\varepsilon]$. In turn, this is equivalent to the equality $\psi(xy, z) = x\psi(y, z)$ for any algebra R, any $x, y \in G(R)$ and any $z \in \mathrm{Spec}(k[\varepsilon])(R) = \mathrm{Hom}_{\mathrm{alg}}(R, k[\varepsilon])$. But this amounts to the equality $\psi(x, z) = x\psi(e, z)$ for any R and $x \in G(R)$, where $\psi(e, z) \in \mathrm{Hom}(\mathrm{Spec}(k[\varepsilon]), G) = G(k[\varepsilon])$. As $\psi \circ (\mathrm{id}, \pi^*) = \mathrm{id}$, we have $\psi(e, z) \in \mathrm{Ker}(G(\pi))$. So $\psi(e, z) \in T_e(G)$; this yields the assertion. □

The above proposition is a key ingredient in the proof of a central result of the theory:

Theorem 1.4.13 *If $p = 0$ then every finite group scheme is reduced.*

Proof By Theorem 1.3.13, it suffices to show that every infinitesimal group scheme G is trivial. Let $\mathfrak{m}$ be the maximal ideal of $\mathcal{O}(G) = A$; then $T_e(G) = (\mathfrak{m}/\mathfrak{m}^2)^*$. By Proposition 1.4.12, there exist $D_1, \ldots, D_m \in \mathrm{Der}^G(A)$ such that $\mathrm{ev}_e(D_1), \ldots, \mathrm{ev}_e(D_m)$ form a basis of $T_e(G)$. So we may choose $a_1, \ldots, a_m \in \mathfrak{m}$ such that the images $\bar{a}_1, \ldots, \bar{a}_m \in \mathfrak{m}/\mathfrak{m}^2$ form the dual basis. Equivalently, $D_i(a_j)(e) = \delta_{i,j}$ for $1 \leq i, j \leq m$.

Consider the algebra $R = k[[T_1, \ldots, T_m]]$ consisting of the formal power series $\sum_{i_1,\ldots,i_m} c_{i_1,\ldots,i_m} T_1^{i_1} \cdots T_m^{i_m}$, where $i_1, \ldots, i_m$ run over the nonnegative integers, and $c_{i_1,\ldots,i_m} \in k$. Then R is a local k-algebra with maximal ideal $(T_1, \ldots, T_m)$ (consisting of the series with constant term 0) and residue field k. Each quotient $R/(T_1, \ldots, T_m)^n$ is isomorphic to the truncated polynomial ring $k[T_1, \ldots, T_m]/(T_1, \ldots, T_m)^n$, a local finite-dimensional algebra.

We have a linear map defined via "Taylor series expansion"

$$u : A \longrightarrow R, \quad a \longmapsto \sum_{i_1,\dots,i_m \geq 0} (D_1^{i_1} \cdots D_m^{i_m} a)(e) \frac{T_1^{i_1} \cdots T_m^{i_m}}{i_1! \cdots i_m!}.$$

Using the formula (1.14), one may check that u is an algebra homomorphism. Also, we have $u(a_i) \equiv T_i \bmod (T_1, \dots, T_m)^2$ for $i = 1, \dots, m$. In view of Lemma 1.2.12, it follows that the composition of u with the quotient map $R \to R/(T_1, \dots, T_m)^n$ is surjective for all $n \geq 0$. But A is finite-dimensional, and the dimension of $R/(T_1, \dots, T_m)^n$ is arbitrarily large if $m \geq 1$ and $n \gg 0$. So we must have $m = 0$, i.e., $A = k$. □

Theorem 1.4.13 is a special case of **Cartier's theorem**: if $p = 0$ then every algebraic group is reduced (see [26, p. 101, Thm.], from which the above proof is borrowed).

We now return to a finite group scheme G in arbitrary characteristic.

Definition 1.4.14 The **Lie algebra** $\mathrm{Lie}(G)$ is the Zariski tangent space $T_e(G)$ equipped with the restricted Lie algebra structure obtained via the isomorphism (1.18).

Example 1.4.15 Let $G = \mathbb{G}_a$. Then $A = k[T]$ and $\mathrm{Der}(A) = A\, d/dT$. One may readily check that $\mathrm{Der}^G(A) = k\, d/dT$. Thus, $\mathrm{Lie}(G) \simeq k$ with trivial pth power map.

Next, let $G = \mathbb{G}_m$. Then $A = k[T, T^{-1}]$ and $\mathrm{Der}(A) = A\, d/dT = A\, T d/dT$; moreover, $\mathrm{Der}^G(A) = k\, T d/dT$. So $\mathrm{Lie}(G) \simeq k$ with pth power map $x \mapsto x^p$.

Every homomorphism of affine group schemes $f : G \to H$ induces a linear map

$$\mathrm{Lie}(f) = df_e : \mathrm{Lie}(G) = T_e(G) \longrightarrow T_e(H) = \mathrm{Lie}(H).$$

Moreover, every G-action α on an affine scheme $X = \mathrm{Spec}(A)$ induces an action of the group $G(k[\varepsilon])$ on the algebra $A[\varepsilon]$ by $k[\varepsilon]$-algebra automorphisms, which lifts the $G(k)$-action on A by algebra automorphisms (Remark 1.3.10). Thus, for any $\xi \in T_e(G) = \mathrm{Ker}(G(\pi) : G(k[\varepsilon]) \to G(k))$, we obtain a $k[\varepsilon]$-automorphism of $A[\varepsilon]$ lifting id, and hence an algebra homomorphism $A \to A[\varepsilon]$, $a \mapsto a + \varepsilon D_\xi$. Then D_ξ is a derivation (Lemma 1.4.8), so that we obtain a map

$$\alpha' : T_e(G) \longrightarrow \mathrm{Der}(A), \quad \xi \longmapsto D_\xi.$$

Proposition 1.4.16 *With the above notation,* $\mathrm{Lie}(f)$ *and* α' *are homomorphisms of restricted Lie algebras.*

This follows from an alternative description of the restricted Lie algebra structure on $T_e(G)$ that we now sketch. The G-action on itself by conjugation fixes e, and hence induces a linear representation

$$\mathrm{Ad} : G \longrightarrow \mathrm{GL}_{T_e(G)},$$

the **adjoint representation** (see [25, Cor. 8.10]). The differential of Ad at e may be identified with a linear map

$$\mathrm{ad} : T_e(G) \longrightarrow \mathfrak{gl}(T_e(G)).$$

In fact, this map equips $T_e(G)$ with a restricted Lie algebra structure such that α' is a homomorphism of restricted Lie algebras for any G-action α (see [7, II.4.4.4, II.4.4.5, II.7.3.4]). In particular, the G-action on itself by right multiplication induces a homomorphism of restricted Lie algebras $T_e(G) \to \mathrm{Der}(\mathcal{O}(G))$; its image is $\mathrm{Der}^G(\mathcal{O}(G))$ by [7, II.4.4.6]). This completes the proof of Proposition 1.4.16 in view of the equivariance of f relative to the conjugation actions.

If $G = \mathrm{GL}_n$, then $T_e(G)$ is the vector space $\mathfrak{gl}_n$ of $n \times n$ matrices, and the adjoint representation is the conjugation action again. Its differential at e is the adjoint representation of $\mathfrak{gl}_n$ (see e.g. [25, Thm. 10.23]) and hence $\mathrm{Lie}(\mathrm{GL}_n) = \mathfrak{gl}_n$ as restricted Lie algebras.

Also, the Lie algebra of any closed subgroup scheme $H \leq G$ is a restricted Lie subalgebra of $\mathrm{Lie}(G)$. We illustrate this on the following:

Example 1.4.17 Let $G = \mu_n$, so that $A = k[T](T^n - 1)$. If $p = 0$ or n is prime to p, then A is reduced and hence $\mathrm{Lie}(G) = 0$ (e.g. by Lemma 1.4.4). On the other hand, if $p > 0$ divides n, then the inclusion $G \leq \mathbb{G}_m$ yields the equality $\mathrm{Lie}(G) = \mathrm{Lie}(\mathbb{G}_m)$ for dimension reasons. So $\mathrm{Lie}(G) = k$ with pth power map $x \mapsto x^p$.

Next, let $G = \alpha_p$. Then similarly, the inclusion $G \leq \mathbb{G}_a$ yields that $\mathrm{Lie}(G) = k$ with trivial pth power map.

Example 1.4.18 Let B be a finite-dimensional algebra. By assigning to any algebra R the automorphism group of the R-algebra $R \otimes B$, one obtains a group functor Aut_B. One may readily check that this group functor is represented by a closed subgroup scheme of GL_B, that we will still denote by Aut_B. We have $\mathrm{Aut}_B(k) = \mathrm{Aut}(B)$ (the automorphism group of B), and $\mathrm{Lie}(\mathrm{Aut}_B) = \mathrm{Der}(B)$ as follows from Lemma 1.4.8.

If B is reduced of dimension n, then $B \simeq k^n$ and hence $\mathrm{Aut}(B)$ is isomorphic to the symmetric group S_n (permuting the idempotents of B). Together with Lemma 1.4.4 (iii), it follows that $\mathrm{Aut}_B \simeq S_n$ as well.

On the other hand, if B is nonreduced, then $\mathrm{Aut}(B)$ is infinite. Indeed, using Theorem 1.2.13, we may assume that B is local with maximal ideal $\mathfrak{m}$; then $\mathrm{Aut}(B) \xrightarrow{\sim} \mathrm{Aut}(\mathfrak{m})$. Denote by n the smallest positive integer such that $\mathfrak{m}^n = 0$. If $n = 2$ then $\mathrm{Aut}(\mathfrak{m}) \simeq \mathrm{GL}(\mathfrak{m})$ is infinite. Otherwise, consider a linear map $f : \mathfrak{m}/\mathfrak{m}^2 \to \mathfrak{m}^{n-1} \subset \mathfrak{m}^2$ as in the proof of Lemma 1.4.4 (iii). Then one may check that the assignment $\mathfrak{m} \to \mathfrak{m}$, $x \mapsto x + f(\bar{x})$ yields an automorphism u_f of $\mathfrak{m}$. Moreover, the assignment $f \mapsto u_f$ yields an injective homomorphism $\mathrm{Hom}(\mathfrak{m}/\mathfrak{m}^2, \mathfrak{m}^{n-1}) \to \mathrm{Aut}(\mathfrak{m})$, where the left-hand side is viewed as an additive group. Thus, $\mathrm{Aut}(\mathfrak{m})$ is infinite in this case too, and hence Aut_B is infinite as well.

Also, Aut_B may be nonreduced if $p > 0$. Take indeed $B = k[T]/(T^p)$ and denote by t the image of T in B, as in Example 1.4.3. Then for any algebra R, the group $\mathrm{Aut}_B(R)$ consists of the maps

$$t \longmapsto a_0 + a_1 t + \cdots + a_{p-1} t^{p-1},$$

where $a_0, \ldots, a_{p-1} \in R$ satisfy $a_0^p = 0$ and $a_1 \in R^\times$. In particular, $\mathrm{Aut}_B(k) = \mathrm{Aut}(B)$ consists of the maps $t \mapsto a_1 t + \cdots + a_{p-1} t^{p-1}$, where $a_1 \in k^\times$ and $a_2, \ldots, a_{p-1} \in k$. So the reduced subgroup scheme $\mathrm{Aut}_{B,\mathrm{red}}$ is the proper closed subscheme of Aut_B with ideal (a_0).

1.4.5 The relative Frobenius morphism

In this subsection, we assume that $\mathrm{char}(k) = p > 0$. Let A be an algebra; then the Frobenius map

$$F = F_A : A \longrightarrow A, \quad a \longmapsto a^p.$$

is a ring endomorphism. But F is not an algebra endomorphism, since $F(ta) = t^p F(a)$ for all $t \in k$, $a \in A$. To correct this, we define a new algebra structure on A via $t \cdot a = t^{1/p} a$. The resulting algebra will be denoted by $A^{(p)}$. Then we obtain an algebra homomorphism

$$F_{A/k} : A^{(p)} \longrightarrow A, \quad a \longmapsto a^p,$$

or equivalently a morphism of affine schemes

$$F_{X/k} : X \longrightarrow X^{(p)},$$

where $X = \mathrm{Spec}(A)$ and $X^{(p)} = \mathrm{Spec}(A^{(p)})$. We say that $F_{X/k}$ is the **relative Frobenius morphism**. If A is finite-dimensional, then $A^{(p)}$ is finite-dimensional as well, and $\dim(A^{(p)}) = \dim(A)$; thus, $F_{X/k}$ is a morphism of finite schemes.

Next, let $f : X \to Y$ be a morphism of affine schemes and denote by $u : \mathcal{O}(Y) = B \to A$ the corresponding algebra homomorphism. Then we have $F_A \circ u = u \circ F_B$. As a consequence, u induces a homomorphism $u^{(p)} : B^{(p)} \to A^{(p)}$ such that $F_{A/k} \circ u^{(p)} = u \circ F_{B/k}$. Equivalently, we have a commutative diagram of affine schemes

$$\begin{array}{ccc} X & \xrightarrow{F_{X/k}} & X^{(p)} \\ {\scriptstyle f}\downarrow & & \downarrow{\scriptstyle f^{(p)}} \\ Y & \xrightarrow{F_{Y/k}} & Y^{(p)} \end{array}$$

Also, we have a natural isomorphism of algebras

$$A^{(p)} \otimes B^{(p)} \xrightarrow{\sim} (A \otimes B)^{(p)},$$

and hence the relative Frobenius morphism commutes with products. Thus, for any affine group scheme G, there exists a natural structure of affine group scheme on $G^{(p)}$ such that $F_{G/k}$ is a homomorphism. Its kernel is the **Frobenius kernel**; we denote it by G_1. If G is finite, then $G^{(p)}$ is finite and we have $|G| = |G^{(p)}|$.

This construction can be iterated: given a positive integer n, we replace p with p^n and F with $F^n : A \to A$, $a \mapsto a^{p^n}$. We then denote by $A^{(p^n)}$ the ring A equipped with an algebra structure via $t \cdot a = t^{1/p^n} a$. This yields a homomorphism $F^n_{A/k} : A^{(p^n)} \to A$, and hence a morphism

$$F^n_{X/k} : X \longrightarrow X^{(p^n)},$$

the n**th iterated relative Frobenius morphism**. The above properties of the relative Frobenius morphism extend to this setting. So for any affine group scheme G, we obtain a homomorphism

$$F^n_{G/k} : G \longrightarrow G^{(p^n)},$$

where $G^{(p^n)}$ is an affine group scheme; if G is finite, then $G^{(p^n)}$ is finite of the same order. The kernel of $F^n_{G/k}$ is the n**th Frobenius kernel** G_n.

This notion gives back some of the examples of §1.2.1:

Example 1.4.19 If G is the multiplicative group $\mathbb{G}_m$, then G_n is the group scheme μ_{p^n} (the kernel of the p^nth power map), of order p^n.

Also, if G is the additive group $\mathbb{G}_a$, then $G_1 = \alpha_p$. More generally, G_n is denoted by α_{p^n}; it has order p^n as well.

Finally, the nth Frobenius kernel of an elliptic curve E has order p^n again. Moreover, we have

$$E_n \simeq \begin{cases} \mu_{p^n} & \text{if } E \text{ is ordinary,} \\ \alpha_{p^n} & \text{if } E \text{ is supersingular} \end{cases}$$

(see [26, §22, Thm.]).

In these three examples, G_n is contained in the kernel of the multiplication by p^n. This is a general fact for commutative group schemes, see [7, IV.3.4.10].

We now record some properties of the iterated relative Frobenius morphism of affine schemes:

Lemma 1.4.20 *Let X be an affine scheme, and n a positive integer.*

(i) The morphism $F^n_{X/k}$ is bijective on k-points.
(ii) For any $x \in X(k)$, the fiber of $F^n_{X/k}$ at $y = F^n_{X/k}(x)$ satisfies

$$\mathcal{O}(X_y) = \mathcal{O}(X)/(f^{p^n}, f \in \mathfrak{m}_x),$$

where $\mathfrak{m}_x$ denotes the maximal ideal of x.
(iii) Assume in addition that X is finite. Then $F^n_{X/k}$ is an isomorphism if and only if X is reduced.

Proof (i) Let $A = \mathcal{O}(X)$. We have to show that every algebra homomorphism $f : A^{(p)} \to k$ extends uniquely to an algebra homomorphism $A \to k$. The uniqueness follows from the fact that every element of k has a unique pth root. For the existence, let $\mathfrak{m} = \mathrm{Ker}(f)$ and $B = A/\mathfrak{m}A$. Then the algebra B has a maximal ideal, and hence there exists an algebra homomorphism $g : B \to K$, where K is a field extension of k. By construction, we have $K^p \subset k$ and hence $K = k$. So $g : B \to k$ yields the desired extension.

(ii) This follows readily from the fact that the ideal of X_y in $\mathcal{O}(X)$ is generated by $(F^n_{X/k})^*(\mathfrak{m}_x)$.

(iii) Using Theorem 1.2.13, we may assume that A is local. If $A = k$ then $(F^n_{X/k})^* = \mathrm{id}$. Otherwise, A has nonzero nilpotent elements, and hence there exists $a \in A$ such that $a^p = 0 \neq a$. Thus, $(F^n_{X/k})^*$ is not injective. □

Lemma 1.4.21 *Let G be an affine group scheme. Then the iterated Frobenius kernels G_n form an increasing sequence of infinitesimal subgroup schemes of G. Moreover,* $\mathrm{Lie}(G_1) = \mathrm{Lie}(G_2) = \cdots = \mathrm{Lie}(G)$.

Assuming in addition that G is finite, we have:

(i) G_n is trivial if and only if G is reduced.
(ii) $G_n \leq G^0$ with equality if and only if $f^{p^n} = 0$ for all $f \in \mathfrak{m}$, where $\mathfrak{m}$ denotes the maximal ideal of e in $\mathcal{O}(G)$.
(iii) $G_n = G^0$ for $n \gg 0$.

Proof Recall that G_n is the fiber of $F^n_{G/k}$ at e. By Lemma 1.4.20, it follows that e is the unique k-point of G_n, and hence G_n is infinitesimal.

Using Lemma 1.4.20 again, we have

$$\mathcal{O}(G_n) = \mathcal{O}(G)/(f^{p^n}, f \in \mathfrak{m}). \tag{1.19}$$

Moreover, the ideals $(f^{p^n}, f \in \mathfrak{m})$ form a decreasing sequence, and hence the closed subschemes G_n form an increasing sequence.

Also, recall that $\mathrm{Lie}(G) = T_e(G) = T_e(G^0) = (\mathfrak{m}/\mathfrak{m}^2)^*$. Likewise, we have $\mathrm{Lie}(G_n) = (\bar{\mathfrak{m}}/\bar{\mathfrak{m}}^2)^*$, where $\bar{\mathfrak{m}}$ denotes the maximal ideal of e in G_n. So

$$\bar{\mathfrak{m}} = \mathfrak{m}/(f^{p^n}, f \in \mathfrak{m}) \subset \mathcal{O}(G)/(f^{p^n}, f \in \mathfrak{m}) = \mathcal{O}(G_n).$$

As $f^{p^n} \in \mathfrak{m}^2$ for all $f \in \mathfrak{m}$, the natural map $\mathfrak{m}/\mathfrak{m}^2 \to \bar{\mathfrak{m}}/\bar{\mathfrak{m}}^2$ is an isomorphism. Thus, $\mathrm{Lie}(G_n) = \mathrm{Lie}(G)$; this yields the first assertion.

We now assume that G is finite. Since G_n is infinitesimal, it is a subgroup scheme of G^0 by Theorem 1.3.13. So we may replace G with G^0, i.e., assume that G is infinitesimal.

To show (i), it suffices to check that G is trivial if G_n is trivial. But then the ideal $\mathfrak{m}$ is generated by its p^nth powers, and hence is zero by Lemma 1.2.12.

The remaining assertions from (ii) and (iii) follow from the isomorphism (1.19) and the vanishing of $\mathfrak{m}^n$ for $n \gg 0$ (Lemma 1.2.11). □

Definition 1.4.22 Let G be an infinitesimal group scheme. If G is nontrivial, then its **height** $\mathrm{ht}(G)$ is the smallest integer $n \geq 1$ such that $G_n = G$. We set $\mathrm{ht}(G) = 0$ if G is trivial.

By Lemma 1.4.21, the height of G exists and satisfies

$$\mathrm{ht}(G) = \min\{n \mid f^{p^n} = 0 \text{ for all } f \in \mathfrak{m}\},$$

where $\mathfrak{m}$ denotes the maximal ideal of the local algebra $\mathcal{O}(G)$.

Theorem 1.4.23 *Let G be an infinitesimal group scheme. Then we have $\mathrm{ht}(G) \leq 1$ if and only if there exists an isomorphism of algebras*

$$\mathcal{O}(G) \simeq k[T_1, \ldots, T_n]/(T_1^p, \cdots, T_n^p). \tag{1.20}$$

Moreover, the assignment $G \mapsto \mathrm{Lie}(G)$ yields an equivalence between the categories of infinitesimal group schemes of height at most 1, and of finite-dimensional restricted Lie algebras.

We refer to [25, Chap. 11.h] for the broad lines of the proof, and to [7, II.7.4] for the full details. We will only sketch the construction of a quasi-inverse functor: given a finite-dimensional restricted Lie algebra $\mathfrak{g}$, one considers its restricted enveloping algebra $\mathrm{U} = \mathrm{U}^{[p]}(\mathfrak{g})$ (§1.4.2), and equips it with algebra homomorphisms

$$\Delta : \mathrm{U} \longrightarrow \mathrm{U} \otimes \mathrm{U}, \quad x \in \mathfrak{g} \longmapsto x \otimes 1 + 1 \otimes x,$$

$$S : \mathrm{U} \longrightarrow \mathrm{U}, \quad x \in \mathfrak{g} \longmapsto -x,$$

$$\varepsilon : \mathrm{U} \longrightarrow k, \quad x \in \mathfrak{g} \longmapsto 0.$$

Then U is a finite-dimensional Hopf algebra, which is co-commutative (i.e., the image of Δ is fixed pointwise by the involution $x \otimes y \mapsto y \otimes x$ of $U \otimes U$), but not necessarily commutative. Thus, the dual vector space is a finite-dimensional commutative Hopf algebra, which yields a finite group scheme $G(\mathfrak{g})$. One then checks that $G(\mathfrak{g})$ is infinitesimal with Lie algebra $\mathfrak{g}$.

The equivalence of categories of Theorem 1.4.23 extends to actions of group schemes: given an infinitesimal group scheme G of height 1 and a scheme $X = \mathrm{Spec}(A)$, the G-actions on X correspond bijectively to the homomorphisms of restricted Lie algebras $\mathrm{Lie}(G) \to \mathrm{Der}(A)$ via $\alpha \mapsto \alpha'$ (see [7, II.7.3.10]). In view of Example 1.4.17, it follows that the actions of μ_p (resp. α_p) correspond bijectively to the vector fields D such that $D^p = D$ (resp. $D^p = 0$).

We now present applications of Theorem 1.4.23 to the structure of finite group schemes:

Corollary 1.4.24 *Let G be an infinitesimal group scheme of height* 1. *Then we have $|G| = p^n$, where $n = \dim(\mathrm{Lie}(G))$.*

Proof The isomorphism (1.20) implies that $|G| = \dim(\mathcal{O}(G)) = p^n$, since the monomials $T_1^{i_1} \cdots T_n^{i_n}$, where $0 \leq i_1, \ldots, i_n \leq p-1$, yield a basis of $k[T_1, \ldots, T_n]/(T_1^p, \ldots, T_n^p)$. Also, this isomorphism identifies the maximal ideal $\mathfrak{m}$ of $\mathcal{O}(G)$ with the image of the maximal ideal $(T_1, \ldots, T_n)$ of $k[T_1, \ldots, T_n]$. Since $(T_1^p, \ldots, T_n^p) \subset (T_1, \ldots, T_n)^2$, this yields in turn an isomorphism $\mathfrak{m}/\mathfrak{m}^2 \simeq (T_1, \ldots, T_n)/(T_1, \ldots, T_n)^2$. Thus, $\dim(\mathrm{Lie}(G)) = \dim(\mathfrak{m}/\mathfrak{m}^2) = n$. □

Corollary 1.4.25 *Every nonreduced finite group scheme contains a subgroup scheme isomorphic to μ_p or α_p.*

Proof This follows by combining Theorem 1.4.23 with Lemmas 1.4.7 and 1.4.21. □

Corollary 1.4.26 *Every finite group scheme of prime order ℓ is isomorphic to $\mathbb{Z}/\ell\mathbb{Z}$ if $\ell \neq p$, and to $\mathbb{Z}/p\mathbb{Z}$, μ_p or α_p if $\ell = p$.*

Proof By Theorem 1.3.13, we have $\mathcal{O}(G) \simeq \mathcal{O}(G^0) \otimes \mathcal{O}(G_{\mathrm{red}})$ as algebras. Counting dimensions, this yields $|G| = |G^0| \cdot |G_{\mathrm{red}}|$. Thus, the assumption that $|G| = \ell$ implies that G is infinitesimal or reduced. In the latter case, we have $G \simeq \mathbb{Z}/\ell\mathbb{Z}$. In the former case, G contains a subgroup scheme H isomorphic to μ_p or α_p by Corollary 1.4.25. Thus, we have $\ell = p$; moreover, the resulting algebra homomorphism $\mathcal{O}(G) \to \mathcal{O}(H)$ is surjective, and hence bijective for dimension reasons. □

Finally, we mention a remarkable structure result for the algebras of infinitesimal group schemes (see [7, Cor. III.3.6.3] or [25, Thm. 11.29]):

Theorem 1.4.27 *Let G be an infinitesimal group scheme. Then there exists an isomorphism of algebras*

$$\mathcal{O}(G) \simeq k[T_1, \ldots, T_n]/(T_1^{p^{a_1}}, \ldots, T_n^{p^{a_n}}),$$

where $a_1, \ldots, a_n$ are positive integers.

By considering a monomial basis of $\mathcal{O}(G)$ as in the proof of Corollary 1.4.24, this yields $|G| = \dim(\mathcal{O}(G)) = p^{a_1 + \cdots + a_n}$. In particular, every infinitesimal group scheme is a p-group. We will obtain an alternative proof of this result in Corollary 1.5.15.

1.5 Quotients

Throughout this section, we consider a finite group scheme G.

Definition 1.5.1 Let X be an affine G-scheme, and $f : X \to Y$ a morphism. We say that f is **G-invariant** if $f(g \cdot x) = f(x)$ for any algebra R and any $g \in G(R)$, $x \in X(R)$.

The invariance condition is equivalent to the equality $f \circ \alpha = f \circ \mathrm{pr}_X$, where $\alpha : G \times X \to X$ denotes the action, and $\mathrm{pr}_X : G \times X \to X$ the projection. In turn, this is equivalent to the equality $\alpha^* \circ f^* = (1 \otimes \mathrm{id})^* \circ f^*$ for the algebra homomorphism $f^* : \mathcal{O}(Y) \to \mathcal{O}(X)$.

Remark 1.5.2 Assume that $p > 0$ and consider an action $G \times X \to X$. Then the induced morphism $G^{(p^n)} \times X^{(p^n)} \to X^{(p^n)}$ is an action as well, and the nth relative Frobenius morphism $F^n_{X/k} : X \to X^{(p^n)}$ is equivariant relative to the homomorphism $F^n_{G/k} : G \to G^{(p^n)}$. Indeed, this follows readily from the functorial properties of relative Frobenius morphisms.

If in addition G is infinitesimal of height n, then $F^n_{X/k}$ is G-invariant.

Observe that every affine G-scheme X is equipped with a G-invariant morphism $q : X \to Y$ which satisfies the following universal property: for any invariant morphism of affine schemes $\varphi : X \to Z$, there exists a unique morphism $\psi : Y \to Z$ such that $\varphi = \psi \circ q$. Indeed, let $A = \mathcal{O}(X)$ and consider the **algebra of invariants**

$$A^G = \{a \in A \mid \alpha^*(a) = \mathrm{pr}_X^*(a)\}.$$

Then the inclusion of A^G into A corresponds to a morphism of affine schemes $q : X \to Y$. Moreover, the morphism φ corresponds to an algebra homomorphism $\mathcal{O}(Z) \to A$ with image contained in A^G. So q satisfies the above universal property. We say that q is the **categorical quotient** of X by G and we denote Y by X/G.

The following properties of the categorical quotient are easily verified:

Lemma 1.5.3 *Let G be a finite group scheme, and X, Y two affine G-schemes with categorical quotients $q_X : X \to X/G$, $q_Y : Y \to Y/G$.*

(i) Every G-equivariant morphism $f : X \to Y$ induces a morphism $f/G : X/G \to Y/G$ such that $f/G \circ q_X = q_Y \circ f$.
(ii) If G acts trivially on Y, then there exists a natural isomorphism

$$(X \times Y)/G \xrightarrow{\sim} X/G \times Y. \tag{1.21}$$

(iii) The finite group $G(k)$ acts on X/G^0 and there exists a natural isomorphism

$$(X/G^0)/G(k) \xrightarrow{\sim} X/G. \tag{1.22}$$

Also, for any algebraic affine G-scheme X, the scheme X/G is algebraic in view of the following:

Theorem 1.5.4 *Let A be a finitely generated G-algebra. Then the algebra of invariants A^G is finitely generated and the A^G-module A is finitely generated as well.*

Proof We may choose $a_1, \ldots, a_m \in A$ such that $A = k[a_1, \ldots, a_m]$.

We first treat the case where G is reduced. Consider the polynomials

$$P_i(T) = \prod_{g \in G} (T - g \cdot a_i) \in A[T] \quad (i = 1, \ldots, m).$$

Then $P_i(a_i) = 0$ and P_i has degree $|G| = N$. Moreover, the coefficients of $P_i(T)$ are G-invariant, since G permutes the $g \cdot a_i$. In other words, $P_i(T) \in A^G[T]$ for all i. Let B denote the subalgebra of A generated by the coefficients of the P_i. Then B is finitely generated and contained in A^G. Moreover, we have $a_i^N \in B+Ba_i+\cdots+Ba_i^{N-1}$ for $i = 1, \ldots, m$, since $P_i(a_i) = 0$. By induction, it follows that $a_i^r \in B + Ba_i + \cdots + Ba_i^{N-1}$ for $r \geq N$ and $i = 1, \ldots, m$. As a consequence, the B-module A is generated by the monomials $a_1^{i_1} \cdots a_m^{i_m}$, where $0 \leq i_1, \ldots, i_m \leq N - 1$. In particular, A is a finite B-module. Using again the finite generation of B, this yields that $A^G \subset A$ is a finite B-module as well. In turn, this yields the assertions in this case.

Next, we treat the case where G is infinitesimal. Let $n = \mathrm{ht}(G)$, then $a^{p^n} \in A^G$ for any $a \in A$ in view of Remark 1.5.2. Thus, $C = k[a_1^{p^n}, \ldots, a_m^{p^n}]$ is a subalgebra of A^G. Also, the C-module A is generated by the monomials $a_1^{i_1} \cdots a_m^{i_m}$, where $0 \leq i_1, \ldots, i_m \leq p - 1$. This yields the assertions by arguing as in the first case.

For an arbitrary finite group scheme G, the algebra A^{G^0} is equipped with an action of $G(k)$ such that $(A^{G^0})^{G(k)} = A^G$. So we conclude by combining the two above cases. □

Remark 1.5.5 Given an affine algebraic G-scheme X, the categorical quotient is an orbit space in the following sense: the natural map $X(k)/G(k) \to (X/G)(k)$ is bijective. Indeed, this is a consequence of [11, Ex. 13.4] if G is reduced. Also, q is bijective if G is infinitesimal (use Lemma 1.4.20 and Remark 1.5.2). This yields the assertion for an arbitrary G in view of Lemma 1.5.3 (iii).

In other words, the set-theoretic fibers of q at k-rational points are exactly the $G(k)$-orbits. But the natural map $X(R)/G(R) \to (X/G)(R)$ is not necessarily bijective for an algebra R. For example, consider the action of μ_n on $\mathbb{A}^1 = \mathrm{Spec}(k[T])$ by multiplication: $x \cdot y = xy$. Then $k[T]^{\mu_n} = k[T^n]$ and hence q is not surjective on (say) $k(T)$-valued points. Also, note that the fiber of q at 0 is the nonreduced scheme $\mathrm{Spec}(k[T]/(T^n))$.

Remark 1.5.6 Invariant theory of finite groups is related to Galois theory as follows. Assume that G is a finite group of automorphisms of an algebra A, which is an integral domain with fraction field K. This yields a G-action on K by field automorphisms; the invariant subfield K^G is easily seen to be the fraction field of A^G. By Galois theory, we have $G = \mathrm{Aut}_{K^G}(K)$ and K is a finite extension of K^G of degree $|G|$. As a consequence, $G = \mathrm{Aut}_{A^G}(A)$ and the A^G-module A has rank $|G|$.

This does not extend to finite group schemes if $p > 0$. For example, consider the action of μ_p on $\alpha_p^n = \alpha_p \times \cdots \times \alpha_p$ (n times) via $y \cdot (x_1, \ldots, x_n) = (yx_1, \ldots, yx_n)$ and form the corresponding semi-direct product $G = \alpha_p^n \rtimes \mu_p$. Then G acts on $\mathbb{A}^1$ via

$$(x_1, \ldots, x_n, y) \cdot t = yt + x_1 + x_2 t^p + \cdots + x_n t^{p^{n-1}}$$

and the algebra of G-invariants of $\mathcal{O}(\mathbb{A}^1) = k[T]$ is $k[T^p]$. So $|G| = p^{n+1}$ and $[K : K^G] = p$ for the induced G-action on $K = k(T)$.

This failure is explained by the fact that the extension K/K^G is purely inseparable for a (functorial) action of an infinitesimal group scheme G on a field K. More specifically, if $\mathrm{ht}(G) \leq n$ then $f^{p^n} \in K^G$ for any $f \in K$ (Remark 1.5.2). In particular, the extension K/K^G has exponent p if $\mathrm{ht}(G) = 1$.

Definition 1.5.7 An action of G on an affine scheme X is **free** if the group $G(R)$ acts freely on the set $X(R)$ for any algebra R.

More specifically, if $g \in G(R)$ and $x \in X(R)$ satisfy $g \cdot x = x$, then $g = e$.

Example 1.5.8 The group scheme G acts freely on itself by left multiplication, and also by right multiplication. But the G-action on itself by conjugation is not free (if G is nontrivial), since this action fixes e.

Remark 1.5.9 Assume that G is a finite group acting faithfully on an affine variety X. Then there exists a nonempty open subset $U \subset X$ which is G-stable (i.e., $g \cdot x \in U$ for all $g \in G$ and $x \in U$) and on which G acts freely. Indeed, the fixed point locus $X^g = \{x \in X \mid g \cdot x = x\}$ is a proper closed subset of X for any $g \in G$. Thus, we may take for U the free locus $X \setminus \bigcup_{g \in G} X^g$.

This does not extend to actions of finite group schemes in view of the example in Remark 1.5.6 (exercise).

Definition 1.5.10 A morphism of affine schemes $f : X \to Y$ is **finite locally free (of rank n)** if the $\mathcal{O}(Y)$-module $\mathcal{O}(X)$ is finitely generated and projective (of constant rank n).

By [11, Ex. 4.12], this is equivalent to the existence of $g_1, \ldots, g_m \in \mathcal{O}(Y)$ such that $(g_1, \ldots, g_m) = \mathcal{O}(Y)$ and the $\mathcal{O}(Y)[\frac{1}{g_i}]$-module $\mathcal{O}(X)[\frac{1}{g_i}]$ is free (of rank n) for $i = 1, \ldots, m$. Here $\mathcal{O}(Y)[\frac{1}{g_i}]$ denotes the localization of $\mathcal{O}(Y)$ by g_i, and likewise for $\mathcal{O}(X)[\frac{1}{g_i}]$.

For example, the projection $\mathrm{pr}_Y : X \times Y \to Y$ is finite locally free if and only if X is finite; then pr_Y has rank $n = \dim(\mathcal{O}(X))$. Also, the immersion $i : X \to X \sqcup Y$ is finite locally free, but not of constant rank if X, Y are nonempty.

Remark 1.5.11 The notion of a finite locally free morphism extends to (not necessarily affine) schemes as follows: a morphism of schemes $f : X \to Y$ is finite locally free if for any affine open subset V of Y, the preimage $U = f^{-1}(V)$ is affine and $f|_U : U \to V$ is finite locally free. For example, given an elliptic curve E with origin 0 and a positive integer n, the morphism $n_E : E \to E$, $x \mapsto nx$ of Example 1.2.3 is finite locally free of rank n^2 (as follows from [15, Ex. IV.4.2]). Thus, the schematic kernel $E[n]$ is a finite group scheme of order n^2.

Theorem 1.5.12 *Let $X = \mathrm{Spec}(A)$ be a scheme of finite type equipped with a free action of G, and $q : X \to X/G = \mathrm{Spec}(A^G)$ the categorical quotient.*

(i) The morphism q is finite locally free of rank $|G|$.
(ii) The morphism $G \times X \to X \times X$, $(g, x) \mapsto (x, g \cdot x)$ induces an isomorphism $G \times X \xrightarrow{\sim} X \times_{X/G} X$.

The second assertion means that the categorical quotient by a free action of G is a **principal G-bundle** in the sense of topologists.

We refer to [26, §12, Thm. 1] or [25, Thm. B.18] for the proof of the above result, and record an important consequence:

Corollary 1.5.13 *Let $H \leq G$ be a subgroup scheme, and $q : G \to G/H$ the categorical quotient by the action via right multiplication.*

(i) The morphism q is finite locally free of rank $|H|$.
(ii) We have $|G| = [G : H]\,|H|$, where $[G : H] = \dim(\mathcal{O}(G/H)) = \dim(\mathcal{O}(G)^H)$.
(iii) The G-action on itself by left multiplication yields a unique action on G/H such that q is equivariant.
(iv) If H is normal in G, then G/H has a unique structure of a finite group scheme such that q is a homomorphism.

Proof (i) This follows from Example 1.5.8 and Theorem 1.5.12.

(ii) By (i), the algebra $\mathcal{O}(G)$ is a projective module of constant rank $|H|$ over its subalgebra $\mathcal{O}(G)^H$. This yields the assertion by counting dimensions.

(iii) Consider the H-action on $G \times G$ via $h \cdot (g_1, g_2) = (g_1, g_2 h^{-1})$. Then the multiplication map $\mu : G \times G \to G$ is equivariant, and hence induces a morphism $\alpha : G \times G/H \to G/H$ as categorical quotients commute with products by schemes with a trivial action (Lemma 1.5.3 (i) and (ii)). Using this commutation property again, one may check that α is an action.

(iv) This is proved by a similar argument. □

With the above assumptions, the quotient G/H is called a **homogeneous space**. The structure theorem for the algebras of infinitesimal group schemes (Theorem 1.4.27) extends unchanged to their homogeneous spaces in view of [7, III.3.6.2].

In turn, this yields further structure results for finite group schemes:

Corollary 1.5.14 *Every finite group scheme G admits a canonical sequence of normal subgroup schemes*

$$e = N_0 \leq N_1 \leq \cdots \leq N_n = G^0$$

such that every quotient N_i/N_{i-1} has height 1.

Proof Take $N_1 = G_1$ (the Frobenius kernel) and argue by induction on $|G|$ by using Corollary 1.5.13. □

Corollary 1.5.15 *If G is infinitesimal, then $|G| = p^n$ for some $n \geq 0$.*

Proof This follows readily by combining Corollaries 1.4.24, 1.5.13 (ii) and 1.5.14. □

Corollary 1.5.16 *The simple finite group schemes are exactly the finite simple groups and the infinitesimal group schemes of height* 1 *associated with the simple finite-dimensional restricted Lie algebras.*

There is a well-known and widely used classification of finite simple groups. The simple finite-dimensional restricted Lie algebras have also been classified, except in small characteristic. More specifically, these Lie algebras are in bijective correspondence with the simple finite-dimensional Lie algebras (see e.g. [36, Thm. 4.1]). These have been classified by Block, Wilson, Strade and Premet: in characteristic $p \geq 7$, they are either of classical type (e.g., $\mathfrak{gl}_n/kI_n$ if n is prime to p) or of Cartan type (e.g., the Jacobson-Witt algebra $\mathrm{Der}(k[T_1, \ldots, T_n]/(T_1^p, \ldots, T_n^p))$ unless $n = 1$ and $p = 2$). If $p = 5$, one gets in addition the Melikian algebras. We refer to [33] for a full account of these developments, and [36] for a nice survey. In characteristics $p = 2, 3$, there are many additional simple (restricted) Lie algebras, see e.g. the recent preprint [5].

1.6 The inverse Galois problem for group schemes

In its original form, the inverse Galois problem may be stated as follows:

Question 1.6.1 Given a finite group G, does there exist a Galois extension of the field of rational numbers with Galois group G?

This classical problem is unsolved, even if many finite groups have been realized as Galois groups over $\mathbb{Q}$; this includes all solvable groups by a theorem of Shafarevich (see [27, (9.5.1)]).

A fruitful approach, initiated by Hilbert, consists in realizing G as a Galois group over the field of rational functions $\mathbb{Q}(T_1, \ldots, T_n)$. Then G can be realized as a Galois group over $\mathbb{Q}$ by specializing $T_1, \ldots, T_n$ appropriately (as a consequence of Hilbert's irreducibility theorem; see e.g. [31, §3.4]). This applies for example to the symmetric group S_n: consider its action on the field of rational functions $k(U_1, \ldots, U_n)$ by permuting the variables. Then the field of invariants $k(U_1, \ldots, U_n)^{S_n}$ is generated by the elementary symmetric functions, and hence is isomorphic to $k(T_1, \ldots, T_n)$.

We refer to [21] for a recent survey of the inverse Galois problem over an arbitrary field K, which asks which finite groups occur as Galois groups over K. It is known that every finite group G can be realized as a Galois group over any **algebraic function field of one variable** K over k, i.e., K/k is a finitely generated field extension of transcendence degree 1. More specifically, there exist infinitely many Galois extensions L/K such that the equalities $G = \mathrm{Aut}_K(L) = \mathrm{Aut}_k(L)$ hold (see [13] for $k = \mathbb{C}$, and [19] for the general case).

In view of the correspondence between algebraic function fields of one variable and algebraic curves (see [15, Cor. I.6.12]), it follows that every nonsingular projective curve X admits a **ramified Galois covering** $q : Y \to X$ with group G, i.e., Y is a nonsingular projective curve equipped with a faithful action of G, and q is the categorical quotient. Moreover, G is the full automorphism group $\mathrm{Aut}(Y)$.

The automorphism group of a nonsingular projective curve X can be described in terms of the genus $g = g(X)$ as follows. If $g = 0$ then X is isomorphic to the projective line $\mathbb{P}^1$, and hence $\mathrm{Aut}(X) \simeq \mathrm{Aut}(\mathbb{P}^1) = \mathrm{PGL}_2(k)$. If $g = 1$ then choosing a point $0 \in X(k)$, we get a commutative algebraic group structure on the elliptic curve X, with neutral element 0. Thus, X acts on itself by translations. One may check that $\mathrm{Aut}(X) = X \rtimes \mathrm{Aut}(X, 0)$, where X denotes the subgroup of translations, and $\mathrm{Aut}(X, 0)$ stands for the subgroup fixing the origin. Moreover, $\mathrm{Aut}(X, 0)$ is finite of order dividing 24 (see [32, Thm. III.10.1]). Finally, if $g \geq 2$ then $\mathrm{Aut}(X)$ is finite of order at most $84(g-1)$ (see [15, Ex. IV.2.5]). In particular, $\mathrm{Aut}(X)$ is the group of k-points of an algebraic group for any nonsingular projective curve X.

More generally, one associates to any projective scheme X the **automorphism group scheme** Aut_X. Its points with values in an algebra R are the automorphisms of $X \times \mathrm{Spec}(R)$ of the form $(x, y) \mapsto (f(x, y), y)$, where $f : X \times \mathrm{Spec}(R) \to X$ is a morphism. (We may view f as a family of automorphisms of X parameterized by $\mathrm{Spec}(R)$). In particular, $\mathrm{Aut}_X(k) = \mathrm{Aut}(X)$. Also, the Lie algebra $\mathrm{Lie}(\mathrm{Aut}_X)$ (the kernel of the group homomorphism $\mathrm{Aut}_X(k[\varepsilon]) \to \mathrm{Aut}_X(k)$, $\varepsilon \mapsto 0$) is identified with the Lie algebra $\mathrm{Vect}(X)$ of vector fields on X. The scheme Aut_X is **locally of finite type**, i.e., it admits an open covering by affine schemes of finite type (but Aut_X is not necessarily an algebraic group, see Example 1.6.4 below). Given a group scheme G, the G-actions on X correspond bijectively to the homomorphisms of group schemes $G \to \mathrm{Aut}_X$.

The construction of Aut_X is due to Grothendieck in a much more general setting, see [14]; it has been extended to proper schemes over an arbitrary field by Matsumura and Oort (see [24]). If X is a finite scheme, this gives back the group scheme Aut_B of Example 1.4.18, where $B = \mathcal{O}(X)$; in particular, Aut_X is not necessarily reduced if $p > 0$. Also, if X is a nonsingular projective curve, then Aut_X is a reduced algebraic group (as a consequence of the above description of $\mathrm{Aut}(X)$ together with a Lie algebra argument). In view of the above discussion, this yields:

Proposition 1.6.2 *Every finite group can be realized as the automorphism group scheme of a nonsingular projective curve.*

By contrast, there are strong restrictions on infinitesimal subgroup schemes of nonsingular projective curves. For example, their Lie algebra has dimension at most 3. If one considers a possibly singular projective curve X, then there are still strong restrictions on the Lie algebra of Aut_X as follows from [30, Thm. 12.1]. The same holds for several classes of nonsingular projective surfaces, see [35, 22, 23].

By analogy with the inverse Galois problem, one may ask:

Question 1.6.3 Which group schemes can be realized as the automorphism group scheme of a projective scheme X? One may further impose geometric conditions on X, e.g., restrict to nonsingular varieties.

This question is wide open, even if our understanding of automorphism group schemes has improved significantly during the last years. Before mentioning some recent developments, we review a classical structure result on the group scheme Aut_X, where X is a proper scheme. Since this group scheme is locally of finite type, its connected component of the identity is a normal subgroup scheme of finite type, denoted by Aut_X^0 and called the **connected automorphism group scheme**. Moreover, the quotient group scheme

$$\mathrm{Aut}_X/\mathrm{Aut}_X^0 = \pi_0(\mathrm{Aut}_X)$$

exists; it is a discrete group scheme which parameterizes the connected components of Aut_X (see [7, II.5.1]). Thus, we may view $\pi_0(\mathrm{Aut}_X)$ as an abstract group. This **component group** is countable (as follows from [14]), and generally infinite in view of the following:

Example 1.6.4 Let E be an elliptic curve with origin 0, and $X = E \times E$. One may check that $\mathrm{Aut}_X = X \rtimes \mathrm{Aut}_{X,(0,0)}$, where X acts on itself by translation. Moreover, $\mathrm{Aut}_X^0 = X$ and $\pi_0(\mathrm{Aut}_X) \simeq \mathrm{Aut}_{X,(0,0)}$. In particular, $\pi_0(\mathrm{Aut}_X)$ contains the group $\mathrm{GL}_2(\mathbb{Z})$ acting on X via $\begin{pmatrix} a & b \\ c & d \end{pmatrix} \cdot (z, w) = (az + bw, cz + dw)$.

In this example, one may easily show that $\pi_0(\mathrm{Aut}_X)$ is an arithmetic group; as a consequence, it admits a finite presentation. More generally, the group $\pi_0(\mathrm{Aut}_X)$ is arithmetic for any **abelian variety**, i.e., a projective group variety. But there exist complex nonsingular projective varieties X such that Aut_X is discrete and non-finitely generated; see [16] for the first example of such a variety, in dimension 6, and [9] for further examples in dimension 2. Such examples also exist when p is odd and k is not the algebraic closure of its prime field, see [28].

So the component group of Aut_X is quite mysterious. By contrast, the identity component can be any prescribed connected algebraic group:

Theorem 1.6.5 *Let G be a connected algebraic group. Then there exists a projective variety X such that $G \simeq \mathrm{Aut}_X^0$. If $p = 0$, one may further take X nonsingular.*

Taking for G an infinitesimal group scheme of height 1 and using the equivalence of categories of Theorem 1.4.23 and its version for actions of group schemes (see [7, II.7.3.10]), this yields:

Corollary 1.6.6 *Assume that $p > 0$ and let $\mathfrak{g}$ be a finite-dimensional restricted Lie algebra. Then there exists a projective variety X such that $\mathfrak{g} \simeq \mathrm{Vect}(X)$.*

Also, Question 1.6.3 has been answered for abelian varieties:

Theorem 1.6.7 *Let A be an abelian variety with origin 0. Then there exists a projective variety X such that $A \simeq \mathrm{Aut}_X$ if and only if the group $\mathrm{Aut}(A, 0)$ is finite. Under these conditions, one may further take X nonsingular.*

This result was first obtained by Lombardo and Maffei over the field of complex numbers (see [18]), and then extended in [1] to an algebraically closed ground field. As a consequence, every elliptic curve E can be realized as the automorphism group scheme of a nonsingular projective variety, but $E \times E$ admits no such realization.

Part of Theorem 1.6.7 has been generalized in [3, Thm. 2]; this yields a necessary condition for a reduced connected algebraic group G to be the full automorphism group scheme of a nonsingular projective variety. But this only gives restrictions when G is not affine. So Question 1.6.3 is still unanswered for linear algebraic groups.

One may also consider Question 1.6.3 over an arbitrary ground field k (not necessarily algebraically closed). Here again, some results have been obtained recently: Proposition 1.6.2 extends to an arbitrary field, see [2]. It also extends to a finite field k and a finite commutative group scheme of order prime to the characteristic, as a consequence of the main result of [6]. Also, Theorem 1.6.5 holds over an arbitrary field (see [4]), as well as Theorem 1.6.7 by a result of Florence (see [12]).

Acknowledgements Many thanks to Sami Al–Asaad, Pascal Fong, Matilde Maccan, David Stewart and an anonymous referee, for their careful reading of preliminary versions of this text and their valuable comments.

References

[1] J. Blanc, M. Brion, "Abelian varieties as automorphism groups of smooth projective varieties in arbitrary characteristics," Ann. Fac. Sci. Toulouse Math. (6) **32** (2023), 607–622.

[2] D. Bragg, "Automorphism groups of curves over arbitrary fields," preprint, `https://arxiv.org/abs/2304.02778`

[3] M. Brion, "Automorphism groups of almost homogeneous varieties," in: Facets of algebraic geometry. A collection in honor of William Fulton's 80th birthday. Volume 1, London Math. Soc. Lecture Note Series **472**, pp. 54–76, Cambridge Univ. Press, 2022.

[4] M. Brion, S. Schröer, "The inverse Galois problem for connected algebraic groups," preprint, `https://arxiv.org/abs/2205.08117`

[5] D. Cushing, G. Stagg, D. Stewart, "A Prolog assisted search for new simple Lie algebras," Math. Comp. **93** (2024), 1473–1495.

[6] R. Darda, T. Yasuda, "Inverse Galois problem for semicommutative finite group schemes," preprint, `https://arxiv.org/abs/2210.01495`

[7] M. Demazure, P. Gabriel, "Groupes algébriques," Masson, Paris, 1970.

[8] M. Demazure, A. Grothendieck, "Séminaire de Géométrie Algébrique du Bois Marie, 1962–64, Schémas en groupes (SGA3)," Tome I. Propriétés générales des schémas en groupes, Doc. Math. **7**, Soc. Math. France, Paris, 2011.

[9] T. Dinh, K. Oguiso, "A surface with discrete and nonfinitely generated automorphism group," Duke Math. J. **168** (2019), 941–966.

[10] D. Eisenbud, J. Harris, "The geometry of schemes," Graduate Texts Math. **197**, Springer, New York, 2000.

[11] D. Eisenbud, "Commutative algebra with a view towards algebraic geometry," Graduate Texts Math. **150**, Springer, New York, 1996.

[12] M. Florence, "Realization of Abelian varieties as automorphism groups," preprint, `https://arxiv.org/abs/2102.02581`

[13] L. Greenberg, "Maximal groups and signatures," in: Discontinuous groups and Riemann surfaces, pp. 207–226. Princeton Univ. Press, Princeton, N.J., 1974.

[14] A. Grothendieck, "Techniques de construction et théorèmes d'existence en géométrie algébrique IV: les schémas de Hilbert," Sém. Bourbaki, Vol. **6** (1960–1961), Exp. 221, 249–276.

[15] R. Hartshorne, "Algebraic geometry," Graduate Texts Math. **52**, Springer, New York, 1977.

[16] J. Lesieutre, "A projective variety with discrete, non-finitely generated automorphism group," Invent. Math. **212** (2018), 189—211.

[17] C. Liedtke, "A McKay correspondence in positive characteristic," preprint, `https://arxiv.org/abs/2207.06286`

[18] D. Lombardo, A. Maffei, "Abelian varieties as automorphism groups of smooth projective varieties," Int. Math. Res. Not. (2020), 1942—1956.

[19] M. Madan, M. Rosen, "The group of automorphisms of a function field," Proc. Amer. Math. Soc. **115** (1992), 923–929.

[20] D. Madden, R. Valentini, "The group of automorphisms of algebraic function fields," J. Reine Angew. Math. **343** (1983), 162–168.

[21] G. Malle, B. Matzat, "Inverse Galois theory. 2nd edition," Springer Monographs in Mathematics, Springer, Berlin, 2018.

[22] G. Martin, "Infinitesimal automorphisms of algebraic varieties and vector fields on elliptic surfaces," Algebra Number Theory **16** (2022), 1655–1704.

[23] G. Martin, "Automorphism group schemes of bielliptic and quasi-bielliptic surfaces," Épijournal Géom. Algébrique **6** (2022), Article no. 9.

[24] H. Matsumura, F. Oort, "Representability of group functors, and automorphisms of algebraic schemes," Invent. Math. **4** (1967), 1–25.

[25] J. S. Milne, "Algebraic groups. The theory of group schemes of finite type over a field," Cambridge Stud. Adv. Math. **170**, Cambridge Univ. Press, Cambridge, 2017.
[26] D. Mumford, "Abelian varieties," Oxford Univ. Press, Oxford, 1970.
[27] J. Neukirch, A. Schmidt, K. Wingberg, "Cohomology of number fields. 2nd ed.," Grundlehren Math. Wiss. **323**, Springer, Berlin, 2008.
[28] K. Oguiso, "A surface in odd characteristic with discrete and non-finitely generated automorphism group," Adv. Math. **375** (2020) 107397, 20 pp.
[29] R. Pink, "Finite group schemes," course notes available at `https://people.math.ethz.ch/~pink/ftp/FGS/CompleteNotes.pdf`
[30] S. Schröer, N. Tziolas, "The structure of Frobenius kernels for automorphism group schemes," Algebra Number Theory **17** (2023), 1637–1680,
[31] J.-P. Serre, "Topics in Galois theory. Notes written by Henri Darmon. 2nd ed.," Research Notes in Math. **1**, Wellesley, MA, 2007.
[32] J. H. Silverman, "The arithmetic of elliptic curves. Second edition," Graduate Texts Math.**106**, Springer, New York, 2009.
[33] H. Strade, "Simple Lie algebras over fields of positive characteristic. I, II, III," de Gruyter Expositions in Mathematics **38**, **42**, **57**, de Gruyter, Berlin, 2013–2017.
[34] J. Tate, "Finite flat group schemes," in: Modular forms and Fermat's last theorem, pp. 121–154, Springer, New York, 1997.
[35] N. Tziolas, "Automorphisms of smooth canonically polarized surfaces in positive characteristic," Adv. Math. **310** (2017), 235–289; corrigendum ibid., 585–593.
[36] F. Viviani, " Simple finite group schemes and their infinitesimal deformations," Rend. Sem. Mat. Univ. Politec. Torino **68** (2010), 171–182.
[37] W. Waterhouse, "Introduction to affine group schemes," Graduate Texts Math. **66**, Springer, New York, 1979.

2
Algorithms for polycyclic groups

Bettina Eick[a]

Abstract

This paper gives a survey about the currently used methods for computing with polycyclic groups. It discusses the different representations for polycyclic groups, gives a brief outline of many existing methods and considers two algorithms in a little more detail: the Frattini subgroup algorithm and the methods for solving the conjugacy problem. The final section of the paper exhibits some open problems.

2.1 Introduction

A group is polycyclic if it is solvable and every subgroup is finitely generated. Equivalently, a group is polycyclic if it has a subnormal series with cyclic quotients. Polycyclic groups are finitely presented and this makes them interesting from a computational point of view.

The class of polycyclic groups is closed with respect to forming subgroups, quotient groups and extensions. Polycyclic groups are residually finite and the number of infinite cyclic quotients in a subnormal series with cyclic quotients is an invariant; it is called the *Hirsch length* nowadays. These and many other results have been proved in a sequence of papers by Hirsch [30, 31, 32, 33, 34] and have led to a rich structure theory of these groups. A first introduction to polycyclic groups can be found in Section 5.4 of [48] and a more detailed account in the book by Segal [50].

a Institut für Analysis und Algebra, Technische Universität Braunschweig, Braunschweig, Germany.
b.eick@tu-braunschweig.de

Examples of polycyclic groups arise in many different areas. For example, every finitely generated nilpotent group and every finite solvable group is polycyclic. A celebrated result by Malcev [44] asserts that every solvable subgroup of $GL(n, \mathbb{Z})$ is polycyclic. Further, every algebraic number field can be encoded into a polycyclic group in some sense: if K is an algebraic number field with maximal order $\mathcal{O}_K$ and unit group $\mathcal{U}_K$, then the split extension $\mathcal{O}_K \rtimes \mathcal{U}_K$ is polycyclic. Moreover, many crystallographic groups are polycyclic, in particular, all of those in dimension at most 3.

Some historical comments

Each finite p-group is polycyclic and the polycyclic structure of finite p-groups has been exploited in many algorithms for this type of group. Early examples of this are the enumerations by Higman [28, 29] and Sims [52] and in the algorithms to construct finite p-groups by Newman & O'Brien [45].

The methods for finite p-groups have been generalised in different directions. Nickel [46] used the polycyclic structure of finitely generated nilpotent groups to design a nilpotent quotient algorithm. Laue, Neubüser & Schoenwaelder [37] described various algorithms for finite solvable groups again by using their polycyclic structure. Arbitrary polycyclic groups have then been considered by Eick in joint work with Nickel and Ostheimer; we refer to [13] for an overview.

All of these algorithms are based on the fundamental observation that polycyclic groups can be described by a particular type of finite presentation: the so-called *polycyclic presentations*. These presentations allow effective computations with the groups they define. The Handbook [35] gives an introduction to this type of presentation and some algorithms based on them, see also Section 2.3.3 below for a brief introduction.

A different line of algorithmic theory for arbitrary polycyclic groups has been introduced by Baumslag, Cannonito, Robinson & Segal [6]. They proved the decidability of many algorithmic questions around polycyclic groups, but without a view towards practicability.

Aims of this survey

The first aim of this survey is to summarise the state of the art of computations with polycyclic groups. We discuss the different ways in which polycyclic groups can be represented on a computer, give a brief outline of the existing methods and exhibit two example algorithms in a little more detail: computing the Frattini subgroup of a polycyclic group (joint work with Neumann-Brosig [19]) and solving the conjugacy problem for elements in a polycyclic group (based on joint work with Ostheimer [21]).

Throughout, we are going to focus on *practical* algorithms. This terminology is not very precisely defined. We use it to indicate *algorithms that can be implemented and produce results on interesting examples.* Our algorithms have been implemented in the computer algebra system GAP [56] and are publically available in the Polycyclic Package [20].

2.2 Preliminaries

This section recalls some of the well-known structure properties of polycyclic groups. First, we exhibit three equivalent characterisations for them. For a proof we refer to [48, p. 152].

Theorem 2.2.1 *Let G be a group. The following are equivalent:*

(a) G has a polycyclic series, *i.e. a subnormal series with cyclic factors.*
(b) G is solvable and every subgroup of G is finitely generated.
(c) G has a normal series whose factors are finitely generated abelian.

If one of these conditions holds, then G is polycyclic.

It is easy to observe that the class of polycyclic groups is closed under taking subgroups, factor groups and extensions. We collect some further interesting structural properties of polycyclic groups in the following.

Theorem 2.2.2 *Let G be polycyclic.*

(a) The number of infinite cyclic factors in a polycyclic series of G is an invariant; it is called the Hirsch length.
(b) G has a maximal nilpotent normal subgroup, its Fitting subgroup *$Fit(G)$. The quotient $G/Fit(G)$ is abelian-by-finite.*
(c) G has finitely many conjugacy classes of finite subgroups. If G is nilpotent, then the elements of finite order form a subgroup $T(G)$.
(d) Each maximal subgroup of G has finite prime-power index in G. If G is non-trivial, then G has proper maximal subgroups.
(e) G is residually finite.

Proof We include proofs or references for completeness.
(a) Given two different polycyclic series, they have isomorphic refinements. Each refinement of a polycyclic series has the same number of infinite cyclic quotients as the original series.
(b) The polycyclic group G satisfies the maximal condition on subgroups and hence has a maximal nilpotent normal subgroup. This is its Fitting subgroup $Fit(G)$. The quotient $G/Fit(G)$ is abelian-by-finite due to a theorem by Malcev [44].
(c) Both parts can be proved using induction along a normal series with elementary or free abelian factors. For the first part consider $N \trianglelefteq G$ free abelian with U/N finite. Then there are only finitely many conjugacy classes of complements to N in U, since $H^1(U/N, N)$ is finite. For the second part consider an infinite cyclic central subgroup $N \trianglelefteq G$. If U/N is finite, then U is abelian of the form $A \times B$ with A infinite cyclic and B finite. In this case B is the unique torsion subgroup of U.
(d) Follows readily, see also [48, 5.4.3].
(e) This is due to Hirsch, see also [48, 5.4.17]. □

The Frattini subgroup $\Phi(G)$ of a group G is the intersection of all maximal subgroups of G. (If G has no maximal subgroups, then $\Phi(G) = \{1\}$). The following is a celebrated result by Hirsch [34].

Theorem 2.2.3 *The Frattini subgroup of a polycyclic group is nilpotent.*

Despite this very significant structure result, the practical determination of the Frattini subgroup of an infinite polycyclic group remained difficult for a long time. We discuss this in Section 2.9 below.

2.3 Representations of polycyclic groups

In applications, polycyclic groups arise in various different forms: they arise as matrix groups over $\mathbb{Q}$ or over a finite field, they arise as permutation groups or they arise as quotients of finitely presented groups. For algorithmic purposes it is most useful to describe them via polycyclic presentations; that is, a special type of finite presentation that is associated with the polycyclic structure of the group and facilitates effective computations. The Handbook [35, Chap. 8] contains an introduction to this topic. We recall the main features here for completeness.

Definition 2.3.1 Let G be polycyclic with polycyclic series $G = G_1 >$

$G_2 > \ldots > G_n > G_{n+1} = \{1\}$. For $1 \leq i \leq n$ choose $g_i \in G$ so that $g_i G_{i+1}$ generates G_i/G_{i+1} and write $r_i = [G_i : G_{i+1}]$ if this is finite and let $r_i = 0$ otherwise. Then

(a) $(g_1, \ldots, g_n)$ is a *polycyclic generating sequence* (pcgs).
(b) $(r_1, \ldots, r_n)$ are the associated *relative orders.*

The following lemma is central in working with polycyclic groups. Its proof is elementary and omitted here.

Lemma 2.3.2 *Let G be polycyclic with pcgs $(g_1, \ldots, g_n)$. Then each $g \in G$ has a* unique normal form

$$g = g_1^{e_1} \cdots g_n^{e_n}$$

with $e_i \in \mathbb{Z}$ and $0 \leq e_i < r_i$ if $r_i \neq 0$.

Let G be polycyclic. Each pcgs of G induces a finite presentation of G whose relations form a rewriting system that facilitates the determination of normal forms of elements. We exhibit a more precise definition.

Lemma 2.3.3 *Let G be polycyclic with pcgs $(g_1, \ldots, g_n)$. For $1 \leq i, j \leq n$ with $j < i$ and $x, y \in \{\pm 1\}$ write*

(a) $g_i^x g_j^y = g_j^y g_{j+1}^{t_{i,j,j+1}(x,y)} \cdots g_n^{t_{i,j,n}(x,y)}$
(b) $g_i^{r_i} = g_{i+1}^{s_{i,i+1}} \cdots g_n^{s_{i,n}}$

where the left hand sides are the corresponding normal forms for the words on the right hand side. Then the generators $(g_1, \ldots, g_n)$ and these relations form a finite presentation for G.

The presentation determined in Lemma 2.3.3 is the *consistent polycyclic presentation* induced by the pcgs $(g_1, \ldots, g_n)$ and its associated relative orders. It is a confluent rewriting system for G.

The word problem can be solved readily in a polycyclic group given by a consistent polycyclic presentation, since the relations of the presentation allow one to rewrite an arbitrary word into its normal form as described in Lemma 2.3.2. The algorithm performing such a rewriting process is called *Collection.* The Collection Algorithm is known to work well in practice; this has been investigated in detail by Leedham-Green & Soicher [38]. An improved novel version has been described by Assmann & Linton [4]. A main drawback of the Collection algorithm is that it does not have polynomial complexity, see Gebhardt [23] for details.

2.4 Arithmetic in polycyclic groups

Let G be a polycyclic group with pcgs $(g_1, \ldots, g_n)$. The multiplication in G has the form

$$(g_1^{x_1} \cdots g_n^{x_n})(g_1^{y_1} \cdots g_n^{y_n}) = g_1^{f_1(x,y)} \cdots g_n^{f_n(x,y)},$$

where $x = (x_1, \ldots, x_n)$ and $y = (y_1, \ldots, y_n)$ and $f_i(x, y)$ is a function in $2n$ variables. A central question in this setting is *What type of functions are $f_1, \ldots, f_n$?* Hall [27] proved the following partial answer:

Theorem 2.4.1 *Let G be polycyclic and torsion free nilpotent.*

(a) There exists a polycyclic series for G which is a central series with infinite cyclic factors.
(b) Let $(g_1, \ldots, g_n)$ be a pcgs for G associated with a polycyclic series as in (a). Then $f_1, \ldots, f_n$ can be represented by polynomials in $\mathbb{Q}[x, y]$.

Polynomials describing the arithmetic in polycyclic and torsion free nilpotent groups can be computed; there is an algorithm introduced by Leedham-Green & Soicher [39] and an alternative method by Cant & Eick [9] based on work by Sims [54, page 443].

2.5 Translation to polycyclic presentations

Polycyclic groups arise in various different forms in group theory. A first step in computations with polycyclic groups is to determine a consistent polycyclic presentation for a given group. We discuss this briefly for the most often occuring types of representations.

Permutation groups and finite matrix groups: Given a subgroup G of a symmetric group $Sym(n)$, there is an algorithm by Sims [53] to check if G is polycyclic and, if so, then to determine a consistent polycyclic presentation for G. This algorithm is highly practical and an implementation is available in GAP [56]. Finite matrix groups can be treated in a similar way using the bound given in [10, Theorem 6.2A].

Rational matrix groups: Given a finitely generated subgroup G of $GL(n, \mathbb{Q})$, there is a method available to check if G is polycyclic and, if so, then to determine a consistent polycyclic presentation for G. We refer to the work of Assmann & Eick [2, 3] with an implementation in [1] and we note that these methods are moderately practical.

Finitely presented groups: Let G be a finitely presented group. A consistent polycyclic presentation for G/G' can be determined readily using the Smith normal form algorithm for integral matrices. A consistent polycyclic presentation for a nilpotent quotient $G/\gamma_c(G)$, where $\gamma_c(G)$ is the c-th term of the lower central series of G, can be obtained using the method by Nickel [46] with an implementation in [47]. The nilpotent quotient algorithm extends the abelian quotient algorithm and is highly practical. Further, Lo [42] invented a method to determine a consistent polycyclic presentation for $G/G^{(n)}$, where $G^{(n)}$ is the n-th term of the derived series of G, provided that this quotient is polycyclic. If the quotient is not polycyclic, then Lo's method may not terminate. Lo's algorithm is the most general of the available quotient methods. Its range of application is often limited.

2.6 Survey of algorithms

Suppose that a group G is given by a consistent polycyclic presentation. What can we compute for G? Many practical algorithms are available for such groups, see [35, Chapter 8] for an introduction. We give a brief overview here. Throughout, let G be a polycyclic group given by a consistent polycyclic presentation on the generators $(g_1, \ldots, g_n)$.

Subgroups, factor groups and homomorphisms: Let $G = G_1 > \ldots > G_n > G_{n+1} = \{1\}$ be the polycyclic series defined by $G_i = \langle g_i, \ldots, g_n \rangle$. Suppose that a subgroup U of G is given via generators. Then the subgroups $U_i = U \cap G_i$ yield a polycyclic series $U = U_1 \geq \ldots \geq U_n \geq U_{n+1} = \{1\}$. This is a natural choice for a polycyclic series for U. An *induced pcgs* $(u_1, \ldots, u_m)$ for U is a pcgs associated with this series (after possibly eliminating subgroups with $U_i = U_{i+1}$). An induced pcgs for a subgroup U can be computed from a set of generators of U with an algorithm called the *non-commutative Gauss-algorithm*. This has been described for finite groups by Laue, Neubüser & Schoenwaelder [37]. It is not difficult to extend this to infinite groups, see [13]. An induced pcgs facilitates an effective membership test and comparision of subgroups.

Computations with factor groups and homomorphisms can be facilitated in a similar fashion. We omit the details here and refer to [35, Section 8.4] instead.

Orbit-stabilizer methods: Let G act on a set X. Given $x, y \in X$, the aim of the orbit methods include to decide if there exists $g \in G$

with $x^g = y$ and, if so, then to determine one such g. The aim of the stabilizer methods is to determine a finite generating set for the stabilizer $Stab_G(x)$. If the orbit x^G is finite, then there is a highly effective method known to determine the full orbit and, simultaneously, an induced pcgs for $Stab_G(x)$; we refer to [35, Section 8.6] for a description. The method works by induction along a polycyclic series $G = G_1 > \ldots > G_n > G_{n+1} = \{1\}$ extending orbits and stabilisers from G_{i+1} to G_i.

A significant improvement over this method can be achieved in the special case that G is a finite p-group acting linearly on the vector space $X = \mathbb{F}_p^d$. This has first been described by Schwingel [49] and can also be found in [18].

If the orbit x^G is infinite, then it may not be possible to list it and solving the orbit-stabilizer problem may not be possible in this case. It is possible in certain special cases. In particular, it can be solved if G is acting linearly on $X = \mathbb{Z}^d$, see the work by Eick & Ostheimer [21]. An extension of this method is used by Eick [16]. These methods rely significantly on the structure of polycyclic subgroups of $GL(d, \mathbb{Z})$.

Commutators and subgroup series: Commutator subgroups in G and series based on commutator subgroups can be computed readily for G. Thus, for example, the derived series and the lower central series of G are straightforward to determine. This also allows one to check if G is nilpotent. Based on the computation of the derived series one can determine a normal series whose factors are either free abelian or elementary abelian. This is frequently used in algorithms as it facilitates induction approaches.

Complements and Extensions: Let A be an abelian normal subgroup in G. A complement to A in G is a subgroup K of G with $AK = G$ and $A \cap K = \{1\}$. The complement problem asks if such a complement exists, and if so, then to compute one complement. A related question is if there are finitely many complements or finitely many conjugacy classes of complements and, if so, then to determine them. There are effective algorithms to solve all of these problems. They translate the problem to an application of linear algebra and solve it in this form. We refer to [35, Sec. 8.7.1] for details.

Let A be a finitely generated abelian group and suppose that a homomorphism $G \to Aut(A)$ is given. An extension of A by G is a group H together with an embedding $\iota : A \to H$ so that $H/\iota(A) \cong G$. The extension problem asks to determine one or all extensions of A by G up to equivalence. There are effective algorithms available to solve these problems. As in the complement case, the extension algorithms translate the problem into an application of linear algebra and then solve this. A difference to the complement case is that the translation is significantly more expensive to compute. We refer to [35, Sec. 8.7.1] for details and to [7] for an example application of these methods.

Intersections of subgroups: Intersections of subgroups of G can be computed readily. A first description for finite polycyclic groups was introduced by Glasby & Slattery [24] and for finitely generated nilpotent groups by Lo [41]. An algorithm for the general case is exhibited by Eick [13].

Centralizers and Normalizers: Given $g, h \in G$, the conjugacy problem asks whether there exists $x \in G$ with $g^x = h$ and, if so, then to determine one such x. Dual to this problem is the computation of the centralizer $C_G(g) = \{y \in G \mid g^y = g\}$. Algorithms to solve these problems use induction along a normal series with elementary or free abelian factors. Each induction step translates to an orbit-stabilizer problem for a polycyclic group acting linearly. In turn, this can be solved with the available orbit-stabilizer methods.

Given $U, V \leq G$, the subgroup conjugacy problem asks whether there exists $x \in G$ with $U^x = V$ and, if so, then to determine one such x. Dual to this problem is the computation of the normalizer $N_G(U) = \{y \in G \mid U^y = U\}$. Again, algorithms to solve these problems use induction along a normal series with elementary or free abelian factors. A description of this algorithm for finite groups was given by Glasby & Slattery [24], for finitely generated nilpotent groups by Lo [41] and for arbitrary polycyclic groups by Eick [16].

Finite subgroups: A polycyclic group has finitely many conjugacy classes of finite subgroups. Computing them is an interesting task. Also, it is of interest to check if G has a torsion subgroup; that is, if the elements of finite order form a subgroup. The later is always the case if G is nilpotent, but not necessarily in general, as the infinite dihedral group $D_\infty = \langle a, b \mid a^2, b^a = b^{-1} \rangle$ shows. The conjugacy classes of finite subgroups can be determined using induction along a normal series with elementary abelian or free abelian factors, see [14].

Maximal subgroups: For finite polycyclic groups there is a highly effective algorithm available to compute the maximal subgroups and the Sylow subgroups of the considered group, see Cannon, Eick & Leedham-Green [8]. This algorithm determines a *special pcgs* for the considered group that allows reading off these subgroups. The special pcgs algorithm does not generalize to the infinite case. In the infinite case there is also a maximal subgroup algorithm available, see [14], but this is less effective than the finite group method.

Fitting Subgroup, Centre and FC-Centre: The Fitting subgroup $Fit(G)$ is the maximal nilpotent normal subgroup of G. An algorithm for computing $Fit(G)$ is outlined in [15]. This algorithm is based on the characterisation of $Fit(G)$ as the centralizer of a specific series of G. The centre $Z(G)$ can then be determined as the centralizer in $Z(Fit(G))$ under G and this allows one to determine $Z(G)$ from $Fit(G)$. Further, the FC-Centre of G is the set of all elements $g \in G$ so that the conjugacy class g^G is finite. This also can be determined based on the Fitting subgroup as exhibited in [15].

2.7 Representations for polycyclic groups

Integral matrix representations play a major role in the algorithmic theory of polycyclic groups and they turn up naturally in many algorithms. This is underlined by the following two famous theorems.

Theorem 2.7.1 (Auslander [5]) *Every polycyclic group embeds into $GL(n, \mathbb{Z})$ for some n.*

Theorem 2.7.2 (Malcev [44]) *Every solvable subgroup of $GL(n, \mathbb{Z})$ is polycyclic.*

2.7.1 Polycyclic rational matrix groups

We investigate the structure of a polycyclic subgroup G of $GL(n, \mathbb{Q})$ further. First, such a group G is finitely generated. Hence there exists a finite set of primes π so that the denominators of the entries of the elements of G are not divisible by any prime outside π. Hence for $p \notin \pi$, the computation modulo p extends to a natural homomorphism

$$\varphi_p : G \to GL(n, \mathbb{F}_p) : g \mapsto (g \bmod p).$$

We call the primes $p \notin \pi$ *admissible primes* for G and we denote with G_p the kernel of φ_p. The group G_p is also called a *congruence subgroup* of G. The following is due to Malcev [44], see also [10, Page 111].

Theorem 2.7.3 *Let G be a polycyclic group and let G_p be a congruence subgroup of G associated with the admissible prime p. Then G_p' is torsion-free nilpotent and conjugate to a subgroup of the upper unitriangular matrices.*

The structure exhibited in Theorem 2.7.3 can be determined computationally. First, the homomorphism φ_p can be explicitly computed and generators $k_1, \ldots, k_m$ for its kernel G_p can be obtained using Schreier generators. Let $V = \mathbb{Q}^n$ denote the natural vector space associated with G. Using the generator for G_p, we determine the subspace $U = \langle v(k_i k_j - k_j k_i) \mid v \in V, 1 \leq i, j \leq m \rangle$. This is the smallest subspace of V on which G_p acts as an abelian group. Theorem 2.7.3 asserts that U is strictly smaller than V. We iterate this construction and obtain a series of subspaces $V = V_1 > V_2 > \ldots > V_k > V_{k+1} = \{0\}$ so that G_p acts as an abelian group on each quotient V_i/V_{i+1}. The next lemma investigates the action on each of the quotients.

Lemma 2.7.4 *Let G be an abelian subgroup of $GL(n, \mathbb{Q})$ and A the matrix subalgebra of $\mathbb{Q}^{n \times n}$ generated by the elements of G. Let $V = \mathbb{Q}^n$.*

(a) For $a \in A$ define $ker(a) = \{v \in V \mid va = 0\}$. Then $ker(a)$ is a G-invariant subspace of V.
(b) For $v \in V$ define $vA = \{va \mid a \in A\}$. Then aV is a G-invariant subspace of V.
(c) If V is irreducible as a G-module, then $dim(V) = dim(A)$ holds.

Proof (a) The kernel $ker(a)$ is a subspace of V by construction. If $h \in G$ and $v \in ker(a)$, then $(vh)g = (vg)h = 0h = 0$ and hence $ker(a)$ is G-invariant.
(b) The image vA is a subspace of V by construction. If $h \in G$, then $(va)g = v(ag) \in vA$ and hence vA is G-invariant.
(c) Suppose that V is irreducible as a G-module. Then V splits into the direct sum of $dim(V)$ different irreducible modules over $\mathbb{C}$. With respect to this splitting, G is a group of diagonal matrices and thus $dim(A) = dim(V)$ follows. □

Lemma 2.7.4 yields two easy constructions for G-submodules if G is an abelian subgroup of $GL(n, \mathbb{Q})$. They can be used to refine a G-invariant series $V = V_1 > \ldots > V_k > V_{k+1} = \{0\}$ with quotients V_i/V_{i+1} on which G acts as abelian group. The irreducibility criterion can often be used to show that the series is maximal refined.

2.7.2 From integral to modular representations

Let G be a polycyclic subgroup of $GL(n, \mathbb{Z})$. Then for every prime p there is a natural homomorphism

$$\varphi_p : G \to GL(n, \mathbb{F}_p) : g \mapsto (g \bmod p).$$

This induces that G has various interesting modules: the natural module $L = \mathbb{Z}^n$, the module $L_{\mathbb{Q}} = L \otimes_{\mathbb{Z}} \mathbb{Q} = \mathbb{Q}^n$ obtained by extending scalars and the induced modules $L_p = \mathbb{F}_p^n$ for every prime p. These are connected via by the following.

Theorem 2.7.5 (Eick & Neumann-Brosig [19]) *Let G be a polycyclic subgroup of $GL(n, \mathbb{Z})$ with natural modules L, $L_{\mathbb{Q}}$ and L_p defined above.*

(a) *Either L_p is semisimple as G-module for almost all primes p or L_p is semisimple as G-module for almost no prime p.*
(b) *L_p is semisimple as G-module for almost all primes p if and only if $L_{\mathbb{Q}}$ is semisimple as G-module.*

There exists a practical algorithm to determine which of the cases in Theorem 2.7.5 arises and to determine the associated finite set of primes. This algorithm uses the structure theory for polycyclic matrix groups as introduced in Section 2.7.1. Note that there are algorithms available to compute the radical of $L_{\mathbb{Q}}$ or to check that $L_{\mathbb{Q}}$ is semisimple; we refer to [13] for details. There are also methods known to compute the radical of L_p for a fixed prime p; see the description of the Meataxe in [35].

2.8 Cohomology for polycyclic groups

Let G be a group and M an additive abelian group with a G-module structure. For $i \in \mathbb{N}$ we write $G^i = G \times \ldots \times G$ for the i-fold direct product of G and we define

$$C^i(G, M) = \{\gamma : G^i \to M \mid (g_1, \ldots, g_i)^\gamma = 0 \text{ if some } g_j = 1\}.$$

Then $C^i(G, M)$ has the structure of an abelian group. The i-th cohomology map $\alpha_i : C^{i-1}(G, M) \to C^i(G, M)$ is defined by

$$\begin{aligned}(g_1, \ldots, g_i)^{\gamma^{\alpha_i}} &= (g_2, \ldots, g_i)^\gamma \\ &\quad + \sum_{j=1}^{i-1} (-1)^j (g_1, \ldots, g_{j-1}, g_j g_{j+1}, g_{j+2}, \ldots, g_i)^\gamma \\ &\quad + (-1)^i ((g_1, \ldots, g_{i-1})^\gamma)^{g_i}.\end{aligned}$$

This is a homomorphism of abelian groups. It is used to define the groups of i-th cocycles and i-coboundaries, respectively, by

$$Z^i(G,M) = ker(\alpha_{i+1}) \leq C^i(G,M), \text{ and}$$
$$B^i(G,M) = im(\alpha_i) \leq C^i(G,M).$$

One can observe that $B^i(G,M) \leq Z^i(G,M)$ holds and this allows one to consider the factor group

$$H^i(G,M) = Z^i(G,M)/B^i(G,M).$$

We denote $H^i(G,M)$ as the i-th cohomology group. The following is a folklore observation.

Lemma 2.8.1 *Let G be a group and M a G-module. Write $H = M \rtimes G$.*

(a) There is a one-to-one correspondence between the conjugacy classes of complements to M in H and the elements of $H^1(G,M)$.
(b) There is a one-to-one correspondence between the equivalence classes of extensions of M by G and the elements of $H^2(G,M)$.

More background on the explicit correspondences and other details can be found in Chapter 6 of [13]. This also contains methods to determine the groups $Z^i(G,M)$ and $B^i(G,M)$ for $i = 1,2$ provided that M is either elementary or free abelian. In either case let d denote the rank of M as abelian group so that $M \cong \mathbb{F}_p^d$ in the elementary abelian case and $M \cong \mathbb{Z}^d$ in the free abelian case. Then there are matrices A_0, A_1, A_2 so that

- $H^1(G,M) \cong Ker(A_1)/Im(A_0)$, and
- $H^2(G,M) \cong Ker(A_2)/Im(A_1)$.

If $M \cong \mathbb{F}_p^d$, then all three matrices are matrices over $\mathbb{F}_p$ and if $M \cong \mathbb{Z}^d$, then all three matrices are matrices over $\mathbb{Z}$. We discuss the latter case in more detail.

Theorem 2.8.2 (Eick & Neumann-Brosig [19]) *Let G be polycyclic and $N \trianglelefteq G$ free abelian. Then either*

(a) N/N^p is complemented in G for almost all primes p, or,
(b) N/N^p is complemented in G for almost no prime p.

We exhibit briefly how it can be determined which of the two cases arises and how the associated finite set of primes can be determined. Let $g_1, \ldots, g_m, t_1, \ldots, t_d$ be a pcgs for G so that $t_1, \ldots, t_d$ is a pcgs of N

with $d = rk(N)$. Then for each complement K to N in G there exist $x_1, \ldots, x_n \in N$ so that $g_1x_1, \ldots, g_nx_n$ is a pcgs for K. As $K \cong G/N : g_ix_i \mapsto g_iN$ is an isomorphism, it follows that the pcgs $g_1x_1, \ldots g_nx_n$ satisfies a defining set of relations of G/N. Write $x_i = t_1^{v_{i1}} \cdots t_d^{v_{id}}$ and let $v = (v_{11}, \ldots, v_{nd})$. Then this construction yields that the complements to N in G correspond one-to-one to the solutions v of a matrix equation

$$vA_1 = w$$

where $A \in \mathbb{Z}^{nd \times ld}$ and $w \in \mathbb{Z}^{ld}$ and n and l are the numbers of generators and relations of a consistent polycyclic presentation of G/N, respectively. Clearly, each complement K to N in G also induces a complement to N/N^p in G for each prime p, but these may not be all complements. The complements to N/N^p in G correspond to the solutions v with

$$vA \equiv w \bmod p.$$

The Smith normal form theorem for integral matrices asserts that there are invertible integer matrices P and Q and a diagonal matrix D so that $A_1 = PDQ$. Thus $vA_1 \equiv w \bmod p$ translates to

$$(vP)D \equiv (wQ^{-1}) \bmod p.$$

Write $vP = (x_1, \ldots, x_l)$ and $wQ^{-1} = (y_1, \ldots, y_m)$ and let $(d_1, \ldots, d_m)$ denote the diagonal entries of D. Then the single equation $(vP)D \equiv (wQ^{-1}) \bmod p$ is equivalent to the list of equations $x_id_i \equiv y_i \bmod p$ for $1 \leq i \leq l$. It is well known how to solve a linear congruence $x_id_i \equiv y_i \bmod p$: this has $gcd(d_i, p)$ solutions in the range $\{0, \ldots, p-1\}$ if $gcd(d_i, p) \mid y_i$ and it has no solution otherwise. This proves the following:

Lemma 2.8.3 *Using the above notation:*

(a) Suppose that $y_i = 0$ implies $d_i = 0$ for all i. Then N/N^p is complemented for all primes except the finitely many primes p with $gcd(d_i, p) \nmid y_i$ for some i.

(b) Suppose there exists i with $y_i \neq 0$ and $d_i = 0$. Then N/N^p is complemented for finitely many primes p only (the primes p have to satisfy $p \mid y_i$).

2.9 The Frattini subgroup

The Frattini subgroup $\Phi(G)$ of a polycyclic group G is the intersection of all its maximal subgroups. It is well known that $\Phi(G)$ is also the set of non-generators of G; that is the set of all $g \in G$ so that if $\langle S, g \rangle = G$, then $\langle S \rangle = G$ holds. Hirsch proved the celebrated result that $\Phi(G)$ is nilpotent for a polycyclic group G. Hence it follows that

$$\Phi(Fit(G)) \leq \Phi(G) \leq Fit(G).$$

2.9.1 Special case: finitely generated nilpotent groups

Burnside's basis theorem asserts that $\Phi(G) = G'G^p$ for a finite p-group G. This allows one to determine $\Phi(G)$ readily in this case. If G is a finite nilpotent group, then $G = G_1 \times \ldots \times G_l$ is the direct product of its Sylow subgroups and $\Phi(G) = \Phi(G_1) \times \ldots \times \Phi(G_l)$ can also be determined readily. We go one step further and consider finitely generated nilpotent groups. The following is elementary, see also [19].

Lemma 2.9.1 *Let G be finitely generated nilpotent with $G/G' = F \times S_1 \times \ldots \times S_l$, where F is free abelian and $S_1, \ldots, S_l$ are the Sylow subgroups of the torsion subgroup of G/G'. Then $G' \leq \Phi(G)$ and*

$$\Phi(G)/G' = \Phi(S_1) \ldots \times \ldots \Phi(S_l).$$

2.9.2 Special case: finite solvable groups

Gaschütz [22] invented a method to determine the Frattini subgroup of a finite solvable group G. For this consider a chief series $G = C_1 > \ldots > C_{l+1} = \{1\}$. Each quotient of this series is either complemented in G or it is *frattinian*; that is, it satisfies $C_i/C_{i+1} \leq \Phi(G/C_{i+1})$. Gaschütz proved that there exists a subgroup $U \leq G$ (nowadays called *Präfrattini-subgroup*) so that U covers all frattinian chief factors, it complements all complemented chief factors and it has the property that

$$\Phi(G) = \cap_{g \in G} U^g.$$

The thesis [12] translates Gaschütz idea into an effective algorithm for computing $\Phi(G)$ in the case that G is finite solvable.

2.9.3 The general case

The general case remained difficult for a long time as far as practical methods goes. The fundamental paper by Baumslag, Cannonito, Robinson & Segal [6] shows that many problems are decidable for polycyclic groups, including the computation of the Frattini subgroup. But the methods of this paper did not seem to translate to practical methods.

Eick & Neumann-Brosig [19] described a new approach towards computing $\Phi(G)$ for an arbitrary polycyclic group G. This uses the computations modulo primes of Sections 2.7 and 2.8 to design an induction approach. It proceeds in the following steps; details of the steps are discussed below.

(1) Compute $N = \Phi(Fit(G))$.
If $N \neq \{1\}$, then let $H/N = \Phi(G/N)$ and return H.
(2) Compute the maximal finite normal subgroup N of G.
Determine $\Phi(G)$ from $\Phi(G/N)$.
(3) Compute $N = Z(G)$.
Determine $\Phi(G)$ from $\Phi(G/N)$.
(4) Compute $N = J(Fit(G))$.
If $N \neq \{1\}$, then let $H/N = \Phi(G/N)$ and return H.
(5) *Now $Fit(G)$ is free abelian, semisimple and $Z(G) = \{1\}$.*
Return $\Phi(G) = \{1\}$.

Step (1) is not difficult, as $Fit(G)$ can be determined and $\Phi(Fit(G))$ can then be computed, since $Fit(G)$ is finitely generated nilpotent. Step (2) considers the finite normal subgroup N and uses finite group methods. Step (3) considers a central normal subgroup N and uses the p-modular methods of Sections 2.7 and 2.8. In Step (4) we determine a subgroup $J(Fit(G))$ which corresponds to the Jacobson Radical of the group algebra of $Fit(G)$ and is a subgroup of $\Phi(G)$; hence it is sufficient to consider its quotient. Step (5) finally is based on a theorem by Eick & Neumann-Brosig [19].

2.10 The conjugacy problem

We first list three of the fundamental problems related to the conjugacy problem. Let G be an arbitrary group.

(1) **The Conjugacy Problem:** Given $g, h \in G$, decide if there exists $x \in G$ with $g^x = h$. If so, then determine one such x.
(2) **The Centralizer Problem:** Given $g \in G$, compute a finite set of generators for $C_G(g) = \{k \in G \mid gk = kg\}$.
(3) **Multiple Conjugacy Problem:** Given $g_1, \ldots, g_n, h_1, \ldots, h_n \in G$, determine $x \in G$ with $g_i^x = h_i$ for $1 \leq i \leq n$ (knowing that such an element x exists).

The Multiple Conjugacy Search Problem has applications in Cryptography, the Anshel-Anshel-Goldfeld cryptosystem is based on groups with difficult to solve multiple conjugacy problem. In a polycyclic group, the multiple conjugacy problem can be solved by an iterated application of the conjugacy problem.

The Conjugacy Problem and the Centralizer Problem are directly connected to each other. They can be solved simultaneously using induction along a normal series with elementary abelian or free abelian factors. In each step, the relevant computations translate to solving an orbit-stabilizer problem in a setting where a group acts as a matrix group over $\mathbb{F}_p$ or over $\mathbb{Z}$ on vectors in $\mathbb{F}_p^d$ or $\mathbb{Z}^d$. We refer to [35] for details.

The arising orbit-stabilizer problem for $\mathbb{F}_p$ can be solved using the methods for finite orbits. The arising orbit-stabilizer problem over $\mathbb{Z}$ has first been considered by Dixon [11]. This work contains the important idea to use the structure of polycyclic matrix groups to solve the problem. These ideas have been picked up by Eick & Ostheimer [21] and have been complemented by practical approaches towards computing with polycyclic matrix groups. The result was a first practical method to solve the orbit-stabilizer over $\mathbb{Z}^d$ and this yields a practical solution of the conjugacy and the centralizer problems.

2.11 Open Problems

In this final section we exhibit some open problems in the area of finding practical algorithms to work with polycyclic groups.

2.11.1 Minimal Generating Set

The *rank* $rk(G)$ of a group G is the smallest cardinality of a generating set of G. A *minimal generating set* of a group G is a generating set with $rk(G)$ elements.

If G is finitely generated nilpotent, then $G' \leq \Phi(G)$ and thus $rk(G) = rk(G/G')$. As G/G' is finitely generated abelian, it is not difficult to compute $rk(G/G')$. A minimal generating set of G can be obtained from one for G/G'.

If G is finite solvable, then Lucchini & Menegazzo [43] proposed a highly effective method to compute a minimal generating set of the considered group G. This uses induction along a normal series with elementary abelian factors and proposes a method to lift a minimal generating set of a quotient G/N_i to the next larger quotient G/N_{i+1}.

The general case of an arbitrary polycyclic group is wide open in this problem. It has been considered by Linnell & Warhurst [40] and by Kassabov & Nikolov [36]. As a result they obtained that $rk(\hat{G}) = rk(G) + \epsilon$ with $\epsilon \in \{0, 1\}$, where

$$rk(\hat{G}) = \max\{rk(G/N) \mid N \text{ normal of finite index in } G\}.$$

This feature eliminates the possibility of a straightforward approach to computing $rk(G)$. So far, it is not even known if the problem of determining $rk(G)$ is decidable in the class of polycyclic groups.

2.11.2 Automorphism group

The automorphism group $Aut(G)$ of a polycyclic group G is finitely generated. One of the challenging aims is to determine an explicit set of generators for $Aut(G)$ if G is polycyclic. Dual to this problem is to decide whether two given polycyclic groups are isomorphic.

For finite p-groups there is a method by Eick, Leedham-Green & O'Brien [18] that provides a practical approach. If G is finite solvable, then the method by Smith [55] can be used. But for infinite groups this is a still very challenging problem.

Grunewald & Segal [25, 26] proved that the isomorphism problem is decidable for finitely generated nilpotent groups. Segal [51] extended this to arbitrary polycyclic groups. The resulting algorithms seem not suitable for implementations.

For finitely generated torsion free nilpotent groups of Hirsch length at most 5 there is a practical solution for the isomorphism problem available in [17]. This gives an explicit classification for the considered groups and also exhibits the shape of the automorphism group.

References

[1] B. Assmann. *Polenta - computing presentations for polycyclic matrix groups*, 2003. A refereed GAP 4 package, see [56].

[2] B. Assmann and B. Eick. Computing polycyclic presentations of polycyclic matrix groups. *J. Symb. Comput.*, 40:1269 – 1284, 2005.

[3] B. Assmann and B. Eick. Testing polycyclicity of finitely generated rational matrix groups. *Math. Comp.*, 76(259):1669–1682 (electronic), 2007.

[4] B. Assmann and S. Linton. Using the Malcev correspondence for collection in polycyclic groups. *J. Algebra*, 316(2):828–848, 2007.

[5] L. Auslander. The automorphism group of a polycyclic group. *Ann. of Math. (2)*, 89:314–322, 1969.

[6] G. Baumslag, F. B. Cannonito, D. J. S. Robinson, and D. Segal. The algorithmic theory of polycyclic-by-finite groups. *J. Algebra*, 142:118–149, 1991.

[7] H. U. Besche and B. Eick. Construction of finite groups. *J. Symb. Comput.*, 27:387–404, 1999.

[8] J. Cannon, B. Eick, and C. R. Leedham-Green. Special polycyclic generating sequences for finite soluble groups. *J. Symb. Comput.*, 38:1445–1460, 2004.

[9] A. Cant and B. Eick. Polynomials describing the multiplication in finitely generated torsion-free nilpotent groups. *J. Symb. Comput.*, 92:203–210, 2019.

[10] J. D. Dixon. *The structure of linear groups.* Van Nostrand Reinhold Company, London, 1971.

[11] J. D. Dixon. The orbit-stabilizer problem for linear groups. *Canad. J. Math.*, 37(2):238–259, 1985.

[12] B. Eick. Spezielle PAG Systeme im Computeralgebra System GAP. Diplomarbeit, RWTH Aachen, 1993.

[13] B. Eick. Algorithms for polycyclic groups. Habilitationsschrift, Universität Kassel, 2001.

[14] B. Eick. Computing with infinite polycyclic groups. In A. Seress and W. M. Kantor, editors, *Groups and Computation III*, pages 139–153. (DIMACS, 1999), 2001.

[15] B. Eick. On the Fitting subgroup of a polycyclic-by-finite group and its applications. *J. Algebra*, 242:176–187, 2001.

[16] B. Eick. Orbit-stabilizer problems and computing normalizers for polycyclic groups. *J. Symb. Comput.*, 34:1–19, 2002.

[17] B. Eick and A.-K. Engel. The isomorphism problem for torsion free nilpotent groups of Hirsch length at most 5. *Groups Complex. Cryptol.*, 9(1):55–75, 2017.

[18] B. Eick, C. R. Leedham-Green, and E. A. O'Brien. Constructing automorphism groups of p-groups. *Comm. Algebra*, 30:2271–2295, 2002.

[19] B. Eick and M. Neumann-Brosig. On the Frattini subgroup of a polycyclic group. *J. Algebra*, 591:523–537, 2022.

[20] B. Eick and W. Nickel. *Polycyclic - computing with polycyclic groups*, 2005. A refereed GAP 4 package, see [56].

[21] B. Eick and G. Ostheimer. On the orbit stabilizer problem for integral matrix actions of polycyclic groups. *Math. Comp*, 72:1511–1529, 2003.

[22] W. Gaschütz. Über die ϕ-Untergruppe endlicher Gruppen. *Math. Z.*, 58:160–170, 1953.

[23] V. Gebhardt. Efficient collection in infinite polycyclic groups. *J. Symb. Comput.*, 34:213–228, 2002.

[24] S. P. Glasby and M. C. Slattery. Computing intersections and normalizers in soluble groups. *J. Symb. Comput.*, 9:637–651, 1990.

[25] F. Grunewald and D. Segal. Some general algorithms, I: Arithmetic groups. *Ann. Math.*, 112:531–583, 1980.

[26] F. Grunewald and D. Segal. Some general algorithms, II: Nilpotent groups. *Ann. Math.*, 112:585–617, 1980.

[27] P. Hall. *The Edmonton notes on nilpotent groups.* Queen Mary College Mathematics Notes. Queen Mary College, Mathematics Department, London, 1969.

[28] G. Higman. Enumerating p-groups. I: Inequalities. *Proc. London Math. Soc.*, 10:24–30, 1960.

[29] G. Higman. Enumerating p-groups. II: Problems whose solution is PORC. *Proc. London Math. Soc.*, 10:566–582, 1960.

[30] K. A. Hirsch. On infinite soluble groups (I). *Proc. London Math. Soc.*, 44(2):53–60, 1938.

[31] K. A. Hirsch. On infinite soluble groups (II). *Proc. London Math. Soc.*, 44(5):336–344, 1938.

[32] K. A. Hirsch. On infinite soluble groups (III). *J. London Math. Soc.*, 49(2):184–94, 1946.

[33] K. A. Hirsch. On infinite soluble groups (IV). *J. London Math. Soc.*, 27:81–85, 1952.

[34] K. A. Hirsch. On infinite soluble groups. V. *J. London Math. Soc.*, 29:250–251, 1954.

[35] D. F. Holt, B. Eick, and E. A. O'Brien. *Handbook of computational group theory.* Discrete Mathematics and its Applications (Boca Raton). Chapman & Hall/CRC, Boca Raton, FL, 2005.

[36] M. Kassabov and N. Nikolov. Generation of polycyclic groups. *J. Group Theory*, 12(4):567–577, 2009.

[37] R. Laue, J. Neubüser, and U. Schoenwaelder. Algorithms for finite soluble groups and the SOGOS system. In *Computational Group Theory*, pages 105–135, London, New York, 1984. (Durham, 1982), Academic Press.

[38] C. R. Leedham-Green and L. H. Soicher. Collection from the left and other strategies. *J. Symb. Comput.*, 9:665–675, 1990.

[39] C. R. Leedham-Green and L. H. Soicher. Symbolic collection using Deep Thought. *LMS J. Comput. Math.*, 1:9–24, 1998.

[40] P. A. Linnell and D. Warhurst. Bounding the number of generators of a polycyclic group. *Arch. Math. (Basel)*, 37(1):7–17, 1981.

[41] E. H. Lo. Finding intersection and normalizer in finitely generated nilpotent groups. *J. Symb. Comput.*, 25:45–59, 1998.

[42] E. H. Lo. A polycyclic quotient algorithm. *J. Symb. Comput.*, 25:61–97, 1998.

[43] A. Lucchini and F. Menegazzo. Computing a set of generators of minimal cardinality in a solvable group. *J. Symb. Comput.*, 17:409–420, 1994.

[44] A. L. Malcev. On certain classes of infinite solvable groups. *Amer. Math. Soc. Transl. (2)*, 2:1–21, 1956.

[45] M. F. Newman and E. A. O'Brien. Application of computers to questions like those of Burnside, II. *Internat. J. Algebra Comput.*, 6:593–605, 1996.

[46] W. Nickel. Computing nilpotent quotients of finitely presented groups. In *Geometric and computational perspectives on infinite groups*, Amer. Math. Soc. DIMACS Series, pages 175–191, 1996.

[47] W. Nickel. *NQ*, 1998. A refereed GAP 4 package, see [56].

[48] D. J. S. Robinson. *A Course in the Theory of Groups*, volume 80 of *Graduate Texts in Math.* Springer-Verlag, New York, Heidelberg, Berlin, 1982.

[49] R. Schwingel. Two matrix group algorithms with applications to computing the automorphism group of a finite p-group. PhD Thesis, QMW, University of London, 2000.

[50] D. Segal. *Polycyclic Groups.* Cambridge University Press, Cambridge, 1983.

[51] D. Segal. Decidable properties of polycyclic groups. *Proc. London Math. Soc.*, 61:497–528, 1990.

[52] C. C. Sims. Enumerating p-groups. *Proc. London Math. Soc.*, 15:151–166, 1965.

[53] C. C. Sims. Computing the order of a solvable permutation group. *J. Symb. Comput.*, 9:699–705, 1990.

[54] C. C. Sims. *Computation with finitely presented groups.* Cambridge University Press, Cambridge, 1994.

[55] M. J. Smith. Computing automorphisms of finite soluble groups. Ph.D. Thesis, Australian National University, 1995.

[56] The GAP Group. GAP – *Groups, Algorithms and Programming, Version 4.10.* Available from http://www.gap-system.org, 2019.

3
The spread of finite and infinite groups

Scott Harper[a]

Abstract

It is well known that every finite simple group has a generating pair. Moreover, Guralnick and Kantor proved that every finite simple group has the stronger property, known as $\frac{3}{2}$-generation, that every nontrivial element is contained in a generating pair. More recently, this result has been generalised in three different directions, which form the basis of this survey article. First, we look at some stronger forms of $\frac{3}{2}$-generation that the finite simple groups satisfy, which are described in terms of spread and uniform domination. Next, we discuss the recent classification of the finite $\frac{3}{2}$-generated groups. Finally, we turn our attention to infinite groups, and we focus on the recent discovery that the finitely presented simple groups of Thompson are also $\frac{3}{2}$-generated, as are many of their generalisations. Throughout the article we pose open questions in this area, and we highlight connections with other areas of group theory.

3.1 Introduction

Every finite simple group can be generated by two elements. This well-known result was proved for most finite simple groups by Steinberg in 1962 [95] and completed via the Classification of Finite Simple Groups (see [2]). Much more is now known about generating pairs for finite simple groups. For instance, for any nonabelian finite simple group G, almost all pairs of elements generate G [73, 79], G has an invariable

[a] School of Mathematics and Statistics, University of St Andrews, St Andrews, KY16 9SS, UK.
scott.harper@st-andrews.ac.uk

generating pair [61, 74], and, with only finitely many exceptions, G can be generated by a pair of elements where one has order 2 and the other has order either 3 or 5 [80, 83].

The particular generation property of finite simple groups that this survey focuses on was established by Guralnick and Kantor [59] and independently by Stein [94]. They proved that if G is a finite simple group, then every nontrivial element of G is contained in a generating pair. Groups with this property are said to be $\frac{3}{2}$*-generated*. We will survey the recent work (mostly from the past five years) that addresses natural questions arising from this theorem.

Section 3.2 focuses on finite groups and considers recent progress towards answering two natural questions. Do the finite simple groups satisfy stronger versions of $\frac{3}{2}$-generation? Which other finite groups are $\frac{3}{2}$-generated? Regarding the first, in Sections 3.2.2 and 3.2.3, we will meet two strong versions of $\frac{3}{2}$-generation, namely (uniform) spread and total/uniform domination. Regarding the second, Section 3.2.4 presents the recent classification of the finite $\frac{3}{2}$-generated groups established by Burness, Guralnick and Harper in 2021 [29]. All these ideas are brought together as we discuss the generating graph in Section 3.2.5. Section 3.2.6 rounds off the first half by highlighting applications of spread to word maps, the product replacement graph and the soluble radical of a group.

Section 3.3 focuses on infinite groups and, in particular, whether any results on the $\frac{3}{2}$-generation of finite groups extend to the realm of infinite groups. After discussing this in general terms in Sections 3.3.1 and 3.3.2, our focus shifts to the finitely presented infinite simple groups of Richard Thompson in Sections 3.3.3 to 3.3.6. Here we survey the ongoing work of Bleak, Donoven, Golan, Harper, Hyde and Skipper, which reveals strong parallels between the $\frac{3}{2}$-generation of these infinite simple groups and the finite simple groups. Section 3.3.3 serves as an introduction to Thompson's groups for any reader unfamiliar with them.

This survey is based on my one-hour lecture at *Groups St Andrews 2022* at the University of Newcastle, and I thank the organisers for the opportunity to present at such an enjoyable and interesting conference. I have restricted this survey to the subject of spread and have barely discussed other aspects of generation. Even regarding the spread of finite simple groups, much more could be said, especially regarding the methods involved in proving the results. Both of these omissions from this survey are discussed amply in Burness' survey article from *Groups St Andrews 2017* [27], which is one reason for deciding to focus in this article on the progress made in the past five years.

Acknowledgements. The author wrote this survey when he was first a Heilbronn Research Fellow and then a Leverhulme Early Career Fellow, and he thanks the Heilbronn Institute for Mathematical Research and the Leverhulme Trust. He thanks Tim Burness, Charles Cox, Bob Guralnick, Jeremy Rickard and a referee for their helpful comments, and he also thanks Guralnick for his input on Application 3, especially his suggested proof of Theorem 3.2.42.

3.2 Finite Groups

3.2.1 Generating pairs

It is easy to write down a pair of generators for each alternating group A_n: for instance, if n is odd, then

$$A_n = \langle (1\,2\,3), (1\,2\,\ldots\,n) \rangle.$$

In 1962, Steinberg [95] proved that every finite simple group of Lie type is 2-generated, by exhibiting an explicit pair of generators. In light of the Classification of Finite Simple Groups, once the sporadic groups were all shown to be 2-generated, it became known that every finite simple group is 2-generated [2]. Since then, numerous stronger versions of this theorem have been proved (see Burness' survey [27]).

Even as early as 1962, Steinberg raised the possibility of stronger versions of his 2-generation result [95]:

"It is possible that one of the generators can be chosen of order 2, as is the case for the projective unimodular group, or even that one of the generators can be chosen as an arbitrary element other than the identity, as is the case for the alternating groups. Either of these results, if true, would quite likely require methods much more detailed than those used here."

That is, Steinberg is suggesting the possibility that for a finite simple group G one might be able to replace just the existence of $x, y \in G$ such that $\langle x, y \rangle = G$, with the stronger statement that for all nontrivial elements $x \in G$ there exists $y \in G$ such that $\langle x, y \rangle = G$. He alludes to the fact that this much stronger condition is known to hold for the alternating groups, which was shown by Piccard in 1939 [91]. In the following example, we will prove this result on alternating groups, but with different methods than Piccard used.

Example 3.2.1 Let $G = A_n$ for $n \geqslant 5$. We will focus on the case $n \equiv 0 \pmod 4$ and then address the remaining cases at the end.

Write $n = 4m$ and let s have cycle shape $[2m-1, 2m+1]$, that is, let s be a product of disjoint cycles of lengths $2m-1$ and $2m+1$. Visibly, s is contained in a maximal subgroup $H \leqslant G$ of type

$$(S_{2m-1} \times S_{2m+1}) \cap G.$$

We claim that no further maximal subgroups of G contain s. Imprimitive maximal subgroups are ruled out since $2m-1$ and $2m+1$ are coprime. In addition, a theorem of Marggraf [100, Theorem 13.5] ensures that no proper primitive subgroup of A_n contains a k-cycle for $k < \frac{n}{2}$, so s is contained in no primitive maximal subgroups as a power of s is a $(2m-1)$-cycle.

Now let x be an arbitrary nontrivial element of G. Choosing g such that x moves some point from the $(2m-1)$-cycle of s^g to a point in the $(2m+1)$-cycle of s^g gives $x \notin H^g$. This means that no maximal subgroup of G contains both x and s^g, so $\langle x, s^g \rangle = G$. In particular, every nontrivial element of G is contained in a generating pair.

We now address the other cases, but we assume that $n \geqslant 25$ for clarity of exposition. If $n \equiv 2 \pmod 4$, then we choose s with cycle shape $[2m-1, 2m+3]$ (where $n = 4m+2$) and proceed as above but now the unique maximal overgroup has type $(S_{2m-1} \times S_{2m+3}) \cap G$. A similar argument works for odd n. Here s has cycle shape $[m-2, m, m+2]$ if $n = 3m$, $[m+1, m+1, m-1]$ if $n = 3m+1$ and $[m+2, m, m]$ if $n = 3m+2$, and the only maximal overgroups of s are the three obvious intransitive ones. For each $1 \neq x \in G$, it is easy to find $g \in G$ such that x misses all three maximal overgroups of s^g and hence deduce that $\langle x, s^g \rangle = G$.

In 2000, Guralnick and Kantor [59] gave a positive answer to the longstanding question of Steinberg by proving the following.

Theorem 3.2.2 *Let G be a finite simple group. Then every nontrivial element of G is contained in a generating pair.*

We say that a group G is $\frac{3}{2}$-*generated* if every nontrivial element of G is contained in a generating pair. The author does not know the origin of this term, but it indicates that the class of $\frac{3}{2}$-generated groups includes the class of 1-generated groups and is included in the class of 2-generated groups. This is somewhat analogous to the class of $\frac{3}{2}$-transitive permutation groups introduced by Wielandt [100, Section 10], which is included in the class of 1-transitive groups and includes the class of 2-transitive groups.

Let us finish this section by briefly turning from simple groups to simple Lie algebras. Here we have a theorem of Ionescu [72], analogous to Theorem 3.2.2.

Theorem 3.2.3 *Let $\mathfrak{g}$ be a finite dimensional simple Lie algebra over $\mathbb{C}$. Then for all $x \in \mathfrak{g} \setminus 0$ there exists $y \in \mathfrak{g}$ such that x and y generate $\mathfrak{g}$ as a Lie algebra.*

In fact, Bois [13] proved that every classical finite dimensional simple Lie algebra in characteristic other than 2 or 3 has this $\frac{3}{2}$-generation property, but Goldstein and Guralnick [56] have proved that $\mathfrak{sl}_n$ in characteristic 2 does not.

3.2.2 Spread

Let us now introduce the concept that gives this article its name.

Definition 3.2.4 Let G be a group.

(i) The *spread* of G, written $s(G)$, is the supremum over integers k such that for any k nontrivial elements $x_1, \ldots, x_k \in G$ there exists $y \in G$ such that $\langle x_1, y \rangle = \cdots = \langle x_k, y \rangle = G$.
(ii) The *uniform spread* of G, written $u(G)$, is the supremum over integers k for which there exists $s \in G$ such that for any k nontrivial elements $x_1, \ldots, x_k \in G$ there exists $y \in s^G$ such that $\langle x_1, y \rangle = \cdots = \langle x_k, y \rangle = G$.

The term spread was introduced by Brenner and Wiegold in 1975 [15], but the term uniform spread was not formally introduced until 2008 [16].

Note that $s(G) > 0$ if and only if every nontrivial element of G is contained in a generating pair. Therefore, spread gives a way of quantifying how strongly a group is $\frac{3}{2}$-generated. Uniform spread captures the idea that the complementary element y, while depending on the elements $x_1, \ldots, x_k$, can be chosen somewhat uniformly for all choices of $x_1, \ldots, x_k$: it can always be chosen from the same prescribed conjugacy class. In Section 3.2.3, we will see a way of measuring how much more uniformity in the choice of y we can insist on. Observe that Example 3.2.1 actually shows that $u(A_n) \geqslant 1$ for all $n \geqslant 5$.

By Theorem 3.2.2, every finite simple group G satisfies $s(G) > 0$. What more can be said about the (uniform) spread of finite simple groups? The main result is the following proved by Breuer, Guralnick and Kantor [16].

Theorem 3.2.5 *Let G be a nonabelian finite simple group. Then $s(G) \geqslant u(G) \geqslant 2$. Moreover, $s(G) = 2$ if and only if $u(G) = 2$ if and only if*

$$G \in \{A_5, A_6, \Omega_8^+(2)\} \cup \{\mathrm{Sp}_{2m}(2) \mid m \geqslant 3\}.$$

The asymptotic behaviour of (uniform) spread is given by the following theorem of Guralnick and Shalev [65, Theorem 1.1]. The version of this theorem stated in [65] is given just in terms of spread, but the result given here follows immediately from their proof (see [65, Lemma 2.1–Corollary 2.3]).

Theorem 3.2.6 *Let (G_i) be a sequence of nonabelian finite simple groups such that $|G_i| \to \infty$. Then $s(G_i) \to \infty$ if and only if $u(G_i) \to \infty$ if and only if (G_i) has no infinite subsequence consisting of either*

(i) alternating groups of degree all divisible by a fixed prime
(ii) symplectic groups over a field of fixed even size or odd-dimensional orthogonal groups over a field of fixed odd size.

Given that $s(G_i) \to \infty$ if and only if $u(G_i) \to \infty$, we ask the following. (Note that $s(G) - u(G)$ can be arbitrarily large, see Theorem 3.2.9(iv) for example.)

Question 3.2.7 Does there exist a constant c such that for all nonabelian finite simple groups G we have $s(G) \leqslant c \cdot u(G)$?

There are explicit upper bounds that justify the exceptions in parts (i) and (ii) of Theorem 3.2.6. Indeed, $s(\mathrm{Sp}_{2m}(q)) \leqslant q$ for even q and $s(\Omega_{2m+1}(q)) \leqslant \frac{1}{2}(q^2 + q)$ for odd q (see [65, Proposition 2.5] for a geometric proof). For alternating groups of composite degree $n > 6$, if p is the least prime divisor of n, then $s(A_n) \leqslant \binom{2p+1}{3}$ (see [65, Proposition 2.4] for a combinatorial proof). For even-degree alternating groups, the situation is clear: $s(A_n) = 4$, but much less is known in odd degrees (see [65, Section 3.1] for partial results).

Question 3.2.8 What is the (uniform) spread of A_n when n is odd?

The spread of even-degree alternating groups was determined by Brenner and Wiegold in the paper where they first introduced the notion of spread. They also studied the spread of two-dimensional linear groups, but their claimed value for $s(\mathrm{PSL}_2(q))$ was only proved to be a lower bound. Further work by Burness and Harper demonstrates that this is not an upper bound when $q \equiv 3 \pmod 4$, where they prove the following (see [31, Theorem 5 & Remark 5]).

Theorem 3.2.9 *Let $G = \mathrm{PSL}_2(q)$ with $q \geqslant 11$.*

(i) If q is even, then $s(G) = u(G) = q - 2$.
(ii) If $q \equiv 1 \pmod 4$, then $s(G) = u(G) = q - 1$.
(iii) If $q \equiv 3 \pmod 4$, then $s(G) \geqslant q - 3$ and $u(G) \geqslant q - 4$.
(iv) If $q \equiv 3 \pmod 4$ is prime, then $s(G) \geqslant \frac{1}{2}(3q-7)$ and $s(G)-u(G) = \frac{1}{2}(q+1)$.

Question 3.2.10 What is the (uniform) spread of the group $\mathrm{PSL}_2(q)$ when $q \equiv 3 \pmod 4$?

In short, determining the spread of simple groups is difficult. We conclude by commenting that the precise value of the spread of only two sporadic groups is known, namely $s(\mathrm{M}_{11}) = 3$ [101] (see also [14]) and $s(\mathrm{M}_{23}) = 8064$ [14, 47].

In contrast, the exact spread and uniform spread of symmetric groups is known. In a series of papers in the late 1960s [3, 5, 4, 6], Binder determined the spread of S_n and also showed that $u(S_n) \geqslant 1$ unless $n \in \{4, 6\}$ (Binder used different terminology). However, the uniform spread of symmetric groups was only completely determined in a 2021 paper of Burness and Harper [31]; indeed, showing that $u(S_n) \geqslant 2$ for even $n > 6$ involves both a long combinatorial argument and a CFSG-dependent group theoretic argument (see [31, Theorem 3 & Remark 3]). We say more on S_6 in Example 3.2.31.

Theorem 3.2.11 *Let $G = S_n$ with $n \geqslant 5$. Then*

$$s(G) = \begin{cases} 2 & \text{if } n \text{ is even} \\ 3 & \text{if } n \text{ is odd} \end{cases} \quad \text{and} \quad u(G) = \begin{cases} 0 & \text{if } n = 6 \\ 2 & \text{otherwise.} \end{cases}$$

Methods. A probabilistic approach. As we turn to discuss the key method behind these results, we return to Example 3.2.1 where we proved that $u(G) \geqslant 1$ when $G = A_n$ for even $n > 6$. We found an element $s \in G$ contained in a unique maximal subgroup H of G. Since G is simple, H is corefree, so $\bigcap_{g \in G} H^g = 1$, which means that for each nontrivial $x \in G$ there exists $g \in G$ such that $x \notin H^g$. This implies that $\langle x, s^g \rangle = G$, so s^G witnesses $u(G) \geqslant 1$. This argument can be generalised in two ways: one yields Lemma 3.2.12, giving a better lower bound on the uniform spread of G and the other yields Lemma 3.2.21, pertaining to the uniform domination number of G, which we will meet in the next section.

Lemma 3.2.12 takes a probabilistic approach, so we need some notation. For a finite group G and elements $x, s \in G$, we write

$$Q(x,s) = \frac{|\{y \in s^G \mid \langle x, y\rangle \neq G\}|}{|s^G|}, \tag{2.1}$$

which is the probability a uniformly random conjugate of s does not generate with x, and write $\mathcal{M}(G,s)$ for the set of maximal subgroups of G that contain s.

Lemma 3.2.12 *Let G be a finite group and let $s \in G$.*

(i) For $x \in G$,

$$Q(x,s) \leqslant \sum_{H \in \mathcal{M}(G,s)} \frac{|x^G \cap H|}{|x^G|}.$$

(ii) For a positive integer k, if $Q(x,s) < \frac{1}{k}$ for all prime order elements $x \in G$, then $u(G) \geqslant k$ is witnessed by s^G.

Proof For (i), let $x \in G$. Then $\langle x, s^g\rangle \neq G$ if and only if $x \in H^g$, or equivalently $x^{g^{-1}} \in H$, for some $H \in \mathcal{M}(G,s)$. Therefore,

$$Q(x,s) = \frac{|\{y \in s^G \mid \langle x, y\rangle \neq G\}|}{|s^G|} \leqslant \sum_{H \in \mathcal{M}(G,s)} \frac{|x^G \cap H|}{|x^G|}.$$

For (ii), fix k. To prove that $u(G) \geqslant k$ is witnessed by s^G, it suffices to prove that for all elements $x_1, \dots, x_k \in G$ of prime order there exists $y \in s^G$ such that $\langle x_i, y\rangle = G$ for all $1 \leqslant i \leqslant k$. Therefore, let $x_1, \dots, x_k \in G$ have prime order. If $Q(x_i, s) < \frac{1}{k}$ for all $1 \leqslant i \leqslant k$, then

$$\frac{|\{y \in s^G \mid \langle x_i, y\rangle = G \text{ for all } 1 \leqslant i \leqslant k\}|}{|s^G|} \geqslant 1 - \sum_{i=1}^{k} Q(x_i, s) > 0,$$

so there exists $y \in s^G$ such that $\langle x_i, y\rangle = G$ for all $1 \leqslant i \leqslant k$. □

Therefore, to obtain lower bounds on the uniform spread (and hence spread) of a finite group, it is enough to (a) identify an element whose maximal overgroups H are tightly constrained, and then (b) for each such H and for all prime order $x \in G$, bound the quantity $\frac{|x^G \cap H|}{|x^G|}$.

The ratio $\frac{|x^G \cap H|}{|x^G|}$ is the well-studied *fixed point ratio*. More precisely, $\frac{|x^G \cap H|}{|x^G|}$ is nothing other than the proportion of points in G/H fixed by x in the natural action of G on G/H. These fixed point ratios, in the context of primitive actions of almost simple groups, have seen many applications via probabilistic methods, not just to spread, but also to base sizes (e.g. the Cameron–Kantor conjecture) and monodromy groups (e.g. the Guralnick–Thompson conjecture), see Burness' survey article [26].

To address task (a), one applies the well-known and extensive literature on the subgroup structure of almost simple groups. For (b), one appeals to the bounds on fixed point ratios of primitive actions of almost simple groups, the most general of which is [78, Theorem 1] of Liebeck and Saxl. This states that

$$\frac{|x^G \cap H|}{|x^G|} \leqslant \frac{4}{3q} \tag{2.2}$$

for any almost simple group of Lie type G over $\mathbb{F}_q$, maximal subgroup $H \leqslant G$ and nontrivial element $x \in G$, with known exceptions. This is essentially best possible, since $\frac{|x^G \cap H|}{|x^G|} \approx q^{-1}$ when q is odd, $G = \mathrm{PGL}_n(q)$, H is the stabiliser of a 1-space of $\mathbb{F}_q^n$ and x lifts to the diagonal matrix $[-1, 1, 1, \dots, 1] \in \mathrm{GL}_n(q)$. However, there are much stronger bounds that take into account the particular group G, subgroup H or element x (see [26, Section 2] for a survey).

Bounding uniform spread via Lemma 3.2.12 was the approach introduced by Guralnick and Kantor in their 2000 paper [59] where they prove that $u(G) \geqslant 1$ for all nonabelian finite simple groups G. Clearly this approach also easily yields further probabilistic information and we refer the reader to Burness' survey article [27] for much more on this approach. We will give just one example, which we will return to later in the article (see [27, Example 3.9]).

Example 3.2.13 Let $G = E_8(q)$ and let s generate a cyclic maximal torus of order $\Phi_{30}(q) = q^8 + q^7 - q^5 - q^4 - q^3 + q + 1$. Weigel proved that $\mathcal{M}(G, s) = \{H\}$ where $H = N_G(\langle s \rangle) = \langle s \rangle : 30$ (see [99, Section 4(j)]). Applying Lemma 3.2.12 with the bound in (2.2), for all nontrivial $x \in G$ we have $u(G) \geqslant 1$ since

$$\sum_{H \in \mathcal{M}(G,s)} \frac{|x^G \cap H|}{|x^G|} \leqslant \frac{4}{3q} \leqslant \frac{2}{3} < 1.$$

However, we can do better: $|x^G \cap H| \leqslant |H| \leqslant q^{14}$ and $|x^G| > q^{58}$ for all nontrivial elements $x \in G$, so $u(G) \geqslant q^{44}$ since

$$\sum_{H \in \mathcal{M}(G,s)} \frac{|x^G \cap H|}{|x^G|} < \frac{1}{q^{44}}.$$

While the overwhelming majority of results on (uniform) spread are established via the probabilistic method encapsulated in Lemma 3.2.12, there are cases where this approach fails, as the following example shows.

Example 3.2.14 Let $m \geqslant 3$ and let $G = \mathrm{Sp}_{2m}(2)$. By Theorem 3.2.5, we know that $u(G) = 2$. However, if x is a transvection, then $Q(x, s) > \frac{1}{2}$ for all $s \in G$. This is proved in [16, Proposition 5.4], and we give an indication of the proof. Every element of $G = \mathrm{Sp}_{2m}(2)$ is contained in a subgroup of type $\mathrm{O}^+_{2m}(2)$ or $\mathrm{O}^-_{2m}(2)$ (see [45], for example).

Assume that s is contained in a subgroup $H \cong \mathrm{O}^-_{2m}(2)$. The groups $\mathrm{Sp}_{2m}(2)$ and $\mathrm{O}^{\pm}_{2m}(2)$ contain $2^{2m} - 1$ and $2^{2m-1} \mp 2^{m-1}$ transvections, respectively, so

$$Q(x, s) \geqslant \frac{2^{2m-1} + 2^{m-1}}{2^{2m} - 1} = \frac{2^{m-1}}{2^m - 1} > \frac{1}{2}.$$

A more involved argument gives $Q(x, s) > \frac{1}{2}$ if s is contained in a subgroup of type $\mathrm{O}^+_{2m}(2)$ but none of type $\mathrm{O}^-_{2m}(2)$, relying on s being reducible here.

3.2.3 Uniform domination

We began by observing that any finite simple group G is $\frac{3}{2}$-generated, that is

$$\text{for all } x \in G \setminus 1 \text{ there exists } y \in G \text{ such that } \langle x, y \rangle = G. \tag{2.3}$$

We then looked to strengthen (2.3) by increasing the scope of the first quantifier. Recall that the *spread* of G, denoted $s(G)$, is the greatest k such that

$$\text{for all } x_1, \ldots, x_k \in G \text{ there exists } y \in G \text{ such that}$$
$$\langle x_1, y \rangle = \cdots = \langle x_k, y \rangle = G.$$

We also have a related notion: the *uniform spread* of G, denoted $u(G)$, is the greatest k for which there exists an element $s \in G$ such that

$$\text{for all } x_1, \ldots, x_k \in G \text{ there exists } y \in s^G \text{ such that}$$
$$\langle x_1, y \rangle = \cdots = \langle x_k, y \rangle = G.$$

Uniform spread inspires us to strengthen (2.3) by narrowing the range of the second quantifier. That is, we say that the *total domination number* of G, denoted $\gamma_t(G)$, is the least size of a subset $S \subseteq G$ such that

$$\text{for all } x \in G \setminus 1 \text{ there exists } y \in S \text{ such that } \langle x, y \rangle = G.$$

Again we have a related notion: the *uniform domination number* of G, denoted $\gamma_u(G)$, is the least size of a subset $S \subseteq G$ of conjugate elements such that

$$\text{for all } x \in G \setminus 1 \text{ there exists } y \in S \text{ such that } \langle x, y \rangle = G.$$

These latter two concepts were introduced by Burness and Harper in [30] and studied further in [31]. The terminology is motivated by the generating graph (see Section 3.2.5).

Let G be a nonabelian finite simple group. Clearly $2 \leqslant \gamma_t(G) \leqslant \gamma_u(G)$, and since $u(G) \geqslant 1$, there exists a class s^G such that $\gamma_u(G) \leqslant |s^G|$. However, the class exhibited in Guralnick and Kantor's proof of $u(G) \geqslant 1$ is typically very large (for groups of Lie type, s is usually a regular semisimple element), so it is natural to seek tighter upper bounds on $\gamma_u(G)$. The following result of Burness and Harper does this [30, Theorems 2, 3 & 4] (see [31, Theorem 4(i)] for the refined upper bound in (iii)).

Theorem 3.2.15 *Let G be a nonabelian finite simple group.*

(i) If $G = A_n$, then $\gamma_u(G) \leqslant 77 \log_2 n$.
(ii) If G is classical of rank r, then $\gamma_u(G) \leqslant 7r + 70$.
(iii) If G is exceptional, then $\gamma_u(G) \leqslant 5$.
(iv) If G is sporadic, then $\gamma_u(G) \leqslant 4$.

In this generality, these bounds are optimal up to constants. If $n \geqslant 6$ is even, then $\log_2 n \leqslant \gamma_t(A_n) \leqslant \gamma_u(A_n) \leqslant 2\log_2 n$, and if G is $\mathrm{Sp}_{2r}(q)$ with q even or $\Omega_{2r+1}(q)$ with q odd, then $r \leqslant \gamma_t(G) \leqslant \gamma_u(G) \leqslant 7r$ [31, Theorems 3(i) & 6.3(iii)]. Regarding the bounds in (iii) and (iv), for sporadic groups, $\gamma_u(G) = 4$ is witnessed by $G = \mathrm{M}_{11}$ [30, Theorem 3], but the best lower bound for exceptional groups is $\gamma_u(G) \geqslant 3$ given by $G = F_4(q)$ [31, Lemma 6.17].

Question 3.2.16 Does there exist a constant c such that for all $n \geqslant 5$ we have $\log_p n \leqslant \gamma_t(A_n) \leqslant \gamma_u(A_n) \leqslant c \log_p n$ where p is the least prime divisor of n?

By [31, Theorem 4(ii)], we know that $\gamma_t(A_n) \geqslant \log_p n$, so to provide an affirmative answer to Question 3.2.16, it suffices to prove the upper bound $\gamma_u(A_n) \leqslant c \log_p n$.

Question 3.2.17 Does there exist a constant c such that for all finite simple groups of Lie type G other than $\mathrm{Sp}_{2r}(q)$ with q even and $\Omega_{2r+1}(q)$ with q odd, we have $\gamma_u(G) \leqslant c$?

By Theorem 3.2.15, to answer Question 3.2.17, it suffices to consider classical groups of large rank, and it was shown in [30, Theorem 6.3(ii)] that $c = 15$ suffices for some families of these groups. Affirmative answers to Questions 3.2.16 and 3.2.17 would answer Question 3.2.18 too.

Question 3.2.18 Does there exist a constant c such that for all nonabelian finite simple groups G we have $\gamma_u(G) \leqslant c \cdot \gamma_t(G)$?

The smallest possible value of $\gamma_u(G)$ is 2 (since G is not cyclic), and an almost complete classification of when this is achieved was given in [31, Corollary 7].

Theorem 3.2.19 *Let G be a nonabelian finite simple group. Then $\gamma_u(G) = 2$ only if G is one of the following*

(i) A_n *for prime* $n \geqslant 13$
(ii) $\mathrm{PSL}_2(q)$ *for odd* $q \geqslant 11$
$\mathrm{PSL}_n^{\varepsilon}(q)$ *for odd* n, $(q, \varepsilon) \notin \{(2,+),(4,+),(3,-),(5,-)\}$ *if* $n = 3$
$\mathrm{PSp}_{4m+2}(q)^*$ *for odd* q *and* $m \geqslant 2$, *and* $\mathrm{P\Omega}^{\pm}_{4m}(q)^*$ *for* $m \geqslant 2$
(iii) ${}^2B_2(q)$, ${}^2G_2(q)$, ${}^2F_4(q)$, ${}^3D_4(q)$, ${}^2E_6(q)$, $E_6(q)$, $E_7(q)$, $E_8(q)$
(iv) M_{23}, J_1, J_4, Ru, Ly, O′N, Fi_{23}, Th, $\mathbb{B}$, $\mathbb{M}$ *or* J_3^*, He^*, Co_1^*, HN^*.

Moreover, $\gamma_u(G) = 2$ in all the cases without an asterisk.

We will say that a subset $S \subseteq G$ of conjugate elements of G is a *uniform dominating set* of G if for all nontrivial $x \in G$ there exists $y \in S$ such that $\langle x, y\rangle = G$, so $\gamma_u(G)$ is the smallest size of a uniform dominating set of G. For groups G such that $\gamma_u(G) = 2$, we know that there exists a uniform dominating set of size two. How abundant are such subsets? To this end, let $P(G, s, 2)$ be the probability that two random conjugates of s form a uniform dominating set for G, and let $P(G) = \max\{P(G, s, 2) \mid s \in G\}$. Then we have the following probabilistic result [31, Corollary 8 & Theorem 9].

Theorem 3.2.20 *Let (G_i) be a sequence of nonabelian finite simple groups such that $|G_i| \to \infty$. Assume that $\gamma_u(G_i) = 2$, and $G_i \notin \{\mathrm{PSp}_{4m+2}(q) \mid \text{odd } q,\ m \geqslant 2\} \cup \{\mathrm{P\Omega}^{\pm}_{4m}(q) \mid m \geqslant 2\} \cup \{\mathrm{J}_3, \mathrm{He}, \mathrm{Co}_1, \mathrm{HN}\}$. Then*

$$P(G_i) \to \begin{cases} \frac{1}{2} & \text{if } G = \mathrm{PSL}_2(q) \\ 1 & \text{otherwise.} \end{cases}$$

Moreover, $P(G_i) \leqslant \frac{1}{2}$ if and only if $G_i = \mathrm{PSL}_2(q)$ for $q \equiv 3 \pmod 4$ or $G_i \in \{A_{13}, \mathrm{PSU}_5(2), \mathrm{Fi}_{23}\}$.

Methods. Bases of permutation groups. Let us now discuss the methods used in [30, 31] to bound $\gamma_u(G)$. Here there is a very pleasing connection with an entirely different topic in permutation group theory: bases. For a group G acting faithfully on a set Ω, a subset $B \subseteq \Omega$ is a *base* if the pointwise stabiliser $G_{(B)}$ is trivial. Since G acts faithfully, the entire domain Ω is a base, so we naturally ask for the smallest size of a base, which we call the *base size* $b(G, \Omega)$. To turn this combinatorial notion into an algebraic one, we observe that when G acts on G/H, a subset $\{Hg_1, \dots, Hg_c\}$ is a base if and only if $\cap_{i=1}^{c} H^{g_i} = 1$, so $b(G, G/H)$ is the smallest number of conjugates of H whose intersection is trivial.

Bases have been studied for over a century, and the base size has been at the centre of several recently proved conjectures, such as Pyber's conjecture that there is a constant c such that

$$\frac{\log |G|}{\log |\Omega|} \leqslant b(G, \Omega) \leqslant c \frac{\log |G|}{\log |\Omega|}$$

for all primitive groups $G \leqslant \mathrm{Sym}(\Omega)$ (see [43]), and Cameron's conjecture that $b(G, \Omega) \leqslant 7$ for nonstandard primitive almost simple groups $G \leqslant \mathrm{Sym}(\Omega)$ (see [32]). There is an ambitious ongoing programme of work, initiated by Saxl, to provide a complete classification of the primitive groups $G \leqslant \mathrm{Sym}(\Omega)$ with $b(G, \Omega) = 2$. There are numerous partial results in this direction, and we give just one, as we will use it below. Burness and Thomas [33] proved that if G is a simple group of Lie type and T is a maximal torus, then $b(G, G/N_G(T)) = 2$ apart from a few known low rank exceptions.

The following result is the bridge that connects bases with uniform domination (see [30, Corollaries 2.2 & 2.3]).

Lemma 3.2.21 *Let G be a finite group and let $s \in G$.*

(i) Assume that $\mathcal{M}(G, s) = \{H\}$ and H is corefree. Then the smallest uniform dominating set $S \subseteq s^G$ satisfies $|S| = b(G, G/H)$.

(ii) Assume that $H \in \mathcal{M}(G, s)$ is corefree. Then every uniform dominating set $S \subseteq s^G$ satisfies $|S| \geqslant b(G, G/H)$.

Proof For (i), note that $x \in H$ if and only if $\langle x, s \rangle \neq G$. Hence, $\{s^{g_1}, \dots, s^{g_c}\}$ is a uniform dominating set if and only if $\bigcap_{i=1}^{c} H^{g_i} = 1$, or said otherwise, if and only if $\{g_1, \dots, g_c\}$ is a base for G acting on G/H. The result follows.

For (ii), if $x \in H$, then $\langle x, s \rangle \neq G$. Therefore, if $\{s^{g_1}, \dots, s^{g_c}\}$ is a uniform dominating set, then $\bigcap_{i=1}^{c} H^{g_i} = 1$, so $\{g_1, \dots, g_c\}$ is a base for G acting on G/H and, consequently, $c \geqslant b(G, G/H)$. □

Let us explain how Lemma 3.2.21 applies. Part (i) gives an upper bound: if we can find $s \in G$ such that $\mathcal{M}(G,s) = \{H\}$ and $b(G,G/H) \leqslant c$, then $\gamma_u(G) \leqslant c$. Part (ii) gives a lower bound: if we can show that for all $s \in G$ there exists $H \in \mathcal{M}(G,s)$ with $b(G,G/H) \geqslant c$, then $\gamma_u(G) \geqslant c$. We give two examples to show how we do this in practice.

Example 3.2.22 Let $G = E_8(q)$ and let s generate a cyclic maximal torus of order $\Phi_{30}(q) = q^8 + q^7 - q^5 - q^4 - q^3 + q + 1$. As noted in Example 3.2.13, $\mathcal{M}(G,s) = \{H\}$ where H is the normaliser of the torus $\langle s \rangle$. Now, applying Burness and Thomas' result [33, Theorem 1] mentioned above, we see that $b(G,G/H) = 2$, so Lemma 3.2.21 implies that $\gamma_u(G) = 2$.

Example 3.2.23 Let $n > 6$ be even and let $G = A_n$. We will give upper and lower bounds on $\gamma_u(G)$ via Lemma 3.2.21.

Seeking an upper bound on $\gamma_u(G)$, let $s = (1\,2\,\dots\,l)(l+1\,l+2\,\dots\,n)$ where $l \in \{\frac{n}{2}-1, \frac{n}{2}-2\}$ is odd. As we showed in Example 3.2.1, $\mathcal{M}(G,s) = \{H\}$ where $H \cong (S_l \times S_{n-l}) \cap A_n$. The action of A_n on A_n/H is just the action of A_n on the set of l-subsets of $\{1,2,\dots,n\}$. The base size of this action was studied by Halasi, and by [66, Theorem 4.2], we have $b(G,G/H) \leqslant \left\lceil \log_{\lceil n/l \rceil} n \right\rceil \cdot (\lceil n/l \rceil - 1) \leqslant 2\log_2 n$. Applying Lemma 3.2.21(i) gives $\gamma_u(G) \leqslant 2\log_2 n$.

Turning to a lower bound, note that every element of G is contained in a subgroup K of type $(S_k \times S_{n-k}) \cap A_n$ for some $0 < k < n$. By [66, Theorem 3.1], we have $b(G,G/K) \geqslant \log_2 n$. Applying Lemma 3.2.21(ii) gives $\gamma_u(G) \geqslant \log_2 n$.

We now address the general case where s is not contained in a unique maximal subgroup of G. In the spirit of how uniform spread was studied, a probabilistic approach is adopted. Write $Q(G,s,c)$ for the probability that a random c-tuple of elements of s^G does not give a uniform dominating set of G and write $\mathcal{P}(G)$ for the set of prime order elements of G. The main lemma is [30, Lemma 2.5].

Lemma 3.2.24 *Let G be a finite group and let $s \in G$.*

(i) For all positive integers c, we have

$$Q(G,s,c) \leqslant \sum_{x \in \mathcal{P}(G)} \left(\sum_{H \in \mathcal{M}(G,s)} \frac{|x^G \cap H|}{|x^G|} \right)^c.$$

(ii) For a positive integer c, if $Q(G,s,c) < 1$, then $\gamma_u(G) \leqslant c$.

Proof Part (ii) is immediate. For part (i), $\{s^{g_1}, \dots, s^{g_c}\}$ is not a uniform dominating set of G if and only if there exists a prime order element $x \in G$ such that $\langle x, s^{g_i} \rangle \neq G$ for all $1 \leqslant i \leqslant c$. Since $Q(x,s)$ is the probability that x does not generate G with a random conjugate of s (see (2.1)), this implies that $Q(G,s,c) \leqslant \sum_{x \in \mathcal{P}(G)} Q(x,s)^c$. The result follows from Lemma 3.2.12(i). □

Could Lemma 3.2.24 yield a better bound than Lemma 3.2.21 when s satisfies $\mathcal{M}(G,s) = \{H\}$? In this case, $Q(G,s,c)$ is nothing other than the probability that a random c-tuple of elements of G/H form a base and $\sum_{x \in \mathcal{P}(G)} \left(\frac{|x^G \cap H|}{|x^G|}\right)^c$ is the upper bound for $Q(G,s,c)$ used, first by Liebeck and Shalev [82] and then by numerous others since, to obtain upper bounds on the base size $b(G, G/H)$. Therefore, Lemma 3.2.24 has nothing new to offer in this special case.

We conclude with an example of Lemma 3.2.24 in action. This establishes a (typical) special case of Theorem 3.2.15(ii).

Example 3.2.25 Let $n \geqslant 10$ be even and let $G = \mathrm{PSL}_n(q)$. We proceed similarly to Example 3.2.23, by fixing odd $l \in \{\frac{n}{2} - 1, \frac{n}{2} - 2\}$ and then letting s lift to a block diagonal matrix $\left(\begin{smallmatrix} A & 0 \\ 0 & B \end{smallmatrix}\right)$ where $A \in \mathrm{SL}_l(q)$ and $B \in \mathrm{SL}_{n-l}(q)$ are irreducible. The order of s is divisible by a *primitive prime divisor* of $q^{n-l} - 1$ (a prime divisor coprime to $q^k - 1$ for each $1 \leqslant k < n - l$). Using the framework of Aschbacher's theorem on the subgroup structure of classical groups [1], Guralnick, Penttila, Praeger and Saxl, classify the subgroups of $\mathrm{GL}_n(q)$ that contain an element whose order is a primitive prime divisor of $q^m - 1$ when $m > \frac{n}{2}$ [62]. With this we deduce that $\mathcal{M}(G,s) = \{G_U, G_V\}$ where U and V are the obviously stabilised subspaces of dimension l and $n - l$, respectively. The fixed point ratio for classical groups acting on the set of k-subspaces of their natural module was studied by Guralnick and Kantor, and by [59, Proposition 3.1], if $G = \mathrm{PSL}_n(q)$ and H is the stabiliser of a k-subspace of $\mathbb{F}_q^n$, then $\frac{|x^G \cap H|}{|x^G|} < \frac{2}{q^k}$. Applying Lemma 3.2.24 gives $\gamma_u(G) \leqslant 2n + 15$ since

$$\begin{aligned} Q(G,s,2n+15) &\leqslant \sum_{x \in \mathcal{P}(G)} \left(\sum_{H \in \mathcal{M}(G,s)} \frac{|x^G \cap H|}{|x^G|} \right)^{2n+15} \\ &\leqslant q^{n^2-1} \left(2 \cdot \frac{2}{q^{\frac{n}{2}-2}} \right)^{2n+15} < 1. \end{aligned}$$

3.2.4 The spread of a finite group

We now look beyond finite simple groups and ask the general question: for which finite groups G is every nontrivial element contained in a generating pair? Brenner and Wiegold's original 1975 paper gives a comprehensive answer for finite soluble groups (see [15, Theorem 2.01] for even more detail).

Theorem 3.2.26 *Let G be a finite soluble group. The following are equivalent:*

(i) $s(G) \geqslant 1$
(ii) $s(G) \geqslant 2$
(iii) every proper quotient of G is cyclic.

The equivalence of (i) and (ii) shows that $s(G) = 1$ for no finite soluble groups G, so Brenner and Wiegold asked the following question [15, Problem 1.04].

Question 3.2.27 Which finite groups G satisfy $s(G) = 1$? In particular, are there perhaps only finitely many such groups?

The condition in (iii) is necessary for every nontrivial element of G to be contained in a generating pair, and this is true for an arbitrary group G. To see this, assume that every nontrivial element of G is contained in a generating pair. Let $1 \neq N \trianglelefteqslant G$ and let $1 \neq n \in N$. Then there exists $g \in G$ such that $G = \langle n, g \rangle$, so $G/N = \langle Nn, Ng \rangle = \langle Ng \rangle$, which is cyclic. In 2008, Breuer, Guralnick and Kantor conjectured that this condition is also sufficient for all finite groups [16, Conjecture 1.8].

Conjecture 3.2.28 *Let G be a finite group. Then $s(G) \geqslant 1$ if and only if every proper quotient of G is cyclic.*

Completing a long line of research in this direction, both Question 3.2.27 and Conjecture 3.2.28 were settled by Burness, Guralnick and Harper in 2021 [29].

Theorem 3.2.29 *Let G be a finite group. Then $s(G) \geqslant 2$ if and only if every proper quotient of G is cyclic.*

Corollary 3.2.30 *No finite group G satisfies $s(G) = 1$.*

The next example (which is [29, Remark 2.16]) shows that Theorem 3.2.29 does not hold if spread is replaced with uniform spread (recall Theorem 3.2.11).

Example 3.2.31 Let $G = S_n$ where $n \geqslant 6$ is even. Suppose that $u(G) > 0$ is witnessed by the class s^G. Since a conjugate of s generates with $(1\,2\,3)$, s must be an odd permutation. Since a conjugate of s generates with $(1\,2)$, s must have at most two cycles. Since n is even, it follows that s is an n-cycle. However, if $n = 6$ and $a \in \mathrm{Aut}(G) \setminus G$, then $s^a \in (1\,2\,3)(4\,5)^G$ also witnesses $u(G) > 0$, which is a contradiction. Therefore, $u(S_6) = 0$.

However, Example 3.2.31 is essentially the only obstacle to a result for uniform spread analogous to Theorem 3.2.29 on spread. Indeed, Burness, Guralnick and Harper gave the following complete description of the finite groups G with $u(G) < 2$ [29, Theorem 3]. (Note that the uniform spread of abelian groups G is not interesting: $u(G) = \infty$ if G is cyclic and $u(G) = 0$ otherwise.)

Theorem 3.2.32 *Let G be a nonabelian finite group such that every proper quotient is cyclic. Then*

(i) $u(G) = 0$ if and only if $G = S_6$

(ii) $u(G) = 1$ if and only if the group G has a unique minimal normal subgroup

$$N = T_1 \times \cdots \times T_k$$

where $k \geqslant 2$ and where $T_i = A_6$ and

$$N_G(T_i)/C_G(T_i) = S_6$$

for all $1 \leqslant i \leqslant k$.

Theorem 3.2.32 emphasises the anomalous behaviour of S_6: it is the only almost simple group G where every proper quotient of G is cyclic but $u(G) < 2$.

Methods. A reduction theorem and Shintani descent. We now outline the proof of Theorems 3.2.29 and 3.2.32 in [29]. We need to consider the finite groups G all of whose proper quotients are cyclic. In light of Theorem 3.2.26, we will assume that G is insoluble, so G has a unique minimal normal subgroup T^k for a nonabelian simple group T, and we can assume that $G = \langle T^k, s\rangle$ where $s = (a, 1, \dots, 1)\sigma \in \mathrm{Aut}(T^k)$ for $a \in \mathrm{Aut}(T)$ and $\sigma = (1\,2\,\dots\,k) \in S_k$. Let us ignore $T = A_6$ due to the complications we have already seen in this case.

The first major step in the proof of Theorems 3.2.29 and 3.2.32 is the following reduction theorem [29, Theorem 2.13].

Theorem 3.2.33 *Fix a nonabelian finite simple group T and assume that $T \neq A_6$. Fix $s = (a, 1, \ldots, 1)\sigma \in \mathrm{Aut}(T^k)$ with $a \in \mathrm{Aut}(T)$ and $\sigma = (1\,2\,\ldots\,k) \in S_k$. Then s witnesses $u(\langle T^k, s\rangle) \geqslant 2$ if both of the following hold:*

(i) a witnesses $u(\langle T, a\rangle) \geqslant 2$
(ii) $\langle a\rangle \cap T \neq 1$, and if a is square in $\mathrm{Aut}(T)$, then $|\langle a\rangle \cap T|$ does not divide 4.

The second major step is to generalise Breuer, Guralnick and Kantor's result Theorem 3.2.5 that $u(T) \geqslant 2$ for all nonabelian finite simple groups T to all almost simple groups $A = \langle T, a\rangle$. This long line of research was initiated by Burness and Guest for $T = \mathrm{PSL}_n(q)$ [28], continued by Harper for the remaining classical groups T [68, 70] and completed by Burness, Guralnick and Harper for exceptional groups T [29]. (We have already noted that the proof of $u(S_n) \geqslant 2$ for $n \neq 6$ was completed by Burness and Harper in [31], and the result for sporadic groups follows from computational work in [16]).

We conclude by highlighting the major obstacle that this body of work faced and then outlining the technique that overcame this obstacle.

Let $G = \langle T, g\rangle$ where T is a finite simple group of Lie type and $g \in \mathrm{Aut}(T)$. Suppose s^G witnesses $u(G) \geqslant 2$. A conjugate of s generates with any element of T, so $G = \langle T, s\rangle$. By replacing s with a power if necessary, we may assume that $s \in Tg$. How do we describe elements of Tg and their overgroups? Suppose that $T = \mathrm{PSL}_n(q)$ with $q = p^f$. If $g \in \mathrm{PGL}_n(q)$, then there are geometric techniques available, but what if g is, say, the field automorphism $(a_{ij}) \mapsto (a_{ij}^p)$?

The technique that was used to answer these questions is known as *Shintani descent.* This was introduced by Shintani in 1976 [93] (and generalised by Kawanaka in [75]) to study irreducible characters of almost simple groups. However, as first exploited by Fulman and Guralnick in their work on the Boston–Shalev conjecture [51], Shintani descent also provides a fruitful way of studying the conjugacy classes of almost simple groups. The main theorem is the following, and we follow Desphande's proof [41]. (Here σ_i is considered as an element of $\langle X, \sigma_i\rangle$ for $i \in \{1, 2\}$.)

Theorem 3.2.34 *Let X be a connected algebraic group, and let $\sigma_1, \sigma_2\colon X \to X$ be commuting Steinberg endomorphisms. Then there is a bijection*

$$F\colon \{X_{\sigma_1}\text{-classes in } X_{\sigma_1}\sigma_2\} \to \{X_{\sigma_2}\text{-classes in } X_{\sigma_2}\sigma_1\}.$$

Proof Let S be the orbits of $\{(g,h) \in X\sigma_2 \times X\sigma_1 \mid [g,h] = 1\}$ under the conjugation action of X. By the Lang–Steinberg theorem, S is in bijection with the orbits of $\{(x\sigma_2, \sigma_1) \mid x \in X_{\sigma_1}\}$ under conjugation by X_{σ_1} and also with the orbits of $\{(\sigma_2, y\sigma_1) \mid x \in X_{\sigma_2}\}$ under conjugation by X_{σ_2}. This provides a bijection between the X_{σ_1}-classes in $X_{\sigma_1}\sigma_2$ and the X_{σ_2}-classes in $X_{\sigma_2}\sigma_1$. □

The bijection in the proof of Theorem 3.2.34, known as the *Shintani map* of (X, σ_1, σ_2), has desirable properties. For instance, if $\sigma_1 = \sigma_2^e$ for $e \geqslant 1$, then

$$F((x\sigma_2)^{X_{\sigma_1}}) = (a^{-1}(x\sigma_1)^{-e}a)^{X_{\sigma_2}} \tag{2.4}$$

for some $a \in X$. Moreover, if $F(g^{X_{\sigma_1}}) = h^{X_{\sigma_2}}$, then it is easy to show that $C_{X_{\sigma_1}}(g) \cong C_{X_{\sigma_2}}(h)$, and, by now, extensive information is also available about how maximal overgroups of g in $\langle X_{\sigma_1}, \sigma_2\rangle$ relate to the maximal overgroups of h in $\langle X_{\sigma_2}, \sigma_1\rangle$. These latter results are crucial to proving Theorem 3.2.29 for almost simple groups of Lie type. The first result in this direction is due to Burness and Guest [28, Corollary 2.15], and the subsequent developments are unified by Harper in [69], to which we refer the reader for further detail.

Example 3.2.35 We sketch how $u(G) \geqslant 2$ was proved for $G = \langle T, g\rangle$ when $T = \Omega^+_{2m}(q)$ with $q = 2^f$ and g is the field automorphism $\varphi\colon (a_{ij}) \mapsto (a_{ij}^2)$. We will not make further assumptions on q, but we will assume that m is large.

Let X be the simple algebraic group $\mathrm{SO}_{2m}(\overline{\mathbb{F}}_2)$ and let F be the Shintani map of (X, φ^f, φ). Then $G = \langle X_{\varphi^f}, \varphi\rangle$, and writing $G_0 = X_\varphi = \Omega^+_{2m}(2)$, we observe that F gives a bijection between the conjugacy classes in Tg and those in G_0.

We define $s \in Tg$ such that $F(s^G) = s_0^{G_0}$ for a well chosen element $s_0 \in G_0$. In particular, (2.4) implies that s_0 is X-conjugate to a power of s. To define s_0, fix k such that $m-k$ is even and $\frac{\sqrt{2m}}{4} < 2k < \frac{\sqrt{2m}}{2}$, fix $A \in \Omega^-_{2k}(2)$ and $B \in \Omega^-_{2m-2k}(2)$ of order 2^k+1 and $2^{m-k}+1$ and let $s_0 = \begin{pmatrix} A & 0 \\ 0 & B \end{pmatrix} \in \Omega^+_{2m}(2)$.

We now study $\mathcal{M}(G,s)$. First, a power of s_0 (and hence s) has a 1-eigenspace of codimension $2k < \frac{\sqrt{2m}}{2}$, so [64, Theorem 7.1] implies that s is not contained in any local or almost simple maximal subgroup of G.

Next, a power of s_0 has order $2^{m-k}+1$, which is divisible by the *primitive part* of $2^{2m-2k}-1$ (the largest divisor of $2^{2m-2k}-1$ that is prime to $2^l - 1$ for all $0 < l < 2m-2k$). For sufficiently large m, we can apply the main theorem of [62] to deduce that all of the maximal overgroups of s_0 in $G_0 = \Omega^+_{2m}(2)$ are reducible. In particular, the only maximal overgroup of s_0 in G_0 arising as the set of fixed points of a closed positive-dimensional φ-stable subgroup of X is the obvious reducible subgroup of type $(\mathrm{O}^-_{2k}(2) \times \mathrm{O}^-_{2m-2k}(2)) \cap G_0$. Now the theory of Shintani descent [69, Theorem 4] implies that the only such maximal overgroup of s in G is one subgroup H of type $(\mathrm{O}^\pm_{2k}(q) \times \mathrm{O}^\pm_{2m-2k}(q)) \cap G$.

Drawing these observations together and using Aschbacher's subgroup structure theorem [1], we deduce that $\mathcal{M}(G,s) = \{H\} \cup \mathcal{M}'$, where $\mathcal{M}'$ consists of subfield subgroups. There are at most $\log_2 f + 1 = \log_2 \log_2 q + 1$ classes of maximal subfield subgroups of G, and by [28, Lemma 2.19], s is contained in at most $|C_{G_0}(s_0)| = (2^k+1)(2^{2m-2k}+1)$ conjugates of a fixed maximal subgroup.

Using the fixed point ratio bound proved for reducible subgroups in [59] and irreducible subgroups in [22, 23, 24, 25], for all nontrivial $x \in G$ we have

$$\begin{aligned} Q(x,s) &< \sum_{H \in \mathcal{M}(G,s)} \frac{|x^G \cap H|}{|x^G|} \\ &< \frac{5}{q^{\sqrt{n}/4}} + (\log_2 \log_2 q + 1)(2^k+1)(2^{2m-2k}+1)\frac{2}{q^{n-3}}. \end{aligned}$$

In particular, for sufficiently large m, $Q(x,s) < \frac{1}{2}$ and $u(G) \geqslant 2$. Moreover, $Q(x,s) \to 0$ and $u(G) \to \infty$, as $m \to \infty$ or, for sufficiently large m, as $q \to \infty$.

3.2.5 The generating graph

In this section, we introduce a combinatorial object that gives a way to visualise the concepts we have seen. The *generating graph* of a group G, denoted $\Gamma(G)$, is the graph whose vertex set is $G \setminus 1$ and where two vertices $g, h \in G$ are adjacent if $\langle g, h \rangle = G$. See Figure 2.1 for examples.

A question that quickly comes to mind is: when is $\Gamma(G)$ is connected? This question has a remarkably straightforward answer.

Theorem 3.2.36 *Let G be a finite group. The following are equivalent:*

(i) $\Gamma(G)$ has no isolated vertices
(ii) $\Gamma(G)$ is connected
(iii) $\Gamma(G)$ has diameter at most two
(iv) every proper quotient of G is cyclic.

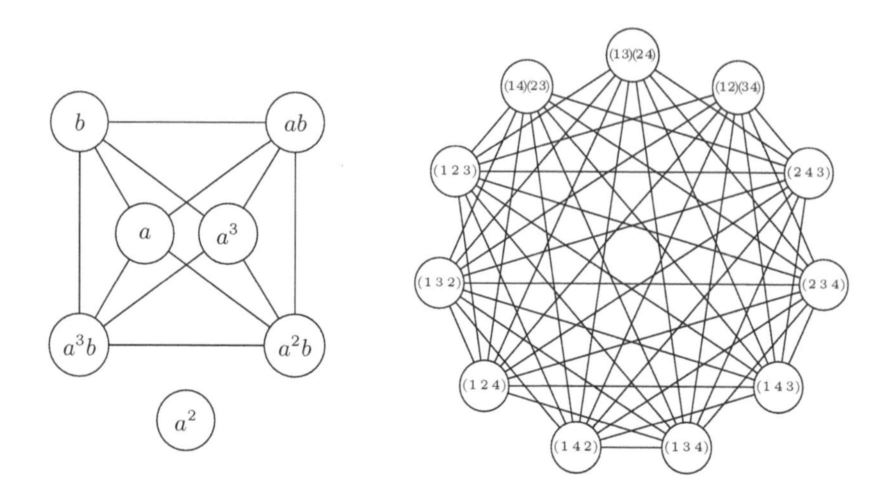

Figure 2.1 Generating graphs of $D_8 = \langle a, b \mid a^4 = b^2 = 1, a^b = a^{-1} \rangle$ and the alternating group A_4.

Of course, Theorem 3.2.36 is simply a reformulation of Theorem 3.2.29 due to Burness, Guralnick and Harper. Indeed, the generating graph gives an enlightening perspective on the concepts introduced so far:

(i) G is $\frac{3}{2}$-generated if and only if $\Gamma(G)$ has no isolated vertices
(ii) $s(G)$ is the greatest k such that any k vertices of $\Gamma(G)$ have a common neighbour, which means that $s(G) \geqslant 2$ if and only if $\mathrm{diam}(\Gamma(G)) \leqslant 2$
(iii) $\gamma_t(G)$ is the total domination number of $\Gamma(G)$: the least size of a set of vertices whose neighbours cover $\Gamma(G)$.

Regarding (iii), the total domination number is a well studied graph invariant and was the inspiration for the group theoretic term (and the symbol γ_t is the graph theoretic notation). Regarding (ii), as far as the author is aware, the graph invariant corresponding to the spread of a group, while natural, does not have a canonical name, but, inspired by the recent work on the spread of finite groups, some authors have started to use the term *spread* for this graph invariant (that is, the *spread* of a graph is the greatest k such that any k vertices have a common neighbour), see for example [34, Section 2.5].

What are the connected components of $\Gamma(G)$ are in general? The following conjecture (first posed as a question in [39]) proposes a straightforward answer.

Conjecture 3.2.37 *Let G be a finite group. Then the graph obtained from $\Gamma(G)$ by removing the isolated vertices is connected.*

Conjecture 3.2.37 is known to be true if G is soluble or characteristically simple by work of Crestani and Lucchini [39, 40] or if every proper quotient of G is cyclic as a consequence of Theorem 3.2.29. Otherwise, Conjecture 3.2.37 remains an intriguing open question about the generating sets of finite groups.

There is now a vast literature on the generating graph and surveying it is beyond the scope of this survey article, so we will make only a few remarks. The first paper to study $\Gamma(G)$ in its own right, and call it the *generating graph*, was [86] by Lucchini and Maróti, who then studied various aspects of this graph in subsequent papers (for example, [17, 85]). However, $\Gamma(G)$ first appeared in the literature, indirectly, as a construction in a proof of Liebeck and Shalev in [81]. A major result of that paper is that there exist constants $c_1, c_2 > 0$ such that for all finite simple groups G, the probability that two randomly chosen elements generate G, denoted $P(G)$, satisfies

$$1 - \frac{c_1}{m(G)} \leqslant P(G) \leqslant 1 - \frac{c_2}{m(G)} \tag{2.5}$$

where $m(G)$ is the smallest index of a subgroup of G (for example, $m(A_n) = n$). From this, Liebeck and Shalev deduce that there is a constant $c > 0$ such that every finite simple group G contains at least $c \cdot m(G)$ elements that pairwise generate G. The proof of this corollary simply involves applying Turán's theorem to $\Gamma(G)$, exploiting (2.5). A couple of subsequent papers [7, 20] continued the study of cliques in $\Gamma(G)$, partly motivated by the observation that the largest size of a clique in $\Gamma(G)$, denoted $\mu(G)$, is a lower bound for the smallest number of proper subgroups whose union is G, denoted $\sigma(G)$. Returning to spread, it is easy to see that $s(G) < \mu(G) \leqslant \sigma(G)$, so these results give upper bounds on spread, which are otherwise difficult to find. Indeed, the best upper bounds for the smaller 14 sporadic groups (including the two where the spread is known exactly) were established in [14] by a clever refinement of this bound (for the larger 12 sporadic groups, different methods were used in [48]).

3.2.6 Applications

We conclude Section 3.2 with three applications of spread. The first shows how $\frac{3}{2}$-generation naturally arises in a completely different context. The second highlights the benefit of studying spread, not just $\frac{3}{2}$-generation. The third moves beyond simple groups and applies the classification of finite $\frac{3}{2}$-generated groups.

Application 1. Word maps. From a word $w = w(x,y)$ in the free group F_2, we obtain a *word map* $w\colon G \times G \to G$ and we write $w(G) = \{w(g,h) \mid g,h \in G\}$.

Let G be a nonabelian finite simple group. For natural choices of w, there has been substantial recent progress showing that $w(G) = G$. For example, solving the Ore Conjecture, Liebeck, O'Brien, Shalev and Tiep [77] proved that $w = x^{-1}y^{-1}xy$ is surjective (that is, every element is a commutator).

Now consider the converse question: which subsets $S \subseteq G$ arise as images of some word $w \in F_2$? Such a subset S must satisfy $1 \in S$ (as $w(1,1) = 1$) and $S^a = S$ for all $a \in \mathrm{Aut}(G)$ (as $w(x^a, y^a) = w(x,y)^a$), and Lubotzky [84] proved that these conditions are sufficient.

Theorem 3.2.38 *Let G be a finite simple group and let $S \subseteq G$. Then S is the image of a word $w \in F_2$ if and only if $1 \in S$ and $S^a = S$ for all $a \in \mathrm{Aut}(G)$.*

The proof of Theorem 3.2.38 is short and fairly elementary, except it uses Theorem 3.2.2. Indeed, Theorem 3.2.2 is the only way it depends on the CFSG.

Proof outline of Theorem 3.2.38 Let $G^2 = \{(a_i, b_i) \mid 1 \leqslant i \leqslant |G|^2\}$ be ordered such that $\langle a_i, b_i \rangle = G$ if and only if $i \leqslant \ell$. Fix the free group $F_2 = \langle x, y \rangle$ and let $\varphi\colon F_2 \to G^{|G|^2}$ be defined as $\varphi(x) = (a_1, \dots, a_{|G|^2})$ and $\varphi(y) = (b_1, \dots, b_{|G|^2})$. Let $z = (z_1, \dots, z_{|G|^2})$ where $z_i = a_i$ if $i \leqslant \ell$ and $a_i \in S$ and $z_i = 1$ otherwise.

By Theorem 3.2.2, every nontrivial element of G is contained in a generating pair, so, in particular, $\{z_i \mid 1 \leqslant i \leqslant |G|^2\} = S \cup \{1\} = S$.

By an elementary argument, Lubotzky shows that $\varphi(F_2) = H \times K$ where H and K are the projections of $\varphi(F_2)$ onto the first ℓ factors of $G^{|G|^2}$ and the remaining $|G|^2 - \ell$ factors, respectively. Moreover, using a theorem of Hall [67], H is the subgroup of G^ℓ isomorphic to $G^{\ell/|\mathrm{Aut}(G)|}$ with the defining property that $(g_1, \dots, g_\ell) \in H$ if and only if for all $a \in \mathrm{Aut}(G)$ and $1 \leqslant i,j \leqslant \ell$ we have $g_i = g_j^a$ whenever $\varphi_i = \varphi_j\, a$. In particular, since $S^a = S$ for all $a \in \mathrm{Aut}(G)$, we deduce that $z \in \varphi(F_2)$. Therefore, there exists $w \in F_2$ such that for all $1 \leqslant i \leqslant |G|^2$ we have $w(a_i, b_i) = \varphi(w)_i = z_i$.

Combining the conclusions of the previous two paragraphs, we deduce that

$$w(G) = \{w(a_i, b_i) \mid 1 \leqslant i \leqslant |G|^2\} = \{z_i \mid 1 \leqslant i \leqslant |G|^2\} = S,$$

as required. □

Application 2. The product replacement graph. For an integer $k \geqslant 0$, the vertices of the product replacement graph $\Gamma_k(G)$ are the generating k-tuples of G, and the neighbours of $(x_1, \dots, x_i, \dots, x_k)$ in $\Gamma_k(G)$ are $(x_1, \dots, x_i x_j^{\pm}, \dots, x_k)$ and $(x_1, \dots, x_j^{\pm} x_i, \dots, x_k)$ for each possible $1 \leqslant i \neq j \leqslant k$.

The product replacement graph arises in a number of contexts, most notably, the product replacement algorithm for computing random elements of G, which involves a random walk on $\Gamma_k(G)$, see [36]. Thus, the connectedness of $\Gamma_k(G)$ is of particular interest. Specifically, Pak [90, Question 2.1.33] asked whether $\Gamma_k(G)$ is connected whenever k is strictly greater than $d(G)$, the smallest size of a generating set for G. This question is open, even for finite simple groups where Wiegold conjectured that the answer is true. Nevertheless, the following lemma of Evans [46, Lemma 2.8] shows the usefulness of spread. (Here a generating tuple of G is said to be *redundant* if some proper subtuple also generates G.)

Lemma 3.2.39 *Let $k \geqslant 3$ and let G be a group such that $s(G) \geqslant 2$. Then all of the redundant generating k-tuples are connected in $\Gamma_k(G)$.*

Proof Let $x = (x_1, \dots, x_k)$ and $y = (y_1, \dots, y_k)$ be two redundant generating k-tuples. By an elementary observation of Pak, it is sufficient to show that x and y are connected after permuting of the entries of x and y. In particular, since x and y are redundant, we may assume that $\langle x_1, \dots, x_{k-1} \rangle = \langle y_1, \dots, y_{k-1} \rangle = G$ and also that $x_1 \neq 1 \neq y_2$. Since $s(G) \geqslant 2$, there exists $z \in G$ such that $\langle x_1, z \rangle = \langle y_2, z \rangle = G$. We now make a series of connections. First, $x = x^{(1)}$ is connected to $x^{(2)} = (x_1, \dots, x_{k-1}, z)$ as $\langle x_1, \dots, x_{k-1} \rangle = G$. Next, $x^{(2)}$ is connected to $x^{(3)} = (x_1, y_2, \dots, y_{k-1}, z)$ as $\langle x_1, z \rangle = G$. Now, $x^{(3)}$ is connected to $x^{(4)} = (y_1, \dots, y_{k-1}, z)$ as $\langle y_2, z \rangle = G$. Finally, $x^{(4)}$ is connected to $(y_1, \dots, y_k) = y$ as $\langle y_1, \dots, y_{k-1} \rangle = G$. This shows that x is connected to y. □

Combining Lemma 3.2.39 with Theorem 3.2.5 shows that to prove Wiegold's conjecture, it suffices to show that for each finite simple group G every irredundant generating k-tuple is connected in $\Gamma_k(G)$ to a redundant one.

As further evidence for the relevance of spread in this area, we note that Wiegold's original conjecture was (the a priori weaker claim) that a related graph $\Sigma_k(G)$ is connected for all finite simple groups G and $k > d(G)$ (see [90, Conjecture 2.5.4]), but a short argument of Pak [90, Proposition 2.5.13] shows that $\Sigma_k(G)$ is connected if and only if $\Gamma_k(G)$ is connected, for all groups G such that $s(G) \geqslant 2$, which by Theorem 3.2.5 includes all finite simple groups G.

Application 3. The $\mathcal{X}$-radical of a group. In 1968, Thompson proved that a finite group is soluble if and only if all of its 2-generated subgroups are soluble [96, Corollary 2]. The result follows from Thompson's classification of the finite insoluble groups all of whose proper subgroups are soluble, but, in 1995, Flavell gave a direct proof of this result [49]. Confirming a conjecture of Flavell [50, Conjecture B], Guralnick, Kunyavskiĭ, Plotkin and Shalev proved the following result about the *soluble radical* of G, written $R(G)$, which is the largest normal soluble subgroup of G [60, Theorem 1.1].

Theorem 3.2.40 *Let G be a finite group. Then*

$$R(G) = \{x \in G \mid \langle x, y \rangle \textit{ is soluble for all } y \in G\}.$$

The key (and only CFSG-dependent) element of the proof of Theorem 3.2.40 is a strong version Theorem 3.2.2 on the $\frac{3}{2}$-generation of finite simple groups. To paint a picture of how the $\frac{3}{2}$-generation of simple groups plays the starring role, we first present the analogue for simple Lie algebras [60, Theorem 2.1]. Recall that the *radical* of a Lie algebra $\mathfrak{g}$, denoted $R(\mathfrak{g})$, is the largest soluble ideal of $\mathfrak{g}$.

Theorem 3.2.41 *Let $\mathfrak{g}$ be a finite-dimensional Lie algebra over $\mathbb{C}$. Then*

$$R(\mathfrak{g}) = \{x \in \mathfrak{g} \mid \langle x, y \rangle \textit{ is soluble for all } y \in \mathfrak{g}\}.$$

Proof Let $x \in \mathfrak{g}$. First assume that $x \in R(\mathfrak{g})$. Let $y \in G$ and let $\mathfrak{h}$ be the smallest ideal of $\langle x, y \rangle$ containing x. Now $\mathfrak{h}$ is soluble as it is a Lie subalgebra of $R(\mathfrak{g})$, and $\langle x, y \rangle / \mathfrak{h}$ is soluble as it is 1-dimensional, so $\langle x, y \rangle$ is soluble.

Now assume that $\langle x, y \rangle$ is soluble for all $y \in G$. We will prove that $x \in R(\mathfrak{g})$. For a contradiction, assume that $\mathfrak{g}$ is a minimal counterexample (by dimension). Consider $\overline{\mathfrak{g}} = \mathfrak{g}/R(\mathfrak{g})$. Then $\langle \overline{x}, \overline{y} \rangle$ is soluble for all $y \in \mathfrak{g}$, and $R(\overline{\mathfrak{g}})$ is trivial, so $\overline{x} \notin R(\overline{\mathfrak{g}})$. Therefore, $R(\mathfrak{g}) = 0$, by the minimality of $\mathfrak{g}$. This means that $\mathfrak{g}$ is semisimple, so we may write $\mathfrak{g} = \mathfrak{g}_1 \oplus \cdots \oplus \mathfrak{g}_k$ where $\mathfrak{g}_1, \ldots, \mathfrak{g}_k$ are simple. Writing $x = (x_1, \ldots, x_k)$, fix $1 \leqslant i \leqslant k$ such that $x_i \neq 0$. Then by Theorem 3.2.3, there exists $y \in \mathfrak{g}_i$ such that $\langle x_i, y \rangle = \mathfrak{g}_i$. In particular, $\mathfrak{g}_i$ is a quotient of $\langle x, y \rangle$, so $\langle x, y \rangle$ is not soluble, which is a contradiction. □

Returning to groups, again using variants of Theorem 3.2.2, Guralnick, Plotkin and Shalev set Theorem 3.2.40 in a more general context [63, Theorem 6.1]. Here we give a short proof of their result by applying Theorem 3.2.29.

Let $\mathcal{X}$ be a class of finite groups that is closed under subgroups, quotients and extensions. The $\mathcal{X}$-radical of a group G, denoted $\mathcal{X}(G)$, is the largest normal $\mathcal{X}$-subgroup of G. For instance, $\mathcal{X}(G) = R(G)$ if $\mathcal{X}$ is the class of soluble groups.

Theorem 3.2.42 *Let $\mathcal{X}$ be a class of finite groups that is closed under subgroups, quotients and extensions. Then*

$$\mathcal{X}(G) = \{x \in G \mid \langle x^{\langle y \rangle} \rangle \textit{ is an } \mathcal{X}\textit{-group for all } y \in G\}.$$

Corollary 3.2.43 *Let $\mathcal{X}$ be a class of finite groups that is closed under subgroups, quotients and extensions. Then G is an $\mathcal{X}$-group if and only if every 2-generated subgroup of G is an $\mathcal{X}$-group.*

Proof If G is an $\mathcal{X}$-group, then every 2-generated subgroup is. Conversely, if for all $x, y \in G$ the subgroup $\langle x, y \rangle$ is an $\mathcal{X}$-group, then so is $\langle x^{\langle y \rangle} \rangle$, so, by Theorem 3.2.42, $x \in \mathcal{X}(G)$, which shows $G = \mathcal{X}(G)$, which is an $\mathcal{X}$-group. □

Corollary 3.2.44 *Let $\mathcal{X}$ be a class of finite groups that is closed under subgroups, quotients and extensions. Assume that $\mathcal{X}$ contains all soluble groups. Then*

$$\mathcal{X}(G) = \{x \in G \mid \langle x, y \rangle \textit{ is an } \mathcal{X}\textit{-group for all } y \in G\}.$$

Proof Let $x \in G$. If for all $y \in G$, $\langle x, y \rangle$ is an $\mathcal{X}$-group, then so is $\langle x^{\langle y \rangle} \rangle$, so, by Theorem 3.2.42, $x \in \mathcal{X}(G)$. Conversely, if $x \in \mathcal{X}(G)$, then $\langle x^{\langle y \rangle} \rangle \leqslant \langle x^G \rangle \leqslant \mathcal{X}(G)$ is an $\mathcal{X}$-group, so $\langle x, y \rangle$, an extension of $\langle x^{\langle y \rangle} \rangle$ by a cyclic group, is an $\mathcal{X}$-group. □

Proof of Theorem 3.2.42 Let $x \in G$. First assume that $x \in \mathcal{X}(G)$. For all $y \in G$, we have $\langle x^{\langle y \rangle} \rangle \leqslant \langle x^G \rangle \leqslant \mathcal{X}(G)$, so $\langle x^{\langle y \rangle} \rangle$ is an $\mathcal{X}$-group.

Now assume that $\langle x^{\langle y \rangle} \rangle$ is an $\mathcal{X}$-group for all $y \in G$. We will prove that $x \in \mathcal{X}(G)$. For a contradiction, assume that G is a minimal counterexample. Consider $\overline{G} = G/\mathcal{X}(G)$. Then $\langle \overline{x}^{\langle \overline{y} \rangle} \rangle$ is an $\mathcal{X}$-group for all $y \in G$ (as $\mathcal{X}$ is closed under quotients), and $\mathcal{X}(\overline{G})$ is trivial (as $\mathcal{X}$ is closed under extensions), so $\overline{x} \notin \mathcal{X}(\overline{G})$. Therefore, $\mathcal{X}(G) = 1$, by the minimality of G. Now consider $H = \langle x^G \rangle$. Then $\langle x^{\langle h \rangle} \rangle$ is an $\mathcal{X}$-group for all $h \in H$, and $\mathcal{X}(H) \leqslant \mathcal{X}(G)$ (as $\mathcal{X}(H)$ is characteristic in H so normal in G), so $x \notin \mathcal{X}(H)$. Therefore, $\langle x^G \rangle = G$, by the minimality of G. Consider a power x' of x of prime order. Then $\langle x'^{\langle y \rangle} \rangle$ is an $\mathcal{X}$-group for all $y \in G$ (as $\mathcal{X}$ is closed under subgroups), and $x' \notin 1 = \mathcal{X}(G)$. Therefore, it suffices to consider the case where x has prime order.

Let N be a minimal normal subgroup of G and write $N = T^k$ where T is simple. Observe that N, or equivalently T, is not an $\mathcal{X}$-group, since $\mathcal{X}(G) = 1$.

Suppose that $x \in N$. Then $G = N$ since $\langle x^G \rangle = G$. In particular, $k = 1$, so, by Theorem 3.2.2, there exists $y \in G$ such that $\langle x, y \rangle = G$, which is not an $\mathcal{X}$-group, so $\langle x^{\langle y \rangle} \rangle$ is not an $\mathcal{X}$-group either: a contradiction. Therefore, $x \notin N$.

Suppose that x centralises N. Then N is central since $\langle x^G \rangle = G$, so $N \cong C_p$ for a prime p. In $\widetilde{G} = G/N$, $\widetilde{x}$ is nontrivial as $x \notin N$ and $\langle \widetilde{x}^{\langle \widetilde{y} \rangle} \rangle$ is an $\mathcal{X}$-group for all $y \in G$ (as $\langle x^{\langle y \rangle} \rangle$ is). The minimality of G means $\widetilde{x} \in \mathcal{X}(\widetilde{G})$, but $\langle \widetilde{x}^{\widetilde{G}} \rangle = \widetilde{G}$ (as $\langle x^G \rangle = G$), so $\widetilde{G}$ is an $\mathcal{X}$-group. Since $N \cong C_p$ is not an $\mathcal{X}$-group, p does not divide $|\widetilde{G}|$. By the Schur–Zassenhaus Theorem, $G = N \times H$ for some $H \cong \widetilde{G}$. Since $N \cong C_p$ is not an $\mathcal{X}$-group and $\langle x \rangle$ is an $\mathcal{X}$-group, p does not divide $|x|$, so $x \in H$, contradicting $\langle x^G \rangle = G$. Therefore, x acts nontrivially on N.

Suppose that N is abelian. As x acts nontrivially on N, there exists $n \in N$ with $[x, n] \neq 1$. Now $\langle [x, n] \rangle$ is isomorphic to T, as $[x, n] \in N$, so it is not an $\mathcal{X}$-group, implying that $\langle x^{\langle n \rangle} \rangle$ is not an $\mathcal{X}$-group either. Therefore, N is nonabelian.

Now x permutes the k factors of $N \cong T^k$ and let $M \cong T^l$ be a nontrivial subgroup of N whose factors are permuted transitively by x. If $l = 1$, then $\langle M, x \rangle$ is almost simple, and if $l > 1$, then, recalling that x has prime order, $\langle x \rangle \cong C_l$ acts regularly on the factors of M. In either case, M is the unique minimal normal subgroup of $\langle M, x \rangle$, so every proper quotient of $\langle M, x \rangle$ is cyclic. Therefore, by Theorem 3.2.29, there exists $m \in \langle M, x \rangle$ such that $\langle m, x \rangle = \langle M, x \rangle$. Moreover, $\langle x^{\langle m \rangle} \rangle$, being a normal subgroup of $\langle M, x \rangle$ containing x, is also $\langle M, x \rangle$. However, $\langle x^{\langle m \rangle} \rangle = \langle M, x \rangle$ is not an $\mathcal{X}$-group since the subgroup T is not an $\mathcal{X}$-group. This contradiction completes the proof. □

3.3 Infinite Groups

3.3.1 Generating infinite groups

We now turn to infinite groups and their generating pairs. Do the results from Section 3.2 on the spread of finite groups extend to infinite groups? Let us recall that the motivating theorem for finite groups is the landmark result that every finite simple group is 2-generated. This result is easily seen to be false when the assumption of finiteness is removed. For example, the alternating group $\mathrm{Alt}(\mathbb{Z})$ is simple but is not even finitely generated since every finite subset of $\mathrm{Alt}(\mathbb{Z})$ is supported on finitely many points and therefore generates a finite subgroup. The problem persists even if we restrict to finitely generated simple groups. Answering a question of Wiegold in the Kourovka Notebook [76, Problem 6.44], in 1982, Guba constructed a finitely generated infinite simple group that is not 2-generated (in fact, the group constructed has the property that every 2-generated subgroup is free) [57]. More recently, Osin and Thom, by studying the ℓ^2-Betti number of groups, proved that for every $k \geqslant 2$ there exists an infinite simple group that is k-generated but not $(k-1)$-generated [89, Corollary 1.2].

With these results in mind, it makes sense to focus on 2-generated groups and ask whether the results about the spread of finite 2-generated groups extend to general 2-generated groups. Recall that Theorem 3.2.29 gives a characterisation of the finite $\frac{3}{2}$-generated groups: a finite group G is $\frac{3}{2}$-generated if and only if every proper quotient of G is cyclic. In particular, every finite simple group is $\frac{3}{2}$-generated. The following example due to Cox in 2022 [38], highlights that this characterisation does not extend to general 2-generated groups (that is, there exists a infinite 2-generated group G that is not $\frac{3}{2}$-generated but for which every proper quotient of G is cyclic).

Example 3.3.1 For each positive n, let G_n be the subgroup of $\mathrm{Sym}(\mathbb{Z})$ defined as $\langle \mathrm{Alt}(\mathbb{Z}), t^n \rangle$ where $t\colon \mathbb{Z} \to \mathbb{Z}$ is the translation $x \mapsto x+1$.

It is straightforward to show that $\mathrm{Alt}(\mathbb{Z})$ is the unique minimal normal subgroup of G_n, so every proper quotient of G is cyclic.

In addition, G_n is 2-generated. Indeed, by [38, Lemma 3.7], $G_n = \langle a_n, t^n \rangle$ for $a_n = \prod_{i=0}^{m+1} x_i^{t^{3ni}}$ where $x_0, \dots, x_{m+1} \in A_{3n}$ satisfy $x_0 = (1\,3)$, $x_{m+1} = (2\,3)$ and $\{(1\,2\,3)^{x_1}, \dots, (1\,2\,3)^{x_m}\} = (1\,2\,3)^{A_{3n}}$. To see this, we note that $[a_n^{t^{-3n(m+1)}}, a_n] = [x_{m+1}, x_0] = (1\,2\,3)$ and $(1\,2\,3)^{a_n^{t^{-3ni}}} = (1\,2\,3)^{x_i}$, so $\langle a_n, t^n \rangle \geqslant \langle (1\,2\,3)^{A_{3n}}, t^n \rangle = \langle A_{3n}, t^n \rangle$, which is simply $\langle \mathrm{Alt}(\mathbb{Z}), t^n \rangle$ (compare with Lemma 3.3.7 below).

However, in [38, Theorem 4.1], Cox proves that if $n \geqslant 3$, then $(1\,2\,3)$ is not contained in a generating pair for G_n, so G_n gives an example of a 2-generated group all of whose proper quotients are cyclic but which is not $\frac{3}{2}$-generated. To simplify the proof, we will assume that $n \geqslant 4$. Let $g \in G_n$. If $g \in \mathrm{Alt}(\mathbb{Z})$, then $\langle (1\,2\,3), g \rangle \leqslant \mathrm{Alt}(\mathbb{Z}) < G_n$. Now assume that $g \notin \mathrm{Alt}(\mathbb{Z})$, so $\langle g \rangle = \langle ht^k \rangle$ where $h \in \mathrm{Alt}(\mathbb{Z})$ and $k \geqslant 4$. It is easy to show that g has exactly k infinite orbits $O_1, \dots, O_k$: indeed, if $\mathrm{supp}(h) \subseteq [a, b]$, then we quickly see that for a suitable permutation $\pi \in S_k$, we can find k orbits $O_1, \dots, O_k$ of g satisfying

$$O_i \setminus [a, b] = \{x > b \mid x \equiv i \pmod{k}\} \cup \{x < a \mid x \equiv i\pi \pmod{k}\}.$$

Since $k \geqslant 4$, we can fix i such that $O_i \cap \{1, 2, 3\} = \emptyset$, so O_i is an orbit of $\langle (1\,2\,3), g \rangle$, which implies that $\langle (1\,2\,3), g \rangle \neq G_n$ in this case too.

In contrast, in [38, Theorem 6.1], Cox shows that G_1 and G_2 are $\frac{3}{2}$-generated, and, in fact, $2 \leqslant u(G_i) \leqslant s(G_i) \leqslant 9$ for $i \in \{1, 2\}$.

The groups in Example 3.3.1 are not simple, so we ask the following.

Question 3.3.2 Is there a 2-generated simple group G with $s(G) = 0$?

For finite groups G, Theorem 3.2.29 also establishes that $s(G) \geqslant 1$ if and only if $s(G) \geqslant 2$, so there are no finite groups G satisfying $s(G) = 1$.

Question 3.3.3 Is there a 2-generated simple group G with $s(G) = 1$?

There is a clear difference between generating finite and infinite groups, and straightforward analogues of the theorems for finite groups do not hold for infinite groups. Nevertheless, do the results on the spread of finite simple groups extend to important classes of infinite simple groups? The investigation of the infinite simple groups of Richard Thompson (and their many generalisations) in Sections 3.3.3–3.3.6 demonstrates that the answer is a resounding yes! However, before turning to these infinite simple groups, in Section 3.3.2, we look at the other important special case we considered for finite groups: soluble groups.

3.3.2 Soluble groups

In the opening to Section 3.2.4, we noted that when Brenner and Wiegold introduced the notion of spread, they proved that for a finite soluble group G, we have $s(G) \geqslant 1$ if and only if $s(G) \geqslant 2$ if and only if every proper quotient of G is cyclic (see Theorem 3.2.26). By Theorem 3.2.5, "soluble" can be removed from the hypothesis (while keeping "finite").

The following theorem establishes that "finite" can be removed from the hypothesis (while keeping "soluble"), in a strong sense. (Theorem 3.3.4 is due to the author, and this is the first appearance of it in the literature.)

Theorem 3.3.4 *Let G be an infinite soluble group such that every proper quotient is cyclic. Then G is cyclic.*

Proof It suffices to show that G is abelian, because an infinite abelian group where every proper quotient is cyclic is itself cyclic. For a contradiction, suppose that G is nonabelian. If $1 \neq N \trianglelefteqslant G$, then G/N is cyclic, so $G' \leqslant N$. Therefore, G' is the unique minimal normal subgroup of G. In particular, G'' is G' or 1, but G is soluble, so $G'' = 1$, which implies that G' is abelian. Therefore, G' is an abelian characteristically simple group, so it is isomorphic to the additive group of a vector space V over a field F, and we may assume that $F = \mathbb{F}_p$ or $F = \mathbb{Q}$.

Let $g \in G$ such that $G/V = \langle Vg \rangle$ and write $H = \langle g \rangle$, so $G = VH$. Observe that $Z(G) = 1$, for otherwise $G/Z(G)$ is cyclic, so G is abelian, a contradiction. Now, if $g^i \in V$, then $g^i \in Z(G) = 1$, so $V \cap H = 1$. Hence, G is a semidirect product $V{:}H$. For all nontrivial $v \in V$, we have $V = \langle v^G \rangle$ since V is a minimal normal subgroup and $\langle v^G \rangle = \langle v^H \rangle$ since V is abelian. Therefore, V is an irreducible FH-module, so V is finite-dimensional since H is cyclic. (To see this, suppose that V is infinite-dimensional, so H is infinite. We give a proper nonzero submodule U, contradicting the irreducibility of V. For $0 \neq v \in V$, either $\{vg^i \mid i \in \mathbb{Z}\}$ is linearly independent, and U is the kernel of $\sum_{i\in\mathbb{Z}} a_i v g^i \mapsto \sum_{i\in\mathbb{Z}} a_i$, or for some $u = vg^i$ we have $a_0 u + a_1 ug + \cdots + a_k ug^k = 0$ and $U = \langle u, ug, \ldots, ug^{k-1} \rangle$.) If $g^i \in C_G(V)$, then $g^i \in Z(G) = 1$, so V is a faithful FH-module. In particular, if F is finite, then so is $G = F^n{:}H \leqslant F^n{:}\mathrm{GL}_n(F)$, so we must have $F = \mathbb{Q}$.

Let $\chi = X^n + a_{n-1}X^{n-1} + \cdots + a_1 X + a_0 \in \mathbb{Q}[X]$ be the characteristic polynomial of g, and let $(e_1, \ldots, e_n)$ be a basis for V with respect to which the matrix A of g is the companion matrix of χ. Let P be the set of prime divisors appearing in the reduced forms of $a_0, \ldots, a_{n-1}$ and note that P is finite. For all $i \in \mathbb{Z}$, write $e_1 A^i$ as a linear combination $\lambda_{i1} e_1 + \cdots + \lambda_{in} e_n$. Any prime that divides the denominator of the reduced form of one of the λ_{ij} is contained in P. Hence, only finitely many primes appear in the denominators of the reduced forms of any element in the subgroup N generated by $\{e_1 A^i \mid i \in \mathbb{Z}\}$. Since N is $\langle e_1^G \rangle$, it is a proper nontrivial subgroup of V that is normal in G, which contradicts V being a minimal normal subgroup of G. Therefore, G is abelian and hence cyclic. □

With a much shorter proof, one can obtain an analogous result for the class of residually finite groups (this was observed by Cox in [38, Lemma 1.1]).

Theorem 3.3.5 *Let G be an infinite residually finite group such that every proper quotient is cyclic. Then G is cyclic.*

Proof Suppose that G is nonabelian. Fix $x, y \in G$ with $[x, y] \neq 1$. Since G is residually finite, G has a finite index normal subgroup N such that $[Nx, Ny]$ is nontrivial in G/N (so, Nx and Ny are nontrivial in G/N). Since G is infinite and N has finite index, we know that N is nontrivial, so G/N is cyclic, which contradicts G/N being nonabelian. Therefore, G is abelian and hence cyclic. □

3.3.3 Thompson's groups: an introduction

In 1965, Richard Thompson introduced three finitely generated infinite groups $F < T < V$ [97]. Among other interesting properties of these groups, V and T were the first known examples of finitely presented infinite simple groups and for 35 years (until the work of Burger and Mozes [21]) all known examples of such groups were closely related to T and V. For an indication of other interesting properties of these groups, we record that F is finitely presented yet it contains a copy of $F \times F$ and is an HNN extension of itself. Moreover, F has exponential growth but contains no nonabelian free groups, and one of the most famous open questions in geometric group theory is whether F is amenable [37]. However, these three groups not only raise interesting group theoretic questions, but they have also played a role in a whole range of mathematical areas such as the word problem for groups, homotopy theory and dynamical systems (see [44, 52, 98] for example). We refer the reader to Canon, Floyd and Parry's introduction to these groups [35].

An appealing feature of Thompson's groups is that they admit concrete representations as transformation groups, which we outline now. Let $X = \{0,1\}^*$ be the set of all finite words over $\{0,1\}$, and let $\mathfrak{C} = \{0,1\}^{\mathbb{N}}$ be *Cantor space*, the set of all infinite sequences over $\{0,1\}$ with the usual topology. For $u \in X$, we write $u\mathfrak{C} = \{uw \mid w \in \mathfrak{C}\}$, and we say a finite set $A \subseteq X$ is a *basis* of $\mathfrak{C}$ if $\{u\mathfrak{C} \mid u \in A\}$ is a partition of $\mathfrak{C}$. Thompson's group V is the group of homeomorphisms $g \in \mathrm{Homeo}(\mathfrak{C})$ for which there exists a *basis pair*, namely a bijection $\sigma \colon A \mapsto B$ between two bases A and B of $\mathfrak{C}$ such that $(uw)g = (u\sigma)w$ for all $u \in A$ and all $w \in \mathfrak{C}$. In other words, V is the group of homeomorphisms of $\mathfrak{C}$ that act

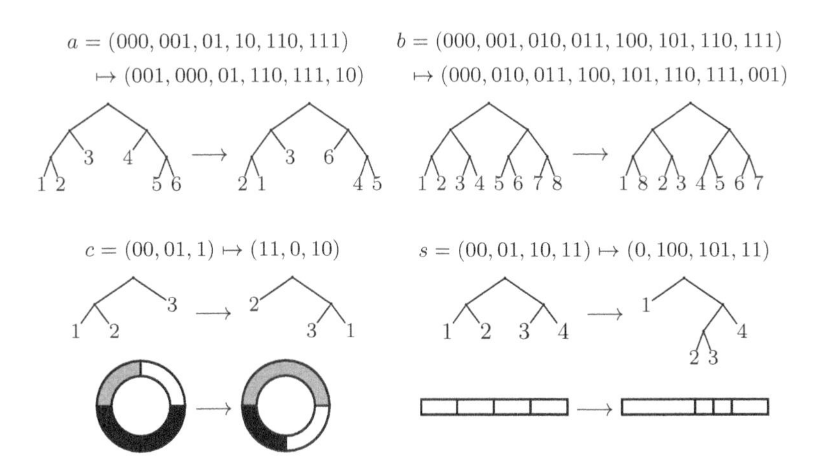

Figure 3.2 Four elements of Thompson's group V.

by prefix substitutions. For instance, $c = (00, 01, 1) \mapsto (11, 0, 10)$ is an element of V that, for example, maps $01010101\ldots$ to $0010101\ldots$. The selfsimilarity of $\mathfrak{C}$ means that there is not a unique choice of basis pair; indeed, by subdividing $\mathfrak{C}$ further c is also represented by $(000, 001, 01, 1) \mapsto (110, 111, 0, 10)$. By identifying elements of X with vertices of the infinite binary rooted tree, we can represent bases as binary rooted trees and elements of V by the familiar *tree pairs*, as shown in Figure 3.2.

A motivating perspective in the study of generating sets of Thompson's groups is that V combines the selfsimilarity of the Cantor space with permutations from the symmetric group. Indeed, to $g \in V$ we may associate (not uniquely) a permutation as follows. Let $\sigma\colon A \to B$ be a basis pair for g and write $A = \{a_1, \ldots, a_n\}$ and $B = \{b_1, \ldots, b_n\}$ where $a_1 < \cdots < a_n$ and $b_1 < \cdots < b_n$ in the lexicographic order. Then the permutation associated to g is the element $\pi_g \in S_n$ satisfying $a_i g = b_{i\pi_g}$. For the elements in Figure 3.2, for instance,

$$\pi_a = (1\,2)(4\,5\,6), \quad \pi_b = (2\,3\,4\,5\,6\,7\,8), \quad \pi_c = (1\,2\,3), \quad \pi_s = 1.$$

This perspective gives an easy way to define F and T. Thompson's groups F and T are the subgroups of V of elements whose associated permutation is trivial and cyclic, respectively (this is well defined). Clearly $F < T < V$ and, referring to Figure 3.2, we see that $s \in F$, $c \in T \setminus F$ and $a, b \in V \setminus T$.

Given a binary word u, we can associate a subset I_u of the unit interval $[0,1]$ (or unit circle $\mathbb{S}^1$) inductively as follows: the empty word corresponds to $(0,1)$ and for any binary word u, the words $u0$ and $u1$ correspond to the open left and right halves of u. In this way, a basis for $\mathfrak{C}$ can be interpreted as a sequence of disjoint open intervals whose closures cover $[0,1]$ (or $\mathbb{S}^1$), and an element of V, as a basis pair, $\sigma\colon \{a_1,\dots,a_n\} \to \{b_1,\dots,b_n\}$ defines a bijection $g\colon [0,1] \to [0,1]$ (or $g\colon \mathbb{S}^1 \to \mathbb{S}^1$) by specifying that $I_{a_i}g = I_{b_i}$ and $g|_{I_{a_i}}$ is affine for all $1 \leqslant i \leqslant n$. Under this correspondence, F is a group of piecewise linear homeomorphisms of $[0,1]$ and T is a group of piecewise linear homeomorphisms of $\mathbb{S}^1$. See Figure 3.2 for some examples.

3.3.4 Thompson's group V

The final three sections of this survey address generating sets for Thompson's groups, which have seen lots of very recent progress. We begin with V.

The group V is 2-generated. Indeed, referring to Figure 3.2, $V = \langle a, b\rangle$. Given this infinite 2-generated simple group V, we are naturally led to ask: is V $\frac{3}{2}$-generated? An answer was given by Donoven and Harper in 2020 [42].

Theorem 3.3.6 *Thompson's group V is $\frac{3}{2}$-generated.*

Theorem 3.3.6 gave the first example of a noncyclic infinite $\frac{3}{2}$-generated group, other than the pathological *Tarski monsters*: the infinite groups whose only proper nontrivial subgroups have order p for a fixed prime p, which are clearly simple and $\frac{3}{2}$-generated and were proved to exist for all $p > 10^{75}$ by Olshanskii in [88]. (Note that the groups G_1 and G_2 in Example 3.3.1 were found later by Cox in [38], motivated by a question posed in [42].) In particular, V was the first finitely presented example of a noncyclic infinite $\frac{3}{2}$-generated group.

Methods. A parallel with symmetric groups. Let us write $G = S_n$ and $\Omega = \{1,\dots,n\}$. It is well known that G is generated by the set of transpositions $\{(i\,j) \mid \text{distinct } i,j \in \Omega\}$ which yields a natural presentation for G, namely

$$\langle t_{i,j} \mid t_{i,j}^2,\ t_{i,j}^{t_{k,l}} = t_{i(k\,l),j(k\,l)}\rangle \tag{3.6}$$

Moreover, using the fact that G is generated by transpositions, we obtain the following *covering lemma* (here $G_{[A]}$ is the subgroup of G supported on $A \subseteq \Omega$).

Lemma 3.3.7 *Let $A_1, \ldots, A_k \subseteq \Omega$ satisfy both $A_i \cap A_{i+1} \neq \emptyset$ and $\bigcup_{i=1}^k A_i = \Omega$. Then $G_{[A_i]} \cong S_{|A_i|}$ for all i, and $G = \langle G_{[A_1]}, \ldots, G_{[A_k]} \rangle$.*

These results for $G = S_n$ have analogues for V. Here, *transpositions* are elements of the following form: for $u, v \in X$ such that the corresponding subsets of $\mathfrak{C}$ are disjoint, we write $(u\,v)$ for the element given as $(u, v, w_1, \ldots, w_k) \mapsto (v, u, w_1, \ldots, w_k)$ where $\{u, v, w_1, \ldots, w_k\}$ is a basis for $\mathfrak{C}$. Brin [18] proved that V is generated by $\{(u\,v) \mid \text{disjoint } u, v \in X\}$.

As an aside, let us point out why the subgroup of elements that are a product of an even number of transpositions is not a proper nontrivial normal subgroup of V (as with the symmetric and alternating groups): this subgroup is not proper. Indeed, every element of V is a product of an even number of transpositions as the selfsimilarity of $\mathfrak{C}$ shows that $(u\,v)$ can be rewritten as $(u0\,v0)(u1\,v1)$.

Bleak and Quick [12, Theorem 1.1] demonstrated how this generating set gives a presentation for V combining the corresponding presentation for the symmetric group in (3.6) with the selfsimilarity of $\mathfrak{C}$, namely

$$\langle t_{u,v} \mid t_{u,v}^2,\ t_{u,w}^{t_{x,y}} = t_{u(x\,y),v(x\,y)},\ t_{u,v} = t_{u0,v0}t_{u1,v1} \rangle \tag{3.7}$$

(see [12, (1.1)] for a full explanation of the notation used in the relations).

We will say no more about presentations, save that Bleak and Quick found a presentation for V with 2 generators and 7 relations [12, Theorem 1.3], which they derived from another, more intuitive, presentation based on the analogy with S_n which has 3 generators and 8 relations [12, Theorem 1.2].

As with the symmetric group, the fact that V is generated by transpositions yields an easy proof of the following.

Lemma 3.3.8 *Let $U_1, \ldots, U_k \subseteq \mathfrak{C}$ be clopen subsets satisfying both $U_i \cap U_{i+1} \neq \emptyset$ and $\bigcup_{i=1}^k U_i = \mathfrak{C}$. Then $V_{[U_i]} \cong V$ for all i, and we have $V = \langle V_{[U_1]}, \ldots, V_{[U_k]} \rangle$.*

With Lemma 3.3.8 in place, we now highlight the main ideas in the proof of Theorem 3.3.6 by way of an example (this is [42, Example 4.1]). We will see an alternative approach in Theorem 3.3.12

Example 3.3.9 Let $x = (00\ \ 01) \in V$. We will construct $y \in V$ such that $\langle x, y \rangle = V$. Let $y_1 = a_{[00]}$ and $y_2 = b_{[01]}$, where for $g \in V$ and clopen $A \subseteq \mathfrak{C}$ we write $g_{[A]}$ for the image of g under the canonical isomorphism $V \to V_{[A]}$. Let $y_3 = (00\ \ 01\ \ 10\ \ 11)_{[0^3 10]} \cdot (0^3 10^3\ \ 010^3)$ and $y_4 = (0000\ \ 0001\ \ \cdots\ \ 1010)_{[0^3 1^2]} \cdot (0^3 1^2 0^4\ \ 1)$, and define $y = y_1 y_2 y_3 y_4$.

Note that y_1, y_2, y_3 and y_4 have coprime orders (6, 7, 5 and 11, respectively). Moreover, these elements have disjoint support, so they commute. Consequently, all four elements are suitable powers of y and are, thus, contained in $\langle x, y\rangle$. We claim that $\langle x, y\rangle = V$. Recall that $V = \langle a, b\rangle$, so $V_{[00]} = \langle a_{[00]}, b_{[00]}\rangle = \langle y_1, y_2^x\rangle \leqslant \langle x, y\rangle$. In addition, $V_{[01]} = (V_{[00]})^x \leqslant \langle x, y\rangle$. Using appropriate elements from $V_{[00]}$ and $V_{[01]}$ we can show that $(000 \ \ 01) \in \langle V_{[00]}, V_{[01]}, y_3\rangle \leqslant \langle x, y\rangle$ and $(000 \ \ 1) \in \langle V_{[00]}, y_4\rangle \leqslant \langle x, y\rangle$. Therefore, $\langle x, y\rangle \geqslant \langle V_{[00]}, V_{[00]}^{(000 \ \ 01)}, V_{[00]}^{(000 \ \ 1)}\rangle = \langle V_{[000\cup 001]}, V_{[01\cup 001]}, V_{[1\cup 001]}\rangle$. Now applying Lemma 3.3.8 twice gives $\langle x, y\rangle = V$.

3.3.5 Generalisations of V

The *Higman–Thompson group* V_n, for $n \geqslant 2$, is an infinite finitely presented group, introduced by Higman in [71]. There is a natural action of V_n on n-ary Cantor space $\mathfrak{C}_n = \{0, 1, \dots, n-1\}^{\mathbb{N}}$, and V_2 is nothing other than V. The derived subgroup of V_n equals V_n for even n and has index two for odd n. In both cases, V_n' is simple and both V_n and V_n' are 2-generated [87].

The *Brin–Thompson group* nV, for $n \geqslant 1$, acts on $\mathfrak{C}^n$ and was defined by Brin in [18]. The groups $V = 1V, 2V, 3V, \dots$ are pairwise nonisomorphic [11], simple [19] and 2-generated [92, Corollary 1.3].

The results about generating V by transpositions have analogues for V_n and nV (see [42, Section 3]), and, in [92, Theorem 1.1], Quick gives a presentation for nV analogous to the one for V in (3.7). Theorem 3.3.6 extends to all of these groups too [42, Theorems 1 & 2].

Theorem 3.3.10 *For all $n \geqslant 2$, the Higman–Thompson groups V_n and V_n' are $\frac{3}{2}$-generated, and for all $n \geqslant 1$, the Brin–Thompson group nV is $\frac{3}{2}$-generated.*

In particular, the groups V_n when n is odd give infinitely many examples of infinite $\frac{3}{2}$-generated groups that are not simple.

As we introduced them, the Higman–Thompson group V_n' is a simple subgroup of $\mathrm{Homeo}(\mathfrak{C}_n)$ and the Brin–Thompson group nV is a simple subgroup of $\mathrm{Homeo}(\mathfrak{C}^n)$. Since $\mathfrak{C}^n$ (nth power of $\mathfrak{C}$) and $\mathfrak{C}_n$ (n-ary Cantor space) are both homeomorphic to $\mathfrak{C}$, all of these groups can be viewed as subgroups of $\mathrm{Homeo}(\mathfrak{C})$. Recent work of Bleak, Elliott and Hyde [9], highlights that these groups, and numerous others (such as Nekrashevych's simple groups of dynamical origin), can be viewed within one unified dynamical framework.

A group $G \leqslant \mathrm{Homeo}(\mathfrak{C})$ is said to be *vigorous* if for any clopen subsets $\emptyset \subsetneq B, C \subsetneq A \subseteq \mathfrak{C}$ there exists $g \in G$ supported on A such that $Bg \subseteq C$. In [9], Bleak, Elliot and Hyde study vigorous groups and, among much else, prove that a perfect vigorous group $G \leqslant \mathrm{Homeo}(\mathfrak{C})$ is simple if and only if it is generated by its elements of *small support* (namely, elements supported on a proper clopen subset of $\mathfrak{C}$). To give a flavour of how these dynamical properties suitably capture the ideas we have seen in this section, compare the following, which is [9, Lemma 2.18 & Proposition 2.19], with Lemma 3.3.8.

Lemma 3.3.11 *Let G be a vigorous group that is generated by its elements of small support. Let $U_1, \dots, U_k \subseteq \mathfrak{C}$ be clopen subsets satisfying $U_i \cap U_{i+1} \neq \emptyset$ and $\bigcup_{i=1}^k U_i = \mathfrak{C}$. Then $G = \langle G_{[U_1]}, \dots, G_{[U_k]} \rangle$. Moreover, if G is simple, then for each i the group $G_{[U_i]}$ is a simple vigorous group.*

Bleak, Elliott and Hyde go on to prove that every finitely generated simple vigorous group is 2-generated [9, Theorem 1.12]. Are all such groups $\frac{3}{2}$-generated? Bleak, Donoven, Harper and Hyde [8] recently proved that $u(G) \geqslant 1$.

Theorem 3.3.12 *Let $G \leqslant \mathrm{Homeo}(\mathfrak{C})$ be a finitely generated simple vigorous group. Then there exists an element $s \in G$ of small support and order 30 such that for every nontrivial $x \in G$ there exists $y \in s^G$ such that $\langle x, y \rangle = G$.*

Theorem 3.3.12 gives $u(G) \geqslant 1$ for all the simple groups G in Theorem 3.3.10. As a special case, we obtain a strong version of Theorem 3.3.6 on Thompson's group V, improving $s(V) \geqslant 1$ to $u(V) \geqslant 1$. It is possible to obtain stronger results on the (uniform) spread of V and its generalisations (and T, discussed below), and this is the subject of current work of the author and others (e.g. [8]).

3.3.6 Thompson's groups T and F

In this final section, we discuss generating sets of Thompson's groups T and F. We begin with T, which is a simple 2-generated group, so it is natural to study its (uniform) spread. In 2022, Bleak, Harper and Skipper [10] proved $u(T) \geqslant 1$.

Theorem 3.3.13 *There exists an element $s \in T$ such that for every nontrivial $x \in T$ there exists $y \in s^T$ such that $\langle x, y \rangle = T$.*

Corollary 3.3.14 *Thompson's group T is $\frac{3}{2}$-generated.*

The element s in Theorem 3.3.13 can be chosen as the one in Figure 3.2. Moreover, in [10, Proposition 3.1], it is shown that if we restrict to elements x of infinite order, then we can choose s to be any infinite order element, that is to say, for any two infinite order elements $x, s \in T$ there exists $g \in T$ such that $\langle x, s^g \rangle = T$. This naturally raises the question of whether an arbitrary infinite order element can be chosen for s in Theorem 3.3.13 (see [10, Question 1]).

We now turn to Thompson's group F. This is 2-generated since if we write $x_0 = (00, 01, 1) \mapsto (0, 10, 11)$ and $x_1 = (0, 100, 101, 1) \mapsto (0, 10, 110, 111)$, then, by [35, Theorem 3.4] for example, $F = \langle x_0, x_1 \rangle$. Moreover, if we inductively define $x_{i+1} = x_i^{x_0}$ for all $i \geqslant 1$, then the elements $x_0, x_1, x_2, \dots$ witness the following well-known presentation

$$F = \langle x_0, x_1, x_2, \cdots \mid x_j^{x_i} = x_{j+1} \text{ for } i < j \rangle.$$

However, F is not a simple group. Considering F in its natural action on $[0,1]$, the homomorphism $\pi \colon F \to \mathbb{Z}^2$ defined as $f \mapsto (\log_2 f'(0^+), \log_2 f'(1^-))$ is surjective and the kernel of π is the derived subgroup F', which is simple. Moreover, F' is the unique minimal normal subgroup of F, so the nontrivial normal subgroups of F are in bijection with normal subgroups of $F/F' = \mathbb{Z}^2$ (see [35, Section 4] for proofs of these claims). In particular, F is not $\frac{3}{2}$-generated since it has a proper noncyclic quotient.

Now F' is not $\frac{3}{2}$-generated for a different reason: it is not finitely generated. Indeed, for any nontrivial normal subgroup $N = \pi^{-1}(\langle (a_0, a_1), (b_0, b_1) \rangle)$, if $\{a_0, b_0\} = \{0\}$ or $\{a_1, b_1\} = \{0\}$, then N is not finitely generated. To see this in the former case, for finitely many elements each of which acts as the identity on an interval containing 0, there exists an interval containing 0 on which they all act as the identity, so they generate a proper subgroup of N (for the latter case, replace 0 with 1). However, the following recent theorem of Golan [55, Theorem 2] shows that these are the only obstructions to $\frac{3}{2}$-generation.

Theorem 3.3.15 *Let $(a_0, a_1), (b_0, b_1) \in \mathbb{Z}^2$ with $\{a_0, b_0\} \neq \{0\}$ and $\{a_1, b_1\} \neq \{0\}$. Let $x \in F$ be a nontrivial element such that $\pi(x) = (a_0, a_1)$. Then there exists $y \in F$ such that $\pi(y) = (b_0, b_1)$ and $\langle x, y \rangle = \pi^{-1}(\langle (a_0, a_1), (b_0, b_1) \rangle)$.*

Theorem 3.3.15 has the following consequence, which asserts that F is almost $\frac{3}{2}$-generated [55, Theorem 1].

Corollary 3.3.16 *Let $f \in F$ and assume that $\pi(f)$ is contained in a generating pair of $\pi(F)$. Then f is contained in a generating pair of F.*

Theorem 3.3.15 also implies that every finitely generated normal subgroup of F is 2-generated. In particular, every finite index subgroup of F is 2-generated.

Methods. Covering lemmas and a generation criterion. We conclude the survey by discussing how Theorems 3.3.13 and 3.3.15 are proved in [10] and [55], respectively. Covering lemmas (analogues of Lemma 3.3.7), again, play a role. For F and T, these results are well known, see [10, Corollary 2.6 & Lemma 2.7] for example. (We call an interval $[a,b]$ *dyadic* if $a,b \in \mathbb{Z}[\frac{1}{2}]$.)

Lemma 3.3.17 *Let $[a_1,b_1],\dots,[a_k,b_k] \subseteq [0,1]$ be dyadic intervals satisfying $\bigcup_{i=1}^{k}(a_i,b_i) = (0,1)$. Then $F_{[a_i,b_i]} \cong F$ for all i, and we have $F = \langle F_{[a_1,b_1]},\dots,F_{[a_k,b_k]}\rangle$.*

Lemma 3.3.18 *Let $[a_1,b_1],\dots,[a_k,b_k] \subseteq \mathbb{S}^1$ be dyadic intervals satisfying $\bigcup_{i=1}^{k}(a_i,b_i) = \mathbb{S}^1$. Then $T_{[a_i,b_i]} \cong F$ for all i, and we have $T = \langle T_{[a_1,b_1]},\dots,T_{[a_k,b_k]}\rangle$.*

Another key ingredient is a criterion due to Golan, for which we need some further notation. Fix a subgroup $H \leqslant F$. An element $f \in F \leqslant \mathrm{Homeo}([0,1])$ is *piecewise-H* if there is a finite subdivision of $[0,1]$ such that on each interval in the subdivision, f coincides with an element of H. The closure of H, written $\mathrm{Cl}(H)$, is the subgroup of F containing all elements that are piecewise-H. The following result combines [53, Theorem 1.3] with [54, Theorem 1.3].

Theorem 3.3.19 *Let $H \leqslant F$. Then the following hold:*

(i) $H \geqslant F'$ if and only if $\mathrm{Cl}(H) \geqslant F'$ and there exist $f \in H$ and a dyadic $\omega \in (0,1)$ such that $f'(\omega^+) = 2$ and $f'(\omega^-) = 1$
(ii) $H = F$ if and only if $\mathrm{Cl}(H) \geqslant F'$ and there exist $f,g \in H$ such that $f'(0^+) = g'(1^-) = 2$ and $f'(1^-) = g'(0^+) = 1$.

We now discuss the proof of Theorem 3.3.13 on T given by Bleak, Harper and Skipper [10]. By Lemma 3.3.18, for each nontrivial $x \in T$ it suffices to find a dyadic interval $[a,b] \subseteq \mathbb{S}^1$ and $y \in s^T$ such that $\bigcup_{g \in \langle x,y\rangle}(a,b)g = \mathbb{S}^1$ and $T_{[a,b]} \leqslant \langle x,y\rangle$. If $|x|$ is infinite, a dynamical argument is used (for any infinite order element s), see [42, Proposition 3.1]. The key ingredients for finite $|x|$ are highlighted in the following example.

Example 3.3.20 Let $x \in T$ be a nontrivial torsion element. We will prove that there exists $y \in s^T$ (for s as in Figure 3.2) such that $\langle x, y \rangle = T$. By replacing x by a power if necessary, x has rotation number $\frac{1}{p}$ for prime p. For exposition, we only discuss the case $p \geqslant 5$. Since any two torsion elements of T with the same rotation number are conjugate, by replacing x by a conjugate if necessary, $x = (00, 01, 10, 110, \ldots, 1^{p-3}0, 1^{p-2}) \mapsto (01, 10, 110, \ldots, 1^{p-3}0, 1^{p-2}, 00)$.

We claim that $T = \langle x, s \rangle$. By Lemma 3.3.18, since $\mathbb{S}^1 = \bigcup_{i \in \mathbb{Z}} (0, \frac{7}{8})x^i$, it suffices to prove that $T_{[0,\frac{7}{8}]} \leqslant \langle x, s \rangle$. Indeed, we claim that $T_{[0,\frac{7}{8}]} = \langle y_0, y_1 \rangle$ for $y_0 = s$ and $y_1 = s^x$. Defining $t\colon (0, \frac{7}{8}) \to (0, 1)$ as $\omega t = \omega$ if $\omega \leqslant \frac{3}{4}$ and $\omega t = 2\omega - \frac{3}{4}$ if $\omega > \frac{3}{4}$, it suffices to prove that $\langle y_0^t, y_1^t \rangle = (T_{[0,\frac{7}{8}]})^t = F$.

To do this, we apply Theorem 3.3.19(ii). To verify the second condition, choose $f = y_0^t$ and $g = (y_1^t)^{-1}$, so $f'(0^+) = g'(1^-) = 2$ and $f'(1^-) = g'(0^+) = 1$. It remains to prove that $\mathrm{Cl}(\langle y_0^t, y_1^t \rangle) \geqslant F'$. Here we apply another criterion: for $g_1, \ldots, g_k \in F$ we have $\langle g_1, \ldots, g_k \rangle \geqslant F'$ if and only if the Stallings 2-core of $\langle g_1, \ldots, g_k \rangle$ equals the Stallings 2-core of F [53, Lemma 7.1 & Remark 7.2]. The *Stallings 2-core* is a directed graph associated to a diagram group introduced by Guba and Sapir [58]. Given elements $g_1, \ldots, g_k \in F$ represented as tree pairs, there is a short combinatorial algorithm to find the Stallings 2-core of $\langle g_1, \ldots, g_k \rangle$, and it is straightforward to compute the Stallings 2-core of $\langle y_0^t, y_1^t \rangle$ and note that it is the Stallings 2-core of F (see the proof of [10, Proposition 3.2]). Therefore, $F = \langle y_0^t, y_1^t \rangle$, completing the proof that $T = \langle x, s \rangle$.

We conclude by briefly outlining the proof of Theorem 3.3.15 on F given by Golan [55], which uses similar methods to those in [10] on T and [42] on V. Let $(a_0, a_1), (b_0, b_1) \in \mathbb{Z}^2$ with $\{a_0, b_0\} \neq \{0\}$ and $\{a_1, b_1\} \neq \{0\}$, and let $x \in F \setminus 1$ with $\pi(x) = (a_0, a_1)$. Observe that it suffices to find an element y such that $\pi(y) = (b_0, b_1)$ and $\langle x, y \rangle \geqslant F'$. In [55], an explicit choice of y, based on x, is given and the condition $\langle x, y \rangle \geqslant F'$ is verified via Theorem 3.3.19(i).

References

[1] M. Aschbacher, *On the maximal subgroups of the finite classical groups*, Invent. Math. **76** (1984), 469–514.

[2] M. Aschbacher and R. Guralnick, *Some applications of the first cohomology group*, J. Algebra **90** (1984), 446–460.

[3] G. J. Binder, *The bases of the symmetric group*, Izv. Vyssh. Uchebn. Zaved. Mat. **78** (1968), 19–25.

[4] G. J. Binder, *Certain complete sets of complementary elements of the symmetric and the alternating group of the nth degree*, Mat. Zametiki **7** (1970), 173–180.

[5] G. J. Binder, *The two-element bases of the symmetric group*, Izv. Vyssh. Uchebn. Zaved. Mat. **90** (1970), 9–11.

[6] G. J. Binder, *The inclusion of the elements of an alternating group of even degree in a two-element basis*, Izv. Vyssh. Uchebn. Zaved. Mat. **135** (1973), 15–18.

[7] S. Blackburn, *Sets of permutations that generate the symmetric group pairwise*, J. Combin. Theory Ser. A **113** (2006), 1572–1581.

[8] C. Bleak, C. Donoven, S. Harper and J. Hyde, *Generating simple vigorous groups*, in preparation.

[9] C. Bleak, L. Elliott and J. Hyde, *Sufficient conditions for a group of homeomorphisms of the Cantor set to be two-generated*, J. Inst. Math. Jussieu, to appear.

[10] C. Bleak, S. Harper and R. Skipper, *Thompson's group T is $\frac{3}{2}$-generated*, Israel J. Math., to appear.

[11] C. Bleak and D. Lanoue, *A family of non-isomorphic results*, Geom. Dedicata **146** (2010), 21–26.

[12] C. Bleak and M. Quick, *The infinite simple group V of Richard J. Thompson: presentations by permutations*, Groups Geom. Dyn. **11** (2017), 1401–1436.

[13] J.-M. Bois, *Generators of simple Lie algebras in arbitrary characteristics*, Math. Z. **262** (2009), 715–741.

[14] J. D. Bradley and P. E. Holmes, *Improved bounds for the spread of sporadic groups*, LMS J. Comput. Math. **10** (2007), 132–140.

[15] J. L. Brenner and J. Wiegold, *Two generator groups, I*, Michigan Math. J. **22** (1975), 53–64.

[16] T. Breuer, R. M. Guralnick and W. M. Kantor, *Probabilistic generation of finite simple groups, II*, J. Algebra **320** (2008), 443–494.

[17] T. Breuer, R. M. Guralnick, A. Lucchini, A. Maróti and G. P. Nagy, *Hamiltonian cycles in the generating graphs of finite groups*, Bull. Lond. Math. Soc. **42** (2010), 621–633.

[18] M. G. Brin, *Higher dimensional Thompson groups*, Geom. Dedicata **108** (2004), 163–192.

[19] M. G. Brin, *On the baker's maps and the simplicity of the higher dimensional Thompson's groups nV*, Publ. Mat. **54** (2010), 433–439.

[20] J. R. Britnell, A. Evseev, R. M. Guralnick, P. E. Holmes and A. Maróti, *Sets of elements that pairwise generate a linear group*, J. Combin. Theory Ser. A **115** (2008), 442–465.

[21] M. Burger and S. Mozes, *Lattices in product of trees*, Inst. Hautes Études Sci. Publ. Math. **92** (2000), 151–194.

[22] T. C. Burness, *Fixed point ratios in actions of finite classical groups, I*, J. Algebra **309** (2007), 69–79.

[23] T. C. Burness, *Fixed point ratios in actions of finite classical groups, II*, J. Algebra **309** (2007), 80–138.

[24] T. C. Burness, *Fixed point ratios in actions of finite classical groups, III*, J. Algebra **314** (2007), 693–748.

[25] T. C. Burness, *Fixed point ratios in actions of finite classical groups, IV*, J. Algebra **314** (2007), 749–788.

[26] T. C. Burness, *Simple groups, fixed point ratios and applications*, in *Local Representation Theory and Simple Groups*, EMS Series of Lectures in Mathematics, European Mathematical Society, 2018, 267–322.

[27] T. C. Burness, *Simple groups, generation and probabilistic methods*, in *Proceedings of Groups St Andrews 2017*, London Math. Soc. Lecture Note Series, vol. 455, Cambridge University Press, 2019, 200–229.

[28] T. C. Burness and S. Guest, *On the uniform spread of almost simple linear groups*, Nagoya Math. J. **209** (2013), 35–109.

[29] T. C. Burness, R. M. Guralnick and S. Harper, *The spread of a finite group*, Ann. of Math. **193** (2021), 619–687.

[30] T. C. Burness and S. Harper, *On the uniform domination number of a finite simple group*, Trans. Amer. Math. Soc. **372** (2019), 545–583.

[31] T. C. Burness and S. Harper, *Finite groups, 2-generation and the uniform domination number*, Israel J. Math. **239** (2020), 271–367.

[32] T. C. Burness, M. .W. Liebeck and A. Shalev, *Base sizes for simple groups and a conjecture of Cameron*, Proc. Lond. Math. Soc. **98** (2009), 116–162.

[33] T. C. Burness and A. R. Thomas, *Normalisers of maximal tori and a conjecture of Vdovin*, J. Algebra **619** (2023), 459–504.

[34] P. J. Cameron, *Graphs defined on groups*, Int. J. Group Theory **11** (2022), 53–107.

[35] J. W. Cannon, W. J. Floyd and W. R. Parry, *Introductory notes on Richard Thompson's groups*, Enseign. Math. **42** (1996), 1–44.

[36] F. Celler, C. R. Leedham-Green, S. H. Murray, A. C. Niemeyer and E. A. O'Brien, *Generating random elements of a finite group*, Comm. Algebra **23** (1995), 4391–4948.

[37] S. Cleary, *Thompson's group*, in *Office Hours with a Geometric Group Theorist*, Princeton University Press, 2017, 331–357.

[38] C. G. Cox, *On the spread of infinite groups*, Proc. Edinb. Math. Soc. **65** (2022), 214–228.

[39] E. Crestani and A. Lucchini, *The generating graph of finite soluble groups*, Israel J. of Math. **198** (2013), 63–74.

[40] E. Crestani and A. Lucchini, *The non-isolated vertices in the generating graph of direct powers of simple groups*, J. Algebraic Combin. **37** (2013), 249–263.

[41] T. Deshpande, *Shintani descent for algebraic groups and almost simple characters of unipotent groups*, Compos. Math. **152** (2016), 1697–1724.

[42] C. Donoven and S. Harper, *Infinite $\frac{3}{2}$-generated groups*, Bull. Lond. Math. Soc. **52** (2020), 657–673.

[43] H. Duyan, Z. Halasi and A. Maróti, *A proof of Pyber's base size conjecture*, Adv. Math. **331** (2018), 720–747.

[44] J. Dydak, *1-movable continua need not be pointed 1-movable*, Bull. Acad. Polon. Sci. Sér. Sci. Math. Astronom. Phys. **25** (1977), 559–562.

[45] R. H. Dye, *Interrelations of symplectic and orthogonal groups in characteristic two*, J. Algebra **59** (1979), 202–221.

[46] M. J. Evans, *T-systems of certain finite simple groups*, Math. Proc. Camb. Phil. Soc. **113** (1993), 9–22.

[47] B. Fairbairn, *The exact spread of* M_{23} *is* 8064, Int. J. Group Theory **1** (2012), 1–2.

[48] B. Fairbairn, *New upper bounds on the spreads of sporadic simple groups*, Comm. Algebra **40** (2012), 1872–1877.

[49] P. Flavell, *Finite groups in which every two elements generate a soluble subgroup*, Invent. Math. **121** (1995), 279–285.

[50] P. Flavell, *Generation theorems for finite groups*, in *Groups and combinatorics – in memory of Michio Suzuki*, Adv. Stud. Pure. Math., vol. 32, Math. Soc. Japan, 2001, 291–300.

[51] J. Fulman and R. M. Guralnick, *Bounds on the number and sizes of conjugacy classes in finite Chevalley groups with applications to derangements*, Trans. Amer. Math. Soc. **364** (2012), 3023–3070.

[52] E. Ghys and V. Sergiescu, *Sur un groupe remarquable de difféomorphismes du cercle*, Comment. Math. Helv. **62** (1987), 185–239.

[53] G. Golan Polak, *The generation problem in Thompson group F*, Mem. Amer. Math. Soc. **292** (2023), v+94.

[54] G. Golan Polak, *On maximal subgroups of Thompson's group F*, preprint, arxiv:2209.03244.

[55] G. Golan Polak, *Thompson's group F is almost $\frac{3}{2}$-generated*, Bull. Lond. Math. Soc. **55** (2023), 2144–2157.

[56] D. Goldstein and R. M. Guralnick, *Generation of Jordan algebras and symmetric matrices*, in preparation.

[57] V. S. Guba, *A finite generated simple group with free 2-generated subgroups*, Sibirsk. Mat. Zh. **27** (1986), 50–67.

[58] V. Guba and M. Sapir, *Diagram groups*, Mem. Amer. Math. Soc. **130** (1997), viii+117.

[59] R. M. Guralnick and W. M. Kantor, *Probabilistic generation of finite simple groups*, J. Algebra **234** (2000), 743–792.

[60] R. Guralnick, B. Kunyavskiĭ, E. Plotkin and A. Shalev, *Thompson-like characterizations of the solvable radical*, J. Algebra **300** (2006), 363–375.

[61] R. M. Guralnick and G. Malle, *Simple groups admit Beauville structures*, J. Lond. Math. Soc. **85** (2012), 694–721.

[62] R. M. Guralnick, T. Penttila, C. E. Praeger and J. Saxl, *Linear groups with orders having certain large prime divisors*, Proc. Lond. Math. Soc. **78** (1997), 167–214.

[63] R. Guralnick, E. Plotkin and A. Shalev, *Burnside-type problems related to solvability*, Internat. J. Algebra Comput. **17** (2007), 1033–1048.

[64] R. M. Guralnick and J. Saxl, *Generation of finite almost simple groups by conjugates*, J. Algebra **268** (2003), 519–571.

[65] R. M. Guralnick and A. Shalev, *On the spread of finite simple groups*, Combinatorica **23** (2003), 73–87.

[66] Z. Halasi, *On the base size of the symmetric group acting on subsets*, Stud. Sci. Math. Hung. **49** (2012), 492–500.

[67] P. Hall, *The Eulerian functions of a group*, Quart. J. Math. **7** (1936), 134–151.

[68] S. Harper, *On the uniform spread of almost simple symplectic and orthogonal groups*, J. Algebra **490** (2017), 330–371.

[69] S. Harper, *Shintani descent, simple groups and spread*, J. Algebra **578** (2021), 319–355.

[70] S. Harper, *The spread of almost simple classical groups*, Lecture Notes in Mathematics, vol. 2286, Springer, 2021.

[71] G. Higman, *Finitely presented infinite simple groups*, Notes on Pure Mathematics, vol. 8, Department of Mathematics, I.A.S., Australia National University, Canberra, 1974.

[72] T. Ionescu, *On the generators of semisimple Lie algebras*, Linear Algebra Appl. **15** (1976), 271–292.

[73] W. M. Kantor and A. Lubotzky, *The probability of generating a finite classical group*, Geom. Dedicata **36** (1990), 67–87.

[74] W. M. Kantor, A. Lubotzky and A. Shalev, *Invariable generation and the Chebotarev invariant of a finite group*, J. Algebra **348** (2011), 302–314.

[75] N. Kawanaka, *On the irreducible characters of the finite unitary groups*, J. Math. Soc. Japan **29** (1977), 425–450.

[76] E. I. Khukhro and V. D. Mazurov (editors), *The Kourovka Notebook: Unsolved Problems in Group Theory*, 20th Edition, Novosibirsk, 2022, arxiv:1401.0300.

[77] M. W. Liebeck, E. A. O'Brien, A. Shalev and P. H. Tiep, *The Ore conjecture*, J. Eur. Math. Soc. **12** (2010), 939–1008.

[78] M. W. Liebeck and J. Saxl, *Minimal degrees of primitive permutation groups, with an application to monodromy groups of covers of Riemann surfaces*, Proc. Lond. Math. Soc. **63** (1991), 266–314.

[79] M. W. Liebeck and A. Shalev, *The probability of generating a finite simple group*, Geom. Dedicata **56** (1995), 103–113.

[80] M. W. Liebeck and A. Shalev, *Probabilistic methods, and the* $(2,3)$*-generation problem*, Ann. of Math. **144** (1996), 77–125.

[81] M. W. Liebeck and A. Shalev, *Simple groups, probabilistic methods, and a conjecture of Kantor and Lubotzky*, J. Algebra **184** (1996), 31–57.

[82] M. W. Liebeck and A. Shalev, *Simple groups, permutation groups, and probability*, J. Amer. Math. Soc. **12** (1999), 497–520.

[83] F. Lübeck and G. Malle, *(2,3)-generation of exceptional groups*, J. Lond. Math. Soc. **59** (1999), 109–122.

[84] A. Lubotzky, *Images of word maps in finite simple groups*, Glasg. Math. J. **56** (2014), 465–469.

[85] A. Lucchini and A. Maróti, *On the clique number of the generating graph of a finite group*, Proc. Amer. Math. Soc. **137** (2009), 3207–3217.

[86] A. Lucchini and A. Maróti, *Some results and questions related to the generating graph of a finite group*, in *Ischia Group Theory 2008*, World Scientific Publishing, 2009, 183–208.
[87] D. R. Mason, *On the 2-generation of certain finitely presented infinite simple groups*, J. Lond. Math. Soc. **16** (1977), 229–231.
[88] A. Y. Ol'shanskii, *An infinite group with subgroups of prime orders*, Izv. Akad. Nauk SSSR Ser. Mat. **44** (1980), 309–321.
[89] D. Osin and A. Thom, *Normal generation and ℓ^2-Betti numbers of groups*, Math. Ann. **355** (2013), 1331–1347.
[90] I. Pak, *What do we know about the product replacement algorithm?*, in *Groups and computation, III (Columbus, OH, 1999)*, Ohio State Univ. Math. Res. Inst. Publ., vol. 8, de Gruyter, 2001, 301–347.
[91] S. Piccard, *Sur les bases du groupe symétrique et du groupe alternant*, Math. Ann. **116** (1939), 752–767.
[92] M. Quick, *Permutation-based presentations for Brin's higher-dimensional Thompson groups nV*, J. Aust. Math. Soc. to appear.
[93] T. Shintani, *Two remarks on irreducible characters of finite general linear groups*, J. Math. Soc. Japan **28** (1976), 396–414.
[94] A. Stein, *$1\frac{1}{2}$-generation of finite simple groups*, Contrib. Algebra and Geometry **39** (1998), 349–358.
[95] R. Steinberg, *Generators for simple groups*, Canadian J. Math. **14** (1962), 277–283.
[96] J. G. Thompson, *Nonsolvable groups all of whose local subgroups are solvable*, Bull. Amer. Math. Soc. **74** (1968), 383–437.
[97] R. J. Thompson, widely circulated handwritten notes (1965), 1–11.
[98] R. J. Thompson, *Embeddings into finitely generated simple groups which preserve the word problem*, in *Word problems, II (Conf. on Decision Problems in Algebra, Oxford, 1976)*, North-Holland, 1980, 401–441.
[99] T. S. Weigel, *Generation of exceptional groups of Lie-type*, Geom. Dedicata **41** (1992), 63–87.
[100] H. Wielandt, *Finite Permutation Groups*, Academic Press, 1964.
[101] A. Woldar, *The exact spread of the Mathieu group M_{11}*, J. Group Theory **10** (2007), 167–171.

4
Discrete subgroups of semisimple Lie groups, beyond lattices

Fanny Kassel[a]

Abstract

Discrete subgroups of $\mathrm{SL}(2, \mathbb{R})$ are well understood, and classified by the geometry of the corresponding hyperbolic surfaces. Discrete subgroups of higher-rank semisimple Lie groups, such as $\mathrm{SL}(n, \mathbb{R})$ for $n > 2$, remain more mysterious. While lattices in this setting are rigid, there also exist more flexible, "thinner" discrete subgroups, which may have large and interesting deformation spaces, giving rise in particular to so-called higher Teichmüller theory. We survey recent progress in constructing and understanding such discrete subgroups from a geometric and dynamical viewpoint.

4.1 Introduction

Recall that a Lie group is a group which is also a differentiable manifold. All Lie groups considered in these notes will be assumed to be real linear Lie groups, i.e. closed subgroups of $\mathrm{GL}(N, \mathbb{R})$ for some $N \in \mathbb{N}$, with finitely many connected components. We will be specifically interested in such Lie groups which are *noncompact*, since our goal is to study their *infinite* discrete subgroups.

We say that a Lie group G is *simple* if its Lie algebra is simple, i.e. nonabelian with no nonzero proper ideals; equivalently, all infinite closed normal subgroups of G have finite index in G and are nonabelian. Simple Lie algebras have been completely classified by É. Cartan, leading

a CNRS and Laboratoire Alexander Grothendieck, Institut des Hautes Études Scientifiques, Université Paris-Saclay, 35 route de Chartres, 91440 Bures-sur-Yvette, France.
kassel@ihes.fr

to a classification of simple Lie groups up to local isomorphism. (Recall that two Lie groups G_1 and G_2 are said to be *locally isomorphic* if they have the same Lie algebra; equivalently, some finite cover of the identity component of G_1 is isomorphic to some finite cover of the identity component of G_2.) Noncompact simple Lie groups come in several infinite families, given in Table 4.1, and 17 (up to local isomorphism) additional groups, called *exceptional* (see e.g. [90, Ch. X]).

	Noncompact simple Lie group G	Maximal compact subgroup K	$\mathrm{rank}_{\mathbb{R}}(G)$
A	$\mathrm{SL}(n,\mathbb{C})$	$\mathrm{SU}(n)$	$n-1$
B	$\mathrm{SO}(2n+1,\mathbb{C})$	$\mathrm{SO}(2n+1)$	n
C	$\mathrm{Sp}(2n,\mathbb{C})$	$\mathrm{Sp}(n)$	n
D	$\mathrm{SO}(2n,\mathbb{C})$	$\mathrm{SO}(2n)$	n
A I	$\mathrm{SL}(n,\mathbb{R})$	$\mathrm{SO}(n)$	$n-1$
A II	$\mathrm{SU}^*(2n)$	$\mathrm{Sp}(n)$	$n-1$
A III	$\mathrm{SU}(p,q)$	$\mathrm{S}(\mathrm{U}(p)\times\mathrm{U}(q))$	$\min(p,q)$
BD I	$\mathrm{SO}(p,q)_0$	$\mathrm{SO}(p)\times\mathrm{SO}(q)$	$\min(p,q)$
D III	$\mathrm{SO}^*(2n)$	$\mathrm{U}(n)$	$\lfloor n/2 \rfloor$
C I	$\mathrm{Sp}(2n,\mathbb{R})$	$\mathrm{U}(n)$	n
C II	$\mathrm{Sp}(p,q)$	$\mathrm{Sp}(p)\times\mathrm{Sp}(q)$	$\min(p,q)$

Table 4.1 *List of classical noncompact simple real linear Lie groups, up to local isomorphism. Here $n,p,q \geq 1$ are integers. For types A, A I, and A II we assume $n \geq 2$, for types D and D III we assume $n \geq 3$, and for type BD I we assume $(p,q) \notin \{(1,1),(2,2)\}$.*

We say that a Lie group G is *semisimple* if it is locally isomorphic to a direct product $G_1 \times \cdots \times G_\ell$ of simple Lie groups G_i, called the *simple factors* of G; in that case, if G is connected and simply connected, then it is actually isomorphic to such a direct product $G_1 \times \cdots \times G_\ell$. For instance, $\mathrm{SO}(2,2)$ and $\mathrm{SO}(4,\mathbb{C})$ are semisimple (they are locally isomorphic to $\mathrm{PSL}(2,\mathbb{R}) \times \mathrm{PSL}(2,\mathbb{R})$ and $\mathrm{PSL}(2,\mathbb{C}) \times \mathrm{PSL}(2,\mathbb{C})$, respectively). Any connected semisimple Lie group is the identity component (for the real topology) of the real points of some $\mathbb{R}$-algebraic group (see [45, § 2.14]).

Infinite discrete subgroups of semisimple Lie groups are important objects that appear in various areas of mathematics, such as geometry, complex analysis, differential equations, number theory, mathematical physics, ergodic theory, representation theory, etc. There are many motivations for studying these discrete subgroups. Let us mention three:

(1) *Historical importance:* The study of second-order linear differential equations over $\mathbb{C}$, in particular by Fuchs, naturally led to the study of discrete subgroups of $\mathrm{PSL}(2,\mathbb{C})$, in particular by Poincaré, and to the celebrated Uniformisation Theorem: any closed Riemann surface of genus ≥ 2 is a quotient of the hyperbolic plane $\mathbb{H}^2$ by a discrete subgroup Γ of $\mathrm{PSL}(2,\mathbb{R})$. See e.g. [122] for details.

(2) *Locally symmetric spaces:* Any discrete subgroup Γ of a noncompact semisimple Lie group G defines a Riemannian locally symmetric space $\Gamma\backslash G/K$, where K is a maximal compact subgroup of G. These locally symmetric spaces, which include real hyperbolic manifolds $\Gamma\backslash\mathbb{H}^n$ for $G = \mathrm{PO}(n,1) = \mathrm{O}(n,1)/\{\pm \mathrm{I}\}$, are geometrically important. They naturally appear in representation theory and harmonic analysis, where symmetric spaces G/K play a central role (see e.g. [7]).

(3) *Geometric structures on manifolds:* A modern point of view on geometry, which dates back to Klein's 1872 Erlangen program and which has been much developed in the twentieth century especially through the work of Ehresmann and Thurston, is to study manifolds that "locally look like" some "model spaces" with large "symmetry groups". Model spaces are typically homogeneous spaces $X = G/H$ where G is a real Lie group (often semisimple). Important examples include $X = G/K$ as above, but also $(X,G) = (\mathbb{RP}^n, \mathrm{PGL}(n+1,\mathbb{R}))$ (real projective geometry), $(\mathbb{CP}^n, \mathrm{PGL}(n+1,\mathbb{C}))$ (complex projective geometry), or $(\mathbb{H}^{p,q}, \mathrm{PO}(p,q+1))$ (pseudo-Riemannian hyperbolic geometry in signature (p,q)). See [78] for details.

An important class of discrete subgroups of noncompact semisimple Lie groups is the class of *lattices*, namely discrete subgroups of finite covolume for the Haar measure (see Section 4.2 below). They play an important role in several fields of mathematics, in addition to the above, such as:

- geometric group theory (lattices are finitely presented groups with many desirable properties — e.g. lattices of $\mathrm{SL}(n,\mathbb{R})$ for $n \geq 3$ have Kazhdan's property (T)),
- combinatorics (construction of expander graphs),
- number theory (arithmetic groups),
- ergodic theory (flows on $\Gamma\backslash G$) and homogeneous dynamics.

See e.g. [135] and references therein. In some of these settings (in particular ergodic theory and homogeneous dynamics), there is currently active research aiming to extend, to classes of discrete subgroups of infinite covolume, classical results involving lattices. Infinite-index subgroups of arithmetic groups (and particularly those that are still Zariski-dense, named *thin groups* by Sarnak) have also attracted considerable interest recently, see e.g. [107].

In these notes, we will review a few properties of lattices, and then focus on the problem of finding other large classes of infinite discrete subgroups Γ of semisimple Lie groups G with desirable properties, including:

(1) the existence of examples with interesting geometric interpretations,
(2) a good control of the subgroups' behaviour under deformation,
(3) interesting dynamics of Γ on certain homogeneous spaces of G.

These properties are typically invariant under replacing Γ by a finite-index subgroup. This will allow us to sometimes reduce to *torsion-free* Γ: indeed, the Selberg lemma [125, Lem. 8] states that any finitely generated subgroup of G admits a finite-index subgroup which is torsion-free.

4.2 Lattices

Let G be a noncompact semisimple Lie group. It admits a *Haar measure*, i.e. a nonzero Radon measure which is invariant under left and right multiplication; this measure is unique up to scaling.

Definition 4.2.1 A *lattice* of G is a discrete subgroup Γ of G such that the quotient $\Gamma\backslash G$ has finite volume for the measure induced by the Haar measure of G.

If Γ is a lattice of G, then the quotient $\Gamma\backslash G$ can be compact (in which case we say that Γ is a *cocompact* or *uniform* lattice) or not.

A fundamental result of Borel and Harish-Chandra [29, 30] states that G always admits both cocompact lattices and noncocompact lattices.

Borel's Density Theorem [28] states that lattices are Zariski-dense in G as soon as G is connected and has no compact simple factors. This means that if the set of real points of some $\mathbb{R}$-algebraic group contains a lattice of G, then it actually contains the whole of G.

We say that a lattice Γ of G is *irreducible* if for any noncompact, infinite-index, closed normal subgroup G' of G, the projection of Γ to G/G' is nondiscrete. (This is automatically satisfied if G is simple.)

4.2.1 Geometric interpretation

Lattices of G can be characterised by their action on the *Riemannian symmetric space* of G. Let us recall what this fundamental object is (see e.g. [66, 90] for details).

As mentioned in the introduction, G admits a maximal compact subgroup K. It is unique up to conjugation, and so the quotient G/K is uniquely defined. For instance, if $G = \mathrm{SL}(n, \mathbb{R})$, then $K = \mathrm{SO}(n)$ up to conjugation, and G/K identifies with the space of ellipsoids of $\mathbb{R}^n$ of volume 1; if $G = \mathrm{SL}(n, \mathbb{C})$, then $K = \mathrm{SU}(n)$ up to conjugation. See Table 4.1 for further examples.

The group K is the set of fixed points of some involution θ of G, called a Cartan involution. This yields a splitting of the Lie algebra $\mathfrak{g}$ of G as the direct sum of two linear subspaces, namely the subspace $\mathfrak{g}^{\mathrm{d}\theta}$ of fixed points of $\mathrm{d}\theta$ (which is the Lie algebra of K) and the subspace $\mathfrak{g}^{-\mathrm{d}\theta}$ of anti-fixed points of $\mathrm{d}\theta$. The tangent space $T_{eK}(G/K)$ to G/K at the origin identifies with $\mathfrak{g}^{-\mathrm{d}\theta}$, on which there is a natural K-invariant positive definite symmetric bilinear form, the Killing form. Pushing forward this bilinear form by elements of G yields a G-invariant Riemannian metric on G/K. With this metric, G/K has nonpositive sectional curvature and is a *symmetric space*: at every point, the geodesic symmetry sending $\exp(tv)$ to $\exp(-tv)$ (where v is a tangent vector) is an isometry.

Since K is compact, any discrete subgroup Γ of G acts properly discontinuously on G/K. The subgroup Γ is a lattice if and only if the quotient $\Gamma\backslash G/K$ has finite volume, which is equivalent to the action of Γ on G/K admitting a fundamental domain of finite volume.

4.2.2 Examples

The following fundamental example goes back to Minkowski.

Example 4.2.2 The group $\Gamma = \mathrm{SL}(n, \mathbb{Z})$ is a noncocompact lattice in $G = \mathrm{SL}(n, \mathbb{R})$.

Let us briefly explain how to see this, starting with the case $n = 2$.

For $n = 2$, the Riemannian symmetric space G/K is the hyperbolic plane $\mathbb{H}^2 \simeq \{z = x + iy \in \mathbb{C} \mid y = \mathrm{Im}(z) > 0\}$ with its G-invariant metric $\mathrm{d}s^2 = (\mathrm{d}x^2 + \mathrm{d}y^2)/y^2$, on which $G = \mathrm{SL}(2, \mathbb{R})$ acts by Möbius transformations: $\left(\begin{smallmatrix} a & b \\ c & d \end{smallmatrix}\right) \cdot z = \frac{az+b}{cz+d}$. It is an easy exercise to check that

$$\mathcal{D} := \left\{ z \in \mathbb{H}^2 \;\middle|\; |\mathrm{Re}(z)| \leq \frac{1}{2} \quad \text{and} \quad |z| \geq 1 \right\}$$

(see Figure 4.1) is a finite-volume fundamental domain for the action of $\Gamma = \mathrm{SL}(2,\mathbb{Z})$ on $\mathbb{H}^2$. (Use that Γ is generated by $\left(\begin{smallmatrix} 1 & 1 \\ 0 & 1 \end{smallmatrix}\right)$ and $\left(\begin{smallmatrix} 0 & 1 \\ -1 & 0 \end{smallmatrix}\right)$ and that the G-invariant volume form on $\mathbb{H}^2$ is given by $\mathrm{dvol} = \mathrm{d}x\,\mathrm{d}y/(4y^2)$.) Therefore Γ is a lattice in $G = \mathrm{SL}(2,\mathbb{R})$. This lattice is not cocompact since for any $\gamma = \left(\begin{smallmatrix} \mathsf{a} & \mathsf{b} \\ \mathsf{c} & \mathsf{d} \end{smallmatrix}\right) \in \Gamma$ we have $\mathrm{Im}(\gamma \cdot i) = 1/(\mathsf{c}^2 + \mathsf{d}^2) \leq 1$, hence there exist points of $\mathbb{H}^2$ (e.g. ti with $t > 0$ large) that are arbitrarily far away from any point of the Γ-orbit of i in $\mathbb{H}^2$.

For general $n \geq 2$, we can use the classical *Iwasawa decomposition* $G = NAK$, where N (resp. A) is the subgroup of $G = \mathrm{SL}(n,\mathbb{R})$ consisting of upper triangular unipotent (resp. positive diagonal) matrices and $K = \mathrm{SO}(n)$. This means that any element $g \in G$ can be written in a unique way as $g = nak$ where $n \in N$, $a \in A$, and $k \in K$. A finite-volume fundamental domain for the action of Γ on G/K is given by the *Siegel set* $\mathcal{S}$ consisting of those elements of G/K of the form naK with $n \in N$ having all entries above the diagonal in $[-1/2, 1/2]$ and $a = \mathrm{diag}(a_1, \ldots, a_n) \in A$ satisfying $|a_i/a_{i+1}| \geq \sqrt{3}/2$ for all $1 \leq i \leq n-1$.

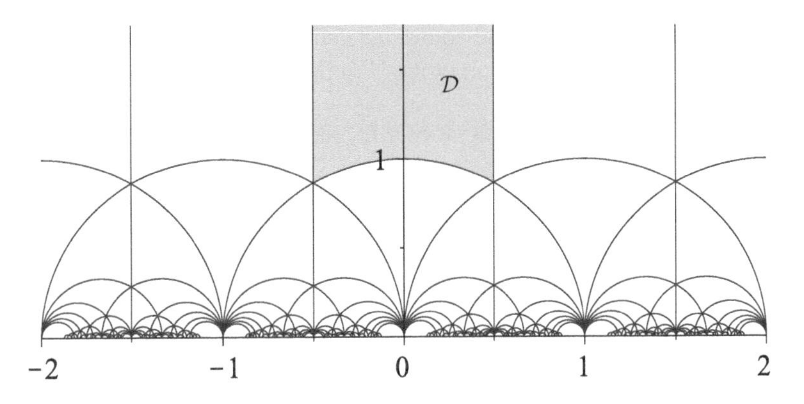

Figure 4.1 Fundamental domains for the action of $\mathrm{SL}(2,\mathbb{Z})$ on the upper half plane model of $\mathbb{H}^2$

Generalising Example 4.2.2, a fundamental result of Borel and Harish-Chandra [30] states that if $\mathbf{G}$ is a semisimple $\mathbb{Q}$-algebraic group, then $\mathbf{G}_{\mathbb{Z}}$ is a lattice in $\mathbf{G}_{\mathbb{R}}$. Godement's cocompactness criterion (see e.g. [19, § 2.8]) states that this lattice is cocompact if and only if it does not contain any nontrivial unipotent elements.

We now give a concrete example of a cocompact lattice (see [19, § 2]).

Example 4.2.3 For $p, q \geq 1$ with $p + q = n \geq 3$, consider the block diagonal matrix

$$J_{p,q} := \begin{pmatrix} \mathrm{I}_p & 0 \\ 0 & -\sqrt{2}\,\mathrm{I}_q \end{pmatrix}$$

and the Lie group $G := \mathrm{SO}(J_{p,q}, \mathbb{R}) \simeq \mathrm{SO}(p,q)$. Then $\Gamma := G \cap \mathrm{SL}(n, \mathbb{Z}[\sqrt{2}])$ is a cocompact lattice in G.

In order to see this, we can apply Weil's trick of "restriction of scalars". Namely, consider the automorphism σ of $\mathrm{SL}(n, \mathbb{Z}[\sqrt{2}])$ obtained by applying the Galois conjugation $x + \sqrt{2}y \mapsto x - \sqrt{2}y$ of $\mathbb{Q}[\sqrt{2}]$ to each entry. Let $J_{p,q}^{\sigma}$ be the image of $J_{p,q}$ under σ, and $\mathbf{H}$ the semisimple algebraic subgroup of $\mathbf{GL}_{2n}$ whose set $\mathbf{H}_{\mathbb{C}}$ of complex points consists of those block matrices of the form

$$h := \begin{pmatrix} a & 2b \\ b & a \end{pmatrix} \in \mathrm{GL}(2n, \mathbb{C})$$

with $\varphi_+(h) := a + \sqrt{2}b \in \mathrm{SO}(J_{p,q}, \mathbb{C})$ and $\varphi_-(h) := a - \sqrt{2}b \in \mathrm{SO}(J_{p,q}^{\sigma}, \mathbb{C})$. An elementary computation (or more abstractly the fact that the family of polynomial equations defining $\mathbf{H}$ is invariant under σ) shows that $\mathbf{H}$ is a $\mathbb{Q}$-algebraic group. We have isomorphisms

$$\begin{cases} \mathbf{H}_{\mathbb{R}} \overset{(\varphi_+,\varphi_-)}{\simeq} \mathrm{SO}(J_{p,q}, \mathbb{R}) \times \mathrm{SO}(J_{p,q}^{\sigma}, \mathbb{R}) = G \times \mathrm{SO}(J_{p,q}^{\sigma}, \mathbb{R}), \\ \mathbf{H}_{\mathbb{Z}} \overset{\varphi_+}{\simeq} \Gamma, \end{cases}$$

where $\mathrm{SO}(J_{p,q}^{\sigma}, \mathbb{R}) \simeq \mathrm{SO}(n)$ is compact. The group $\mathbf{H}_{\mathbb{Z}}$ is a lattice in $\mathbf{H}_{\mathbb{R}}$, hence Γ is a lattice in G. Moreover, $\mathbf{H}_{\mathbb{Z}}$ does not contain any nontrivial unipotent elements since φ_- takes Γ to a subgroup of a *compact* group, hence without nontrivial unipotent elements, and a homomorphism of algebraic groups takes unipotent elements to unipotent elements. Godement's criterion then ensures that $\mathbf{H}_{\mathbb{Z}} \backslash \mathbf{H}_{\mathbb{R}}$ is compact, and so $\Gamma \backslash G$ is compact too.

In both Examples 4.2.2 and 4.2.3, the group Γ is *arithmetic* in G, i.e. there is a homomorphism $\pi : \mathbf{H} \to \mathbf{G}$ of semisimple $\mathbb{Q}$-algebraic groups such that $G = \mathbf{G}_{\mathbb{R}}$, such that the kernel of π in $\mathbf{H}_{\mathbb{R}}$ is compact, and such that Γ is commensurable to $\pi(\mathbf{H}_{\mathbb{Z}})$ (see [135]).

Nonarithmetic lattices are known to exist in $G = \mathrm{SO}(n,1)$ for any $n \geq 2$: examples were constructed by Vinberg [129] for small n using reflection groups, then by Gromov and Piatetski-Shapiro [80] for any n. Later, different examples were constructed by Agol [1] and Belolipetsky–Thomson [13] (see also the very recent work [63]) in the form of lattices of $\mathrm{SO}(n,1)$ whose systole (i.e. length of the shortest closed geodesic) is arbitrarily small. (Due to a separability property later established in [20, Cor. 1.12], Agol's construction [1] actually works for any n.) Finitely many commensurability classes of nonarithmetic lattices are also known in $\mathrm{SU}(2,1)$ and $\mathrm{SU}(3,1)$ by Deligne–Mostow [57, 116] and Deraux–Parker–Paupert [58, 59]. It is an open question whether nonarithmetic lattices exist in $\mathrm{SU}(n,1)$ for $n > 3$.

On the other hand, in noncompact simple Lie groups which are not locally isomorphic to $\mathrm{SO}(n,1)$ or $\mathrm{SU}(n,1)$, all lattices are arithmetic (as a consequence of superrigidity, see Section 4.2.4).

4.2.3 Rank one versus higher rank

The *real rank* of a semisimple Lie group is an integer defined as follows.

Definition 4.2.4 The *real rank* of G, denoted $\mathrm{rank}_{\mathbb{R}}(G)$, is the maximum dimension of a closed connected subgroup of G which is diagonalisable over $\mathbb{R}$; equivalently, for noncompact G, it is the maximum dimension of a totally geodesic subspace of the Riemannian symmetric space G/K which is flat (i.e. of constant zero sectional curvature).

The real rank is invariant under local isomorphism, and the real rank of a product is the sum of the real ranks of the factors. We refer to Table 4.1 for the real ranks of the classical noncompact simple Lie groups. A compact Lie group has real rank 0.

The simple Lie groups of real rank 1 are, up to local isomorphism, $\mathrm{SO}(n,1)$, $\mathrm{SU}(n,1)$, $\mathrm{Sp}(n,1)$ for $n \geq 2$, and the exceptional group $F_{4(-20)}$. (Note that $\mathrm{PSL}(2,\mathbb{R}) \simeq \mathrm{SO}(2,1)_0$ and $\mathrm{PSL}(2,\mathbb{C}) \simeq \mathrm{SO}(3,1)_0$, where the subscript 0 denotes the identity components.)

Semisimple Lie groups G of real rank 1 are characterised by the fact that the sectional curvature of the corresponding Riemannian symmetric space G/K is everywhere < 0. (In fact, the curvature is then *pinched*, i.e. contained in an interval of the form $[\alpha, \beta]$ where $\alpha \leq \beta < 0$.) This implies that the geodesic metric space G/K is *Gromov hyperbolic*, meaning that there exists $\delta \geq 0$ such that all geodesic triangles (a, b, c) of G/K are δ-thin: the side $[a, b]$ lies in the uniform δ-neighbourhood of the union $[b, c] \cup [c, a]$ of the other two sides (see Figure 4.2). On the other hand, when $r := \mathrm{rank}_{\mathbb{R}}(G) \geq 2$, the Riemannian symmetric space G/K is only nonpositively curved, and not Gromov hyperbolic; its geometry is somewhat more complicated due to the presence of *flats* (i.e. isometric copies of Euclidean $\mathbb{R}^r$, where the curvature vanishes).

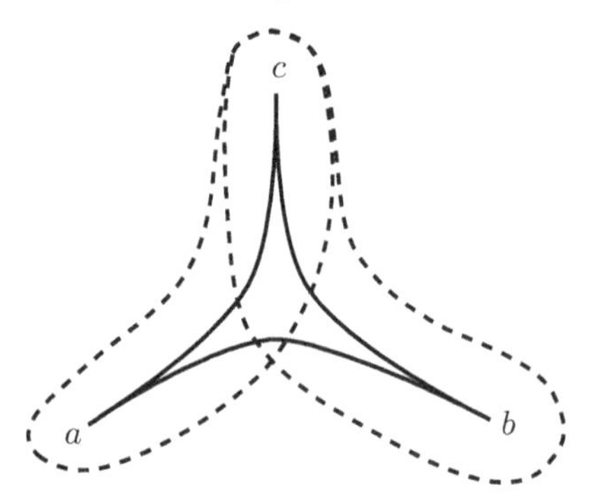

Figure 4.2 A δ-thin triangle in a geodesic metric space. The side $[a, b]$ is contained in the union of the uniform δ-neighbourhoods (indicated by dashes) of the sides $[b, c]$ and $[c, a]$.

There are a number of differences between lattices in real rank one and lattices in higher real rank.

One difference concerns hyperbolicity. Namely, if $\mathrm{rank}_{\mathbb{R}}(G) = 1$, then

- any cocompact lattice Γ of G is *Gromov hyperbolic*, i.e. Γ acts properly discontinuously, by isometries, with compact quotient, on a Gromov hyperbolic proper geodesic metric space X (e.g. $X = G/K$);
- any noncocompact lattice Γ of G is *relatively hyperbolic* with respect to some collection $\mathcal{P}$ of subgroups which are virtually (i.e. up to finite index) nilpotent: this means that Γ acts properly discontinuously by isometries on some visual Gromov hyperbolic proper metric space X (e.g. $X = G/K$), and with compact quotient on some closed subset of X of the form $X \smallsetminus \bigcup_{P\in\mathcal{P}} B_P$ where each B_P is a P-invariant open horoball of X and $B_P \cap B_{P'} = \emptyset$ for $P \neq P'$ (see [89, § 4] for details).

On the other hand, if $\mathrm{rank}_{\mathbb{R}}(G) \geq 2$, then lattices of G are never Gromov hyperbolic, nor relatively hyperbolic with respect to any collection of subgroups [12]. This follows from the fact that these groups are *metrically thick* in the sense of [12] (see [105] for cocompact lattices). In fact, if $\mathrm{rank}_{\mathbb{R}}(G) \geq 2$, then any isometric action of a lattice Γ of G on a Gromov hyperbolic metric space X is "trivial" (i.e. admits a global fixed point in X or its boundary), unless it is obtained by projecting Γ to a rank-one factor of G [3, 87].

More generally, lattices Γ in simple Lie groups G with $\mathrm{rank}_{\mathbb{R}}(G) \geq 2$ tend to have global fixed points when they act on various classes of spaces. For instance, any continuous action by affine isometries of Γ on a Hilbert space has a global fixed point. This property, which is equivalent to Kazhdan's property (T) (see [135, Ch. 13]), is also satisfied by lattices in the rank-one Lie groups $\mathrm{Sp}(n,1)$ with $n \geq 2$ or $F_{4(-20)}$. However, other fixed point properties actually distinguish higher rank from rank one. For instance, for any simple Lie group G with $\mathrm{rank}_{\mathbb{R}}(G) \geq 2$ and any σ-finite positive measure ν on a standard Borel space, any continuous affine isometric action of a lattice of G on $L^p(\nu)$ for $1 < p < +\infty$ has a global fixed point, by Bader–Furman–Gelander–Monod; on the other hand, by Pansu and Bourdon–Pajot, any cocompact lattice Γ in a simple Lie group G with $\mathrm{rank}_{\mathbb{R}}(G) = 1$ (and more generally, any Gromov hyperbolic group Γ) admits fixed-point-free affine isometric actions on $L^p(\Gamma)$ whose linear part is the regular representation, for any $p > 1$ large enough. See [4].

Another difference between real rank one and higher real rank concerns normal subgroups. Namely, if $\mathrm{rank}_{\mathbb{R}}(G) = 1$, then lattices of G have many normal subgroups (see Gromov [79]); in fact, if Γ is a lattice of G, then any countable group can be embedded into a quotient of Γ by some normal subgroup (this "universality" property holds for all relatively hyperbolic groups [2]). On the other hand, if $\mathrm{rank}_{\mathbb{R}}(G) \geq 2$, then all normal subgroups of an irreducible lattice Γ of G are finite or finite-index in Γ (this is Margulis's Normal Subgroups Theorem, see [112]).

Note that for an irreducible lattice Γ of G, the finite normal subgroups of Γ are easy to describe: for connected G, they are the subgroups of the finite abelian group $\Gamma \cap Z(G)$ (using Borel's Density Theorem [28]). On the other hand, much more effort is required to understand the finite-index normal subgroups of Γ. By [11], for $\Gamma = \mathrm{SL}(n,\mathbb{Z})$ with $n \geq 3$, any finite-index normal subgroup of Γ is a *congruence subgroup*, i.e. contains the kernel of the natural projection $\mathrm{SL}(n,\mathbb{Z}) \to \mathrm{SL}(n,\mathbb{Z}/m\mathbb{Z})$ for some $m \geq 1$; this is false for $\Gamma = \mathrm{SL}(2,\mathbb{Z})$. In general, it is conjectured that lattices of G have a slightly weaker form of this "Congruence Subgroup Property" if and only if $\mathrm{rank}_{\mathbb{R}}(G) \geq 2$: see [127].

We now discuss some rigidity results for representations of lattices inside G, which hold in particular for $\mathrm{rank}_{\mathbb{R}}(G) \geq 2$.

4.2.4 Deformations and rigidity

Let Γ be a discrete subgroup of G. We denote by $\mathrm{Hom}(\Gamma, G)$ the space of representations of Γ to G, endowed with the compact-open topology (if Γ admits a finite generating subset F, then this coincides with the topology of pointwise convergence on F).

By a *continuous deformation* of Γ in G we mean a continuous path

$(\rho_t)_{t\in[0,1)}$ in $\mathrm{Hom}(\Gamma, G)$ where ρ_0 is the natural inclusion of Γ in G. Certain continuous deformations of Γ in G are considered *trivial*: namely, those of the form $\rho_t = g_t\,\rho_0(\cdot)\,g_t^{-1}$ where $(g_t)_{t\in[0,1)}$ is a continuous path in G (and g_0 is the identity element). In other words, if $\mathrm{Hom}(\Gamma, G)/G$ denotes the quotient of $\mathrm{Hom}(\Gamma, G)$ by the natural action of G by conjugation at the target, then the trivial deformations are those whose image in $\mathrm{Hom}(\Gamma, G)/G$ are constant.

For $G = \mathrm{PSL}(2,\mathbb{R}) \simeq \mathrm{SO}(2,1)_0$, torsion-free lattices Γ of G admit many nontrivial continuous deformations. Indeed, if Γ is noncocompact in G, then Γ is a nonabelian free group on finitely many generators $\gamma_1,\dots,\gamma_m$, and the natural inclusion $\rho_0 : \Gamma \hookrightarrow G$ can be continuously deformed by deforming independently the image of each γ_i; the map $\rho \mapsto (\rho(\gamma_1),\dots,\rho(\gamma_m))$ yields an isomorphism $\mathrm{Hom}(\Gamma, G) \simeq G^m$. If Γ is cocompact in G, then Γ identifies with the fundamental group of the closed hyperbolic surface $S := \Gamma\backslash\mathbb{H}^2$; the connected component of the natural inclusion ρ_0 in $\mathrm{Hom}(\Gamma, G)$ consists entirely of injective and discrete representations [77], and its image in $\mathrm{Hom}(\Gamma, G)/G$ is homeomorphic to $\mathbb{R}^{6g-6}$: it is the *Teichmüller space* of S.

On the other hand, a number of rigidity results have been proved for lattices in other noncompact semisimple Lie groups G, including local rigidity, Mostow rigidity, and Margulis superrigidity, which we now briefly state and comment on. See [67, 119] for details and references.

Local rigidity (Selberg, Calabi, Weil, Garland–Raghunathan) *Let G be a semisimple Lie group with no simple factors that are compact or locally isomorphic to* $\mathrm{PSL}(2,\mathbb{R})$ *(resp.* $\mathrm{PSL}(2,\mathbb{K})$ *with $\mathbb{K} = \mathbb{R}$ or $\mathbb{C}$). If Γ is a cocompact (resp. noncocompact) irreducible lattice of G, then any continuous deformation of Γ in G is trivial.*

Note that noncocompact lattices of $G = \mathrm{PSL}(2,\mathbb{C})$ are not locally rigid: they can be deformed using Thurston's *hyperbolic Dehn surgery* theory. However, they do not admit nontrivial deformations sending unipotent elements to unipotent elements.

Local rigidity is an important ingredient in the proof of Wang's finiteness theorem, which states that if G is simple and not locally isomorphic to $\mathrm{PSL}(2,\mathbb{K})$ with $\mathbb{K} = \mathbb{R}$ or $\mathbb{C}$, then for any $v > 0$ there are only finitely many conjugacy classes of lattices of G with covolume $\leq v$.

Mostow rigidity (Mostow, Prasad, Margulis) *Let G, G' be connected semisimple Lie groups, with trivial centre, and with no simple factors that are compact or locally isomorphic to* $\mathrm{PSL}(2,\mathbb{R})$*. If Γ and Γ' are* *irreducible lattices of G and G', respectively, then any isomorphism between Γ and Γ' extends to a continuous isomorphism between G and G'.*

This implies (see Section 4.2.1) that the fundamental group of any locally symmetric space $\Gamma\backslash G/K$ completely determines its geometry.

Margulis superrigidity (Margulis, Corlette, Gromov–Schoen, see e.g. [135, Th. 16.1.4]) *Let G be a noncompact semisimple Lie group which is connected, algebraically simply connected, and not locally isomorphic to the product of* $\mathrm{SO}(n,1)$ *or* $\mathrm{SU}(n,1)$ *with a compact Lie group. Then any irreducible lattice Γ of G is* superrigid, *in the sense that any representation $\rho : \Gamma \to \mathrm{GL}(d,\mathbb{R})$ (for any $d \geq 2$) continuously extends to G up to finite index and to bounded error.*

Here "ρ continuously extends to G up to finite index and to bounded error" means that there exist a finite-index subgroup Γ' of Γ, a continuous homomorphism $\rho_G : G \to \mathrm{GL}(d,\mathbb{R})$, and a compact subgroup C of $\mathrm{GL}(d,\mathbb{R})$ centralising $\rho_G(G)$ such that $\rho(\gamma) \in \rho_G(\gamma)C$ for all $\gamma \in \Gamma'$. Under an appropriate assumption on the image of ρ, we can take C to be trivial. "Algebraically simply connected" is a technical assumption which is always satisfied up to passing to a finite cover: see [135, § 16.1].

Margulis used his superrigidity (over $\mathbb{R}$ as above, but also over non-Archimedean local fields) to prove that if G is semisimple with no compact simple factors and if $\mathrm{rank}_{\mathbb{R}}(G) \geq 2$, then all irreducible lattices Γ of G are arithmetic in the sense of Section 4.2.2. The same conclusion holds when G is locally isomorphic to $\mathrm{Sp}(n,1)$ with $n \geq 2$ or $F_{4(-20)}$, as superrigidity holds for these rank-one groups as well.

Margulis superrigidity was further extended by Zimmer into a rigidity result for cocycles, see [69]. This was the starting point of important new directions of research at the intersection of group theory and dynamics (see e.g. [73]), including the so-called *Zimmer program* (see [44, 68]). The idea of this program is the following: for a lattice Γ in a simple Lie group G with $\mathrm{rank}_{\mathbb{R}}(G) \geq 2$, Margulis superrigidity states that any linear representation of Γ essentially comes from a linear representation of G; in particular, the minimal dimension of a finite-kernel linear representation of Γ is equal to the minimal dimension of a finite-kernel linear representation of G. Zimmer asked whether this last property has a nonlinear analogue, for actions by diffeomorphisms of Γ on closed manifolds: namely, is the minimal dimension of a closed manifold on which Γ acts faithfully by diffeomorphisms equal to the minimal dimension of a closed manifold on which G (or a compact form of the complexification of G)

acts faithfully by diffeomorphisms? Brown, Fisher, and Hurtado have recently answered this question positively in many cases, building on new developments in dynamics and on recent strengthenings of Kazhdan's property (T): see [44, 68]. This has led to intense research activity around rigidity questions for actions by diffeomorphisms of higher-rank lattices on manifolds.

4.3 A change of paradigm

We just saw that many important rigidity results have been established for lattices since the 1960s, particularly in higher real rank, and that this topic is still very active. On the other hand, since the 1990s and early 2000s, there has been growing interest in *flexibility*: namely, there has been increasing effort to find and study infinite discrete subgroups of semisimple Lie groups which are more flexible than lattices, and which in certain cases can have large deformation spaces. Such discrete subgroups have been known to exist for a long time in real rank one, whereas the investigation of their analogues in higher real rank has gathered momentum only much more recently. We present a few examples below.

To be more precise, we are interested in infinite discrete subgroups Γ of semisimple Lie groups G that admit continuous deformations $(\rho_t)_{t\in[0,1)} \subset \mathrm{Hom}(\Gamma, G)$ as in Section 4.2.4 which, not only are nontrivial, but also satisfy that each ρ_t is injective with discrete image, so that the $\rho_t(\Gamma)$ for $t > 0$ are still discrete subgroups of G isomorphic (but not conjugate) to Γ. An ideal situation is when the natural inclusion $\rho_0 : \Gamma \hookrightarrow G$ admits a full open neighbourhood in $\mathrm{Hom}(\Gamma, G)$ consisting entirely of injective and discrete representations, with a nonconstant image in $\mathrm{Hom}(\Gamma, G)/G$.

We are thus led, for given discrete subgroups Γ of G, to study subsets of $\mathrm{Hom}(\Gamma, G)$ consisting of injective and discrete representations, and their images in the corresponding character varieties. In this framework, we discuss so-called *higher Teichmüller theory* in Section 4.3.4 below.

Remark 4.3.1 In the sequel, we go back and forth between two equivalent points of view: studying discrete subgroups Γ of G, or fixing an abstract group Γ_0 and studying the injective and discrete representations of Γ_0 into G (corresponding to the various ways of realising Γ_0 as a discrete subgroup of G). We sometimes allow ourselves to weaken "injective" into "finite-kernel".

4.3.1 Examples in real rank one

Examples of flexible discrete subgroups in real rank one include classical Schottky groups (which are nonabelian free groups), quasi-Fuchsian groups (which are closed surface groups), as well as other discrete subgroups which are fundamental groups of higher-dimensional manifolds. We briefly review such examples, referring to [95, 113] for more details.

Schottky groups

For $n \geq 2$, let $X = \mathbb{H}^n$ be the real hyperbolic space of dimension n, with visual boundary $\partial_\infty X \simeq \mathbb{S}^{n-1}$. Concretely, choosing a symmetric bilinear form $\langle \cdot, \cdot \rangle_{n,1}$ of signature $(n, 1)$ on $\mathbb{R}^{n+1}$, we can realise X as the open subset

$$\mathbb{H}^n = \{[v] \in \mathbb{P}(\mathbb{R}^{n+1}) \mid \langle v, v \rangle_{n,1} < 0\} \tag{4.1}$$

of the real projective space $\mathbb{P}(\mathbb{R}^{n+1})$ and $\partial_\infty X$ as the boundary of X in $\mathbb{P}(\mathbb{R}^{n+1})$. The geodesics of X are then the nonempty intersections of X with projective lines of $\mathbb{P}(\mathbb{R}^{n+1})$, the geodesic copies of $\mathbb{H}^{n-1}$ in X are the nonempty intersections of X with projective hyperplanes of $\mathbb{P}(\mathbb{R}^{n+1})$, and the isometry group $G = \mathrm{Isom}(X)$ of X is $\mathrm{PO}(n, 1) = \mathrm{O}(n, 1)/\{\pm \mathrm{I}\}$.

An *open disk* in $\partial_\infty X$ is the boundary at infinity of an open half-space of X, bounded by a geodesic copy of $\mathbb{H}^{n-1}$. (For $n = 2$, open disks are just open intervals in $\partial_\infty X \simeq \mathbb{S}^1$.)

For $m \geq 2$, choose $2m$ pairwise disjoint open disks $B_1^\pm, \ldots, B_m^\pm$ in $\partial_\infty X$, such that $\partial_\infty X \smallsetminus \bigcup_{i=1}^m (B_i^- \cup B_i^+)$ has nonempty interior, and elements $\gamma_1, \ldots, \gamma_m \in G$ such that $\gamma_i \cdot \mathrm{Int}(\partial_\infty X \smallsetminus B_i^-) = B_i^+$ for all i. Let Γ be the subgroup of G generated by $\gamma_1, \ldots, \gamma_m$.

Claim 4.3.2 *The group Γ is a nonabelian free group with free generating subset $\{\gamma_1, \ldots, \gamma_m\}$. It is discrete in G.*

Proof Consider any reduced word $\gamma = \gamma_{i_1}^{\sigma_1} \cdots \gamma_{i_N}^{\sigma_N}$ in the alphabet $\{\gamma_1^{\pm 1}, \ldots, \gamma_m^{\pm 1}\}$, where $1 \leq i_j \leq m$ and $\sigma_j \in \{\pm 1\}$ for all $1 \leq j \leq N$. Since $\gamma_{i_j}^{\sigma_j} \cdot \mathrm{Int}(\partial_\infty X \smallsetminus B_{i_j}^{-\mathrm{sign}(\sigma_j)}) = B_{i_j}^{\mathrm{sign}(\sigma_j)}$ for all j and since $B_{i_j}^{\mathrm{sign}(\sigma_j)} \subset \mathrm{Int}(\partial_\infty X \smallsetminus B_{i_{j-1}}^{-\mathrm{sign}(\sigma_{j-1})})$ for $j \geq 2$, we see that the element of Γ corresponding to γ sends $\partial_\infty X \smallsetminus \bigcup_{i=1}^m (B_i^- \cup B_i^+)$ into the closure of $B_{i_1}^{\mathrm{sign}(\sigma_1)}$ in $\partial_\infty X$. On the other hand, the set of elements $g \in G$ sending $\partial_\infty X \smallsetminus \bigcup_{i=1}^m (B_i^- \cup B_i^+)$ into the closure of $\bigcup_{i=1}^m (B_i^- \cup B_i^+)$ in $\partial_\infty X$ is a closed subset of G that does not contain the identity element. □

Such a group Γ is called a *Schottky group*. The proof of Claim 4.3.2 is based on the so-called *ping pong dynamics* of Γ on $\partial_\infty X$: imagine the ping pong players are the generators $\gamma_1, \gamma_1^{-1}, \ldots, \gamma_m, \gamma_m^{-1}$; the ping pong table is $\partial_\infty X$, which is divided into several open regions, namely the $B_i^\pm$ and the "central region" $\mathrm{Int}(\partial_\infty X \smallsetminus \bigcup_i (B_i^- \cup B_i^+))$; the rules of the game are that each player $\gamma_i^{\pm 1}$ sends all regions but one (namely $B_i^\mp$) into a single region (namely $B_i^\pm$). The ping pong ball is a point which is initially in the central region. For any reduced word in the generators, we successively apply the corresponding ping pong players; the ball ends up in one of the $B_i^\pm$. We deduce that the element of Γ corresponding to this reduced word is nontrivial in Γ, and not too close to the identity in G.

Remark 4.3.3 Let $\mathcal{D} := \partial_\infty X \smallsetminus \bigcup_{i=1}^m (B_i^- \cup B_i^+)$ and $\Omega := \mathrm{Int}(\bigcup_{\gamma \in \Gamma} \gamma \cdot \mathcal{D})$. Then Ω is an open subset of $\partial_\infty X$ on which Γ acts properly discontinuously with fundamental domain $\mathcal{D}$.

(Here we have assumed that $\mathcal{D}$ has nonempty interior; therefore $\Omega \neq \emptyset$ and Γ is *not* a lattice in G: it has infinite covolume for the Haar measure.) See e.g. [117] for beautiful illustrations in dimension two, for $X = \mathbb{H}^3$.

Since Schottky groups Γ are nonabelian free groups, they admit, as in Section 4.2.4, many nontrivial continuous deformations $(\rho_t)_{t \in [0,1)} \subset \mathrm{Hom}(\Gamma, G)$, obtained by independently deforming the image of each generator γ_i. Some of these deformations $(\rho_t)_{t \in [0,1)}$ are "good" in the sense that for every $t \in [0,1)$, the group $\rho_t(\Gamma)$ still has a ping pong configuration analogous to that of Γ, hence ρ_t is injective with discrete image by arguing as in Claim 4.3.2. If the open disks $B_1^\pm, \ldots, B_m^\pm$ in the initial configuration have pairwise disjoint *closures* (i.e. Γ is a "strong" Schottky group), then all small deformations are "good": the natural inclusion $\rho_0 : \Gamma \hookrightarrow G$ admits an open neighbourhood in $\mathrm{Hom}(\Gamma, G)$ consisting entirely of injective and discrete representations, with a ping pong configuration analogous to that of Γ.

Quasi-Fuchsian groups

Quasi-Fuchsian groups are important infinite discrete subgroups of $\mathrm{PSL}(2, \mathbb{C})$ which have been much studied (see [113]), and which are *not* lattices in $\mathrm{PSL}(2, \mathbb{C})$. They are by definition the images of quasi-Fuchsian representations. Let us briefly recall what these are.

Let S be a closed orientable surface of genus $g \geq 2$. By the Uniformisation Theorem (see Section 4.1), there exist injective and discrete representations from the fundamental group $\pi_1(S)$ to $\mathrm{PSL}(2, \mathbb{R})$. These representations form two connected components of $\mathrm{Hom}(\pi_1(S), \mathrm{PSL}(2, \mathbb{R}))$ [77], switched by conjugation by elements of $\mathrm{PGL}(2, \mathbb{R}) \smallsetminus \mathrm{PSL}(2, \mathbb{R})$ (i.e. by orientation-reversing isometries of $\mathbb{H}^2$). The image of either of these connected components in $\mathrm{Hom}(\pi_1(S), \mathrm{PSL}(2,\mathbb{R}))/\mathrm{PSL}(2,\mathbb{R})$ identifies with the *Teichmüller space* of S, which is homeomorphic to $\mathbb{R}^{6g-6}$.

Now view $\mathrm{PSL}(2,\mathbb{R})$ as a subgroup of $\mathrm{PSL}(2,\mathbb{C})$. Recall that $\mathrm{PSL}(2,\mathbb{C}) \simeq \mathrm{PO}(3,1)_0$ acts by isometries on the hyperbolic space $\mathbb{H}^3$; the subgroup $\mathrm{PSL}(2,\mathbb{R}) \simeq \mathrm{PO}(2,1)_0$ preserves an isometric copy of $\mathbb{H}^2$ inside $\mathbb{H}^3$. We see the injective and discrete representations $\rho : \pi_1(S) \to \mathrm{PSL}(2,\mathbb{R})$ as representations with values in $\mathrm{PSL}(2,\mathbb{C})$, called *Fuchsian.* They preserve a circle in $\partial_\infty \mathbb{H}^3$, namely the boundary $\partial_\infty \mathbb{H}^2$ of the isometric copy of $\mathbb{H}^2$ preserved by $\mathrm{PSL}(2,\mathbb{R})$.

The Fuchsian representations admit an open neighbourhood in $\mathrm{Hom}(\pi_1(S), \mathrm{PSL}(2,\mathbb{C}))$ consisting entirely of injective and discrete representations, called *quasi-Fuchsian.* Each quasi-Fuchsian representation preserves a topological circle in $\partial_\infty \mathbb{H}^3$, but which may now be "wiggly" as in Figure 4.3. Quasi-Fuchsian representations form an open subset of $\mathrm{Hom}(\pi_1(S), \mathrm{PSL}(2,\mathbb{C}))$ which is dense in the set of injective and discrete representations; its image in $\mathrm{Hom}(\pi_1(S), \mathrm{PSL}(2,\mathbb{C}))/\mathrm{PSL}(2,\mathbb{C})$ admits a natural parametrisation (due to Bers) by two copies of the Teichmüller space of S (hence by $\mathbb{R}^{12g-12}$). See e.g. [126] for details and references.

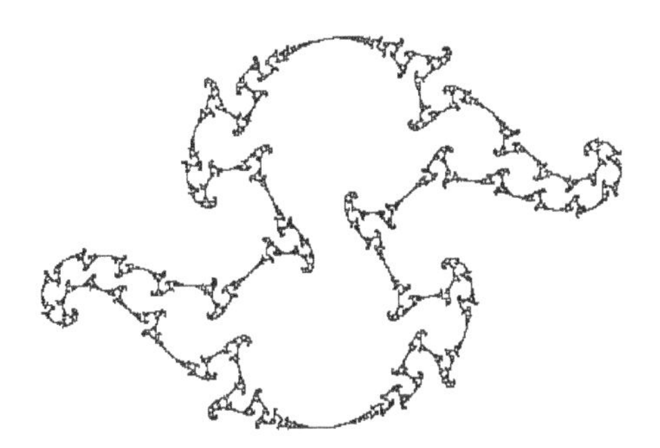

Figure 4.3 The limit set (an invariant topological circle) of a quasi-Fuchsian group in $\partial_\infty \mathbb{H}^3 \simeq \mathbb{C} \cup \{\infty\}$

Deformations of Fuchsian representations for higher-dimensional groups

Recall that $\mathrm{PSL}(2,\mathbb{R}) \simeq \mathrm{SO}(2,1)_0$ and $\mathrm{PSL}(2,\mathbb{C}) \simeq \mathrm{SO}(3,1)_0$. We now consider any integer $n \geq 2$ and let Γ be a cocompact lattice of $\mathrm{SO}(n,1)_0$. As above, we can see Γ as a discrete subgroup of $\mathrm{SO}(n+1,1)$ (which is not a lattice anymore). Interestingly, although all continuous deformations of

Γ in $\mathrm{SO}(n,1)$ are trivial for $n \geq 3$ (by Mostow rigidity, see Section 4.2.4), there can exist nontrivial continuous deformations of Γ in $\mathrm{SO}(n+1,1)$. Such deformations were constructed in [94, 108] based on a construction of Thurston called *bending*.

The idea is the following. The cocompact lattice Γ of $\mathrm{SO}(n,1)_0$ defines a closed hyperbolic manifold $M = \Gamma\backslash\mathbb{H}^n$ whose fundamental group $\pi_1(M)$ identifies with Γ. Suppose that M admits a closed totally geodesic embedded hypersurface N. Its fundamental group $\pi_1(N)$ is a subgroup of Γ contained in a copy of $\mathrm{SO}(n-1,1)$ inside $\mathrm{SO}(n,1)$. In particular, the centraliser of $\pi_1(N)$ in $\mathrm{SO}(n+1,1)$ contains a one-parameter subgroup $(g_t)_{t\in\mathbb{R}}$ which is not contained in $\mathrm{SO}(n,1)$.

If N separates M into two submanifolds M_1 and M_2, then by van Kampen's theorem $\pi_1(M)$ is the amalgamated free product $\pi_1(M_1) *_{\pi_1(N)} \pi_1(M_2)$ of $\pi_1(M_1)$ and $\pi_1(M_2)$ over $\pi_1(N)$. Let $\rho_0 : \Gamma \to \mathrm{SO}(n+1,1)$ be the natural inclusion. A continuous deformation $(\rho_t)_{t\in[0,1)} \subset \mathrm{Hom}(\Gamma, \mathrm{SO}(n+1,1))$ is obtained by defining ρ_t to be ρ_0 when restricted to $\pi_1(M_1)$ and $g_t\rho_0(\cdot)g_t^{-1}$ when restricted to $\pi_1(M_2)$ (these two representations coincide on $\pi_1(N)$).

Otherwise, $M' := M \smallsetminus N$ is connected and $\pi_1(M)$ is an HNN extension of $\pi_1(M')$: it is generated by $\pi_1(M')$ and some element ν with the relations $\nu\, j_1(\gamma)\, \nu^{-1} = j_2(\gamma)$ for all $\gamma \in \pi_1(N)$, where $j_1 : \pi_1(N) \to \pi_1(M)$ and $j_2 : \pi_1(N) \to \pi_1(M)$ are the inclusions in $\pi_1(M)$ of the fundamental groups of the two sides of N. Let $\rho_0 : \Gamma \to \mathrm{SO}(n+1,1)$ be the natural inclusion. A continuous deformation $(\rho_t)_{t\in[0,1)} \subset \mathrm{Hom}(\Gamma, \mathrm{SO}(n+1,1))$ is obtained by defining ρ_t to be ρ_0 when restricted to $\pi_1(M')$ and setting $\rho_t(\nu) := \nu g_t$ (the relations $\nu\, j_1(\gamma)\, \nu^{-1} = j_2(\gamma)$ for $\gamma \in \pi_1(N)$ are preserved since g_t centralises $\pi_1(N)$).

In either case, Johnson and Millson [94] observed that for small enough $t > 0$ the representation ρ_t has Zariski-dense image in $\mathrm{SO}(n+1,1)$; moreover, ρ_t is still injective and discrete for small t (see Section 4.4.2).

Remarks 4.3.4 (1) In this construction, ρ_t is not injective and discrete for all $t \in \mathbb{R}$. Indeed, the one-parameter subgroup $(g_t)_{t\in\mathbb{R}}$ takes values in a copy of $\mathrm{SO}(2)$ in $\mathrm{SO}(n+1,1)$, which centralises $\rho_0(\pi_1(N))$. For $t \in \mathbb{R}$ such that $g_t = -\mathrm{I}$ in $\mathrm{SO}(2)$, the representation ρ_t takes values in $\mathrm{SO}(n,1)$ but is not injective and discrete.

(2) The fact that ρ_t is injective and discrete for small t also follows from Maskit's combination theorems, which generalise the idea of ping pong to amalgamated free products and HNN extensions (see [113, § VIII.E.3]).

4.3.2 Ping pong in higher real rank

Examples of "flexible" discrete subgroups of higher-rank semisimple Lie groups G which are nonabelian free groups can be constructed by generalising the classical Schottky groups of Section 4.3.1 in various ways. Let us mention three geometric constructions.

Ping pong in projective space

The idea of the following construction goes back to Tits [128] in his proof of the Tits alternative. The construction was later studied in a more quantitative way by Benoist [15]. It works in any flag variety G/P where G is a noncompact semisimple Lie group and P a proper parabolic subgroup of G, but for simplicity we consider the projective space $\mathbb{P}(\mathbb{R}^d)$ which is a flag variety of $G = \mathrm{SL}(d, \mathbb{R})$, for $d \geq 3$. We fix a Riemannian metric $\mathsf{d}_{\mathbb{P}(\mathbb{R}^d)}$ on $\mathbb{P}(\mathbb{R}^d)$.

An element $g \in G$ is said to be *biproximal* in $\mathbb{P}(\mathbb{R}^d)$ if it admits a unique complex eigenvalue of highest modulus and a unique complex eigenvalue of lowest modulus, and if these two eigenvalues (which are then necessarily real) have multiplicity 1; equivalently, g is conjugate to a block-diagonal matrix $\mathrm{diag}(t, A, s^{-1})$ where $t, s > 1$ and $A \in \mathrm{GL}(d-2, \mathbb{R})$ is such that the spectral radii of A and A^{-1} are $< t$ and $< s$, respectively (for instance, A could be the identity matrix). In this case, g has a unique attracting fixed point x_g^+ and a unique repelling fixed point x_g^- in $\mathbb{P}(\mathbb{R}^d)$, corresponding to the eigenspaces for the highest and lowest eigenvalues. More precisely, g has the following "North-South dynamics" on $\mathbb{P}(\mathbb{R}^d)$:

- it preserves a unique projective hyperplane X_g^+ (resp. X_g^-) of $\mathbb{P}(\mathbb{R}^d)$ containing x_g^+ (resp. x_g^-), corresponding to the sum of the generalised eigenspaces for the eigenvalues of nonminimal (resp. nonmaximal) modulus,
- for any $x \in \mathbb{P}(\mathbb{R}^d) \smallsetminus X_g^{\mp}$ we have $g^{\pm k} \cdot x \to x_g^{\pm}$ as $k \to +\infty$, uniformly on compact sets.

In particular, if we fix $\varepsilon > 0$, then any large power of g sends the complement of the open uniform ε-neighbourhood $B_{g^{-1}}^{\varepsilon}$ of X_g^- into the closure of the open ball b_g^{ε} of radius ε centred at x_g^+ for $\mathsf{d}_{\mathbb{P}(\mathbb{R}^d)}$, and g^{-1} has a similar behaviour after replacing $(B_{g^{-1}}^{\varepsilon}, X_g^-, b_g^{\varepsilon}, x_g^+)$ by $(B_g^{\varepsilon}, X_g^+, b_{g^{-1}}^{\varepsilon}, x_g^-)$ (see Figure 4.4).

Let $\gamma_1, \ldots, \gamma_m \in G$ be biproximal elements which are "transverse" in the sense that $x_{\gamma_i}^+, x_{\gamma_i}^- \notin X_{\gamma_j}^+ \cup X_{\gamma_j}^-$ for all $1 \leq i \neq j \leq m$ (in other words, the configuration of pairs $(x_{\gamma_i}^{\bullet}, X_{\gamma_i}^{\bullet})_{1 \leq i \leq m,\ \bullet \in \{+,-\}}$ with $x_{\gamma_i}^{\bullet} \in X_{\gamma_i}^{\bullet}$ is

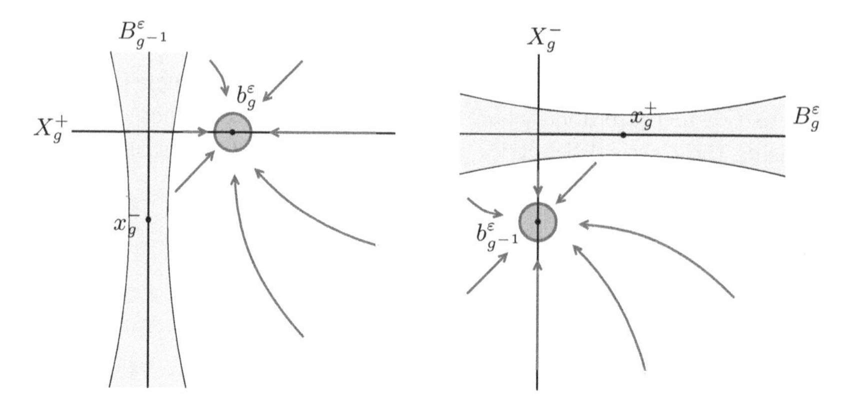

Figure 4.4 Left (resp. right) panel: the dynamics of a large power of g (resp. g^{-1}) on $\mathbb{P}(\mathbb{R}^d)$ for a biproximal element $g \in \mathrm{SL}(d, \mathbb{R})$

generic). Up to replacing each γ_i by a large power, we may assume that there exists $\varepsilon > 0$ such that $\mathbb{P}(\mathbb{R}^d) \smallsetminus \bigcup_{i=1}^{m}(B^{\varepsilon}_{\gamma_i} \cup B^{\varepsilon}_{\gamma_i^{-1}})$ has nonempty interior and such that for any $\alpha \neq \beta$ in $\{\gamma_1, \gamma_1^{-1}, \dots, \gamma_m, \gamma_m^{-1}\}$, the sets b^{ε}_{α} and B^{ε}_{β} have disjoint closures in $\mathbb{P}(\mathbb{R}^d)$ and α sends the interior of $\mathbb{P}(\mathbb{R}^d) \smallsetminus B^{\varepsilon}_{\alpha^{-1}}$ into b^{ε}_{α} (see Figure 4.5 for $m = 2$).

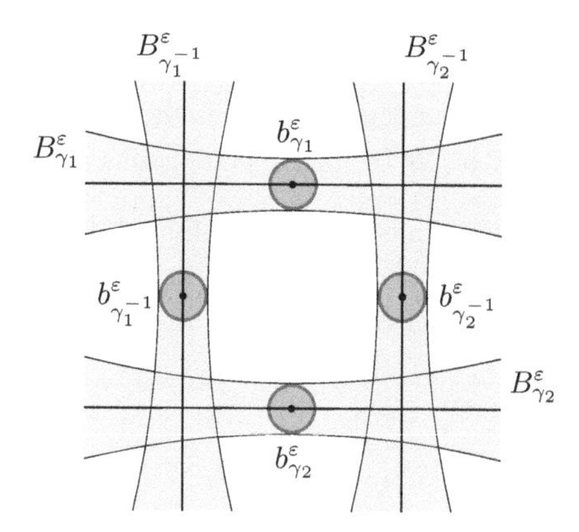

Figure 4.5 A ping pong configuration as in Claim 4.3.5

Let Γ be the subgroup of G generated by $\gamma_1, \dots, \gamma_m$. The following is analogous to Claim 4.3.2.

Claim 4.3.5 *The group Γ is a nonabelian free group with free generating subset $\{\gamma_1, \dots, \gamma_m\}$. It is discrete in G.*

Proof Consider any reduced word $\gamma = \gamma_{i_1}^{\sigma_1} \dots \gamma_{i_N}^{\sigma_N}$ in the alphabet $\{\gamma_1^{\pm 1}, \dots, \gamma_m^{\pm 1}\}$, where $1 \leq i_j \leq m$ and $\sigma_j \in \{\pm 1\}$ for all $1 \leq j \leq N$. Using the inclusions $\alpha \cdot \mathrm{Int}\big(\mathbb{P}(\mathbb{R}^d) \smallsetminus B^{\varepsilon}_{\alpha^{-1}}\big) \subset b^{\varepsilon}_{\alpha}$ for $\alpha = \gamma_{i_j}^{\sigma_j}$ and $b^{\varepsilon}_{\alpha} \subset \mathrm{Int}\big(\mathbb{P}(\mathbb{R}^d) \smallsetminus B^{\varepsilon}_{\beta^{-1}}\big)$ for $(\alpha, \beta) = (\gamma_{i_j}^{\sigma_j}, \gamma_{i_{j-1}}^{\sigma_{j-1}})$ with $j \geq 2$, we see that the element of Γ corresponding to γ sends $\mathbb{P}(\mathbb{R}^d) \smallsetminus \bigcup_{i=1}^m (B^{\varepsilon}_{\gamma_i} \cup B^{\varepsilon}_{\gamma_i^{-1}})$ into the closure of $b^{\varepsilon}_{\gamma_{i_1}^{\sigma_1}}$ (hence of $B^{\varepsilon}_{\gamma_{i_1}^{\sigma_1}}$) in $\mathbb{P}(\mathbb{R}^d)$. On the other hand, the set of elements $g \in G$ sending $\mathbb{P}(\mathbb{R}^d) \smallsetminus \bigcup_{i=1}^m (B^{\varepsilon}_{\gamma_i} \cup B^{\varepsilon}_{\gamma_i^{-1}})$ into the closure of $\bigcup_{i=1}^m (B^{\varepsilon}_{\gamma_i} \cup B^{\varepsilon}_{\gamma_i^{-1}})$ in $\mathbb{P}(\mathbb{R}^d)$ is a closed subset of G that does not contain the identity element. □

Similarly to the classical strong Schottky groups of Section 4.3.1, the group Γ admits nontrivial continuous deformations $(\rho_t)_{t \in [0,1)} \subset \mathrm{Hom}(\Gamma, G)$, obtained by independently deforming the image of each generator γ_i; moreover, there is a neighbourhood of the natural inclusion $\rho_0 : \Gamma \hookrightarrow G$ consisting entirely of injective and discrete representations.

Schottky groups with disjoint ping pong domains

In certain situations it is possible to construct discrete subgroups of G, with ping pong dynamics, for which the ping pong domains are pairwise disjoint, as in the case of the classical rank-one Schottky groups of Section 4.3.1. Achieving this disjointness may require using a slightly modified ping pong compared to Figures 4.4 and 4.5, allowing the attracting and repelling subsets of the generators to be larger than points.

Such a construction has been made in $G = \mathrm{PGL}(2n, \mathbb{K})$, acting on the projective space $\mathbb{P}(\mathbb{K}^{2n})$, for $\mathbb{K} = \mathbb{R}$ or $\mathbb{C}$: the first examples were constructed by Nori in the 1980s, for $\mathbb{K} = \mathbb{C}$, then generalised by Seade and Verjovsky (see [124]); it was observed in [91] that the construction also works for $\mathbb{K} = \mathbb{R}$. The idea is to consider pairwise disjoint $(n-1)$-dimensional projective subspaces $X_1^+, X_1^-, \dots, X_m^+, X_m^-$ of $\mathbb{P}(\mathbb{K}^{2n})$ and elements $\gamma_1, \dots, \gamma_m \in G$ such that for any $1 \leq i \leq m$ we have $\gamma_i^{\pm k} \cdot x \to X_i^{\pm}$ for all $x \in \mathbb{P}(\mathbb{K}^{2n}) \smallsetminus X_i^{\mp}$ as $k \to +\infty$, uniformly on compact sets. Up to replacing each γ_i by a large power, we may assume that there exist tubular neighbourhoods $B_i^{\pm}$ of $X_i^{\pm}$ such that $B_1^+, B_1^-, \dots, B_m^+, B_m^-$ are pairwise disjoint, $\mathbb{P}(\mathbb{K}^{2n}) \smallsetminus \bigcup_{i=1}^m (B_i^- \cup B_i^+)$ has nonempty interior, and $\gamma_i \cdot \mathrm{Int}(\mathbb{P}(\mathbb{K}^{2n}) \smallsetminus B_i^-) = B_i^+$ for all i. Then the subgroup Γ of G generated by $\gamma_1, \dots, \gamma_m$ is a nonabelian free group with free generating subset $\{\gamma_1, \dots, \gamma_m\}$. It is discrete in G, and it acts properly discontinuously on $\Omega := \mathrm{Int}(\bigcup_{\gamma \in \Gamma} \gamma \cdot \mathcal{D})$ with fundamental domain $\mathcal{D}$, as in Remark 4.3.3. As for the classical strong Schottky groups of Section 4.3.1, there is a neighbourhood of the natural inclusion $\rho_0 : \Gamma \hookrightarrow G$ consisting entirely of injective and discrete representations.

Crooked Schottky groups

Here is another ping pong construction, introduced and studied in [36].

Let $G = \mathrm{Sp}(2n, \mathbb{R})$ be the group of elements of $\mathrm{GL}(2n, \mathbb{R})$ that preserve the skew-symmetric bilinear form $\omega(v, v') = \sum_{i=1}^{2n} (-1)^i v_i v'_{2n+1-i}$ on $\mathbb{R}^{2n}$. A *symplectic basis* of $\mathbb{R}^{2n}$ is a basis in which the matrix of ω is antidiagonal with entries $1, -1, \dots, 1, -1$; for instance, the canonical basis is a symplectic basis. To any symplectic basis $(e_1, \dots, e_{2n})$ of $\mathbb{R}^{2n}$ we associate an open simplex $B = \mathbb{P}(\mathbb{R}^{>0}\text{-span}(e_1, \dots, e_{2n}))$ in $\mathbb{P}(\mathbb{R}^{2n})$, which we call a *symplectic simplex*. Its *dual* $B^* := \{[v] \in \mathbb{P}(\mathbb{R}^{2n}) \mid \mathbb{P}(v^\perp) \cap \overline{B} = \emptyset\}$ (where $v^\perp$ denotes the orthogonal of v with respect to ω and $\overline{B}$ the closure of B in $\mathbb{P}(\mathbb{R}^{2n})$) is still a symplectic simplex, associated to the symplectic basis $(e_{2n}, -e_{2n-1}, \dots, e_2, -e_1)$. Note that B and B^* are two of the 2^{2n-1} connected components of $\mathbb{P}(\mathbb{R}^{2n}) \smallsetminus \bigcup_{i=1}^{2n} \mathbb{P}(e_i^\perp)$ (see Figure 4.6). We also make the elementary observation that for any symplectic simplices B_1 and B_2, we have $B_1 \subset B_2^*$ if and only if $B_2 \subset B_1^*$.

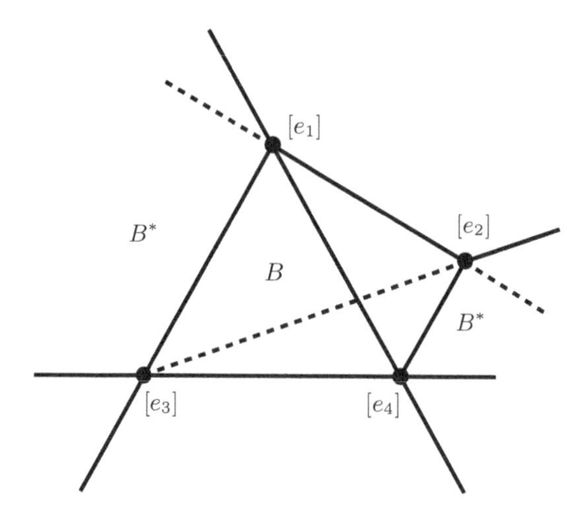

Figure 4.6 A symplectic simplex B of $\mathbb{P}(\mathbb{R}^4)$, associated to a symplectic basis (e_1, e_2, e_3, e_4) of $\mathbb{R}^4$, and its dual B^*

Lemma 4.3.6 *For any $m \geq 2$, there exist $2m$ symplectic simplices $B_1^\pm, \dots, B_m^\pm$ in $\mathbb{P}(\mathbb{R}^{2n})$ such that $B \subset B'^*$ for all $B \neq B'$ in $\{B_1^\pm, \dots, B_m^\pm\}$.*

Proof Choose any symplectic simplex B_1. Note that any nonempty open subset of $\mathbb{P}(\mathbb{R}^{2n})$ contains a symplectic simplex; therefore, we can find a symplectic simplex B_2 such that $\overline{B_2} \subset B_1^*$. Moreover, any neighbourhood of the closure of a symplectic simplex meets the dual of the simplex; therefore $B_1^* \cap B_2^*$ is nonempty. By induction, we construct symplectic simplices $B_1, \ldots, B_{2m}$ such that $\overline{B_j} \subset \bigcap_{i=1}^{j-1} B_i^*$ for all $2 \leq j \leq 2m$. We then have $B_j \subset B_i^*$ for all $1 \leq i < j \leq 2m$. By the elementary observation above, we also have $B_i \subset B_j^*$ for all $1 \leq i < j \leq 2m$. We can then take $(B_i^+, B_i^-) := (B_i, B_{m+i})$ for all $1 \leq i \leq m$. □

For $m \geq 2$, choose symplectic simplices $B_1^\pm, \ldots, B_m^\pm$ as in Lemma 4.3.6, and elements $\gamma_1, \ldots, \gamma_m \in G$ such that $\gamma_i \cdot (B_i^-)^* = B_i^+$ for all $1 \leq i \leq m$ (these exist since G acts transitively on the set of symplectic simplices). A ping pong argument as in Claim 4.3.2 shows that the subgroup Γ of G generated by $\gamma_1, \ldots, \gamma_m$ is a nonabelian free group with free generating subset $\{\gamma_1, \ldots, \gamma_m\}$, and that it is discrete in G.

Moreover, there is an interesting counterpart of Remark 4.3.3, not in the projective space $\mathbb{P}(\mathbb{R}^{2n})$, but in the space $\mathrm{Lag}(\mathbb{R}^{2n})$ of *Lagrangians* of $(\mathbb{R}^{2n}, \omega)$, i.e. of n-dimensional linear subspaces of $\mathbb{R}^{2n}$ which are totally isotropic for ω. For this, we associate to any symplectic simplex $B = \mathbb{P}(\mathbb{R}^{>0}\text{-span}(e_1, \ldots, e_{2n}))$ of $\mathbb{P}(\mathbb{R}^{2n})$ an open subset of $\mathrm{Lag}(\mathbb{R}^{2n})$, namely $\mathcal{H}(B) := \{L \in \mathrm{Lag}(\mathbb{R}^{2n}) \mid L \cap B \neq \emptyset\}$ where we see each $L \in \mathrm{Lag}(\mathbb{R}^{2n})$ as an $(n-1)$-dimensional projective subspace of $\mathbb{P}(\mathbb{R}^{2n})$. In [36] we prove the following remarkable property: for any symplectic simplex B, we have $\mathrm{Lag}(\mathbb{R}^{2n}) = \overline{\mathcal{H}(B)} \sqcup \mathcal{H}(B^*)$. In other words, $\mathcal{H}(B)$ and $\mathcal{H}(B^*)$ are two open "half-spaces" of $\mathrm{Lag}(\mathbb{R}^{2n})$, bounded by their common boundary $\partial\mathcal{H}(B) = \overline{\mathcal{H}(B)} \smallsetminus \mathcal{H}(B) = \overline{\mathcal{H}(B^*)} \smallsetminus \mathcal{H}(B^*)$. We observe [36] that these boundaries $\partial\mathcal{H}(B)$ are nice geometric objects which for $n = 2$ coincide with the *crooked surfaces* of Frances [72] in the Einstein universe $\mathrm{Ein}^3 \simeq \mathrm{Lag}(\mathbb{R}^4)$. Remark 4.3.3 generalises as follows: consider symplectic simplices $B_1^\pm, \ldots, B_m^\pm$ and elements $\gamma_1, \ldots, \gamma_m \in G$ as above. If we set $\mathcal{D} := \mathrm{Lag}(\mathbb{R}^{2n}) \smallsetminus \bigcup_{i=1}^m (\mathcal{H}(B_i^-) \cup \mathcal{H}(B_i^+))$, then the group Γ generated by $\gamma_1, \ldots, \gamma_m$ acts properly discontinuously on $\Omega := \mathrm{Int}(\bigcup_{\gamma \in \Gamma} \gamma \cdot \mathcal{D})$ with fundamental domain $\mathcal{D}$. We call Γ a *crooked Schottky group*.

As in the classical case of Section 4.3.1, continuous deformations of configurations of symplectic simplices yield nontrivial continuous deformations $(\rho_t)_{t \in [0,1)} \subset \mathrm{Hom}(\Gamma, G)$ of crooked Schottky groups Γ for which each ρ_t is injective with discrete image. If the symplectic simplices $B_1^\pm, \ldots, B_m^\pm$ in the initial configuration have pairwise disjoint *closures* (i.e. Γ is a "strong" crooked Schottky group), then the natural inclusion

$\rho_0 : \Gamma \hookrightarrow G$ admits an open neighbourhood in $\mathrm{Hom}(\Gamma, G)$ consisting entirely of injective and discrete representations.

4.3.3 Higher-rank deformations of Fuchsian representations

Inspired by the quasi-Fuchsian representations and their higher-dimensional analogues from Section 4.3.1, here is one strategy for constructing "flexible" infinite discrete subgroups, beyond nonabelian free groups, in semisimple Lie groups G of higher real rank. Consider a finitely generated group Γ_0, an injective and discrete representation σ_0 of Γ_0 into a simple Lie group G' of real rank one, and a nontrivial Lie group homomorphism $\tau : G' \to G$. Consider the composed representation

$$\rho_0 : \Gamma_0 \overset{\sigma_0}{\hookrightarrow} G' \overset{\tau}{\hookrightarrow} G.$$

In some important cases (see e.g. Facts 4.5.6 and 4.5.7), there will be an open neighbourhood of ρ_0 in $\mathrm{Hom}(\Gamma_0, G)$ consisting entirely of injective and discrete representations. The goal is then to deform ρ_0 nontrivially in $\mathrm{Hom}(\Gamma_0, G)$ outside of $\mathrm{Hom}(\Gamma_0, G')$, so as to obtain discrete subgroups of G that are isomorphic to Γ_0 but not conjugate to Γ_0 or any subgroup of G' (ideally Zariski-dense discrete subgroups of G).

This strategy works well, for instance, for $\Gamma_0 = \pi_1(S)$ where S is a closed orientable surface of genus ≥ 2 as in Section 4.3.1, and $G' = \mathrm{SL}(2,\mathbb{R})$ or $\mathrm{PSL}(2,\mathbb{R})$. Let us give three examples in this setting.

Barbot representations

For $d \geq 2$, consider the standard embedding $\tau : G' = \mathrm{SL}(2,\mathbb{R}) \hookrightarrow G = \mathrm{SL}(d,\mathbb{R})$, acting trivially on a $(d-2)$-dimensional linear subspace of $\mathbb{R}^d$. Then there is a neighbourhood of ρ_0 in $\mathrm{Hom}(\Gamma_0, G)$ consisting entirely of injective and discrete representations (see Section 4.5.1). Nontrivial continuous deformations $(\rho_t)_{t\in[0,1)} \subset \mathrm{Hom}(\Gamma_0, G)$ exist; for $d = 3$, they were studied by Barbot, who particularly investigated [8] the case that the ρ_t take values in $\mathrm{GL}(2,\mathbb{R}) \ltimes \mathbb{R}^2$, seen as the subgroup of G consisting of lower block-triangular matrices with blocks of size $(2,1)$.

Hitchin representations

For $d \geq 2$, consider the irreducible embedding $\tau_d : G' = \mathrm{PSL}(2,\mathbb{R}) \hookrightarrow G = \mathrm{PSL}(d,\mathbb{R})$. It is unique modulo conjugation by $\mathrm{PGL}(d,\mathbb{R})$, and given concretely as follows: identify $\mathbb{R}^d$ with the vector space $\mathbb{R}[X,Y]_{d-1}$ of real polynomials in two variables X, Y which are homogeneous of degree $d-1$. The group $\mathrm{SL}(2,\mathbb{R})$ acts on $\mathbb{R}[X,Y]_{d-1}$ by

$$\begin{pmatrix} \mathtt{a} & \mathtt{b} \\ \mathtt{c} & \mathtt{d} \end{pmatrix} \cdot P\begin{pmatrix} X \\ Y \end{pmatrix} = P\left(\begin{pmatrix} \mathtt{a} & \mathtt{b} \\ \mathtt{c} & \mathtt{d} \end{pmatrix}^{-1} \begin{pmatrix} X \\ Y \end{pmatrix}\right),$$

and this defines an irreducible representation $\mathrm{SL}(2,\mathbb{R}) \to \mathrm{SL}(\mathbb{R}[X,Y]_{d-1}) \simeq \mathrm{SL}(d,\mathbb{R})$, which is injective if d is even, and has kernel $\{\pm \mathrm{I}\}$ if d is odd. It factors into an embedding $\tau_d : \mathrm{PSL}(2,\mathbb{R}) \hookrightarrow \mathrm{PSL}(d,\mathbb{R})$. In this setting, the following result was proved by Choi–Goldman [46] for $d = 3$, and by Labourie [109] and Fock–Goncharov [70] for general d (recall that the case $d = 2$ is due to Goldman [77]).

Theorem 4.3.7 *Let $\Gamma_0 = \pi_1(S)$ be a closed surface group and $\sigma_0 : \Gamma_0 \to \mathrm{PSL}(2,\mathbb{R})$ an injective and discrete representation. For any $d \geq 2$, the connected component of $\rho_0 := \tau_d \circ \sigma_0$ in $\mathrm{Hom}(\Gamma_0, \mathrm{PSL}(d,\mathbb{R}))$ consists entirely of injective and discrete representations.*

The image of this connected component in the $\mathrm{PSL}(d,\mathbb{R})$-character variety of Γ_0 had previously been studied by Hitchin [92], and is now known as the *Hitchin component*. The corresponding representations are called *Hitchin representations*.

Rough sketch of the proofs of Theorem 4.3.7 The proof of Choi–Goldman [46] for $d = 3$ is geometric. The point is that the group $\tau_3(\mathrm{PSL}(2,\mathbb{R})) \simeq \mathrm{SO}(2,1)_0$ preserves a nondegenerate symmetric bilinear form $\langle \cdot, \cdot \rangle_{2,1}$ of signature $(2,1)$ on $\mathbb{R}^3$; in particular, it preserves the open subset

$$\Omega = \{[v] \in \mathbb{P}(\mathbb{R}^3) \mid \langle v, v \rangle_{2,1} < 0\}$$

of the projective plane $\mathbb{P}(\mathbb{R}^3)$, which is a model for the hyperbolic plane $\mathbb{H}^2$ (see (4.1)). This set Ω is *properly convex*: it is convex and bounded in some affine chart of $\mathbb{P}(\mathbb{R}^3)$ (e.g. it is the open unit disk in the affine chart $\{v_3 = 1\}$, see Figure 4.7, left). The group $\mathrm{PSL}(2,\mathbb{R})$ acts properly and transitively on Ω via τ_3, hence Γ_0 acts properly discontinuously with compact quotient on Ω via $\rho_0 = \tau_3 \circ \sigma_0$. By work of Koszul, the set of representations through which Γ_0 acts properly discontinuously with compact quotient on some nonempty properly convex open subset of $\mathbb{P}(\mathbb{R}^3)$ is open in $\mathrm{Hom}(\Gamma_0, \mathrm{PSL}(3,\mathbb{R}))$. Choi and Goldman proved that this set is also closed. Therefore the entire connected component of ρ_0 consists of such representations, and they are injective and discrete.

The proofs of Labourie and Fock–Goncharov for general d are dynamical. They involve two key objects. The first one is the *Gromov boundary*

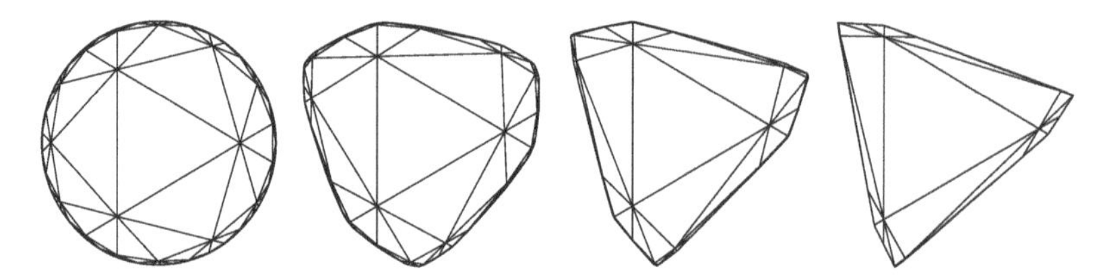

Figure 4.7 The left-most picture shows the projective model of $\mathbb{H}^2$ (an open disk in an affine chart of the projective plane $\mathbb{P}(\mathbb{R}^3)$), tiled by fundamental domains for the action of a triangle group T given by some injective and discrete representation $\rho_0 : T \to \mathrm{PO}(2,1) \subset \mathrm{PGL}(3,\mathbb{R})$. (Note that T admits a finite-index subgroup which is a closed surface group $\Gamma_0 = \pi_1(S)$ as in the proof of Theorem 4.3.7.) The other pictures show the effect of a continuous deformation of ρ_0 in $\mathrm{Hom}(T, \mathrm{PGL}(3,\mathbb{R}))$: the open disk deforms into an invariant properly convex open subset of $\mathbb{P}(\mathbb{R}^3)$ which is not a disk anymore, and the tiling deforms as the action of T remains properly discontinuous and cocompact. These pictures are taken from [16].

$\partial_\infty \Gamma_0$ of Γ_0: by definition, this is the visual boundary of a proper geodesic metric space on which Γ_0 acts properly discontinuously, by isometries, with compact quotient; in our situation, Γ_0 is a closed surface group and $\partial_\infty \Gamma_0$ is the visual boundary of $\mathbb{H}^2$, namely a circle. The second key object is the space $\mathrm{Flags}(\mathbb{R}^d)$ of full flags $(V_1 \subset \cdots \subset V_{d-1} \subset \mathbb{R}^d)$ of $\mathbb{R}^d$ (where each V_i is an i-dimensional linear subspace of $\mathbb{R}^d$); this space $\mathrm{Flags}(\mathbb{R}^d)$ is compact with a transitive action of $G = \mathrm{PSL}(d,\mathbb{R})$, and may be thought of as a kind of "boundary" for G or its symmetric space. The point of the proof is then to show that for any Hitchin representation $\rho : \Gamma_0 \to G$, there exists a continuous, injective, ρ-equivariant "boundary map" $\xi : \partial_\infty \Gamma_0 \to \mathrm{Flags}(\mathbb{R}^d)$. See Figure 4.10. (By *$\rho$-equivariant* we mean that $\xi(\gamma \cdot w) = \rho(\gamma) \cdot \xi(w)$ for all $\gamma \in \Gamma_0$ and all $w \in \partial_\infty \Gamma_0$.) The existence of such a boundary map ξ easily implies that ρ is injective and discrete: see the proof of Lemma 4.5.3.(3). Indeed, the idea is that the continuous, injective, equivariant map ξ "transfers", to $\mathrm{Flags}(\mathbb{R}^d)$, the dynamics of the intrinsic action of Γ_0 on $\partial_\infty \Gamma_0$, which is a so-called *convergence action*: any sequence $(\gamma_k)_{k\in\mathbb{N}}$ of pairwise distinct elements of Γ_0 comes with some contraction in $\partial_\infty \Gamma_0$, hence (using ξ) the sequence $(\rho(\gamma_k))_{k\in\mathbb{N}}$ comes with some contraction in $\mathrm{Flags}(\mathbb{R}^d)$, and this prevents $(\rho(\gamma_k))_{k\in\mathbb{N}}$ from converging to the identity element of G.

We note that the existence of continuous, injective, equivariant boundary maps for Hitchin representations is obtained by an open-and-closed argument, using stronger properties satisfied by these maps (namely, some uniform forms of contraction and transversality for Labourie, and a positivity property for Fock and Goncharov). □

Maximal representations

For $n \geq 2$, consider the embedding $\tau : G' = \mathrm{PSL}(2,\mathbb{R}) \simeq \mathrm{SO}(2,1)_0 \hookrightarrow \mathrm{SO}(2,n)$. Then the entire connected component of ρ_0 in $\mathrm{Hom}(\Gamma_0, G)$ consists of injective and discrete representations, as was proved by Burger, Iozzi, and Wienhard [38]. This is an example of a so-called *maximal component*: it consists of representations (called *maximal representations*) that maximise the *Toledo invariant* (a topological invariant generalising the Euler number, see e.g. [39, § 5.1]).

4.3.4 Higher Teichmüller theory

We already encountered in Sections 4.2.4 and 4.3.1 the Teichmüller space of a closed surface S of genus ≥ 2. It is a fundamental object in many areas of mathematics, which can be viewed both as a moduli space for marked complex structures on S or, via the Uniformisation Theorem, as a moduli space for marked hyperbolic structures on S. In this second point of view, the holonomy representation of the fundamental group $\Gamma_0 = \pi_1(S)$ naturally realises the Teichmüller space of S as a connected component of the G-character variety of Γ_0 for $G = \mathrm{PSL}(2,\mathbb{R})$, corresponding to the image, modulo conjugation by G at the target, of a connected component of $\mathrm{Hom}(\Gamma_0, G)$ consisting entirely of injective and discrete representations.

An interesting and perhaps surprising phenomenon, which has led to a considerable amount of research in the past twenty years, is that for certain semisimple Lie groups G of higher real rank, there also exist connected components of $\mathrm{Hom}(\Gamma_0, G)$ consisting entirely of injective and discrete representations, and which are nontrivial in the sense that they are not reduced to a single representation and its conjugates by G. The images in the G-character variety of these components are now called *higher(-rank) Teichmüller spaces*. We saw two examples in Section 4.3.3:

- *Hitchin components* when G is $\mathrm{PSL}(d,\mathbb{R})$, or more generally a real split simple Lie group; these are by definition components containing a Fuchsian representation $\rho_0 : \Gamma_0 \hookrightarrow \mathrm{PSL}(2,\mathbb{R}) \hookrightarrow G$, where $\mathrm{PSL}(2,\mathbb{R}) \hookrightarrow G$ is the so-called *principal embedding*;
- *maximal components* when G is $\mathrm{SO}(2,n)$, or more generally a simple Lie group of Hermitian type; these are by definition components of representations that maximise the Toledo invariant.

See [39, 41, 103, 120, 134] for details about these examples.

Towards a full list of higher Teichmüller spaces

Recently, new higher Teichmüller spaces were discovered in [22, 33, 83] when G is $\mathrm{O}(p,q)$ with $p \neq q$ or an exceptional simple Lie group whose restricted root system is of type F_4. These higher Teichmüller spaces consist of so-called Θ*-positive* representations, introduced by Guichard and Wienhard [85, 86]. Notions of positivity for Hitchin representations and maximal representations had been previously found by Fock–Goncharov [70] (based on Lusztig's total positivity [111]) and Burger–Iozzi–Wienhard [38]; the notion of Θ-positivity encompasses them both. Together with Hitchin components and maximal components, these new Θ-positive components conjecturally (see [85]) form the full list of higher Teichmüller spaces.

Without entering into technical details, let us mention briefly the role of Higgs bundles in this conjectural classification. See [32, 34, 74] for details.

Let Σ be a Riemann surface homeomorphic to S. By definition, a *G-Higgs bundle* over Σ is a pair (E, φ) where E is a holomorphic $K_{\mathbb{C}}$-bundle over Σ and φ (the *Higgs field*) is a holomorphic section of a certain natural bundle over Σ associated to E. (Here $K_{\mathbb{C}}$ is the complexification of a maximal compact subgroup K of G.) The *non-Abelian Hodge correspondence* of Hitchin, Donaldson, Corlette, Simpson, and others (see [75]), gives a homeomorphism between the G-character variety of $\pi_1(S)$ and the moduli space $\mathcal{M}_G(\Sigma)$ of so-called *polystable* G-Higgs bundles over Σ. This was used by Hitchin to define and study the Hitchin component.

Some of the connected components of the G-character variety of $\pi_1(S)$ can be distinguished using topological invariants. However, such invariants are not sufficient to distinguish them all in general. One fruitful approach is to use the fact, proved by Hitchin, that $(E, \varphi) \mapsto \|\varphi\|^2_{L^2(\Sigma)}$ defines a proper Morse function f from $\mathcal{M}_G(\Sigma)$ to $\mathbb{R}_{\geq 0}$; therefore, the connected components of $\mathcal{M}_G(\Sigma)$ can be studied by examining the local minima of f. The zero locus $f^{-1}(0)$ of f corresponds, in the G-character variety, to representations of $\pi_1(S)$ whose image lies in a compact subgroup of G; in particular, these representations are not injective and discrete, and so connected components for which f has a local minimum of 0 cannot be higher Teichmüller spaces. This approach has already been successfully exploited to find and count almost all connected components of the G-character variety of $\pi_1(S)$ for simple G, including conjecturally all higher Teichmüller spaces: see [32, 33, 34, 74].

Similarities with the classical Teichmüller space

The study of higher Teichmüller spaces, or *higher Teichmüller theory*, has been very active in the past twenty years. In particular, striking similarities have been found between higher Teichmüller spaces and the classical Teichmüller space of S, including:

- associated notions of positivity (see above);
- for Hitchin components: the topology of $\mathbb{R}^{\dim(G)|\chi(S)|}$ (Hitchin);
- the proper discontinuity of the action of the mapping class group (Labourie, Wienhard);
- good systems of coordinates (Goldman, Fock–Goncharov, Bonahon–Dreyer, Strubel, Zhang);
- analytic Riemannian metrics invariant under the mapping class group (Bridgeman–Canary–Labourie–Sambarino, Pollicott–Sharp);
- natural maps to the space of geodesic currents on S (Labourie, Bridgeman–Canary–Labourie–Sambarino, Martone–Zhang, Ouyang–Tamburelli);
- versions of the collar lemma for the associated locally symmetric spaces (Lee–Zhang, Burger–Pozzetti, Beyrer–Pozzetti, Beyrer–Guichard–Labourie–Pozzetti–Wienhard);
- interpretations of higher Teichmüller spaces as moduli spaces of geometric structures on S or on closed manifolds fibering over S (Choi–Goldman, Guichard–Wienhard, Collier–Tholozan–Toulisse).

There are also conjectural interpretations of higher Teichmüller spaces as moduli spaces of "higher complex structures" on S (Fock–Thomas), as well as various approaches to see higher Teichmüller spaces as mapping-class-group-equivariant fiber bundles over the classical Teichmüller space of S (Labourie, Loftin, Alessandrini–Collier, Collier–Tholozan–Toulisse). We refer to [39, 103, 120, 134] for more details and references.

Higher higher Teichmüller spaces

Phenomena analogous to Theorem 4.3.7 have also been uncovered for fundamental groups of higher-dimensional manifolds, in two situations.

The first one is in the context of *convex projective geometry*, which is by definition the study of properly convex open subsets Ω of real projective spaces $\mathbb{P}(\mathbb{R}^d)$, as in Choi–Goldman's proof of Theorem 4.3.7 for $d = 3$. Let $\Gamma_0 = \pi_1(M)$ where M is a closed topological manifold of dimension $n \geq 2$. Generalising Theorem 4.3.7, Benoist [17] proved that if Γ_0 does not contain an infinite nilpotent normal subgroup, then the set of representations through which Γ_0 acts properly discontinuously with compact quotient on some properly convex open subset of $\mathbb{P}(\mathbb{R}^{n+1})$ is closed in $\mathrm{Hom}(\Gamma_0, G)$ for $G = \mathrm{PGL}(n+1, \mathbb{R})$. This set is also open in $\mathrm{Hom}(\Gamma_0, G)$ by Koszul, and so it is a union of connected components of $\mathrm{Hom}(\Gamma_0, G)$. It consists entirely of injective and discrete representations. Recent results of Marseglia and Cooper–Tillman extend this to some cases where M and the quotients of the properly convex sets are not necessarily closed (see [50]).

The second situation is in the context of *pseudo-Riemannian hyperbolic geometry*, which is by definition the study of pseudo-Riemannian manifolds (i.e. smooth manifolds with a smooth assignment, to each tangent space, of a nondegenerate quadratic form) which have constant negative sectional curvature. In signature (p, q), such manifolds are locally modeled on the pseudo-Riemannian symmetric space $\mathbb{H}^{p,q} = \mathrm{PO}(p, q+1)/\mathrm{P}(\mathrm{O}(p) \times \mathrm{O}(q+1))$, which can be realised as an open set in projective space, namely $\{[v] \in \mathbb{P}(\mathbb{R}^{p+q+1}) \mid \langle v, v\rangle_{p,q+1} < 0\}$ where $\langle \cdot, \cdot \rangle_{p,q+1}$ is a symmetric bilinear form of signature $(p, q+1)$ on $\mathbb{R}^{p+q+1}$. For $q = 0$ we recover the real hyperbolic space $\mathbb{H}^p$, with its projective model (4.1), and for $q = 1$ the space $\mathbb{H}^{p,1}$ is the $(p+1)$-dimensional *anti-de Sitter space* (a Lorentzian analogue of the real hyperbolic space). Let $\Gamma_0 = \pi_1(M)$ where M is a closed hyperbolic p-manifold, with holonomy $\sigma_0 : \Gamma_0 \to \mathrm{O}(p, 1)$, and let $\tau : \mathrm{O}(p, 1) \hookrightarrow G = \mathrm{PO}(p, q+1)$ be the standard embedding. For $q = 1$, Barbot [9] proved that the connected component of $\rho_0 = \tau \circ \sigma_0$ in $\mathrm{Hom}(\Gamma_0, G)$ consists entirely of injective and discrete representations (corresponding to holonomies of so-called *globally hyperbolic spatially compact* anti-de Sitter manifolds, studied in [114] for $p = 2$). This was recently extended in [21] to general $p \geq 2$ and $q \geq 1$. In fact, the following more general result is proved in [21]: for $\Gamma_0 = \pi_1(M)$ where M is any closed topological manifold of dimension $p \geq 2$, the set of so-called $\mathbb{H}^{p,q}$*-convex cocompact* representations is a union of connected components in $\mathrm{Hom}(\Gamma_0, G)$. These $\mathbb{H}^{p,q}$-convex cocompact representations are injective and discrete representations with a nice geometric behaviour in $\mathbb{H}^{p,q}$ (see Section 4.5.2); they include the representations $\tau \circ \sigma_0 : \Gamma_0 = \pi_1(M) \to \mathrm{O}(p, 1) \hookrightarrow G = \mathrm{PO}(p, q+1)$ above where M is a closed hyperbolic manifold, but also other examples where M can be quite "exotic" (see [110, 115] for $q = 1$). These representations can have Zariski-dense image in G: see e.g. [21] for a bending argument as in Section 4.3.1.

In these two situations, there are connected components in $\mathrm{Hom}(\Gamma_0, G)$ consisting entirely of injective and discrete representations, where Γ_0 is the fundamental group of an n-dimensional closed manifold with $n > 2$

and G is a semisimple Lie group with $\mathrm{rank}_{\mathbb{R}}(G) \geq 2$. It is natural to call *higher-dimensional higher-rank Teichmüller spaces* (or *higher higher Teichmüller spaces* for short) the images of these components in the G-character variety of Γ_0. It would be interesting in the future to investigate whether these higher higher Teichmüller spaces have any topological or geometric analogies with classical Teichmüller space or its higher-rank counterparts, as above. See also [134, § 14] for some further discussion.

4.4 Classes of discrete subgroups in real rank one

In Section 4.3 we saw various examples of "flexible" infinite discrete subgroups of semisimple Lie groups. We now present some general theory in which these examples fit, first in real rank one (this section), then in higher real rank (Section 4.5).

More precisely, throughout this section we consider a semisimple Lie group G with $\mathrm{rank}_{\mathbb{R}}(G) = 1$. We discuss two important classes of finitely generated discrete subgroups of G that have received considerable attention, namely convex cocompact subgroups and geometrically finite subgroups. The inclusion relations between these classes and lattices of G are shown in Figure 4.8.

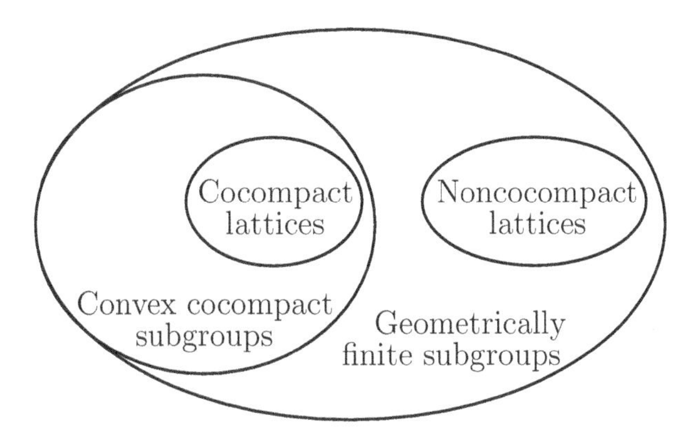

Figure 4.8 Inclusions between four important classes of discrete subgroups of G for $\mathrm{rank}_{\mathbb{R}}(G) = 1$

4.4.1 Definitions

Consider, as in Section 4.2.1, the Riemannian symmetric space $X = G/K$, where K is a maximal compact subgroup of G. If $G = \mathrm{SO}(n,1)$ (resp. $\mathrm{SU}(n,1)$, resp. $\mathrm{Sp}(n,1)$), then X is the n-dimensional hyperbolic space over $\mathbb{R}$ (resp. $\mathbb{C}$, resp. the quaternions). If G is the exceptional group $F_{4(-20)}$, then X is the "hyperbolic plane over the octonions".

There is a natural notion of convexity in X: any two points x, y of X are joined by a unique geodesic segment; we say that a subset $\mathcal{C}$ of X is convex if this segment is contained in $\mathcal{C}$ for all $x, y \in X$. See [66, § 1.6] for more details. For any $\varepsilon > 0$ and any subset $\mathcal{C}$ of X, we denote by $\mathcal{U}_\varepsilon(\mathcal{C})$ the uniform ε-neighbourhood of $\mathcal{C}$ in X.

Definition 4.4.1 Suppose $\mathrm{rank}_{\mathbb{R}}(G) = 1$. A discrete subgroup Γ of G is *convex cocompact* (resp. *geometrically finite*) if it is finitely generated and there is a nonempty Γ-invariant convex subset $\mathcal{C}$ of X such that the quotient $\Gamma\backslash\mathcal{C}$ is compact (resp. the quotient $\Gamma\backslash\mathcal{U}_\varepsilon(\mathcal{C})$ has finite volume for some $\varepsilon > 0$).

Alternatively (see Remark 4.3.1), given a group Γ_0, we say that a representation $\rho : \Gamma_0 \to G$ is *convex cocompact* (resp. *geometrically finite*) if it has finite kernel and discrete, convex cocompact (resp. geometrically finite) image.

Remark 4.4.2 If there is a nonempty Γ-invariant convex subset $\mathcal{C}$ of X such that $\Gamma\backslash\mathcal{C}$ is compact, then Γ is automatically finitely generated, by the Švarc–Milnor lemma (see e.g. [65, Th. 8.37]). Thus the assumption that Γ be finitely generated can be omitted in the definition of convex cocompactness. On the other hand, this assumption cannot be omitted in general in the definition of geometric finiteness: see [88].

Remark 4.4.3 Bowditch [31] gave several equivalent definitions of geometric finiteness. Here we use a variation on his definition F5, where the uniform bound on the orders of finite subgroups of Γ is replaced by the assumption that Γ be finitely generated. The two definitions are equivalent by [31] and the Selberg lemma [125, Lem. 8].

We now explain how Definition 4.4.1 can be rephrased in terms of a specific convex set in G/K. For this, we first recall the important notion of the *limit set* of a discrete subgroup of G.

Limit sets and convex cores

Let $\partial_\infty X$ be the visual boundary of $X = G/K$, i.e. the set of equivalence classes of geodesic rays in X for the equivalence relation "to remain at bounded distance". There is a natural topology on $\overline{X} := X \sqcup \partial_\infty X$ that extends that of X and makes $\overline{X}$ compact, and the action of G on X extends continuously to $\overline{X}$ (see e.g. [66, § 1.7]). For instance, as in Sections 4.3.1 and 4.3.4, if $G = \mathrm{PO}(n,1)$ and $X = \mathbb{H}^n$, then we can realise X as the open subset (4.1) of $\mathbb{P}(\mathbb{R}^{n+1})$ where some quadratic form of signature $(n,1)$ is negative, and $\overline{X}$ is then the closed subset of $\mathbb{P}(\mathbb{R}^{n+1})$ where the quadratic form is nonpositive, endowed with the topology from $\mathbb{P}(\mathbb{R}^{n+1})$ and the natural action of $G = \mathrm{PO}(n,1)$.

Definition 4.4.4 Let Γ be a discrete subgroup of G. The *limit set* of Γ is the set Λ_Γ of accumulation points in $\overline{X}$ of a Γ-orbit of X; it is contained in $\partial_\infty X$ and does not depend on the choice of Γ-orbit. The *convex core* $\mathcal{C}_\Gamma^{\mathsf{cor}} \subset X$ of Γ is the convex hull of Λ_Γ in X (i.e. the smallest closed convex subset of X whose closure in $\overline{X}$ contains Λ_Γ).

Note that Λ_Γ and $\mathcal{C}_\Gamma^{\mathsf{cor}}$ are both invariant under the action of Γ on $\overline{X}$.

The limit set Λ_Γ is nonempty if and only if Γ is infinite. This set has either at most two elements (in which case we say Γ is *elementary*), or infinitely many. If Γ is not elementary, then the action of Γ on Λ_Γ is minimal (all orbits are dense), and any nonempty Γ-invariant closed subset of $\partial_\infty X$ contains Λ_Γ (see e.g. [31, § 3.2]); in particular, any nonempty Γ-invariant closed convex subset of X contains the convex core $\mathcal{C}_\Gamma^{\mathsf{cor}}$. We deduce the following.

Fact 4.4.5 *Suppose* $\mathrm{rank}_{\mathbb{R}}(G) = 1$. *A finitely generated infinite discrete subgroup Γ of G is convex cocompact (resp. geometrically finite) if and only if the quotient $\Gamma\backslash\mathcal{C}_\Gamma^{\mathsf{cor}}$ is compact and nonempty (resp. the quotient $\Gamma\backslash\mathcal{U}_\varepsilon(\mathcal{C}_\Gamma^{\mathsf{cor}})$ has finite volume for some $\varepsilon > 0$).*

Remark 4.4.6 In our setting where $\mathrm{rank}_{\mathbb{R}}(G) = 1$, the group G acts transitively on $\partial_\infty X$. The stabilisers in G of points of $\partial_\infty X$ are the *proper parabolic subgroups* of G. Thus $\partial_\infty X$ is G-equivariantly homeomorphic to G/P where P is a proper parabolic subgroup of G.

4.4.2 Properties

Let us briefly mention a few useful properties of geometrically finite and convex cocompact representations.

Domains of discontinuity

We first observe that any discrete subgroup Γ of G acts properly discontinuously on the open subset $\Omega_\Gamma := \partial_\infty X \smallsetminus \Lambda_\Gamma$ of $\partial_\infty X$, and in fact on

$X \cup \Omega_\Gamma$. Indeed, let $\mathcal{C}$ be a nonempty Γ-invariant closed convex subset of X. One can check that the closest point projection from X to $\mathcal{C}$ extends to a continuous Γ-equivariant map from $X \cup \Omega_\Gamma$ to $\mathcal{C}$. The fact that Γ acts properly discontinuously on $\mathcal{C} \subset X$ then implies that Γ acts properly discontinuously on $X \cup \Omega_\Gamma$.

If Γ is convex cocompact, then the quotient $\Gamma\backslash\Omega_\Gamma$ is compact (possibly empty), and $\Gamma\backslash(X \cup \Omega_\Gamma)$ is a compactification of $\Gamma\backslash X$.

If Γ is geometrically finite, then $\Gamma\backslash(X \cup \Omega_\Gamma)$ is not necessarily compact, but it has only finitely many topological ends, each of which is a "parabolic end"; this actually characterises geometric finiteness: see [31].

Deformations

Convex cocompactness is stable under small deformations:

Fact 4.4.7 *Suppose* $\mathrm{rank}_{\mathbb{R}}(G) = 1$. *For any finitely generated group* Γ_0, *the space of convex cocompact representations is open in* $\mathrm{Hom}(\Gamma_0, G)$.

On the other hand, geometric finiteness is in general not stable under small deformations. If one restricts to small deformations that are *cusp-preserving* (i.e. that keep parabolic elements parabolic), then stability holds for $G = \mathrm{PO}(n,1)$ when $n \leq 3$ or when all cusps have rank $\geq n-2$, but not in general. See e.g. [82, App. B] for more details and references.

Homomorphisms

Convex cocompactness behaves well under Lie group homomorphisms:

Fact 4.4.8 *Suppose* $\mathrm{rank}_{\mathbb{R}}(G) = 1$. *Let* G' *be another semisimple Lie group with* $\mathrm{rank}_{\mathbb{R}}(G') = 1$ *and let* $\tau : G' \to G$ *be a Lie group homomorphism with compact kernel. For any finitely generated group* Γ_0 *and any representation* $\sigma_0 : \Gamma_0 \to G'$, *the composed representation* $\tau \circ \sigma_0 : \Gamma_0 \to G$ *is convex cocompact if and only if* σ_0 *is.*

4.4.3 Examples

- If Γ is a lattice in G, then $\Lambda_\Gamma = \partial_\infty X$ and $\mathcal{C}_\Gamma^{\mathsf{cor}} = X$, and Γ is geometrically finite. If Γ is cocompact in G, then it is convex cocompact.
- Suppose $G = \mathrm{PSL}(2,\mathbb{R}) \simeq \mathrm{PO}(2,1)_0$. Then every finitely generated discrete subgroup Γ of G is geometrically finite; Γ is convex cocompact if and only if the associated hyperbolic surface $\Gamma\backslash\mathbb{H}^2$ has no cusps.

Remark 4.4.9 On the other hand, for $G = \mathrm{PO}(n,1)$ with $n \geq 3$, there exist finitely generated discrete subgroups of G which are *not* geometrically finite. The first examples were given by Bers for $n = 3$ ("singly degenerate" Kleinian groups, for which the domain of discontinuity Ω_Γ is simply connected): see [95, § 2].

- Any discrete subgroup of $G = \mathrm{PO}(n,1)$ generated by the orthogonal reflections in the faces of a finite-sided right-angled polyhedron of $\mathbb{H}^n$ is geometrically finite; it is convex cocompact if and only if no distinct facets of the polyhedron have closures meeting in $\partial_\infty \mathbb{H}^n$ (see [60, § 4]).
- The Schottky groups of Section 4.3.1 are geometrically finite; the strong Schottky groups (for which $B_1^\pm, \dots, B_m^\pm$ have pairwise disjoint closures) are convex cocompact. Their limit sets are Cantor sets. The set Ω of Remark 4.3.3 is the domain of discontinuity $\Omega_\Gamma = \partial_\infty X \smallsetminus \Lambda_\Gamma$ of Γ in $\partial_\infty X$ from Section 4.4.2.
- Any quasi-Fuchsian group $\Gamma = \rho(\pi_1(S))$ as in Section 4.3.1 is convex cocompact. The limit set Λ_Γ is a topological circle in $\partial_\infty \mathbb{H}^3$ (see Figure 4.3). The quotient $\Gamma \backslash \mathcal{C}_\Gamma^{\mathsf{cor}}$ is homeomorphic to $S \times [0,1]$.
- The small deformations of cocompact lattices of $G' = \mathrm{SO}(n,1)$ inside $G = \mathrm{SO}(n+1,1)$ from Section 4.3.1 are convex cocompact by Facts 4.4.7 and 4.4.8 (see also Remark 4.3.4.(2)).

4.4.4 A few characterisations of convex cocompactness

Preliminaries

Given a finitely generated group Γ_0, we choose a finite generating subset F of Γ_0 and denote by $\mathrm{Cay}(\Gamma_0) = \mathrm{Cay}(\Gamma_0, F)$ the corresponding Cayley graph, with its metric $\mathsf{d}_{\mathrm{Cay}(\Gamma_0)}$.

As in Section 4.2.3, a group Γ_0 is called *Gromov hyperbolic* if it is finitely generated and acts properly discontinuously, by isometries, with compact quotient, on some Gromov hyperbolic proper geodesic metric space Y; in that case, we can take Y to be $\mathrm{Cay}(\Gamma_0)$. As in the proof of Theorem 4.3.7, the *Gromov boundary* $\partial_\infty \Gamma_0$ of Γ_0 is then the visual boundary of Y, endowed with the action of Γ_0 extending that on Y. The Gromov boundary $\partial_\infty \Gamma_0$ does not depend on Y up to Γ_0-equivariant homeomorphism. An important property is that the action of Γ_0 on $\partial_\infty \Gamma_0$ is a *convergence action*: for any sequence $(\gamma_k)_{k \in \mathbb{N}}$ of pairwise distinct elements of Γ_0, up to passing to a subsequence, there exist $w^+, w^- \in \partial_\infty \Gamma_0$ such that $\gamma_k \cdot w \to w^+$ for all $w \in \partial_\infty \Gamma_0 \smallsetminus \{w^-\}$, uniformly on compact sets. Moreover, any infinite-order element of Γ_0 has two fixed points in $\partial_\infty \Gamma_0$, one attracting and one repelling. The group Γ_0 is called *elementary* if it is finite (in which case $\partial_\infty \Gamma_0$ is empty) or if it admits a finite-index subgroup which is cyclic (in which case $\partial_\infty \Gamma_0$ consists of two points). If Γ_0 is not elementary, then the set of attracting fixed points of infinite-order elements of Γ_0 is infinite and dense in $\partial_\infty \Gamma_0$. See e.g. [14] for details.

Examples 4.4.10 If Γ_0 is a nonabelian free group with finite free generating subset F, then Γ_0 is Gromov hyperbolic, $\mathrm{Cay}(\Gamma_0)$ is a tree, and $\partial_\infty \Gamma_0$ is a Cantor set. If $\Gamma_0 = \pi_1(M)$ for some closed negatively-curved manifold M, then Γ_0 is Gromov hyperbolic, we can take Y to be the universal cover $\widetilde{M}$ of M, and $\partial_\infty \Gamma_0 = \partial_\infty \widetilde{M}$. In particular, if $\Gamma_0 = \pi_1(S)$ for some closed orientable surface of genus ≥ 2, then Γ_0 is Gromov hyperbolic and $\partial_\infty \Gamma_0$ is a circle (as in the proof of Theorem 4.3.7).

Remark 4.4.11 A Gromov hyperbolic group can never contain a subgroup isomorphic to $\mathbb{Z}^2$ or to a Baumslag–Solitar group $BS(m,n) := \langle a, t \,|\, t^{-1} a^m t = a^n \rangle$. Understanding how close this is to characterising Gromov hyperbolic groups is an important question in geometric group theory: see e.g. [76].

For any isometry g of a metric space (M, d_M), we define the *translation length* of g in M to be

$$\mathrm{transl}_M(g) := \inf_{m \in M} \mathsf{d}_M(m, g \cdot m) \geq 0. \tag{4.2}$$

Finally, we denote by d_X the metric on the Riemannian symmetric space $X = G/K$ (see Section 4.2.1). We fix a basepoint $x_0 \in X$, and a Riemannian metric $\mathsf{d}_{\partial_\infty X}$ on the visual boundary $\partial_\infty X$.

A few classical characterisations

Many interesting characterisations of convex cocompactness have been found by various authors including Beardon, Bowditch, Maskit, Sullivan, Thurston, Tukia, and others. We now give a few. We refer to [95, 96] for more details and references, as well as further characterisations (e.g. in terms of *conical limit points*). We also refer to [31, 95] for characterisations of geometric finiteness.

Theorem 4.4.12 *Suppose* $\mathrm{rank}_{\mathbb{R}}(G) = 1$. *For any infinite group* Γ_0 *and any representation* $\rho : \Gamma_0 \to G$, *the following are equivalent:*

(1) ρ *is convex cocompact (Definition 4.4.1);*
(2) Γ_0 *is finitely generated and* ρ *is a* quasi-isometric embedding*: there exist* $c, c' > 0$ *such that for any* $\gamma \in \Gamma$,

$$\mathsf{d}_X(x_0, \rho(\gamma) \cdot x_0) \geq c\, \mathsf{d}_{\mathrm{Cay}(\Gamma_0)}(e, \gamma) - c'; \tag{4.3}$$

(3) *Γ_0 is Gromov hyperbolic and ρ is* well-displacing*: there exist $c, c'' > 0$ such that for any $\gamma \in \Gamma$,*

$$\mathrm{transl}_X(\rho(\gamma)) \geq c\,\mathrm{transl}_{\mathrm{Cay}(\Gamma_0)}(\gamma) - c''; \tag{4.4}$$

(4) *Γ_0 is Gromov hyperbolic and there exists a ρ-equivariant map*

$$\xi : \partial_\infty \Gamma_0 \longrightarrow \partial_\infty X$$

which is continuous, injective, and dynamics-preserving *(i.e. for any infinite-order element $\gamma \in \Gamma_0$, the image by ξ of the attracting fixed point of γ in $\partial_\infty \Gamma_0$ is an attracting fixed point of $\rho(\gamma)$ in $\partial_\infty X$);*

(5) *Γ_0 is Gromov hyperbolic and there exists a ρ-equivariant map*

$$\xi : \partial_\infty \Gamma_0 \longrightarrow \partial_\infty X$$

which is continuous, injective, and strongly dynamics-preserving *(i.e. for any $(\gamma_k) \in \Gamma_0^{\mathbb{N}}$ and any $w^+, w^- \in \partial_\infty \Gamma_0$, if $\gamma_k \cdot w \to w^+$ for all $w \in \partial_\infty \Gamma_0 \smallsetminus \{w^-\}$, then $\rho(\gamma_k) \cdot z \to \xi(w^+)$ for all $z \in \partial_\infty X \smallsetminus \{\xi(w^-)\}$);*

(6) *ρ has finite kernel, discrete image, and the action of Γ on $\partial_\infty X$ via ρ is* expanding *at $\Lambda_{\rho(\Gamma_0)}$, i.e. for any $z \in \Lambda_{\rho(\Gamma_0)}$, there exist a neighbourhood $\mathcal{U}$ of z in $\partial_\infty X$ and an element $\gamma \in \Gamma_0$ such that*

$$\inf_{z_1 \neq z_2 \text{ in } \mathcal{U}} \frac{\mathsf{d}_{\partial_\infty X}(\rho(\gamma) \cdot z_1, \rho(\gamma) \cdot z_2)}{\mathsf{d}_{\partial_\infty X}(z_1, z_2)} > 1. \tag{4.5}$$

Remarks 4.4.13 • Using the triangle inequality, one sees that condition (2) does not depend on the choice of basepoint $x_0 \in X$ (changing x_0 may change the values of c, c' but not their existence).

- One also sees that for Γ_0 with finite generating subset F, the reverse inequality $\mathsf{d}_X(x_0, \rho(\gamma) \cdot x_0) \leq C\, \mathsf{d}_{\mathrm{Cay}(\Gamma_0)}(e, \gamma)$ to (4.3) holds for any representation $\rho : \Gamma_0 \to G$, with $C := \max_{f \in F} \mathsf{d}_X(x_0, \rho(f) \cdot x_0)$.
- In condition (3) we cannot remove the assumption that Γ_0 be Gromov hyperbolic: for instance, there exist finitely generated infinite groups Γ_0 with only finitely many conjugacy classes [118], and for such Γ_0 any representation $\rho : \Gamma_0 \to G$ is well-displacing.
- In condition (4), *dynamics-preserving* implies that for any $\gamma \in \Gamma_0$ of infinite order, $\rho(\gamma)$ is a *hyperbolic* element of G (i.e. an element with two fixed points in $\partial_\infty X$, one attracting and one repelling). In condition (5), *strongly dynamics-preserving* means that ξ preserves the convergence action of Γ_0 on $\partial_\infty \Gamma_0$ mentioned above.

Sketches of proofs

Proof of (1) ⇒ (2): We may assume that the basepoint x_0 belongs to the convex core $\mathcal{C}^{\mathsf{cor}}_{\rho(\Gamma_0)}$. By the Švarc–Milnor lemma (see e.g. [65, Th. 8.37]), if Γ_0 acts properly discontinuously, by isometries, with compact quotient, on a proper geodesic metric space M, then Γ_0 is finitely generated and any orbital map $\gamma \mapsto \gamma \cdot m$ is a quasi-isometric embedding: there exist $c, c' > 0$ such that $\mathtt{d}_M(m, \gamma \cdot m) \geq c\, \mathtt{d}_{\mathrm{Cay}(\Gamma_0)}(e, \gamma) - c'$ for all $\gamma \in \Gamma_0$. We apply this to the convex core $M = \mathcal{C}^{\mathsf{cor}}_{\rho(\Gamma_0)}$, endowed with the restriction of the metric $\mathtt{d}_X$. □

Proof of (2) ⇒ (1): Since $\mathtt{d}_X(x_0, \rho(\gamma) \cdot x_0) \to +\infty$ as $\mathtt{d}_{\mathrm{Cay}(\Gamma_0)}(e, \gamma) \to +\infty$, the representation ρ has finite kernel and discrete image.

The orbital map $\gamma \mapsto \rho(\gamma) \cdot x_0$ from Γ_0 to X extends to a map from $\mathrm{Cay}(\Gamma_0)$ to X sending edges of $\mathrm{Cay}(\Gamma_0)$ to geodesic segments of X. The fact that ρ is a quasi-isometric embedding implies the existence of $c, c' > 0$ such that any geodesic of $\mathrm{Cay}(\Gamma_0)$ is sent to a (c, c')-quasigeodesic in X, and the Morse lemma (see e.g. [65, Th. 11.40 & 11.105]) states that (c, c')-quasigeodesics are uniformly close to actual geodesics in X. Therefore the orbit $\rho(\Gamma_0) \cdot x_0$ is *quasiconvex*: there exists a uniform neighbourhood $\mathcal{U}$ of $\rho(\Gamma_0) \cdot x_0$ in X such that any geodesic segment between two points of $\rho(\Gamma_0) \cdot x_0$ is contained in $\mathcal{U}$. We conclude using the fact (see [31, Prop. 2.5.4]) that any quasiconvex subset of X lies at finite Hausdorff distance from its convex hull in X. □

In order to prove (2) ⇒ (3), we consider, for any metric space $(M, \mathtt{d}_M)$ and any isometry g of M, the *stable length*

$$\mathrm{length}^{\infty}_M(g) := \lim_k \frac{1}{k}\, \mathtt{d}_M(m, g^k \cdot m) \geq 0$$

of g. It is an easy exercise to check, using the triangle inequality, that this limit exists (because the sequence $(d_M(m, g^k \cdot m))_{k \in \mathbb{N}}$ is subadditive) and that it does not depend on the choice of $m \in M$. Note that

$$\mathrm{length}^{\infty}_M(g) \leq \mathrm{transl}_M(g). \tag{4.6}$$

Indeed, for any $m \in M$ and any $k \geq 1$ we have $\mathtt{d}_M(m, g^k \cdot m) \leq k\, \mathtt{d}_M(m, g \cdot m)$ by the triangle inequality. Dividing by k and passing to the limit yields $\mathrm{length}^{\infty}_M(g) \leq \mathtt{d}_M(m, g \cdot m)$, and we conclude by taking an infimum over all $m \in M$ on the right-hand side.

Proof of (2) ⇒ (3): Applying (4.3) to γ^k instead of γ, dividing by k, and passing to the limit yields $\mathrm{length}^{\infty}_X(\rho(\gamma)) \geq c\, \mathrm{length}^{\infty}_{\mathrm{Cay}(\Gamma_0)}(\gamma)$ for all

$\gamma \in \Gamma$. In order to obtain (4.4), it is sufficient to use (4.6) for $M = X$ and to check that

(i) for $M = \mathrm{Cay}(\Gamma_0)$, the inequality (4.6) is "almost" an equality: $\mathrm{length}^{\infty}_{\mathrm{Cay}(\Gamma_0)}(g) \geq \mathrm{transl}_{\mathrm{Cay}(\Gamma_0)}(g) - 8\delta$ where $\delta \geq 0$ is a hyperbolicity constant for $\mathrm{Cay}(\Gamma_0)$ (i.e. all triangles of $\mathrm{Cay}(\Gamma_0)$ are δ-thin).

Indeed, then (4.4) will hold with $c'' = 8\delta$. We note that actually

(ii) for $M = X = G/K$, the inequality (4.6) is an equality.

Indeed, (ii) is based on the fact that X is a *CAT(0) space*: any geodesic triangle of X is "at least as thin" as a triangle with the same side lengths in the Euclidean plane. Applying this to a geodesic triangle with vertices $m, g \cdot m, g^2 \cdot m$, we see that if m' is the midpoint of the geodesic segment $[m, g \cdot m]$ (so that $g \cdot m'$ is the midpoint of $[g \cdot m, g^2 \cdot m]$), then $\mathtt{d}_X(m', g \cdot m') \leq \mathtt{d}_X(m, g^2 \cdot m)/2$. By induction on k, we obtain that for any $m \in M$ and any $k \geq 1$, there exists $m_k \in M$ such that $\mathtt{d}_X(m, g^{2^k} \cdot m) \geq 2^k\, \mathtt{d}_X(m_k, g \cdot m_k) \geq 2^k\, \mathrm{transl}_M(g)$. We conclude by dividing by 2^k and passing to the limit.

(i) can be proved in a similar way, replacing the CAT(0) inequality $\mathtt{d}_X(m', g \cdot m') \leq \mathtt{d}_X(m, g^2 \cdot m)/2$ by the Gromov hyperbolicity inequality $\mathtt{d}_X(m', g \cdot m') \leq \mathtt{d}_X(m, g^2 \cdot m)/2 + 4\delta$ (see [51, Ch. 10, Prop. 5.1]). ☐

Proof of (3) ⇒ (2)*:* The Gromov hyperbolic group Γ_0 has the following property: there exist a finite subset S of Γ_0 and a constant $C' > 0$ such that for any $\gamma \in \Gamma_0$ we can find $s \in S$ with $\mathrm{transl}_{\mathrm{Cay}(\Gamma_0)}(s\gamma) \geq \mathtt{d}_{\mathrm{Cay}(\Gamma_0)}(e, \gamma) - C'$. (If Γ_0 is nonelementary, then we can take $S = \{\gamma_1^N, \gamma_1^{-N}, \gamma_2^N, \gamma_2^{-N}\}$ for some large N, where $\gamma_1, \gamma_2 \in \Gamma_0$ are infinite-order elements such that the attracting fixed points in $\partial_\infty \Gamma_0$ of γ_1, γ_1^{-1}, γ_2, and γ_2^{-1} are pairwise distinct: see e.g. [139, Lem. B.2].)

Given $\gamma \in \Gamma_0$, consider $s \in S$ as above. Applying (4.4) to $s\gamma$ yields

$$\mathrm{transl}_X(\rho(s\gamma)) \geq c\, \mathrm{transl}_{\mathrm{Cay}(\Gamma_0)}(s\gamma) - c'' \geq c\, \mathtt{d}_{\mathrm{Cay}(\Gamma_0)}(e, \gamma) - (c\,C' + c'').$$

To conclude, we observe that

$$\mathrm{transl}_X(g_1 g_2) \leq \mathtt{d}_X(x_0, g_1 g_2 \cdot x_0) \leq \mathtt{d}_X(x_0, g_1 \cdot x_0) + \mathtt{d}_X(x_0, g_2 \cdot x_0)$$

for all $g_1, g_2 \in G$. Applying this to $(g_1, g_2) = (\rho(s), \rho(\gamma))$, we obtain (4.3) with $c' = c\,C' + c'' + \max_{s' \in S} \mathtt{d}_X(x_0, \rho(s') \cdot x_0)$. ☐

Proof of (1) ⇒ (5)*:* We have seen in the proof of (1) ⇒ (2) that for any $m \in \mathcal{C}^{\mathsf{cor}}_{\rho(\Gamma_0)}$, the orbital map $\gamma \mapsto \rho(\gamma) \cdot m$ is a quasi-isometry from Γ_0 to $\mathcal{C}^{\mathsf{cor}}_{\rho(\Gamma_0)}$. It is a classical result in geometric group theory (see e.g.

[65, Th. 11.108]) that such a quasi-isometry extends to a Γ_0-equivariant homeomorphism ξ from $\partial_\infty\Gamma_0$ to $\partial_\infty\mathcal{C}^{\mathsf{cor}}_{\rho(\Gamma_0)}$. Here $\partial_\infty\mathcal{C}^{\mathsf{cor}}_{\rho(\Gamma_0)}$ is a subset of $\partial_\infty X$ (namely the intersection of $\partial_\infty X$ with the closure of $\mathcal{C}^{\mathsf{cor}}_{\rho(\Gamma_0)}$ in $\overline{X}$). Thus we can view ξ as a ρ-equivariant, continuous, injective map from $\partial_\infty\Gamma_0$ to $\partial_\infty X$, such that for any $(\gamma_k) \in \Gamma_0^{\mathbb{N}}$ and any $w^+, w^- \in \partial_\infty\Gamma_0$, if $\gamma_k \cdot w \to w^+$ for all $w \in \partial_\infty\Gamma_0 \smallsetminus \{w^-\}$, then $\rho(\gamma_k)\cdot z \to \xi(w^+)$ for all $z \in \partial_\infty\mathcal{C}^{\mathsf{cor}}_{\rho(\Gamma_0)} \smallsetminus \{\xi(w^-)\}$.

In order to see that this last convergence holds for all $z \in \partial_\infty X \smallsetminus \{\xi(w^-)\}$, one possibility is to use the fact (Cartan decomposition) that if we choose a point $x \in X$ and a geodesic line $\mathcal{G}$ of X through x with endpoints $z_0^+, z_0^- \in \partial_\infty X$, then any element $\rho(\gamma_k) \in G$ can be written as $\rho(\gamma_k) = \kappa_k a_k \kappa'_k$ where $\kappa_k, \kappa'_k \in G$ fix x and $a_k \in G$ is a pure translation along $\mathcal{G}$ towards z_0^+. The subgroup of G fixing x is compact (it is conjugate to K); therefore, up to passing to a subsequence we may assume that $(\kappa_k)_{k\in\mathbb{N}}, (\kappa'_k)_{k\in\mathbb{N}}$ converge respectively to some $\kappa, \kappa' \in G$. Since ρ has finite kernel and discrete image, we have $a_k \cdot z \to z_0^+$ for all $z \in \partial_\infty X \smallsetminus \{z_0^-\}$. Therefore $\rho(\gamma_k)\cdot z \to \kappa\cdot z_0^+$ for all $z \in \partial_\infty X \smallsetminus \{\kappa'^{-1}\cdot z_0^-\}$. Necessarily $\kappa \cdot z_0^+ = \xi(w^+)$ and $\kappa'^{-1}\cdot z_0^- = \xi(w^-)$. □

The implication (5) ⇒ (4) is immediate by considering, for any infinite-order element $\gamma \in \Gamma_0$, the sequence $(\gamma_k) := (\gamma^k) \in \Gamma_0^{\mathbb{N}}$. The implication (4) ⇒ (2) can be proved using flows as in Section 4.5.1 below (see Remark 4.5.5 and the implication (1) ⇒ (2) in Theorem 4.5.13).

Proof of (1) ⇒ (6)*:* We treat the case that X is $\mathbb{H}^n$, seen as the open unit ball of $\mathbb{R}^n$ for a Euclidean norm $\|\cdot\|$, that 0 belongs to $\mathcal{C}^{\mathsf{cor}}_{\rho(\Gamma_0)}$, and that $\mathsf{d}_{\partial_\infty X}$ is the metric induced by $\|\cdot\|$ on the unit sphere $\partial_\infty X$ of $\mathbb{R}^n$.

We first observe that for any element $g \in G$ that does not fix 0, the closed subset

$$\mathcal{H}_g := \{x \in X \mid \|x\| \leq \|g\cdot x\|\} = \{x \in X \mid \mathsf{d}_X(0,x) \leq \mathsf{d}_X(0, g\cdot x)\}$$

of X is bounded by the bisector between 0 and $g^{-1}\cdot 0$. Moreover, for any neighbourhood $\mathcal{V}$ in $\overline{X}$ of the closure of $\mathcal{H}_g$ in $\overline{X}$, the restriction of g to $\partial_\infty X \smallsetminus \mathcal{V}$ is uniformly expanding in the sense that

$$\inf_{z_1\neq z_2 \text{ in } \partial_\infty X\smallsetminus\mathcal{V}} \frac{\mathsf{d}_{\partial_\infty X}(g\cdot z_1, g\cdot z_2)}{\mathsf{d}_{\partial_\infty X}(z_1,z_2)} > 1.$$

Indeed, one can check this when g is a pure translation along a geodesic of X through 0, and then conclude using the fact (Cartan decomposition) that any $g \in G$ can be written as $g = \kappa a \kappa'$ where $a \in G$ is such a pure translation and $\kappa, \kappa' \in G$ fix 0 and preserve $\|\cdot\|$.

Consider the Dirichlet domain of $\mathcal{C}^{\mathsf{cor}}_{\rho(\Gamma_0)}$ centred at 0:

$$\mathcal{D} = \bigcap_{\gamma\in\Gamma_0} \mathcal{H}_{\rho(\gamma)} \cap \mathcal{C}^{\mathsf{cor}}_{\rho(\Gamma_0)}.$$

It is compact by (1). Since Γ_0 acts properly discontinuously on $\mathcal{C}^{\mathsf{cor}}_{\rho(\Gamma_0)}$ via ρ, the set $\mathcal{F}$ of elements $\gamma \in \Gamma_0$ such that $\mathcal{D} \cap \rho(\gamma) \cdot \mathcal{D} \neq \emptyset$ and $\rho(\gamma) \cdot 0 \neq 0$ is finite. One easily checks that $\mathcal{D} = \bigcap_{\gamma\in\mathcal{F}} \mathcal{H}_{\rho(\gamma)} \cap \mathcal{C}^{\mathsf{cor}}_{\rho(\Gamma_0)}$. For each $\gamma \in \mathcal{F}$, let $\mathcal{V}_{\rho(\gamma)}$ be a closed neighbourhood in $\overline{X}$ of the closure of $\mathcal{H}_{\rho(\gamma)}$ in $\overline{X}$. If we choose these neighbourhoods small enough, then $\mathcal{D}' := \bigcap_{\gamma\in\mathcal{F}} \mathcal{V}_{\rho(\gamma)} \cap \mathcal{C}^{\mathsf{cor}}_{\rho(\Gamma_0)}$ is still a compact subset of X, and so $\Lambda_{\rho(\Gamma_0)} \subset \bigcup_{\gamma\in\mathcal{F}} (\partial_\infty X \smallsetminus \mathcal{V}_{\rho(\gamma)})$. We conclude using the fact, observed above, that (4.5) holds for $\mathcal{U} := \partial_\infty X \smallsetminus \mathcal{V}_{\rho(\gamma)}$ for each $\gamma \in \mathcal{F}$. □

Proof of (6) $\Rightarrow$ (1)*:* We again treat the case that X is $\mathbb{H}^n$, seen as the open unit ball of $\mathbb{R}^n$ for a Euclidean norm $\|\cdot\|$, and that the metric $\mathsf{d}_{\partial_\infty X}$ is induced by $\|\cdot\|$. We denote by $\mathsf{d}_{\mathrm{Euc}}$ the Euclidean distance on $\mathbb{R}^n$ associated to $\|\cdot\|$.

Suppose that (6) holds. Then $\Lambda_{\rho(\Gamma_0)}$ contains at least two points. (Indeed, by assumption $\rho(\Gamma_0)$ is an infinite discrete subgroup of G, hence $\Lambda_{\rho(\Gamma_0)}$ is nonempty; moreover, the expansion assumption prevents $\Lambda_{\rho(\Gamma_0)}$ from being a singleton, as follows e.g. from the classification of elementary discrete subgroups of G: see [31, Prop. 3.2.1].) Therefore $\mathcal{C}^{\mathsf{cor}}_{\rho(\Gamma_0)}$ is nonempty. Moreover, one can check (e.g. using the Cartan decomposition as in the proof of (1) $\Rightarrow$ (6) just above) that for any $z \in \Lambda_{\rho(\Gamma_0)}$, there exist a neighbourhood $\mathcal{U}$ of z in $\mathbb{R}^n$ (rather than just $\partial_\infty X$) and an element $\gamma \in \Gamma_0$ such that (4.5) holds for $\mathsf{d}_{\mathrm{Euc}}$ (rather than $\mathsf{d}_{\partial_\infty X}$).

Suppose by contradiction that the action of Γ_0 on $\mathcal{C}^{\mathsf{cor}}_{\rho(\Gamma_0)}$ via ρ is *not* cocompact. Let $(\varepsilon_m)_{m\in\mathbb{N}}$ be a sequence of positive reals going to 0. For any m, the set $\mathcal{K}_m := \{x \in \mathcal{C}^{\mathsf{cor}}_{\rho(\Gamma_0)} \mid \mathsf{d}_{\mathrm{Euc}}(x, \Lambda_{\rho(\Gamma_0)}) \geq \varepsilon_m\}$ is compact, hence there exists a $\rho(\Gamma_0)$-orbit contained in $\mathcal{C}^{\mathsf{cor}}_{\rho(\Gamma_0)} \smallsetminus \mathcal{K}_m$. By proper discontinuity of the action on $\mathcal{C}^{\mathsf{cor}}_{\rho(\Gamma_0)}$, the supremum of $\mathsf{d}_{\mathrm{Euc}}(\cdot, \Lambda_{\rho(\Gamma_0)})$ on this orbit is achieved at some point $x_m \in \mathcal{C}^{\mathsf{cor}}_{\rho(\Gamma_0)}$, and by construction we have $0 < \mathsf{d}_{\mathrm{Euc}}(\rho(\gamma) \cdot x_m, \Lambda_{\rho(\Gamma_0)}) \leq \mathsf{d}_{\mathrm{Euc}}(x_m, \Lambda_{\rho(\Gamma_0)}) \leq \varepsilon_m$ for all $\gamma \in \Gamma_0$. Up to passing to a subsequence, we may assume that $(x_m)_{m\in\mathbb{N}}$ converges to some $z \in \Lambda_{\rho(\Gamma_0)}$. Consider a neighbourhood $\mathcal{U}$ of z in $\mathbb{R}^n$ and an element $\gamma \in \Gamma_0$ such that (4.5) holds for $\mathsf{d}_{\mathrm{Euc}}$, and let $c > 1$ be the infimum in (4.5). For any $m \in \mathbb{N}$, there exists $z_m \in \Lambda_{\rho(\Gamma_0)}$ such that $\mathsf{d}_{\mathrm{Euc}}(\rho(\gamma) \cdot x_m, \Lambda_{\rho(\Gamma_0)}) = \mathsf{d}_{\mathrm{Euc}}(\rho(\gamma) \cdot x_m, \rho(\gamma) \cdot z_m)$. For large enough m we have $x_m, z_m \in \mathcal{U}$, and so $\mathsf{d}_{\mathrm{Euc}}(\rho(\gamma)\cdot x_m, \Lambda_{\rho(\Gamma_0)}) \geq c\,\mathsf{d}_{\mathrm{Euc}}(x_m, z_m) \geq c\,\mathsf{d}_{\mathrm{Euc}}(x_m, \Lambda_{\rho(\Gamma_0)}) \geq c\,\mathsf{d}_{\mathrm{Euc}}(\rho(\gamma)\cdot x_m, \Lambda_{\rho(\Gamma_0)}) > 0$. This is impossible since $c > 1$. □

4.5 Classes of discrete subgroups in higher real rank

We have seen in Section 4.4 two important classes of discrete subgroups of semisimple Lie groups G with $\mathrm{rank}_{\mathbb{R}}(G) = 1$, namely convex cocompact subgroups and geometrically finite subgroups. These classes have been much studied, although many interesting questions remain open even in the case of $G = \mathrm{PO}(n,1)$ for $n \geq 4$ (see e.g. [95]).

We now turn to infinite discrete subgroups of semisimple Lie groups G for $\mathrm{rank}_{\mathbb{R}}(G) \geq 2$. These discrete subgroups, beyond lattices, remain more mysterious, and very few general results are known (see [71] for a notable exception). Recently, an important class has emerged, namely the class of *Anosov subgroups*, which are by definition the images of the *Anosov representations* of Gromov hyperbolic groups introduced by Labourie [109] as part of his study of Hitchin representations (see Section 4.3.3). In fact, most examples in Section 4.3 are Anosov subgroups. We now discuss these subgroups, make the link with convex cocompactness, and mention some generalisations.

4.5.1 Anosov subgroups

Given a noncompact semisimple Lie group G, there are several possible types of Anosov subgroups of G, depending on the choice of one of the (finitely many) *flag varieties* G/P of G, where P is a proper parabolic subgroup of G. For simplicity, in these notes we consider $G = \mathrm{PGL}(d,\mathbb{K})$ or $\mathrm{SL}^{\pm}(d,\mathbb{K}) = \{g \in \mathrm{GL}(d,\mathbb{K}) \mid \det(g) = \pm 1\}$ where $\mathbb{K} = \mathbb{R}$ or $\mathbb{C}$; we take $P = P_i$ to be the stabiliser in G of an i-plane of $\mathbb{K}^d$, for some $1 \leq i \leq d-1$, so that $G/P_i = \mathrm{Gr}_i(\mathbb{K}^d)$ is the Grassmannian of i-planes of $\mathbb{K}^d$.

Definition and first observations

Here is the original definition from Labourie, which appeared in [109] for surface groups $\Gamma_0 = \pi_1(S)$ and in [84] for general hyperbolic groups.

Definition 4.5.1 Let Γ_0 be an infinite Gromov hyperbolic group and $G = \mathrm{PGL}(d,\mathbb{K})$ or $\mathrm{SL}^{\pm}(d,\mathbb{K})$. For $1 \leq i \leq d-1$, a representation $\rho : \Gamma_0 \to G$ is *P_i-Anosov* if there exist ρ-equivariant maps $\xi_i : \partial_\infty \Gamma_0 \to G/P_i = \mathrm{Gr}_i(\mathbb{K}^d)$ and $\xi_{d-i} : \partial_\infty \Gamma_0 \to G/P_{d-i} = \mathrm{Gr}_{d-i}(\mathbb{K}^d)$ which

- are continuous,
- are *transverse*: $\xi_i(w) \oplus \xi_{d-i}(w') = \mathbb{K}^d$ for all $w \neq w'$ in $\partial_\infty \Gamma_0$;
- satisfy a uniform contraction property (Condition 4.5.4 below) which strengthens the dynamics-preserving condition of Theorem 4.4.12.(4).

By *the dynamics-preserving condition of Theorem 4.4.12.*(4) for ξ_i we mean that for any infinite-order element $\gamma \in \Gamma_0$, the image by ξ_i of the attracting fixed point of γ in $\partial_\infty \Gamma_0$ (see Section 4.4.4) is an attracting fixed point of $\rho(\gamma)$ in $\mathrm{Gr}_i(\mathbb{K}^d)$.

We note that for an element $g \in G$, the property of admitting an attracting fixed point in G/P_i can be characterised in terms of eigenvalues, namely as $(\lambda_i - \lambda_{i+1})(g) > 0$ (Notation 4.5.10). In this case the attracting fixed point is unique and we say that g is *proximal* in $\mathrm{Gr}_i(\mathbb{K}^d)$.

Remark 4.5.2 For our purposes, working with $\mathrm{PGL}(d, \mathbb{K})$ or $\mathrm{SL}^{\pm}(d, \mathbb{K})$ is equivalent. Indeed, a representation $\rho : \Gamma_0 \to \mathrm{SL}^{\pm}(d, \mathbb{K})$ is P_i-Anosov if and only if its composition with the natural projection $\mathrm{SL}^{\pm}(d, \mathbb{K}) \to \mathrm{PGL}(d, \mathbb{K})$ is P_i-Anosov, and up to passing to a finite-index subgroup (which does not change the property of being P_i-Anosov) any representation $\rho : \Gamma_0 \to \mathrm{PGL}(d, \mathbb{K})$ with Γ_0 Gromov hyperbolic lifts to $\mathrm{SL}^{\pm}(d, \mathbb{K})$.

The uniform contraction property in Definition 4.5.1 is reminiscent of the condition defining Anosov flows in dynamics, which explains the terminology *Anosov representation*. Before stating it (Condition 4.5.4), let us make a few elementary observations that already follow from the fact that ξ_i and ξ_{d-i} are continuous, transverse, and dynamics-preserving.

Lemma 4.5.3 *If $\rho : \Gamma \to G$ is P_i-Anosov, then*

(1) the boundary maps ξ_i and ξ_{d-i} are unique, and compatible*: $\xi_{\min(i,d-i)}(w) \subset \xi_{\max(i,d-i)}(w)$ for all $w \in \partial_\infty \Gamma_0$; the image of ξ_i is the* proximal limit set *of $\rho(\Gamma_0)$ in $\mathrm{Gr}_i(\mathbb{K}^d)$, i.e. the closure in $\mathrm{Gr}_i(\mathbb{K}^d)$ of the set of attracting fixed points of proximal elements of $\rho(\Gamma_0)$;*
(2) ξ_i and ξ_{d-i} are injective, hence they are homeomorphisms onto their images;
(3) ρ has finite kernel and discrete image.

By (3), the images of P_i-Anosov representations are infinite discrete subgroups of G; we shall call them *P_i-Anosov subgroups*.

Proof (1) Recall from Section 4.4.4 that the subset of $\partial_\infty \Gamma_0$ consisting of the attracting fixed points of infinite-order elements of Γ_0 is dense in $\partial_\infty \Gamma_0$. Since ξ_i and ξ_{d-i} are dynamics-preserving, they are uniquely determined on this subset, and compatible on this subset. By continuity, they are uniquely determined and compatible on all of $\partial_\infty \Gamma_0$. Moreover, the image of ξ_i is the proximal limit set of $\rho(\Gamma_0)$ in $\mathrm{Gr}_i(\mathbb{K}^d)$.

(2) For any $w \neq w'$ in $\partial_\infty \Gamma_0$, the subspaces $\xi_i(w)$ and $\xi_{d-i}(w')$ are transverse by definition, whereas $\xi_i(w)$ and $\xi_{d-i}(w)$ are not by (1) above.

(3) Suppose Γ_0 is nonelementary. In order to show that ρ has finite kernel and discrete image, it is sufficient to consider an arbitrary sequence $(\gamma_k)_{k\in\mathbb{N}}$ of pairwise distinct points of Γ_0 and to check that $(\rho(\gamma_k))_{k\in\mathbb{N}}$ does not converge to the identity of G. Recall from Section 4.4.4 that the action of Γ_0 on $\partial_\infty \Gamma_0$ is a convergence action. Therefore, up to passing to a subsequence, there exist $w^+, w^- \in \partial_\infty \Gamma_0$ such that $\gamma_k \cdot w \to w^+$ for all $w \in \partial_\infty \Gamma_0 \smallsetminus \{w^-\}$. By ρ-equivariance and continuity of ξ_i, we then have $\rho(\gamma_k) \cdot \xi_i(w) = \xi_i(\gamma_k \cdot w) \to \xi_i(w^+)$ for all $w \in \partial_\infty \Gamma_0 \smallsetminus \{w^-\}$. Since $\partial_\infty \Gamma_0$ is infinite and ξ_i is injective, there exists $w \in \partial_\infty \Gamma_0 \smallsetminus \{w^-\}$ such that $\xi_i(w) \neq \xi_i(w^+)$. The convergence $\rho(\gamma_k) \cdot \xi_i(w) \to \xi_i(w^+)$ then implies that $(\rho(\gamma_k))_{n\in\mathbb{N}}$ does not converge to the identity element of G. This shows that ρ has finite kernel and discrete image.

If Γ_0 is elementary, then it admits a finite-index subgroup Γ_0' which is cyclic. The fact that ξ_i is dynamics-preserving implies that ρ is injective and discrete in restriction to Γ_0'. From this one easily deduces that ρ has finite kernel and discrete image. □

The uniform contraction condition

Let us state this condition in the original case considered by Labourie [109], where $\Gamma_0 = \pi_1(M)$ for some closed negatively curved manifold M. We denote by $\widetilde{M}$ the universal cover of M, by T^1 the unit tangent bundle, and by $(\varphi_t)_{t\in\mathbb{R}}$ the geodesic flow on either $T^1(M)$ or $T^1(\widetilde{M})$. (For a general Gromov hyperbolic group Γ_0, one should replace $T^1(\widetilde{M})$ by a certain *flow space* for Γ_0, see [84] or [26, § 4.1].)

For simplicity, we take $G = \mathrm{SL}^{\pm}(d, \mathbb{K})$ (see Remark 4.5.2). Any representation $\rho : \Gamma_0 \to G$ then determines a flat vector bundle

$$E^\rho = \Gamma_0 \backslash (T^1(\widetilde{M}) \times \mathbb{K}^d)$$

over $T^1(M) = \Gamma_0 \backslash T^1(\widetilde{M})$, where Γ_0 acts on $T^1(\widetilde{M}) \times \mathbb{K}^d$ by $\gamma \cdot (\tilde{x}, v) = (\gamma \cdot \tilde{x}, \rho(\gamma) \cdot v)$. The geodesic flow $(\varphi_t)_{t\in\mathbb{R}}$ on $T^1(M)$ lifts to a flow $(\psi_t)_{t\in\mathbb{R}}$ on E^ρ, given by $\psi_t \cdot [(\tilde{x}, v)] = [(\varphi_t \cdot \tilde{x}, v)]$.

Suppose, as in Definition 4.5.1, that there exist continuous, transverse, ρ-equivariant boundary maps $\xi_i : \partial_\infty \Gamma_0 \to \mathrm{Gr}_i(\mathbb{K}^d)$ and $\xi_{d-i} : \partial_\infty \Gamma_0 \to$

$\mathrm{Gr}_{d-i}(\mathbb{K}^d)$. By transversality, for each $\tilde{x} \in T^1(\widetilde{M})$ we have a decomposition $\mathbb{K}^d = \xi_i(\tilde{x}^+) \oplus \xi_{d-i}(\tilde{x}^-)$, where $\tilde{x}^\pm = \lim_{t\to\pm\infty} \varphi_t \cdot \tilde{x} \in \partial_\infty \widetilde{M} \simeq \partial_\infty \Gamma_0$ are the forward and backward endpoints of the geodesic determined by $\tilde{x}$, and this defines a decomposition of the vector bundle E^ρ into the direct sum of two subbundles $E_i^\rho = \{[(\tilde{x}, v)] \,|\, v \in \xi_i(\tilde{x}^+)\}$ and $E_{d-i}^\rho = \{[(\tilde{x}, v)] \,|\, v \in \xi_{d-i}(\tilde{x}^-)\}$. This decomposition is invariant under the flow (ψ_t). By definition, the representation ρ is P_i-Anosov if the following "dominated splitting" condition is satisfied.

Condition 4.5.4 *The flow $(\psi_t)_{t\in\mathbb{R}}$ uniformly contracts E_i^ρ with respect to E_{d-i}^ρ, i.e. given a continuous family $(\|\cdot\|_x)_{x\in T^1(M)}$ of norms on the fibers $E^\rho(x)$, there exist $C, C' > 0$ such that for any $t \geq 0$, any $x \in T^1(M)$, and any nonzero $v_i \in E_i^\rho(x)$ and $v_{d-i} \in E_{d-i}^\rho(x)$,*

$$\frac{\|\psi_t \cdot v_i\|_{\varphi_t \cdot x}}{\|\psi_t \cdot v_{d-i}\|_{\varphi_t \cdot x}} \leq e^{-Ct+C'} \frac{\|v_i\|_x}{\|v_{d-i}\|_x},$$

By compactness of $T^1(M)$, this condition does not depend on the choice of continuous family of norms $(\|\cdot\|_x)_{x\in T^1(M)}$ (changing the norms may change the values of C, C' but not their existence).

Remark 4.5.5 Guichard and Wienhard [84] showed that if there exist ρ-equivariant maps ξ_i and ξ_{d-i} which are continuous, transverse, and dynamics-preserving, and if the group $\rho(\Gamma_0)$ is *Zariski-dense* in G, then Condition 4.5.4 is automatically satisfied.

Properties

- P_i-Anosov is equivalent to P_{d-i}-Anosov, as the integers i and $d-i$ play a similar role in Definition 4.5.1 and Condition 4.5.4 (up to reversing the flow, which switches contraction and expansion). In particular, we may restrict to P_i-Anosov for $1 \leq i \leq d/2$.
- When $\mathrm{rank}_{\mathbb{R}}(G) = 1$ (i.e. $d = 2$ for $G = \mathrm{PGL}(d, \mathbb{K})$ or $\mathrm{SL}^\pm(d, \mathbb{K})$), there is only one proper parabolic subgroup P of G up to conjugation (see Remark 4.4.6), hence only one notion of Anosov. In that case, an infinite discrete subgroup of G is Anosov if and only if it is convex cocompact in the classical sense of Definition 4.4.1.
- When $\mathrm{rank}_{\mathbb{R}}(G) \geq 2$ (i.e. $d \geq 3$ for $G = \mathrm{PGL}(d, \mathbb{K})$ or $\mathrm{SL}^\pm(d, \mathbb{K})$), Anosov subgroups are *not* lattices of G (since Anosov subgroups are Gromov hyperbolic unlike lattices, see Section 4.2.3).
- Uniform contraction over a compact space as in Condition 4.5.4 is stable under small deformations, which implies the following analogue of Fact 4.4.7.

Fact 4.5.6 *Let G be a noncompact semisimple Lie group and P a proper parabolic subgroup of G. For any infinite Gromov hyperbolic group Γ_0, the space of P-Anosov representations is open in* $\mathrm{Hom}(\Gamma_0, G)$.

- Anosov representations behave well under Lie group homomorphisms: the following holds similarly to Fact 4.4.8. (We refer to Remark 4.4.13 for the notion of a hyperbolic element of G'.)

Fact 4.5.7 (see [84]) *Let G' be a semisimple Lie group with $\mathrm{rank}_{\mathbb{R}}(G') = 1$ and let $\tau : G' \to \mathrm{PGL}(d, \mathbb{K})$ be a Lie group homomorphism with compact kernel. For any Gromov hyperbolic group Γ_0, any representation $\sigma_0 : \Gamma_0 \to G'$, and any $1 \leq i \leq d-1$, the following are equivalent:*

(1) the representation $\tau \circ \sigma_0 : \Gamma_0 \to \mathrm{PGL}(d, \mathbb{K})$ is P_i-Anosov;
(2) σ_0 is convex cocompact (Definition 4.4.1) and $(\lambda_i - \lambda_{i+1})(\tau(g')) > 0$ for some hyperbolic element $g' \in G'$.

In this case, τ induces an embedding $\partial_\infty \tau_i : G'/P' \hookrightarrow \mathrm{Gr}_i(\mathbb{K}^d)$ (where G'/P' is the visual boundary of the symmetric space of G', see Remark 4.4.6) and the boundary map of ρ_0 is the composition of the boundary map $\partial_\infty \Gamma_0 \to G'/P'$ of σ_0 (see Theorem 4.4.12) with $\partial_\infty \tau_i$. Moreover, by Fact 4.5.6 there is in that case a neighbourhood of $\tau \circ \sigma_0$ in $\mathrm{Hom}(\Gamma_0, \mathrm{PGL}(d, \mathbb{K}))$ consisting entirely of P_i-Anosov representations (hence with finite kernel and discrete image — see Lemma 4.5.3.(3)).

Examples in higher real rank

Many of the discrete subgroups in Section 4.3 were Anosov subgroups.

- Section 4.3.2: It follows from the work of Benoist [15] that the ping pong groups of Claim 4.3.5 are quasi-isometrically embedded (see Remark 4.5.11) in $\mathrm{PGL}(d, \mathbb{R})$. They are in fact P_1-Anosov: see [42, 98].
- Section 4.3.2: When they are defined by $B_1^{\pm}, \ldots, B_m^{\pm}$ which have pairwise disjoint closures, the Schottky groups in $\mathrm{PGL}(2n, \mathbb{K})$ of Nori and Seade–Verjovsky are P_n-Anosov (see [84]) and the crooked Schottky groups in $\mathrm{Sp}(2n, \mathbb{R}) \subset \mathrm{SL}(2n, \mathbb{R})$ are P_1-Anosov (see [36]).
- Section 4.3.3: By Facts 4.5.6 and 4.5.7, the Barbot representations of closed surface groups into $\mathrm{SL}(d, \mathbb{R})$ are P_1-Anosov. The Hitchin representations into $\mathrm{PSL}(d, \mathbb{R})$ are P_i-Anosov for all $1 \leq i \leq d-1$: this is Labourie's original result from [109], where he introduced Anosov

representations. The maximal representations of closed surface groups into $\mathrm{SO}(2,n) \subset \mathrm{SL}(n+2,\mathbb{R})$ are P_1-Anosov (see [37, 83]). We refer to Figures 4.9 and 4.10 for some illustrations of boundary maps.

Remark 4.5.8 Being P_i-Anosov for all $1 \leq i \leq d-1$ is the strongest possible form of Anosov; in this case, the various boundary maps $\xi_i : \partial_\infty \Gamma_0 \to \mathrm{Gr}_i(\mathbb{K}^d)$ for $1 \leq i \leq d-1$ combine into a continuous, injective, ρ-equivariant boundary map $\xi : \partial_\infty \Gamma_0 \to \mathrm{Flags}(\mathbb{R}^d)$ as in the proof of Theorem 4.3.7.

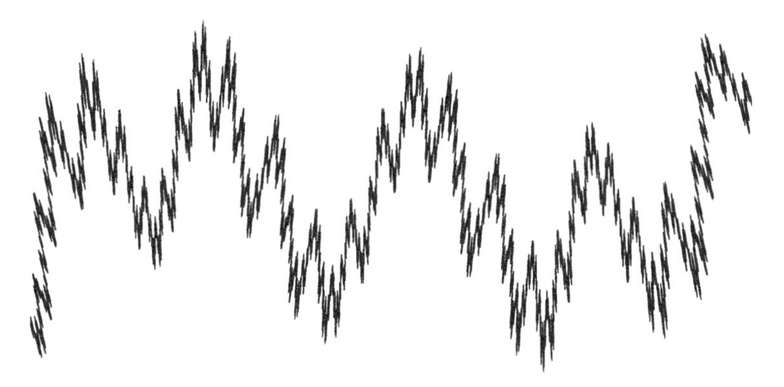

Figure 4.9 The image of the boundary map $\xi_1 : \partial_\infty \Gamma_0 \to \mathbb{P}(\mathbb{R}^3)$ of a representation $\rho : \Gamma_0 = \pi_1(S) \to \mathrm{SL}(3,\mathbb{R})$ which is a small deformation of $\Gamma_0 \overset{\sigma_0}{\hookrightarrow} \mathrm{SL}(2,\mathbb{R}) \overset{\tau}{\hookrightarrow} \mathrm{SL}(3,\mathbb{R})$, where σ_0 is injective and discrete and τ is the standard representation. This image is a topological circle in $\mathbb{P}(\mathbb{R}^3)$ which has Hölder, but not Lipschitz, regularity.

- Section 4.3.4: All (known) higher Teichmüller spaces consist of Anosov representations (see [22, 37, 83, 109]).

Remark 4.5.9 Not all Anosov representations of closed surface groups belong to higher Teichmüller spaces. For instance, the Barbot representations of $\pi_1(S)$ into $\mathrm{SL}(d,\mathbb{R})$ from Section 4.3.3 are P_1-Anosov, but their connected component in $\mathrm{Hom}(\pi_1(S), \mathrm{SL}(d,\mathbb{R}))$ contains representations that are not injective and discrete.

- Section 4.3.4: The two known families of higher-dimensional higher-rank Teichmüller spaces that we mentioned for Gromov hyperbolic groups $\Gamma_0 = \pi_1(M)$ are P_1-Anosov: for holonomies of convex projective structures, see [16], and for $\mathbb{H}^{p,q}$-convex cocompact representations, see [10] (case $q = 1$) and [53, 54] (general case).

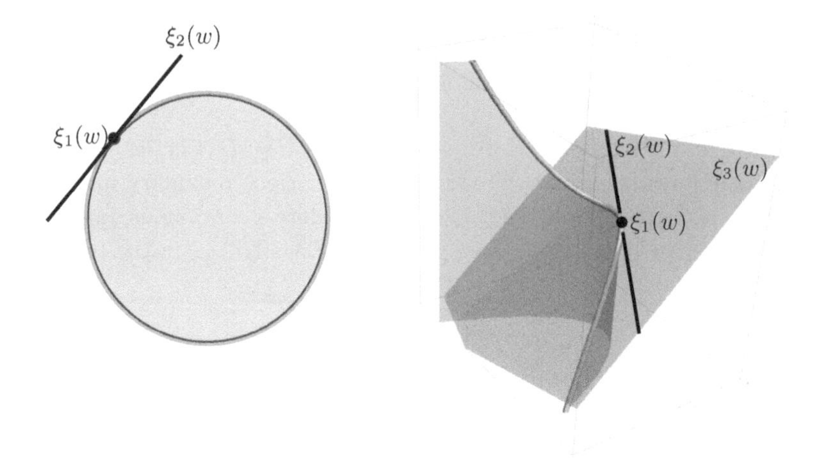

Figure 4.10 If $\xi = (\xi_1, \ldots, \xi_{d-1}) : \partial_\infty \Gamma_0 \to \mathrm{Flags}(\mathbb{R}^d)$ is the boundary map of a Hitchin representation $\rho : \Gamma_0 = \pi_1(S) \to \mathrm{PSL}(d, \mathbb{R})$, then the image of ξ_1 is a C^1 curve in $\mathbb{P}(\mathbb{R}^d)$, and $\xi(w)$ is the osculating flag to this curve at the point $\xi_1(w)$ for all $w \in \partial_\infty \Gamma_0$. For $d = 3$ the curve is the boundary of the properly convex open subset of $\mathbb{P}(\mathbb{R}^3)$ preserved by ρ, while for $d = 4$ the curve is homotopically nontrivial in $\mathbb{P}(\mathbb{R}^4)$. This figure shows the curve $\xi_1(\partial_\infty \Gamma_0)$ and an osculating flag $\xi(w)$ when $\rho : \Gamma_0 \to \mathrm{PSL}(2, \mathbb{R}) \hookrightarrow \mathrm{PSL}(d, \mathbb{R})$ is Fuchsian, for $d = 3$ (left) and $d = 4$ (right); for $d = 4$, the curve is the so-called *twisted cubic* in $\mathbb{P}(\mathbb{R}^4)$, given by $t \mapsto (t, t^2, t^3)$ in some affine chart.

Interlude: eigenvalues and singular values

Before giving (in Theorem 4.5.13 below) some characterisations of Anosov representations that generalise Theorem 4.4.12, we introduce some notation and make a few preliminary observations.

Notation 4.5.10 For any $g \in \mathrm{GL}(d, \mathbb{C})$, we denote by $\lambda_1(g) \geq \cdots \geq \lambda_d(g)$ the logarithms of the moduli of the complex eigenvalues of g, and by $\mu_1(g) \geq \cdots \geq \mu_d(g)$ the logarithms of the *singular values* of g (i.e. of the square roots of the eigenvalues of $\overline{g}^T g$, which are positive numbers). For any $1 \leq i < j \leq d$, this defines functions $\lambda_i - \lambda_j : \mathrm{GL}(d, \mathbb{C}) \to \mathbb{R}_{\geq 0}$ and $\mu_i - \mu_j : \mathrm{GL}(d, \mathbb{C}) \to \mathbb{R}_{\geq 0}$ which factor through $\mathrm{PGL}(d, \mathbb{C})$.

As in Section 4.4.4, for any finitely generated group Γ_0, we choose a finite generating subset of Γ_0 and denote by $\mathrm{Cay}(\Gamma_0)$ the corresponding Cayley graph, with its metric $\mathtt{d}_{\mathrm{Cay}(\Gamma_0)}$. We denote by $X = G/K$ the Riemannian symmetric space of G, with its metric $\mathtt{d}_X$, and fix a basepoint $x_0 \in X$. We denote the translation length as in (4.2).

Our starting point is the following (see Theorem 4.4.12 and its proof).

Remark 4.5.11 Let Γ_0 be a finitely generated group and $\rho : \Gamma_0 \to G$ a representation. Then

- ρ has finite kernel and discrete image if and only if $\mathtt{d}_X(x_0, \rho(\gamma)\cdot x_0) \to +\infty$ as $\mathtt{d}_{\mathrm{Cay}(\Gamma_0)}(e,\gamma) \to +\infty$;
- ρ is called a *quasi-isometric embedding* if there exist $c, c' > 0$ such that $\mathtt{d}_X(x_0, \rho(\gamma)\cdot x_0) \geq c\,\mathtt{d}_{\mathrm{Cay}(\Gamma_0)}(e,\gamma) - c'$ for all $\gamma \in \Gamma$;
- ρ is called *well-displacing* if there exist $c, c'' > 0$ such that $\mathrm{transl}_X(\rho(\gamma)) \geq c\,\mathrm{transl}_{\mathrm{Cay}(\Gamma_0)}(\gamma) - c''$ for all $\gamma \in \Gamma$.

(As in Remarks 4.4.13, the inequality $\mathtt{d}_X(x_0, \rho(\gamma) \cdot x_0) \leq C\,\mathtt{d}_{\mathrm{Cay}(\Gamma_0)}(e,\gamma)$ always holds for $C := \max_{f\in F} \mathtt{d}_X(x_0, \rho(f)\cdot x_0)$; it implies that the inequality $\mathrm{transl}_X(\rho(\gamma)) \leq C\,\mathrm{transl}_{\mathrm{Cay}(\Gamma_0)}(\gamma)$ always holds too: see the proof of the implication (2) $\Rightarrow$ (3) of Theorem 4.4.12.)

We now reinterpret Remark 4.5.11 using Notation 4.5.10. Let $\|\cdot\|_{\mathrm{Euc}}$ be the standard Euclidean norm on $\mathbb{R}^d$. For $G = \mathrm{PGL}(d,\mathbb{K})$ with $\mathbb{K} = \mathbb{R}$ or $\mathbb{C}$, we can take $K = \mathrm{PO}(d)$ or $\mathrm{PU}(d)$ and $x_0 = eK \in G/K = X$, so that for any $g \in G$ lifting to $\hat{g} \in \mathrm{GL}(d,\mathbb{K})$ with $|\det(\hat{g})| = 1$,

$$\mathtt{d}_X(x_0, g\cdot x_0) = \|(\mu_1(\hat{g}),\dots,\mu_d(\hat{g}))\|_{\mathrm{Euc}},$$
$$\mathrm{transl}_X(g) = \|(\lambda_1(\hat{g}),\dots,\lambda_d(\hat{g}))\|_{\mathrm{Euc}}.$$

On the other hand, we have $\sum_{i=1}^d \mu_i(\hat{g}) = \sum_{i=1}^d \lambda_i(\hat{g}) = 0$, and on the linear hyperplane $\{v \in \mathbb{R}^d \mid \sum_{i=1}^d v_i = 0\}$ of $\mathbb{R}^d$ the Euclidean norm $\|\cdot\|_{\mathrm{Euc}}$ is equivalent to $\sum_{i=1}^{d-1} |v_i - v_{i+1}|$. In this setting we can therefore rewrite Remark 4.5.11 as follows.

Remark 4.5.12 Let Γ_0 be a finitely generated group and $\rho : \Gamma_0 \to G = \mathrm{PGL}(d,\mathbb{K})$ a representation. Then

- ρ has finite kernel and discrete image if and only if $\sum_{i=1}^{d-1}(\mu_i - \mu_{i+1})(\rho(\gamma)) \to +\infty$ as $\mathtt{d}_{\mathrm{Cay}(\Gamma_0)}(e,\gamma) \to +\infty$;
- ρ is a quasi-isometric embedding if and only if there exist $c, c' > 0$ such that $\sum_{i=1}^{d-1}(\mu_i - \mu_{i+1})(\rho(\gamma)) \geq c\,\mathtt{d}_{\mathrm{Cay}(\Gamma_0)}(e,\gamma) - c'$ for all $\gamma \in \Gamma$;
- ρ is well-displacing if and only if there exist $c, c'' > 0$ such that $\sum_{i=1}^{d-1}(\lambda_i - \lambda_{i+1})(\rho(\gamma)) \geq c\,\mathrm{transl}_{\mathrm{Cay}(\Gamma_0)}(\gamma) - c''$ for all $\gamma \in \Gamma$.

Remark 4.5.12 should be kept in mind will reading Theorem 4.5.13.(2)–(3) below, as it explains how Anosov representations are refinements of quasi-isometric embeddings and well-displacing representations.

Characterisations

The following characterisations of Anosov representations were established by Kapovich–Leeb–Porti, Guéritaud–Guichard–Kassel–Wienhard, Bochi–Potrie–Sambarino, and Kassel–Potrie. More precisely, (1) $\Rightarrow$ (2) is easy and follows from a property of dominated splittings proved in [25]. The implication (2) $\Rightarrow$ (1) was proved in [101], with an alternative proof later given in [26]. The implication (2) $\Rightarrow$ (3) is easy and similar to the implication (2) $\Rightarrow$ (3) of Theorem 4.4.12 (note that $\lambda_i(\hat{g}) = \lim_k \mu_i(\hat{g}^k)/k$ for all $\hat{g} \in \mathrm{GL}(d, \mathbb{K})$). The implication (3) $\Rightarrow$ (2) was proved in [104]. The implications (1) $\Leftrightarrow$ (4) $\Leftrightarrow$ (5) and (1) $\Rightarrow$ (6) were proved in [81] and [98], and (6) $\Rightarrow$ (1) was proved in [98]. We refer to [96, 99] for further characterisations (e.g. in terms of *conical limit points*).

We fix a basepoint $x_0 \in X = G/K$ and a Riemannian metric on $G/P_{i,d-i} = \mathrm{Flags}_{i,d-i}(\mathbb{R}^d) = \{(V_{\min(i,d-i)} \subset V_{\max(i,d-i)}) \mid \dim(V_\bullet) = \bullet\}$. See Definition 4.5.1 for the notions of transversality and dynamics-preserving, and Theorem 4.4.12.(6) for expansion at the limit set. The notion of limit set that we use is discussed in the next section.

Theorem 4.5.13 *Let $G = \mathrm{PGL}(d, \mathbb{K})$ or $\mathrm{SL}^{\pm}(d, \mathbb{K})$ where $\mathbb{K} = \mathbb{R}$ or $\mathbb{C}$, and let $1 \leq i \leq d-1$. Let Γ_0 be a finitely generated infinite group and $\rho : \Gamma_0 \to G$ a representation. Then the following are equivalent:*

(1) Γ_0 is Gromov hyperbolic and ρ is P_i-Anosov,

(2) ρ is a quasi-isometric embedding "in the i-th direction": there exist $c, c' > 0$ such that for any $\gamma \in \Gamma$,

$$(\mu_i - \mu_{i+1})(\rho(\gamma)) \geq c\, \mathrm{d}_{\mathrm{Cay}(\Gamma_0)}(e, \gamma) - c';$$

(3) Γ_0 is Gromov hyperbolic and ρ is well-displacing "in the i-th direction": there exist $c, c'' > 0$ such that for any $\gamma \in \Gamma$,

$$(\lambda_i - \lambda_{i+1})(\rho(\gamma)) \geq c\, \mathrm{transl}_{\mathrm{Cay}(\Gamma_0)}(\gamma) - c'';$$

(4) Γ_0 is Gromov hyperbolic, there exist ρ-equivariant maps

$$\xi_\bullet : \partial_\infty \Gamma_0 \longrightarrow G/P_\bullet = \mathrm{Gr}_\bullet(\mathbb{K}^d),$$

for $\bullet \in \{i, d-i\}$, which are continuous, transverse, dynamics-preserving, and $(\mu_i - \mu_{i+1})(\rho(\gamma)) \to +\infty$ as $\mathrm{d}_{\mathrm{Cay}(\Gamma_0)}(e, \gamma) \to +\infty$;

(5) Γ_0 is Gromov hyperbolic and there exist ρ-equivariant maps

$$\xi_\bullet : \partial_\infty \Gamma_0 \longrightarrow G/P_\bullet = \mathrm{Gr}_\bullet(\mathbb{K}^d),$$

for $\bullet \in \{i, d-i\}$, which are continuous, transverse, and strongly dynamics-preserving *(i.e. for any* $(\gamma_k) \in \Gamma_0^{\mathbb{N}}$ *and* $w^+, w^- \in \partial_\infty \Gamma_0$*, if* $\gamma_k \cdot w \to w^+$ *for all* $w \in \partial_\infty \Gamma_0 \smallsetminus \{w^-\}$*, then* $\rho(\gamma_k) \cdot z \to \xi_i(w^+)$ *for all* $z \in G/P_i$ *transverse to* $\xi_{d-i}(w^-) \in G/P_{d-i}$*);*

(6) $(\mu_i - \mu_{i+1})(\rho(\gamma)) \to +\infty$ *as* $\mathrm{d}_{\mathrm{Cay}(\Gamma_0)}(e, \gamma) \to +\infty$*, any two points of the limit set of* $\rho(\Gamma_0)$ *in* $G/P_{i,d-i}$ *are transverse, and the action of* Γ_0 *on* $G/P_{i,d-i}$ *via* ρ *is expanding at this limit set.*

Remark 4.5.14 Recall that when $\mathrm{rank}_{\mathbb{R}}(G) = 1$ (i.e. $d = 2$), an infinite discrete subgroup of G is Anosov if and only if it is convex cocompact in the classical sense of Definition 4.4.1. In that case, the flag variety $G/P_i = G/P_{d-i}$ identifies with the visual boundary $\partial_\infty X$ of $X = G/K$ (Remark 4.4.6) and conditions (2), (3), (4), (5), (6) of Theorem 4.5.13 are the same as conditions (2), (3), (4), (5), (6) of Theorem 4.4.12 (see Remark 4.5.12 and Lemma 4.5.3.(3)). On the other hand, when $\mathrm{rank}_{\mathbb{R}}(G) \geq 2$, conditions (2) and (3) of Theorem 4.5.13 are strictly stronger than conditions (2) and (3) of Theorem 4.4.12 (see Remark 4.5.12).

As in Remarks 4.4.13, in condition (3) we cannot remove the assumption that Γ_0 be Gromov hyperbolic. See [104, § 4.4] for further discussion.

Limit sets

We now explain the notion of limit set used in Theorem 4.5.13.(6). It is based on an important decomposition of the noncompact semisimple Lie group G: the *Cartan decomposition* $G = K \exp(\mathfrak{a}^+) K$. We refer to [90] for the general theory for noncompact semisimple Lie groups G. For $G = \mathrm{PGL}(d, \mathbb{K})$, as in Remark 4.5.12, we can take $K = \mathrm{PO}(d)$ or $\mathrm{PU}(d)$, and $\mathfrak{a}^+$ to be the set of diagonal matrices in $\mathfrak{g} = \{y \in M_d(\mathbb{K}) \mid \mathrm{tr}(y) = 0\}$ whose entries $t_1, \dots, t_d \in \mathbb{R}$ are in nonincreasing order, with $t_1 + \cdots + t_d = 0$; the Cartan decomposition can then be stated as follows.

Fact 4.5.15 *Any* $g \in \mathrm{PGL}(d, \mathbb{K})$ *can be written as* $g = \kappa \exp(a) \kappa'$ *for some* $\kappa, \kappa' \in K$ *and a unique* $a \in \mathfrak{a}^+$*; the entries of* a *are* $\mu_1(\hat{g}), \dots, \mu_d(\hat{g})$ *(see Notation 4.5.10) where* $\hat{g} \in \mathrm{GL}(d, \mathbb{K})$ *is any lift of* g *with* $|\det(\hat{g})| = 1$.

Proof By the polar decomposition, any element of $\mathrm{GL}(d, \mathbb{R})$ (resp. $\mathrm{GL}(d, \mathbb{C})$) can be written as the product of an orthogonal (resp. unitary) matrix and a positive semi-definite real symmetric (resp. Hermitian) matrix; on the other hand, any real symmetric (resp. Hermitian) matrix can be diagonalised by an orthogonal (resp. unitary) matrix. □

Here is a useful consequence of the Cartan decomposition.

Lemma 4.5.16 *For $1 \leq i \leq d-1$ and a sequence (g_m) of points of $G = \mathrm{PGL}(d, \mathbb{K})$, consider the following two conditions:*

(a) $(\mu_i - \mu_{i+1})(g_m) \to +\infty$,
(b) there exist $z^+ \in G/P_i$ and $z^- \in G/P_{d-i}$ such that $g_m \cdot z \to z^+$ for all $z \in G/P_i$ transverse to z^-.

If (g_m) satisfies (a), then some subsequence of (g_m) satisfies (b). Conversely, if (g_m) satisfies (b), then it satisfies (a).

Proof Let $z_0^+ := \mathrm{span}(e_1, \ldots, e_i) \in G/P_i$ and $z_0^- := \mathrm{span}(e_{i+1}, \ldots, e_d) \in G/P_{d-i}$, where $(e_1, \ldots, e_d)$ is the canonical basis of $\mathbb{K}^d$. By Fact 4.5.15, for any m we can write $g_m = \kappa_m \exp(a_m) \kappa'_m$ where $\kappa_m, \kappa'_m \in K$ and $a_m \in \mathfrak{a}^+$ is diagonal; the entries of a_m are $\mu_1(\hat{g}_m), \ldots, \mu_d(\hat{g}_m)$ where $\hat{g}_m \in \mathrm{GL}(d, \mathbb{K})$ is any lift of g_m with $|\det(\hat{g}_m)| = 1$.

(a) $\Rightarrow$ (b): If $(\mu_i - \mu_{i+1})(g_m) \to +\infty$, then $a_m \cdot z \to z_0^+$ for all $z \in G/P_i$ transverse to z_0^-. Since K is compact, up to passing to a subsequence, we may assume that $(\kappa_m), (\kappa'_m)$ converge respectively to some $\kappa, \kappa' \in K$. Then $g_m \cdot z \to z^+ := \kappa \cdot z_0^+$ for all $z \in G/P_i$ transverse to $z^- := \kappa'^{-1} \cdot z_0^-$.

(b) $\Rightarrow$ (a): If $(\mu_i - \mu_{i+1})(g_m)$ does not tend to $+\infty$, then up to passing to a subsequence it converges to some nonnegative real number, and one easily sees that the image by a_m of any open subset of G/P_i fails to converge to a point. Up to passing to a subsequence, we may assume that $(\kappa_m), (\kappa'_m)$ converge in K. Then the image by g_m of any open subset of G/P_i fails to converge to a point. □

For (g_m) and z^+ as in condition (b) of Lemma 4.5.16, we say that z^+ is a *contraction point* for (g_m) in G/P_i. We then define the *limit set in* G/P_i of a discrete subgroup Γ of G to be the set of contraction points in G/P_i of sequences of elements of Γ. It is a closed Γ-invariant subset of G/P_i. When Γ is P_i-Anosov, it coincides with the proximal limit set of Γ in G/P_i, which is also the image of the boundary map $\xi_i : \partial_\infty \Gamma \to G/P_i$.

Similarly, sequences $(g_m) \in G^{\mathbb{N}}$ satisfying both $(\mu_i - \mu_{i+1})(g_m) \to +\infty$ and $(\mu_{d-i} - \mu_{d-i+1})(g_m) \to +\infty$ define contraction points in $G/P_{i,d-i} = \mathrm{Flags}_{i,d-i}(\mathbb{R}^d)$. This gives a notion of *limit set in* $G/P_{i,d-i}$ of a discrete subgroup Γ of G, as considered in Theorem 4.5.13.(6).

We note that in the setting of Theorem 4.5.13.(6), the limit set of $\rho(\Gamma_0)$ in $G/P_{i,d-i}$ is nonempty. Indeed, $(\mu_{d-i} - \mu_{d-i+1})(g) = (\mu_i - \mu_{i+1})(g^{-1})$ for all $g \in G$, and so $(\mu_i - \mu_{i+1})(\rho(\gamma)) \to +\infty$ as $d_{\mathrm{Cay}(\Gamma_0)}(e, \gamma) \to +\infty$ implies $(\mu_{d-i} - \mu_{d-i+1})(\rho(\gamma)) \to +\infty$ as $d_{\mathrm{Cay}(\Gamma_0)}(e, \gamma) \to +\infty$.

Cocompact domains of discontinuity

We end this section by briefly mentioning a generalisation to Anosov representations of a nice feature of rank-one convex cocompact representations. Namely, we have seen in Section 4.4.2 that for $\mathrm{rank}_{\mathbb{R}}(G) = 1$, if $X = G/K$ denotes the Riemannian symmetric space of G, then any convex cocompact subgroup Γ of G acts properly discontinuously, with compact quotient, on the open subset $\Omega_\Gamma := \partial_\infty X \smallsetminus \Lambda_\Gamma$ of $\partial_\infty X$. In that case, $\partial_\infty X$ is the unique flag variety G/P of G with P a proper parabolic subgroup of G (Remark 4.4.6).

Guichard and Wienhard [84], inspired by work of Frances, generalised this picture to show that in certain situations, for certain proper parabolic subgroups P and Q of G, any P-Anosov subgroup Γ of G acts properly discontinuously, with compact quotient, on some open subset Ω of G/Q which is obtained by removing all points of G/Q that are "not transverse enough" (in some precise sense) to the limit set of Γ in G/P. This phenomenon was then investigated and described in full generality by Kapovich, Leeb, and Porti [100]. Let us give one concrete example.

Example 4.5.17 Let b be a nondegenerate symmetric bilinear form on $\mathbb{R}^d$ with noncompact automorphism group $G := \mathrm{Aut}(b) \subset \mathrm{SL}^\pm(d, \mathbb{R})$. (If b is symmetric, then $G = \mathrm{O}(p, q)$ for some $p, q \geq 1$; we require p and q to be distinct. If b is skew-symmetric, then $d = 2n$ is even and $G = \mathrm{Sp}(2n, \mathbb{R})$.) Let Γ_0 be an infinite Gromov hyperbolic group, $\rho : \Gamma_0 \to G \subset \mathrm{SL}^\pm(d, \mathbb{R})$ a P_1-Anosov representation, and $\Lambda_{\rho(\Gamma_0)}$ the limit set of $\rho(\Gamma_0)$ in $\mathbb{P}(\mathbb{R}^d)$. Let $\mathcal{L}$ be the space of maximal b-isotropic subspaces of $\mathbb{R}^d$. (It identifies with G/Q where Q is the stabiliser in G of a maximal b-isotropic subspace of $\mathbb{R}^d$.) Then Γ_0 acts properly discontinuously with compact quotient, via ρ, on

$$\Omega_{\rho(\Gamma_0)} := \mathcal{L} \smallsetminus \bigcup_{z \in \Lambda_{\rho(\Gamma_0)}} \mathcal{L}_z,$$

where $\mathcal{L}_z$ is the set of maximal b-isotropic subspaces of $\mathbb{R}^d$ that contain the line z.

When b is skew-symmetric, i.e. $G = \mathrm{Sp}(2n, \mathbb{R})$, the set $\mathcal{L}$ is the space $\mathrm{Lag}(\mathbb{R}^{2n})$ of *Lagrangians* of $\mathbb{R}^{2n}$. In this setting, if $\rho(\Gamma_0)$ is a "strong" crooked Schottky group as in Section 4.3.2, defined by $B_1^\pm, \ldots, B_m^\pm$ with pairwise disjoint closures, then the set $\Omega_{\rho(\Gamma_0)}$ of Example 4.5.17 coincides with the set $\Omega = \mathrm{Int}(\bigcup_{\gamma \in \Gamma_0} \rho(\gamma) \cdot \mathcal{D}) \subset \mathrm{Lag}(\mathbb{R}^{2n})$ of Section 4.3.2.

4.5.2 Anosov representations and convex cocompactness

Recall that when $\mathrm{rank}_{\mathbb{R}}(G) = 1$, Anosov representations coincide with convex cocompact representations in the classical sense of Definition 4.4.1. When $\mathrm{rank}_{\mathbb{R}}(G) \geq 2$, Theorem 4.5.13 shows that Anosov representations have a number of similarities, in terms of their dynamics, with rank-one convex cocompact representations: see Remark 4.5.14. Another similarity, of a more geometric nature, is the existence of cocompact domains of discontinuity as in Example 4.5.17: given an Anosov representation $\rho : \Gamma_0 \to G$, such a domain of discontinuity $\Omega \subset G/Q$ yields, by taking the quotient, a closed manifold $\rho(\Gamma_0)\backslash\Omega$ locally modeled on G/Q, whose geometry can be quite interesting (see [134, § 5]). These manifolds $\rho(\Gamma_0)\backslash\Omega$ do not satisfy any kind of convexity properties in general.

Given these similarities, it is natural to wonder if Anosov representations could also be characterised geometrically in terms of some suitable notion of convex cocompactness. We will see below that this is indeed the case. This will give more geometric intuition about Anosov representations, and yield new examples constructed geometrically.

Two attempts

Our starting point is the following special case of Fact 4.5.7.

Fact 4.5.18 *Let Γ_0 be an infinite group and $\rho : \Gamma_0 \to \mathrm{PO}(n,1) = \mathrm{Isom}(\mathbb{H}^n)$ a representation. Then ρ is convex cocompact (Definition 4.4.1) if and only if Γ_0 is Gromov hyperbolic and $\rho : \Gamma_0 \to \mathrm{PO}(n,1) \hookrightarrow \mathrm{PGL}(n+1,\mathbb{R})$ is P_1-Anosov.*

We would like to generalise this equivalence to higher-rank semisimple Lie groups G.

A natural first attempt would be to replace $\mathbb{H}^n$ by the Riemannian symmetric space of G. However, this turns out to be rather restrictive: Kleiner–Leeb [106] and Quint [121] proved that if G is a real simple Lie group of real rank ≥ 2, with Riemannian symmetric space $X = G/K$, then any Zariski-dense discrete subgroup of G, acting with compact quotient on some nonempty convex subset of X, is a cocompact lattice in G; in particular, Γ is *not* Gromov hyperbolic and ρ is *not* Anosov. Thus this approach does not provide a generalisation of Fact 4.5.18.

Instead, we make a second attempt by viewing $\mathbb{H}^n$ as a properly convex open set in projective space as in (4.1); we can then try to generalise Fact 4.5.18 by replacing $\mathbb{H}^n$ with any properly convex open subset Ω of $\mathbb{P}(\mathbb{R}^{n+1})$.

Recall from the proof of Theorem 4.3.7 that Ω being *properly convex* means that it is convex and bounded in some affine chart of $\mathbb{P}(\mathbb{R}^{n+1})$. In this setting Ω carries a natural proper metric d_Ω, the *Hilbert metric*, which is invariant under $\mathrm{Aut}(\Omega) := \{g \in \mathrm{PGL}(n+1,\mathbb{R}) \mid g \cdot \Omega = \Omega\}$ (see Figure 4.11). In particular, any discrete subgroup of $\mathrm{Aut}(\Omega)$ acts properly discontinuously on Ω.

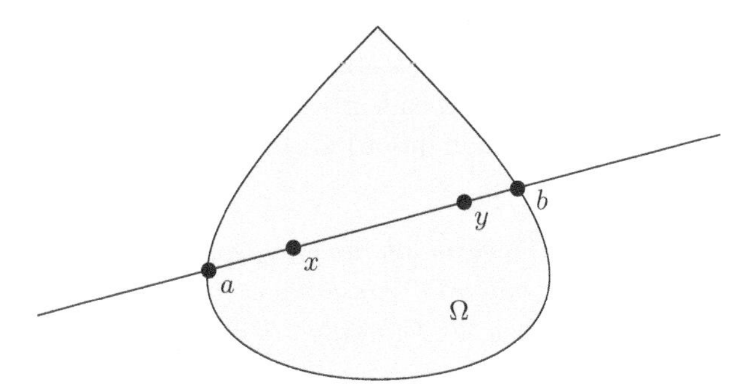

Figure 4.11 In a properly convex open subset Ω of $\mathbb{P}(\mathbb{R}^d)$, the *Hilbert distance* between two distinct points $x, y \in \Omega$ is given by $\mathsf{d}_\Omega(x,y) := \frac{1}{2}\log[a,x,y,b]$, where $[\cdot,\cdot,\cdot,\cdot]$ is the cross-ratio on $\mathbb{P}^1(\mathbb{R})$, normalised so that $[0,1,y,\infty] = y$, and a, b are the intersection points of $\partial\Omega$ with the projective line through x and y, with a, x, y, b in this order. The Hilbert metric d_Ω coincides with the hyperbolic metric when $\Omega = \mathbb{H}^n$ as in (4.1), but in general d_Ω is not Riemannian, only Finsler.

Note that $\mathbb{H}^n$, viewed as a properly convex open subset of $\mathbb{P}(\mathbb{R}^{n+1})$, does not contain any nontrivial projective segments in its boundary. For properly convex open sets Ω with this property (also known as *strictly convex* open sets), we consider the following analogue of Definition 4.4.1.

Definition 4.5.19 Let Ω be a properly convex open subset of $\mathbb{P}(\mathbb{R}^d)$, whose boundary $\partial\Omega$ does not contain any nontrivial projective segments. Let Γ_0 be a group and $\rho : \Gamma_0 \to \mathrm{Aut}(\Omega) \subset \mathrm{PGL}(d,\mathbb{R})$ a representation. We say that the action of Γ_0 on Ω via ρ is *convex cocompact* if it is properly discontinuous and if there exists a nonempty $\rho(\Gamma_0)$-invariant convex subset $\mathcal{C}$ of Ω such that $\rho(\Gamma_0)\backslash\mathcal{C}$ is compact.

In that case, the representation ρ has finite kernel and discrete image and, as in Remark 4.4.2, the group Γ_0 is finitely generated.

Similarly to Fact 4.4.5, we can rephrase convex cocompactness in terms of some specific convex set in Ω. Namely, define the *orbital limit set* $\Lambda^{\mathsf{orb}}_{\rho(\Gamma_0)}(\Omega)$ of $\rho(\Gamma_0)$ in Ω to be the set of accumulation points in $\partial\Omega$ of some $\rho(\Gamma_0)$-orbit of Ω; one easily checks that $\Lambda^{\mathsf{orb}}_{\rho(\Gamma_0)}(\Omega)$ does not depend on the choice of $\rho(\Gamma_0)$-orbit, because $\partial\Omega$ does not contain any nontrivial segments. Define the *convex core* $\mathcal{C}^{\mathsf{cor}}_{\rho(\Gamma_0)}(\Omega) \subset \Omega$ of $\rho(\Gamma_0)$ to be the convex hull of $\Lambda^{\mathsf{orb}}_{\rho(\Gamma_0)}(\Omega)$ in Ω (i.e. the smallest closed convex subset of Ω whose closure in $\mathbb{P}(\mathbb{R}^d)$ contains $\Lambda^{\mathsf{orb}}_{\rho(\Gamma_0)}$). Similarly to Fact 4.4.5, for infinite Γ_0, the action of Γ_0 on Ω via ρ is then convex cocompact if and only if it is properly discontinuous and $\rho(\Gamma_0)\backslash\mathcal{C}^{\mathsf{cor}}_{\rho(\Gamma_0)}(\Omega)$ is compact and nonempty.

The following result is a generalisation of Fact 4.5.18 in this setting. It was first proved in [53] for representations ρ with values in $\mathrm{PO}(p,q)$, and then in general in [54] and independently (in a slightly different form and under some irreducibility assumption) in [141]. See also [10, 16, 52, 114] for related earlier results.

Theorem 4.5.20 *Let Γ_0 be an infinite group and $\rho : \Gamma_0 \to \mathrm{PGL}(d,\mathbb{R})$ a representation. Suppose that $\rho(\Gamma_0)$ preserves a nonempty properly convex open subset of $\mathbb{P}(\mathbb{R}^d)$. Then the following are equivalent:*

(1) Γ_0 is Gromov hyperbolic and $\rho : \Gamma_0 \to \mathrm{PGL}(d,\mathbb{R})$ is P_1-Anosov;
(2) ρ is strongly convex cocompact *in $\mathbb{P}(\mathbb{R}^d)$: the group Γ_0 acts convex cocompactly (Definition 4.5.19) via ρ on some properly convex open subset Ω of $\mathbb{P}(\mathbb{R}^d)$ such that $\partial\Omega$ is C^1 and contains no segments.*

Here we say that $\partial\Omega$ is C^1 if every point of $\partial\Omega$ has a unique supporting hyperplane. The phrase *strongly convex cocompact* is meant to reflect the strong regularity imposed on $\partial\Omega$ (namely, C^1 and no segments).

A few comments on Theorem 4.5.20

In certain situations, the assumption in Theorem 4.5.20 that $\rho(\Gamma_0)$ preserve a properly convex open subset of $\mathbb{P}(\mathbb{R}^d)$ is automatically satisfied for P_1-Anosov representations ρ. For instance, this is the case when $\partial_\infty\Gamma_0$ is connected and ρ takes values in $\mathrm{PO}(p,q) \subset \mathrm{PGL}(d,\mathbb{R})$ for some $p,q \geq 1$ with $p+q=d$, by [53]. In this case, the $\rho(\Gamma_0)$-invariant properly convex open set Ω given by Theorem 4.5.20.(2) can be taken in

$$\{[v] \in \mathbb{P}(\mathbb{R}^{p+q}) \mid \langle v,v\rangle_{p,q} < 0\} = \mathbb{H}^{p,q-1}$$

(we then say that ρ is $\mathbb{H}^{p,q-1}$*-convex cocompact*) or in

$$\{[v] \in \mathbb{P}(\mathbb{R}^{p+q}) \mid -\langle v,v\rangle_{p,q} < 0\} \simeq \mathbb{H}^{q,p-1}$$

(we then say that ρ is $\mathbb{H}^{q,p-1}$*-convex cocompact*), where $\langle\cdot,\cdot\rangle_{p,q}$ is the symmetric bilinear form of signature (p,q) on $\mathbb{R}^{p+q}$ defining $\mathrm{PO}(p,q)$.

On the other hand, there exist P_1-Anosov representations that do not preserve any properly convex open subset of $\mathbb{P}(\mathbb{R}^d)$: e.g. Hitchin representations (see Sections 4.3.3 and 4.5.1) into $\mathrm{PSL}(d, \mathbb{R})$ for even d.

However, one can always reduce to preserving a properly convex open set by considering a larger projective space. Indeed, consider the natural action of $\mathrm{GL}(d, \mathbb{R})$ on the vector space $\mathrm{Sym}_d(\mathbb{R})$ of symmetric $(d \times d)$ real matrices by $g \cdot A = gAg^T$. It induces a representation $\tau : \mathrm{PGL}(d, \mathbb{R}) \to \mathrm{PGL}(\mathrm{Sym}_d(\mathbb{R}))$, which preserves the open subset Ω_{sym} of $\mathbb{P}(\mathrm{Sym}_d(\mathbb{R}))$ corresponding to positive definite symmetric matrices. The set Ω_{sym} is properly convex. One can check (see [84], or use one of the characterisations of Theorem 4.5.13) that a representation $\rho : \Gamma_0 \to \mathrm{PGL}(d, \mathbb{R})$ is P_1-Anosov if and only if $\tau \circ \rho : \Gamma_0 \to \mathrm{PGL}(\mathrm{Sym}_d(\mathbb{R}))$ is P_1-Anosov. Theorem 4.5.20 then implies the following.

Corollary 4.5.21 *For any infinite group Γ_0 and any representation $\rho : \Gamma_0 \to \mathrm{PGL}(d, \mathbb{R})$, the following are equivalent:*

(1) Γ_0 is Gromov hyperbolic and $\rho : \Gamma_0 \to \mathrm{PGL}(d, \mathbb{R})$ is P_1-Anosov;
(2) $\tau \circ \rho$ is strongly convex cocompact in $\mathbb{P}(\mathrm{Sym}_d(\mathbb{R}))$.

This actually yields a characterisation of P-Anosov representations into G for any proper parabolic subgroup P of any noncompact semisimple Lie group G, by considering an appropriate representation of G to some large projective linear group. For instance, for $G = \mathrm{PGL}(d, \mathbb{R})$ and $P = P_i$ with $1 \le i \le d-1$ as in Section 4.5.1, we can consider the natural representation $\tau_i : \mathrm{PGL}(d, \mathbb{R}) \to \mathrm{PGL}(S^2(\Lambda^i \mathbb{R}^d))$ where $S^2(\Lambda^i \mathbb{R}^d)$ is the second symmetric power of the i-th exterior power of the standard representation of $\mathrm{GL}(d, \mathbb{R})$ on $\mathbb{R}^d$. (For $i = 1$, this identifies with $\tau : \mathrm{PGL}(d, \mathbb{R}) \to \mathrm{PGL}(\mathrm{Sym}_d(\mathbb{R}))$ above.) Again, one can check that $\rho : \Gamma_0 \to \mathrm{PGL}(d, \mathbb{R})$ is P_i-Anosov if and only if $\tau_i \circ \rho : \Gamma_0 \to \mathrm{PGL}(S^2(\Lambda^i \mathbb{R}^d))$ is P_1-Anosov. Theorem 4.5.20 then implies the following.

Corollary 4.5.22 *For any infinite group Γ_0, any representation $\rho : \Gamma_0 \to \mathrm{PGL}(d, \mathbb{R})$, and any $1 \le i \le d-1$, the following are equivalent:*

(1) Γ_0 is Gromov hyperbolic and $\rho : \Gamma_0 \to \mathrm{PGL}(d, \mathbb{R})$ is P_i-Anosov;
(2) $\tau_i \circ \rho$ is strongly convex cocompact in $\mathbb{P}(S^2(\Lambda^i \mathbb{R}^d))$.

Sketch of proof of Theorem 4.5.20

Proof of (1) $\Rightarrow$ (2): Suppose that Γ_0 is Gromov hyperbolic, that ρ is P_1-Anosov with boundary maps $\xi_1 : \partial_\infty \Gamma_0 \to \mathrm{Gr}_1(\mathbb{R}^d) = \mathbb{P}(\mathbb{R}^d)$ and $\xi_{d-1} : \partial_\infty \Gamma_0 \to \mathrm{Gr}_{d-1}(\mathbb{R}^d) = \mathbb{P}((\mathbb{R}^d)^*)$, and that $\rho(\Gamma_0)$ preserves a nonempty properly convex open subset Ω of $\mathbb{P}(\mathbb{R}^d)$. Since Ω was chosen without care, it is possible that $\partial\Omega$ contains segments or that the action of Γ_0 on Ω via ρ is not convex cocompact. Therefore, we do not work with Ω itself, but consider instead the connected component $\Omega_{\max}$ of $\mathbb{P}(\mathbb{R}^d) \smallsetminus \bigcup_{w\in\partial_\infty\Gamma_0} \xi_{d-1}(w)$ containing Ω (where we view each $\xi_{d-1}(w)$ as a projective hyperplane in $\mathbb{P}(\mathbb{R}^d)$); it is $\rho(\Gamma_0)$-invariant, open, and convex (not necessarily bounded) in some affine chart of $\mathbb{P}(\mathbb{R}^d)$. Using Lemma 4.5.16, one can show that the action of Γ_0 on $\Omega_{\max}$ via ρ is properly discontinuous, and that the set of accumulation points of any $\rho(\Gamma_0)$-orbit of $\Omega_{\max}$ is $\xi_1(\partial_\infty\Gamma_0)$.

Consider the convex hull $\mathcal{C}$ of $\xi_1(\partial_\infty\Gamma_0)$ in $\Omega_{\max}$. One easily checks, using the transversality of ξ_1 and ξ_{d-1}, that $\xi_1(\partial_\infty\Gamma_0)$ is not contained in a single supporting hyperplane to $\Omega_{\max}$ in $\mathbb{P}(\mathbb{R}^d)$, and therefore that $\mathcal{C}$ is nonempty. Using the expansion property (6) of Theorem 4.5.13 for Anosov representations, a similar reasoning to the proof of (6) $\Rightarrow$ (1) in Section 4.4.4 then shows that $\rho(\Gamma_0)\backslash\mathcal{C}$ is compact: see [54, § 8].

By transversality of ξ_1 and ξ_{d-1}, there are no nontrivial segments in $\partial\Omega_{\max}$ between points of $\xi_1(\partial_\infty\Gamma_0)$. This makes it possible to "smooth out" $\Omega_{\max}$ to obtain a $\rho(\Gamma_0)$-invariant properly convex open subset $\Omega' \subset \Omega_{\max}$ containing $\mathcal{C}$ such that $\partial\Omega'$ is C^1 and contains no segments: see [54, § 9]. The action of Γ_0 on Ω' via ρ is convex cocompact as desired. □

Proof of (2) $\Rightarrow$ (1)*:* Suppose that Γ_0 acts convex cocompactly via ρ on some properly convex open subset Ω of $\mathbb{P}(\mathbb{R}^d)$ such that $\partial\Omega$ is C^1 and contains no segments. Because $\partial\Omega$ contains no segments, the geodesic rays of Ω for the Hilbert metric d_Ω (Figure 4.11) are exactly the projective segments between a point of Ω and a point of $\partial\Omega$, and two such rays remain at bounded Hausdorff distance for d_Ω if and only if their endpoints in $\partial\Omega$ are the same. Therefore the convex core $\mathcal{C}^{\mathsf{cor}}_{\rho(\Gamma_0)}$, endowed with the restriction of d_Ω, is a geodesic metric space whose visual boundary $\partial_\infty\mathcal{C}^{\mathsf{cor}}_{\rho(\Gamma_0)}$ identifies with its ideal boundary $\overline{\mathcal{C}^{\mathsf{cor}}_{\rho(\Gamma_0)}} \cap \partial\Omega$ in $\partial\Omega$.

Using the fact that $\partial\Omega$ contains no segments, one can check by a limiting argument that all triangles in $\mathcal{C}^{\mathsf{cor}}_{\rho(\Gamma_0)}$ must be uniformly thin, i.e. the metric space $(\mathcal{C}^{\mathsf{cor}}_{\rho(\Gamma_0)}, \mathsf{d}_\Omega)$ is Gromov hyperbolic: see [54, Lem. 6.3]. Since the action of Γ_0 on $(\mathcal{C}^{\mathsf{cor}}_{\rho(\Gamma_0)}, \mathsf{d}_\Omega)$ via ρ is properly discontinuous, by isometries, with compact quotient, we deduce that Γ_0 is Gromov hyperbolic and (as in the proof of (1) $\Rightarrow$ (5) in Section 4.4.4) that any orbital map $\Gamma_0 \to \mathcal{C}^{\mathsf{cor}}_{\rho(\Gamma_0)}$ extends to a continuous ρ-equivariant boundary map $\xi_1 : \partial_\infty\Gamma_0 \to \partial_\infty\mathcal{C}^{\mathsf{cor}}_{\rho(\Gamma_0)} \subset \mathbb{P}(\mathbb{R}^d)$.

Consider the dual $\Omega^* = \{H \in \mathbb{P}((\mathbb{R}^d)^*) \mid H \cap \overline{\Omega} = \emptyset\}$ of Ω (where we view $\mathbb{P}((\mathbb{R}^d)^*)$ as the set of projective hyperplanes in $\mathbb{P}(\mathbb{R}^d)$). It is a properly convex open subset of $\mathbb{P}((\mathbb{R}^d)^*)$. The boundary $\partial\Omega^*$ of Ω^* is C^1 (because $\partial\Omega$ contains no segments), and it contains no segments (because $\partial\Omega$ is C^1). One can show that the dual action of Γ_0 on Ω^* via ρ is still convex cocompact: see [54, § 5]. Then the same reasoning as above yields a continuous ρ-equivariant boundary map $\xi_{d-1} : \partial_\infty\Gamma_0 \to \mathbb{P}((\mathbb{R}^d)^*)$.

By construction, ξ_1 and ξ_{d-1} are transverse: indeed, $\xi_{d-1}(w)$ is a supporting hyperplane to Ω at $\xi_1(w)$ for any w, and $\partial\Omega$ contains no segments. One checks that ξ_1 and ξ_{d-1} are dynamics-preserving and (using Lemma 4.5.16) that $(\mu_1 - \mu_2)(\rho(\gamma)) \to +\infty$ as $\mathsf{d}_{\mathrm{Cay}(\Gamma_0)}(e,\gamma) \to +\infty$: see [54, § 7]. We then apply the implication (4) $\Rightarrow$ (1) of Theorem 4.5.13. □

Applications

Theorem 4.5.20 and Corollaries 4.5.21–4.5.22 give geometric interpretations for Anosov representations.

Example 4.5.23 For odd d, any Hitchin representation $\rho : \pi_1(S) \to \mathrm{PSL}(d,\mathbb{R})$ as in Section 4.3.3 preserves a nonempty properly convex open subset of $\mathbb{P}(\mathbb{R}^d)$ (see [53, 54, 141]). Therefore these representations are strongly convex cocompact in $\mathbb{P}(\mathbb{R}^d)$ by Theorem 4.5.20. This extends the case $d = 3$ due to Choi and Goldman (see the proof of Theorem 4.3.7).

Example 4.5.24 For $n \geq 2$, any maximal representation $\rho : \pi_1(S) \to \mathrm{SO}(2,n)$ as in Section 4.3.3 preserves a nonempty properly convex open subset of $\mathbb{P}(\mathbb{R}^{n+2})$, contained in $\mathbb{H}^{2,n-1} = \{[v] \in \mathbb{P}(\mathbb{R}^{n+2}) \mid \langle v, v\rangle_{2,n} < 0\}$ (see [47, 53]). Therefore these representations are strongly convex cocompact in $\mathbb{P}(\mathbb{R}^{n+2})$ by Theorem 4.5.20, and in fact $\mathbb{H}^{2,n-1}$-*convex cocompact* as in Section 4.3.4 (see the comments after Theorem 4.5.20).

Theorem 4.5.20 can also be used to construct new examples of Anosov representations. One source of examples comes from representations of Coxeter groups as linear reflection groups. Recall that a Coxeter group is a group with a presentation by generators and relations of the form

$$W = \langle s_1, \ldots, s_N \mid (s_i s_j)^{m_{i,j}} = e \quad \forall 1 \leq i, j \leq N \rangle \tag{4.7}$$

where $m_{i,i} = 1$ (i.e. s_i is an involution) and $m_{i,j} \in \{2, 3, 4, \ldots\} \cup \{\infty\}$ for all $i \neq j$. (By convention, $(s_i s_j)^\infty = e$ means that $s_i s_j$ has infinite order in the group W.) Vinberg [130] developed a theory of representations of W *as a reflection group* in a finite-dimensional real vector space V: these are by definition representations $\rho : W \to \mathrm{GL}(V)$ such that each $\rho(s_i)$ is a linear reflection in a hyperplane of V and the configuration of these reflections is such that ρ is injective, discrete, and the associated fundamental polytope has nonempty interior. These representations may preserve a nondegenerate quadratic form on V (e.g. the image of ρ could be a discrete subgroup of $\mathrm{O}(n,1)$ generated by orthogonal reflections in the faces of a right-angled polyhedron of $\mathbb{H}^n$ as in Section 4.4.3), but in general they need not preserve any nonzero quadratic form. Representations of W as a reflection group constitute a subset $\mathrm{Hom}_{\mathrm{refl}}(W, \mathrm{GL}(V))$ of $\mathrm{Hom}(W, \mathrm{GL}(V))$ which is semialgebraic (defined by finitely many equalities and inequalities).

Example 4.5.25 ([53, 56, 110]) Let W be a Coxeter group in N generators as in (4.7). Suppose W is infinite and Gromov hyperbolic. Then for any $d \geq N$ there exist representations $\rho : W \to \mathrm{SL}^{\pm}(d, \mathbb{R})$ of W as a reflection group which are strongly convex cocompact in $\mathbb{P}(\mathbb{R}^d)$; for $d \geq 2N-2$, they constitute the full interior of $\mathrm{Hom}_{\mathrm{refl}}(W, \mathrm{GL}(d, \mathbb{R}))$. By Theorem 4.5.20, these representations are P_1-Anosov.

By [64], a conclusion similar to that of Example 4.5.25 holds if W is an infinite Gromov hyperbolic group which is not necessarily a Coxeter group, but which embeds into a right-angled Coxeter group as a so-called *quasiconvex subgroup*. Using celebrated work of Agol and Haglund–Wise, this provides Anosov representations for a large class of infinite Gromov hyperbolic groups, namely all those which admit a properly discontinuous and cocompact action on a CAT(0) cube complex.

One can also use the geometric interpretation of Anosov representations from Theorem 4.5.20 to prove that free products $\Gamma_1 * \Gamma_2$ of Anosov subgroups Γ_1, Γ_2 are Anosov [55], using a generalisation of the ping pong arguments of Sections 4.3.1–4.3.2. For instance, for $1 \leq i \leq d-1$, let $\tau_i : \mathrm{SL}(d, \mathbb{R}) \to \mathrm{SL}(S^2(\Lambda^i \mathbb{R}^d))$ be the second symmetric power of the i-th exterior power of the standard representation as in Corollary 4.5.22, let $V_i' := S^2(\Lambda^i \mathbb{R}^d) \oplus \mathbb{R}$, and let $\tau_i' : \mathrm{SL}(d, \mathbb{R}) \to \mathrm{SL}(V_i')$ be the direct sum of τ_i and of the trivial representation. Then the following holds.

Example 4.5.26 ([55]) Let $1 \leq i \leq d-1$ and let Γ_1, Γ_2 be any discrete subgroups of $\mathrm{SL}(d, \mathbb{R})$. Then there exists $g \in \mathrm{SL}(V_i')$ such that the representation $\rho : \Gamma_1 * g\Gamma_2 g^{-1} \to \mathrm{SL}(V_i')$ induced by the restrictions of τ_i' to Γ_1 and $g\Gamma_2 g^{-1}$ has finite kernel and discrete image. If moreover Γ_1 and Γ_2 are P_i-Anosov, then we can choose g so that ρ is P_1-Anosov.

(Note that beyond Anosov representations, this construction can be used to prove that the free product of two $\mathbb{Z}$-linear groups is $\mathbb{Z}$-linear, and that there exist Zariski-dense discrete subgroups of $\mathrm{SL}(V_i')$ which are not lattices but contain cocompact lattices of $\tau_i'(\mathrm{SL}(d-1, \mathbb{R}))$: see [55].)

We refer to [61, 62] for other combination theorems for Anosov representations which do not use Theorem 4.5.20.

Finally we note that, although we have seen many constructions of Anosov representations above, it is expected that not every linear Gromov hyperbolic group admits an Anosov representation into some noncompact semisimple Lie group; a concrete example remains to be found.

4.5.3 Generalisations of Anosov subgroups

In the past few years, several fruitful generalisations of Anosov subgroups have appeared, which are currently being actively investigated. These generalisations exploit both the dynamical definition of Anosov subgroups from Section 4.5.1 and their geometric characterisation from Section 4.5.2. Let us briefly mention three of these generalisations.

More general convex cocompact subgroups

We just saw in Theorem 4.5.20 and Corollaries 4.5.21 and 4.5.22 that Anosov representations can be characterised geometrically by a strong convex cocompactness condition in projective space. Here *strong* refers to the regularity imposed on the properly convex open set Ω (its boundary $\partial\Omega$ should be C^1 and contain no segments).

It is natural to try to generalise Anosov representations by relaxing this strong regularity requirement. Removing it altogether in Definition 4.5.19 leads to a notion which is not stable under small deformations (see [54, 55]). Instead, we impose the following mild condition, which relies on the notions of *full orbital limit set* and *convex core.*

Definition 4.5.27 ([54]) Let Ω be a properly convex open subset of $\mathbb{P}(\mathbb{R}^d)$. Let Γ_0 be a group and $\rho : \Gamma_0 \to \mathrm{Aut}(\Omega) \subset \mathrm{PGL}(d,\mathbb{R})$ a representation.

- The *full orbital limit set* $\Lambda^{\mathsf{orb}}_{\rho(\Gamma_0)}(\Omega)$ of $\rho(\Gamma_0)$ in Ω is the set of all accumulation points in $\partial\Omega$ of all possible $\rho(\Gamma_0)$-orbits of Ω.
- The *convex core* $\mathcal{C}^{\mathsf{cor}}_{\rho(\Gamma_0)}(\Omega) \subset \Omega$ of $\rho(\Gamma_0)$ is the convex hull of $\Lambda^{\mathsf{orb}}_{\rho(\Gamma_0)}(\Omega)$ in Ω.
- The action of Γ_0 on Ω via ρ is *convex cocompact* if it is properly discontinuous and if there exists a nonempty $\rho(\Gamma_0)$-invariant convex subset $\mathcal{C}$ of Ω such that $\rho(\Gamma_0)\backslash\mathcal{C}$ is compact and $\mathcal{C}$ is "large enough" in the sense that it contains the convex core $\mathcal{C}^{\mathsf{cor}}_{\rho(\Gamma_0)}(\Omega)$.

Note that Definition 4.5.27 coincides with Definition 4.5.19 when $\partial\Omega$

does not contain any nontrivial projective segments. Indeed, in that case the full orbital limit set $\Lambda^{\mathsf{orb}}_{\rho(\Gamma_0)}(\Omega)$ is the set of accumulation points of any single $\rho(\Gamma_0)$-orbit of Ω, hence any nonempty $\rho(\Gamma_0)$-invariant convex subset $\mathcal{C}$ of Ω contains the convex core $\mathcal{C}^{\mathsf{cor}}_{\rho(\Gamma_0)}(\Omega)$ (see the comments after Definition 4.5.19).

Definition 4.5.28 Given a group Γ_0, we say that a representation $\rho : \Gamma_0 \to \mathrm{PGL}(d, \mathbb{R})$ is *convex cocompact in* $\mathbb{P}(\mathbb{R}^d)$ if Γ_0 acts convex cocompactly via ρ on some properly convex open subset Ω of $\mathbb{P}(\mathbb{R}^d)$. In that case, we also say that the image $\rho(\Gamma_0)$ is *convex cocompact in* $\mathbb{P}(\mathbb{R}^d)$.

As above, if ρ is convex cocompact in $\mathbb{P}(\mathbb{R}^d)$, then it has finite kernel and discrete image, and the group Γ_0 is finitely generated.

This notion turns out to be quite fruitful: by [54], the set of convex cocompact representations is open in $\mathrm{Hom}(\Gamma_0, \mathrm{PGL}(d, \mathbb{R}))$, and it is stable under duality and under embedding into a larger projective space; moreover, a representation $\rho : \Gamma_0 \to \mathrm{PGL}(d, \mathbb{R})$ is strongly convex cocompact in $\mathbb{P}(\mathbb{R}^d)$ in the sense of Theorem 4.5.20.(2) if and only if it is convex cocompact in $\mathbb{P}(\mathbb{R}^d)$ and Γ_0 is Gromov hyperbolic. Theorem 4.5.20 then shows that convex cocompact representations are generalisations of P_1-Anosov representations, for finitely generated infinite groups Γ_0 that are not necessarily Gromov hyperbolic, and that may therefore contain subgroups isomorphic to $\mathbb{Z}^2$ (see Remark 4.4.11).

In fact, Weisman [131] has recently given a dynamical characterisation of convex cocompact representations of Γ_0 that extends the characterisation of Anosov representations of Theorem 4.5.13.(6). The expansion now takes place in various Grassmannians (not only projective space): namely, at each face of the full orbital limit set in $\partial\Omega$, there is expansion in the Grassmannian of i-planes of $\mathbb{R}^d$ where $i - 1$ is the dimension of the face.

We conclude this section by mentioning a few examples of convex cocompact groups that are not necessarily Gromov hyperbolic (i.e. that are not necessarily Anosov subgroups).

Example 4.5.29 Let Γ be a discrete subgroup of $\mathrm{PGL}(d, \mathbb{R})$ *dividing* (i.e. acting properly discontinuously with compact quotient on) some properly convex open subset Ω of $\mathbb{P}(\mathbb{R}^d)$. Then $\Lambda^{\mathsf{orb}}_{\rho(\Gamma_0)}(\Omega) = \partial\Omega$ and the action of Γ on Ω is convex cocompact. By [16], the group Γ is Gromov hyperbolic if and only if $\partial\Omega$ contains no segments. Examples where $\partial\Omega$ contains segments include the symmetric divisible convex sets $\Omega_{\mathrm{sym}} \subset \mathbb{P}(\mathrm{Sym}_{d'}(\mathbb{R})) \simeq \mathbb{P}(\mathbb{R}^d)$ with $d = d'(d'+1)/2 \geq 6$ discussed before

Corollary 4.5.21. The first nonsymmetric irreducible examples were constructed in small dimensions ($4 \leq d \leq 7$) by Benoist [18]; examples in all dimensions $d \geq 4$ were recently constructed by Blayac and Viaggi [23].

Example 4.5.30 For Γ dividing Ω as in Example 4.5.29, we can lift Γ to a subgroup $\hat{\Gamma}$ of $\mathrm{SL}^{\pm}(d,\mathbb{R})$ preserving a properly convex cone of $\mathbb{R}^d$ lifting Ω, and then embed $\hat{\Gamma}$ into $\mathrm{PGL}(D,\mathbb{R})$ for some $D \geq d$. By the result of [54] mentioned above, the discrete subgroup of $\mathrm{PGL}(D,\mathbb{R})$ obtained in this way will be convex cocompact in $\mathbb{P}(\mathbb{R}^D)$; moreover, it will remain convex cocompact in $\mathbb{P}(\mathbb{R}^D)$ after any small deformation in $\mathrm{PGL}(D,\mathbb{R})$.

Recall that, given a Coxeter group W as in (4.7), a subgroup of W is called *standard* if it is generated by a subset of the generating set $\{s_1,\dots,s_N\}$. The Coxeter group W is called *affine* if it is irreducible (i.e. it cannot be written as a direct product of two nontrivial Coxeter groups) and if it is virtually (i.e. it admits a finite-index subgroup which is) isomorphic to $\mathbb{Z}^k$ for some $k \geq 1$. Affine Coxeter groups have been completely classified; they include the Coxeter groups of type $\tilde{A}_k$ (which are virtually isomorphic to $\mathbb{Z}^k$), where we say that W is of type $\tilde{A}_1$ if $N = 2$ and $m_{1,2} = \infty$, and W is of type $\tilde{A}_{N-1}$ for $N \geq 3$ if $m_{i,j} = 3$ for all $i \neq j$ with $|i-j| = 1 \bmod N$ and $m_{i,j} = 2$ for all other $i \neq j$.

Example 4.5.31 ([56]) As a generalisation of Example 4.5.25, let W be a Coxeter group in N generators as in (4.7). Suppose W is infinite. Then there exists a representation $\rho \in \mathrm{Hom}_{\mathrm{refl}}(W,\mathrm{GL}(d,\mathbb{R}))$ which is convex cocompact in $\mathbb{P}(\mathbb{R}^d)$ for some d if and only if any affine standard subgroup of W is of type $\tilde{A}_k$ for some $k \geq 1$ and W does not contain a direct product of two infinite standard subgroups. If this holds, then we can take any $d \geq N$ and the convex cocompact representations then constitute a large open subset of $\mathrm{Hom}_{\mathrm{refl}}(W,\mathrm{GL}(d,\mathbb{R}))$: see [56, § 1.5].

Examples 4.5.29, 4.5.30, and 4.5.31 provide many convex cocompact groups which are not Gromov hyperbolic. (In Example 4.5.31, the group W is nonhyperbolic as soon as it contains an affine standard subgroup of type $\tilde{A}_k$ with $k \geq 2$, see Remark 4.4.11.)

Some of these groups are still relatively hyperbolic: e.g. in Example 4.5.31, the group W is relatively hyperbolic with respect to a collection of virtually abelian subgroups of rank ≥ 2 (see [56, Cor. 1.7]). We refer to [93] for general results about the structure of relatively hyperbolic groups which are convex cocompact in $\mathbb{P}(\mathbb{R}^d)$ and about the geometry of the associated convex sets. On the other hand, Example 4.5.29 includes, for $d = d'(d'+1)/2 \geq 6$, discrete subgroups of $\mathrm{PGL}(d,\mathbb{R})$ which divide a symmetric properly convex open set $\Omega_{\mathrm{sym}} \subset \mathbb{P}(\mathbb{R}^d) \simeq \mathbb{P}(\mathrm{Sym}_{d'}(\mathbb{R}))$ and which are isomorphic to cocompact lattices of $\mathrm{PGL}(d',\mathbb{R})$, hence *not* relatively hyperbolic (see Section 4.2.3). Further examples of convex cocompact groups which are not relatively hyperbolic can be constructed e.g. using free products inside larger projective spaces: see [55].

Relatively Anosov subgroups

Kapovich–Leeb [97] and Zhu [137, 138] have developed notions of a *relatively Anosov representation* of a relatively hyperbolic group into a noncompact semisimple Lie group G, which generalise the notion of an Anosov representation of a hyperbolic group into G from Section 4.5.1. They obtain various characterisations similar to those of Theorem 4.5.13. The original definition of Anosov representations using flows (Definition 4.5.1 and Condition 4.5.4) is recovered in this more general setting by recent work of Zhu and Zimmer [139].

Extending Fact 4.5.6, if Γ_0 is relatively hyperbolic with respect to a collection of subgroups (called *peripheral subgroups*), then relatively Anosov representations of Γ_0 into a given G are stable under small deformations that preserve the conjugacy class of the image of each peripheral subgroup [97, 139].

Any relatively Anosov representation $\rho : \Gamma_0 \to G$ has finite kernel and discrete image $\rho(\Gamma_0)$, called a *relatively Anosov subgroup* of G. There are many examples of relatively Anosov subgroups (see [97, 140]), including:

- geometrically finite subgroups of G for $\mathrm{rank}_{\mathbb{R}}(G) = 1$ (Definition 4.4.1),
- some of the Schottky groups of Section 4.3.2,
- the images of certain compositions $\tau\circ\sigma_0 : \Gamma_0 \to G$ where $\sigma_0 : \Gamma_0 \to G'$ is a geometrically finite representation into a semisimple Lie group G' with $\mathrm{rank}_{\mathbb{R}}(G') = 1$ and $\tau : G' \to G$ is a representation with compact kernel (e.g. Fact 4.5.7 generalises to the relative setting);
- similarly to Section 4.3.3, small deformations in G of such $\tau \circ \sigma_0(\Gamma_0)$, preserving the conjugacy class of the image of each peripheral subgroup;
- certain representations of $\mathrm{PSL}(2,\mathbb{Z})$ into $\mathrm{PGL}(3,\mathbb{R})$ constructed by Schwartz [123] by iterating Pappus's theorem (see [140, § 13]);
- for a finite-volume hyperbolic surface S, the images of positive (in the sense of Fock–Goncharov [70]) type-preserving representations of $\Gamma_0 =$

$\pi_1(S)$ into a real split simple Lie group G, see [43] (for closed S, these coincide with the Hitchin representations of Sections 4.3.3–4.3.4);
- discrete subgroups of $\mathrm{PGL}(d,\mathbb{R})$ preserving a properly convex open subset Ω of $\mathbb{P}(\mathbb{R}^d)$ with strong regularity ($\partial\Omega$ is C^1 with no segments), and whose action on Ω is geometrically finite in the sense of [52].

It would be interesting to determine whether relatively Anosov representations of relatively hyperbolic groups can also be fully characterised geometrically similarly to Theorem 4.5.20 and Corollaries 4.5.21–4.5.22.

Extended geometrically finite subgroups

Recently, Weisman [132, 133] has introduced a notion of *extended geometrically finite* (or *EGF* for short) representation of a relatively hyperbolic group. This is a dynamical notion, which extends a dynamical characterisation of Anosov representations in terms of multicones [26]. EGF representations include all Anosov or relatively Anosov representations, all representations of relatively hyperbolic groups which are convex cocompact in the sense of Definition 4.5.28, as well as other examples (see [133, Th. 1.5–1.7] and [23, Prop. 6.5 & Rem. 6.2]). They are stable under certain small deformations, called *peripherally stable*, for which the dynamics of the peripheral subgroups does not degenerate too much.

On the other hand, it would be interesting to define a general notion of geometric finiteness in convex projective geometry (involving properly convex open subsets Ω of $\mathbb{P}(\mathbb{R}^d)$ where $\partial\Omega$ may contain segments or not be C^1), and to make the link with Weisman's EGF representations. A good notion of geometric finiteness should contain as a particular case the notion of convex cocompactness from Definition 4.5.27. More precisely, a convex projective manifold $M = \Gamma\backslash\Omega$ should be geometrically finite if its convex core $\Gamma\backslash\mathcal{C}^{\mathsf{cor}}_{\Gamma}(\Omega)$ (see Definition 4.5.27) is covered by a compact piece and finitely many ends of M, called *cusps*, with a controlled geometry. It is not completely clear what the right definition of a cusp should be. Following Cooper, Long, and Tillmann [48], one could define a (full) cusp to be the image in M of some convex open subset of Ω whose stabiliser in Γ is infinite and does not contain any hyperbolic element (i.e. any element of this stabiliser has all its complex eigenvalues of the same modulus); in that case, the cusp is diffeomorphic to the direct product of $\mathbb{R}$ with an affine $(d-2)$-dimensional manifold called the *cusp cross-section*, and the stabiliser of the cusp is virtually nilpotent [48, Th. 5.3]. The cusp is said to have *maximal rank* if the cross-section is compact. A more general notion of cusp of maximal rank, where the stabiliser may contain hyperbolic elements but is still assumed to be virtually nilpotent, was studied in [5]. A notion of geometric finiteness involving only such generalised cusps of maximal rank was introduced and studied in [136], where it was characterised in dynamical terms. Examples (both of finite and infinite volume) were constructed in [6, 24] as small deformations of finite-volume real hyperbolic manifolds, using a stability result from [49]; the corresponding representations are EGF by [133]. On the other hand, the study of convex projective cusps of nonmaximal rank, possibly allowing for hyperbolic elements, is still at its infancy, and a good general notion of geometric finiteness in this setting still remains to be found, together with appropriate dynamical characterisations.

Acknowledgements. These notes grew out of a minicourse given at Groups St Andrews 2022 in Newcastle. I would like to warmly thank the organisers (Colin Campbell, Martyn Quick, Edmund Robertson, Colva Roney-Dougal, and David Stewart) for a very stimulating and enjoyable meeting.

I am grateful to Pierre-Louis Blayac, Jean-Philippe Burelle, Jeffrey Danciger, Sami Douba, Balthazar Fléchelles, Ilia Smilga, Jérémy Toulisse, and the referee for many useful comments on a preliminary version of this text, and to Florian Stecker for providing Figure 4.9.

This project received funding from the European Research Council (ERC) under the European Union's Horizon 2020 research and innovation programme (ERC starting grant DiGGeS, grant agreement No. 715982). The final proofreading was done at the Institute for Advanced Study in Princeton, supported by the National Science Foundation under Grant No. DMS-1926686.

References

[1] I. Agol, "Systoles of hyperbolic 4-manifolds", arXiv:0612290.

[2] G. Arzhantseva, A. Minasyan, D. Osin, "The SQ-universality and residual properties of relatively hyperbolic groups", J. Algebra **315** (2007), 165–177.

[3] U. Bader, P.-E. Caprace, A. Furman, A. Sisto, "Hyperbolic actions of higher-rank lattices come from rank-one factors", arXiv:2206.06431.

[4] U. Bader, A. Furman, T. Gelander, N. Monod, "Property (T) and rigidity for actions on Banach spaces", Acta Math. **198** (2007), 57–105.

[5] S. Ballas, D. Cooper, A. Leitner, "Generalized cusps on convex projective manifolds: Classification", J. Topol. **13** (2020), 1455–1496.

[6] S. Ballas, L. Marquis, "Convex projective bendings of hyperbolic manifolds", Groups Geom. Dyn. **14** (2020), 653–688.

[7] E. P. van den Ban, M. Flensted-Jensen, H. Schlichtkrull, "Harmonic analysis on semisimple symmetric spaces: A survey of some general results", in "Representation theory and automorphic forms", Proceedings of an instructional conference (Edinburgh, UK, 1996), 191–217, Proceedings of Symposia in Pure Mathematics, vol. 61, American Mathematical Society, Providence, RI, 1997.

[8] T. Barbot, "Flag structures on Seifert manifolds", Geom. Topol. **5** (2001), 227–266.

[9] T. Barbot, "Deformations of Fuchsian AdS representations are quasi-Fuchsian", J. Differential Geom. **101** (2015), 1–46.

[10] T. Barbot, Q. Mérigot, "Anosov AdS representations are quasi-Fuchsian", Groups Geom. Dyn. **6** (2012), 441–483.

[11] H. Bass, J. W. Milnor, J.-P. Serre, "Solution of the congruence subgroup problem for SL_n $(n \geq 3)$ and Sp_{2n} $(n \geq 2)$", Pub. Math. Inst. Hautes Études Sci. **33** (1967), 59–137.

[12] J. Behrstock, C. Druţu, L. Mosher, "Thick metric spaces, relative hyperbolicity, and quasi-isometric rigidity", Math. Ann. **344** (2009), 543–595.

[13] M. V. Belolipetsky, S. A. Thomson, "Systoles of hyperbolic manifolds", Algebr. Geom. Topol. **11** (2010).

[14] N. Benakli, I. Kapovich, "Boundaries of hyperbolic groups", in "Combinatorial and geometric group theory (New York, 2000/Hoboken, NJ, 2001)", 39–93, Contemporary Mathematics, vol. 296, American Mathematical Society, Providence, RI, 2002.

[15] Y. Benoist, "Propriétés asymptotiques des groupes linéaires", Geom. Funct. Anal. **7** (1997), 1–47.

[16] Y. Benoist, "Convexes divisibles I", in "Algebraic groups and arithmetic", Tata Inst. Fund. Res. Stud. Math. **17** (2004), 339–374.

[17] Y. Benoist, "Convex divisibles III", Ann. Sci. Éc. Norm. Supér. **38** (2005), 793–832.

[18] Y. Benoist, "Convexes divisibles IV", Invent. Math. **164** (2006), 249–278.

[19] Y. Benoist, "Five lectures on lattices in semisimple Lie groups", in "Géométries à courbure négative ou nulle, groupes discrets et rigidités", 117–176, Séminaires et Congrès, vol. 18, Société Mathématique de France, Paris, 2009.

[20] N. Bergeron, F. Haglund, D. T. Wise, "Hyperplane sections in arithmetic hyperbolic manifolds", J. Lond. Math. Soc. **83** (2011), 431–448.

[21] J. Beyrer, F. Kassel, "$\mathbb{H}^{p,q}$-convex cocompactness and higher higher Teichmüller spaces", arXiv:2305.15031.

[22] J. Beyrer, M. B. Pozzetti, "Positive surface group representations in PO(p,q)", arXiv:2106.14725.

[23] P.-L. Blayac, G. Viaggi, "Divisible convex sets with properly embedded cones", arXiv:2302.07177.

[24] M. D. Bobb, “Convex projective manifolds with a cusp of any non-diagonalizable type”, J. Lond. Math. Soc. **100** (2019), 183–202.

[25] J. Bochi, N. Gourmelon, “Some characterizations of domination”, Math. Z. **263** (2009), 221–231.

[26] J. Bochi, R. Potrie, A. Sambarino, “Anosov representations and dominated splittings”, J. Eur. Math. Soc. **21** (2019), 3343–3414.

[27] F. Bonahon, G. Dreyer, “Parameterizing Hitchin components”, Duke Math. J. **163** (2014), 2935–2975.

[28] A. Borel, “Density properties for certain subgroups of semi-simple groups without compact components”, Ann. of Math. **72** (1960), 179–188.

[29] A. Borel, “Compact Clifford–Klein forms of symmetric spaces”, Topology **2** (1963), 111–122.

[30] A. Borel, Harish-Chandra, “Arithmetic subgroups of algebraic groups”, Ann. of Math. **75** (1962), 485–535.

[31] B. H. Bowditch, “Geometrical finiteness with variable negative curvature”, Duke Math. J. **77** (1995), 229–274.

[32] S. Bradlow, “Global properties of Higgs bundle moduli spaces”, arXiv:2312.00762, to appear in a volume in honor of Peter Newstead on the occasion of his 80th birthday.

[33] S. Bradlow, B. Collier, O. García-Prada, P. B. Gothen, A. Oliveira, “A general Cayley correspondence and higher Teichmüller spaces”, to appear in Ann. Math.

[34] S. B. Bradlow, O. García-Prada, P. B. Gothen, “Maximal surface group representations in isometry groups of classical Hermitian symmetric spaces”, Geom. Dedicata **122** (2006), 185–213.

[35] M. Bridgeman, R. D. Canary, F. Labourie, A. Sambarino, “The pressure metric for Anosov representations”, Geom. Funct. Anal. **25** (2015), 1089–1179.

[36] J.-P. Burelle, F. Kassel, “Crooked surfaces and symplectic Schottky groups”, in preparation.

[37] M. Burger, A. Iozzi, F. Labourie, A. Wienhard, “Maximal representations of surface groups: Symplectic Anosov structures”, Pure Appl. Math. Q. **1** (2005), special issue in memory of Armand Borel, 543–590.

[38] M. Burger, A. Iozzi, A. Wienhard, “Surface group representations with maximal Toledo invariant”, Ann. of Math. **172** (2010), 517–566.

[39] M. Burger, A. Iozzi, A. Wienhard, “Higher Teichmüller spaces: from $\mathrm{SL}(2,\mathbb{R})$ to other Lie groups”, in “Handbook of Teichmüller theory IV”, 539–618, IRMA Lectures in Mathematics and Theoretical Physics, vol. 19, EMS Publishing House, Zürich, 2014.

[40] M. Burger, M. B. Pozzetti, “Maximal representations, non Archimedean Siegel spaces, and buildings”, Geom. Topol. **21** (2017), 3539–3599.

[41] R. D. Canary, “Hitchin representations of Fuchsian groups”, EMS Surv. Math. Sci. **9** (2022), 355–388, special issue in honor of Dennis Sullivan on the occasion of his 80th birthday.

[42] R. D. Canary, M. Lee, A. Sambarino, M. Stover, "Projective Anosov Schottky groups and strongly amalgam Anosov representations", appendix to "Amalgam Anosov representations" by Canary, Lee and Stover, Geom. Topol. **21** (2017), 240–248.

[43] R. D. Canary, T. Zhang, A. Zimmer, "Cusped Hitchin representations and Anosov representations of geometrically finite Fuchsian groups", Adv. Math. **404** (2022), 1–67.

[44] S. Cantat, "Progrès récents concernant le programme de Zimmer [d'après A. Brown, D. Fisher et S. Hurtado]", Séminaire Bourbaki, Exposé 1136, Astérisque **414** (2019), 1–48.

[45] C. Chevalley, "Théorie des groupes de Lie : II. Groupes algébriques", Actualités Scientifiques et Industrielles, vol. 1152, Hermann & Cie, Paris, 1951.

[46] S. Choi, W. M. Goldman, "Convex real projective structures on closed surfaces are closed", Proc. Amer. Math. Soc. **118** (1993), 657–661.

[47] B. Collier, N. Tholozan, J. Toulisse, "The geometry of maximal representations of surface groups into $SO_0(2,n)$", Duke Math. J. **168** (2019), 2873–2949.

[48] D. Cooper, D. D. Long, S. Tillmann, "On convex projective manifolds and cusps", Adv. Math. **277** (2015), 181–251.

[49] D. Cooper, D. Long, S. Tillmann, "Deforming convex projective manifolds", Geom. Topol. **22** (2018), 1349–1404.

[50] D. Cooper, S. Tillmann, "The space of properly-convex structures", arXiv:2009.06568.

[51] M. Coornaert, T. Delzant, A. Papadopoulos, "Géométrie et théorie des groupes. Les groupes hyperboliques de Gromov", Lecture Notes in Mathematics, vol. 1441, Springer-Verlag, Berlin, 1990.

[52] M. Crampon, L. Marquis, "Finitude géométrique en géométrie de Hilbert", with an appendix by C. Vernicos, Ann. Inst. Fourier **64** (2014), 2299–2377.

[53] J. Danciger, F. Guéritaud, F. Kassel, "Convex cocompactness in pseudo-Riemannian hyperbolic spaces", Geom. Dedicata **192** (2018), 87–126, special issue "Geometries: A Celebration of Bill Goldman's 60th Birthday".

[54] J. Danciger, F. Guéritaud, F. Kassel, "Convex cocompact actions in real projective geometry", Ann. Sci. Éc. Norm. Supér. **57** (2024), 1751–1841.

[55] J. Danciger, F. Guéritaud, F. Kassel, "Combination theorems in convex projective geometry", arXiv:2407.09439.

[56] J. Danciger, F. Guéritaud, F. Kassel, G.-S. Lee, L. Marquis, "Convex cocompactness for Coxeter groups", to appear in J. Eur. Math. Soc.

[57] P. Deligne, G. D. Mostow, "Monodromy of hypergeometric functions and non-lattice integral monodromy", Publ. Math. Inst. Hautes Études Sci. **63** (1986), 5–89.

[58] M. Deraux, "A new non-arithmetic lattice in PU(3,1)", Algebr. Geom. Topol. **20** (2020), 925–963.

[59] M. Deraux, J. R. Parker, J. Paupert, "New non-arithmetic complex hyperbolic lattices", Invent. Math. **203** (2016), 681–771.

[60] M. Desgroseilliers, F. Haglund, "On some convex cocompact groups in real hyperbolic space", Geom. Topol. **17** (2013), 2431–2484.

[61] S. Dey, M. Kapovich, "Klein–Maskit combination theorem for Anosov subgroups: Free products", Math. Z. **305** (2023), article number 35.

[62] S. Dey, M. Kapovich, "Klein–Maskit combination theorem for Anosov subgroups: Amalgams", arXiv:2301.02345.

[63] S. Douba, "Systoles of hyperbolic hybrids", arXiv:2309.16051.

[64] S. Douba, B. Fléchelles, T. Weisman, F. Zhu, "Cubulated hyperbolic groups admit Anosov representations", arXiv:2309.03695.

[65] C. Druţu, M. Kapovich, "Geometric group theory", American Mathematical Society Colloquium Publications, vol. 63, American Mathematical Society, Providence, RI, 2018.

[66] P. B. Eberlein, "Geometry of nonpositively curved manifolds", Chicago Lectures in Mathematics, The University of Chicago Press, Chicago, IL, 1996.

[67] D. Fisher, "Superrigidity, arithmeticity, normal subgroups: results, ramifications and directions", in "Dynamics, geometry, number theory — The impact of Margulis on modern mathematics", 9–46, Chicago University Press, Chicago, IL 2022.

[68] D. Fisher, "Rigidity, lattices and invariant measures beyond homogeneous dynamics", Proceedings of the International Congress of Mathematicians 2022 (ICM 2022), 3484–3507, EMS Publishing House, Berlin, 2023.

[69] D. Fisher, G. A. Margulis, "Local rigidity for cocycles", in "Surveys in differential geometry, vol. VIII" (Boston, MA, 2002), 191–234, Surveys in Differential Geometry, vol. 8, International Press, Somerville, MA, 2003.

[70] V. V. Fock, A. B. Goncharov, "Moduli spaces of local systems and higher Teichmüller theory", Publ. Math. Inst. Hautes Études Sci. **103** (2006), 1–211.

[71] M. Fraczyk, T. Gelander, "Infinite volume and infinite injectivity radius", Ann. of Math. **197** (2023), 389–421.

[72] C. Frances, "The conformal boundary of Margulis space-times", C. R. Acad. Sci. Paris **336** (2003), 751–756.

[73] A. Furman, "A survey of measured group theory", in "Geometry, rigidity, and group actions", 296–374, Chicago Lectures in Mathematics, University of Chicago Press, Chicago, IL, 2011.

[74] O. García-Prada, "Higgs bundles and higher Teichmüller spaces", Handbook of Teichmüller theory VII", 239–285, IRMA Lectures in Mathematics and Theoretical Physics, vol. 30, EMS Publishing House, Zürich, 2020.

[75] O. García-Prada, P. B. Gothen, I. Mundet i Riera, "The Hitchin-Kobayashi correspondence, Higgs pairs and surface group representations", arXiv:0909.4487.

[76] G. Gardam, D. Kielak, A. D. Logan, "Algebraically hyperbolic groups", arXiv:2112.01331.

[77] W. M. Goldman, "Discontinuous groups and the Euler class", PhD thesis, University of California, Berkeley, 1980, see `https://www.math.umd.edu/~wmg/PhDthesis.pdf`

[78] W. M. Goldman, "Geometric structures on manifolds", Graduate Studies in Mathematics, vol. 227, American Mathematical Society, Providence, RI, 2022.

[79] M. Gromov, "Hyperbolic groups", in "Essays in group theory", 75–263, Mathematical Sciences Research Institute Publications, vol. 8, Springer, New York, 1987.

[80] M. Gromov, I. Piatetski-Shapiro, "Nonarithmetic groups in Lobachevsky spaces", Publ. Math. Inst. Hautes Études Sci. **66** (1988), 93–103.

[81] F. Guéritaud, O. Guichard, F. Kassel, A. Wienhard, "Anosov representations and proper actions", Geom. Topol. **21** (2017), 485–584.

[82] F. Guéritaud, F. Kassel, "Maximally stretched laminations on geometrically finite hyperbolic manifolds", Geom. Topol. **21** (2017), 693–840.

[83] O. Guichard, F. Labourie, A. Wienhard, "Positivity and representations of surface groups", arXiv:2106.14584.

[84] O. Guichard, A. Wienhard, "Anosov representations : Domains of discontinuity and applications", Invent. Math. **190** (2012), 357–438.

[85] O. Guichard, A. Wienhard, "Positivity and higher Teichmüller theory", Proceedings of the European Congress of Mathematics (Zürich 2016), 289–310, European Mathematical Society, 2018.

[86] O. Guichard, A. Wienhard, "Generalizing Lusztig's total positivity", arXiv:2208.10114.

[87] T. Haettel, "Hyperbolic rigidity of higher rank lattices", with an appendix by C. Horbez and V. Guirardel, Ann. Sci. Éc. Norm. Supér. **53** (2020), 437–468.

[88] E. Hamilton, "Geometrical finiteness for hyperbolic orbifolds", Topology **37** (1998), 635–657.

[89] B. B. Healy, G. C. Hruska, "Cusped spaces and quasi-isometries of relatively hyperbolic groups", arXiv:2010.09876.

[90] S. Helgason, "Differential geometry, Lie groups, and symmetric spaces", corrected reprint of the 1978 original, Graduate Studies in Mathematics, vol. 34, American Mathematical Society, Providence, RI, 2001.

[91] G. Hinojosa, A. Verjovsky, "Actions of discrete groups on spheres and real projective spaces", Bull. Braz. Math. Soc. (N. S.) **39** (2008), 157–171.

[92] N. J. Hitchin, "Lie groups and Teichmüller space", Topology **31** (1992), 339–365.

[93] M. Islam, A. Zimmer, "Convex co-compact actions of relatively hyperbolic groups", Geom. Topol. **27** (2023), 417–511.

[94] D. Johnson, J. J. Millson, "Deformation spaces associated to compact hyperbolic manifolds", in "Discrete groups in geometry and analysis", 48–106, Progress in Mathematics, vol. 67, Birkhäuser, Boston, MA, 1987.

[95] M. Kapovich, "Kleinian groups in higher dimensions", in "Geometry and dynamics of groups and spaces", 485–562, Progress in Mathematics, vol. 265, Birkhäuser Verlag, Basel, 2007.

[96] M. Kapovich, B. Leeb, "Discrete isometry groups of symmetric spaces", in "Handbook of group actions IV", 191–290, Advanced Lectures in Mathematics, vol. 41, International Press, Boston, MA, 2018.

[97] M. Kapovich, B. Leeb, "Relativizing characterizations of Anosov subgroups, I", Groups Geom. Dyn. **17** (2023), 1005–1071.

[98] M. Kapovich, B. Leeb, J. Porti, "Morse actions of discrete groups on symmetric spaces", arXiv:1403.7671.

[99] M. Kapovich, B. Leeb, J. Porti, "Some recent results on Anosov representations", Transform. Groups **21** (2016), 1105–1121.

[100] M. Kapovich, B. Leeb, J. Porti, "Dynamics on flag manifolds: domains of proper discontinuity and cocompactness", Geom. Topol. **22** (2018), 157–234.

[101] M. Kapovich, B. Leeb, J. Porti, "A Morse Lemma for quasigeodesics in symmetric spaces and euclidean buildings", Geom. Topol. **22** (2018), 3827–3923.

[102] F. Kassel, "Geometric structures and representations of discrete groups", Proceedings of the International Congress of Mathematicians 2018 (ICM 2018), 1113–1150, World Scientific, 2019.

[103] F. Kassel, M. B. Pozzetti, A. Sambarino, A. Wienhard (editors), "Arbeitsgemeinschaft: Higher rank Teichmüller theory", Oberwolfach Rep. **19** (2022), 2687–2740.

[104] F. Kassel, R. Potrie, "Eigenvalue gaps for hyperbolic groups and semigroups", J. Mod. Dyn. **18** (2022), 161–208.

[105] B. Kleiner, B. Leeb, "Rigidity of quasi-isometries for symmetric spaces and Euclidean buildings", Publ. Math. Inst. Hautes Études Sci. **86** (1997), 115–197.

[106] B. Kleiner, B. Leeb, "Rigidity of invariant convex sets in symmetric spaces", Invent. Math. **163** (2006), 657–676.

[107] A. Kontorovich, D. D. Long, A. Lubotzky, A. W. Reid, "What is... a thin group?", Notices Amer. Math. Soc. **66** (2019), 905–910.

[108] C. Kourouniotis, "Deformations of hyperbolic structures on manifolds of several dimensions", Math. Proc. Cambridge Philos. Soc. **98** (1985), 247–261.

[109] F. Labourie, "Anosov flows, surface groups and curves in projective space", Invent. Math. **165** (2006), 51–114.

[110] G.-S. Lee, L. Marquis, "Anti-de Sitter strictly GHC-regular groups which are not lattices", Trans. Amer. Math. Soc. **372** (2019), 153–186.

[111] G. Lusztig, "Total positivity in reductive groups", in "Lie theory and geometry", 531–568, Progress in Mathematics, vol. 123, Birkhäuser, Boston, MA, 1994.

[112] G. A. Margulis, "Discrete subgroups of semisimple Lie groups", Ergebnisse der Mathematik und ihrer Grenzgebiete (3), vol. 17, Springer-Verlag, Berlin, 1991.

[113] B. Maskit, "Kleinian groups", Grundlehren der mathematischen Wissenschaften, vol. 287, Springer-Verlag, Berlin, 1988.

[114] G. Mess, "Lorentz spacetimes of constant curvature" (1990), Geom. Dedicata **126** (2007), 3–45.

[115] D. Monclair, J.-M. Schlenker, N. Tholozan, "Gromov–Thurston manifolds and anti-de Sitter geometry", arXiv:2310.12003.

[116] G. D. Mostow, "On a remarkable class of polyhedra in complex hyperbolic space", Pacific J. Math. **86** (1980), 171–276.

[117] D. Mumford, C. Series, D. Wright, "Indra's pearls. The vision of Felix Klein", Cambridge University Press, 2002.

[118] D. Osin, "Small cancellations over relatively hyperbolic groups and embedding theorems", Ann. of Math. **172** (2010), 1–39.

[119] P. Pansu, "Sous-groupes discrets des groupes de Lie : rigidité, arithméticité", Séminaire Bourbaki, Exposé 778, Astérisque **227** (1995), 69–105.

[120] M. B. Pozzetti, "Higher rank Teichmüller theories", Séminaire Bourbaki, Exposé 1159, Astérisque **422** (2019), 327–354.

[121] J.-F. Quint, "Groupes convexes cocompacts en rang supérieur", Geom. Dedicata **113** (2005), 1–19.

[122] H. P. de Saint-Gervais, "Uniformization of Riemann surfaces. Revisiting a hundred-year old theorem", Heritage of European Mathematics, European Mathematical Society, Zürich, 2016.

[123] R. Schwartz, "Pappus's theorem and the modular group", Publ. Math. Inst. Hautes Études Sci. **78** (1993), 187–206.

[124] J. Seade, A. Verjovsky, "Complex Schottky groups", Astérisque **287** (2003), 251–272.

[125] A. Selberg, "On discontinuous groups in higher-dimensional symmetric spaces" (1960), in "Collected papers", vol. 1, 475–492, Springer-Verlag, Berlin, 1989.

[126] C. Series, "A crash course on Kleinian groups", Rend. Istit. Mat. Univ. Trieste **37** (2005), 1–38.

[127] B. Sury, "The congruence subgroup problem — An elementary approach aimed at applications", Texts and Readings in Mathematics, vol. 24, Hindustan Book Agency, New Delhi, 2003.

[128] J. Tits, "Free subgroups in linear groups", J. Algebra **20** (1972), 250–270.

[129] E. B. Vinberg, "Discrete groups generated by reflections in Lobačevskiĭ spaces, Math. USSR Sb. **1** (1968), 429–444.

[130] E. B. Vinberg, "Discrete linear groups generated by reflections", Math. USSR Izv. **5** (1971), 1083–1119.

[131] T. Weisman, "Dynamical properties of convex cocompact actions in projective space", J. Topol. **16** (2023), 990–1047.

[132] T. Weisman, "An extended definition of Anosov representation for relatively hyperbolic groups", arXiv:2205.07183.

[133] T. Weisman, "Examples of extended geometrically finite representations", arXiv:2311.18653.

[134] A. Wienhard, "An invitation to higher Teichmüller theory", Proceedings of the International Congress of Mathematicians 2018 (ICM 2018), 1007–1034, World Scientific, 2019.

[135] D. Witte Morris, "Introduction to Arithmetic Groups", Deductive Press, 2015.

[136] A. Wolf, "Convex projective geometrically finite structures", PhD thesis, Stanford University, 2020.

[137] F. Zhu, "Relatively dominated representations", Ann. Inst. Fourier **71** (2021), 2169–2235.
[138] F. Zhu, "Relatively dominated representations from eigenvalue gaps and limit maps", Geom. Dedicata **217** (2023), article number 39.
[139] F. Zhu, A. Zimmer, "Relatively Anosov representations via flows I: theory", arXiv:2207.14737.
[140] F. Zhu, A. Zimmer, "Relatively Anosov representations via flows II: examples", arXiv:2207.14738.
[141] A. Zimmer, "Projective Anosov representations, convex cocompact actions, and rigidity", J. Differential Geom. **119** (2021), 513–586.

5

Complete reducibility and subgroups of exceptional algebraic groups

Alastair J. Litterick[a], David I. Stewart[b], and Adam R. Thomas[c]

Abstract

This survey article has two components. The first part gives a gentle introduction to Serre's notion of G-complete reducibility, where G is a connected reductive algebraic group defined over an algebraically closed field. The second part concerns consequences of this theory when G is simple of exceptional type, specifically its role in elucidating the subgroup structure of G. The latter subject has a history going back about sixty years. We give an overview of what is known, up to the present day. We also take the opportunity to offer several corrections to the literature.

Introduction

Let G be an affine algebraic group over an algebraically closed field k of characteristic $p \geq 0$. In this article we will only be interested in the case that G is smooth, connected and reductive. Except in one or two places, the subgroups of G encountered in this article will be smooth, and thus we think of G as a variety as per [49], rather than adopting the scheme-theoretic language of [84]. In any case, we may assume that G is a subgroup of some general linear group defined by the vanishing of

[a] School of Mathematics, Statistics and Actuarial Science, University of Essex, Colchester, Essex, CO4 3SQ, United Kingdom.
a.litterick@essex.ac.uk

[b] Department of Mathematics, University of Manchester, Manchester, M13 9PL, United Kingdom.
david.i.stewart@manchester.ac.uk

[c] Mathematics Institute, University of Warwick, Coventry, CV4 7AL, United Kingdom.
Adam.R.Thomas@warwick.ac.uk

some polynomials in the matrix entries – more specifically by a radical ideal. Throughout, all vector spaces, representations and general linear groups will be of finite dimension.

The idea of G-complete reducibility is due to J-P. Serre [101]. It generalises the property of a representation $\rho : H \to G = \mathrm{GL}(V)$ of a group H being completely reducible, by restating the definition in terms of the relationship between the subgroup $\rho(H) \subseteq G$ and the parabolic subgroups of G. In this way, the same property can be formulated when G is any connected reductive algebraic group and H is one of its subgroups.

The notion of G-complete reducibility appears in many unexpected places, by virtue of the links it offers between representation theory, group theory, algebraic geometry and geometric invariant theory. For the purposes of this article, it offers the cleanest language to talk about the subgroup structure of G.

Unless otherwise mentioned, we only consider closed subgroups of G – those cut out from G by polynomial equations. A linear algebraic group is *unipotent* if it is isomorphic to a subgroup of the upper unitriangular matrices in $\mathrm{GL}(V)$ for some V. As the rank of G grows, one sees that it quickly becomes impossible to say very much about the unipotent subgroups of G, in much the same way that finite p-groups become unmanageable. The same problem arises when H is allowed to have a non-trivial normal connected unipotent subgroup. Therefore we focus on those subgroups H of G which do not contain a non-trivial normal connected unipotent subgroup; in other words H is *reductive.* Reductive subgroups are also interesting from a geometric perspective, as they are precisely those subgroups H such that the coset space G/H is affine [93].

Next, a subgroup $H \subseteq \mathrm{GL}(V)$ is the same thing as a faithful representation of H. Since we do not want to consider the representation theory of all finite groups, we will assume that H is *connected.* Further still, a connected reductive group H takes the form $H = \mathscr{D}(H) \cdot \mathscr{R}(H)$ where $\mathscr{R}(H)$ is a central torus of H and $\mathscr{D}(H)$ is the semisimple derived subgroup of H. It would be unenlightening to enumerate all tori in G which commute with $\mathscr{D}(H)$, and so:

we assume that H is semisimple.

This is still not enough in general to attempt a classification, unless $p = 0$. For, in positive characteristic a non-trivial semisimple group has a rich theory of indecomposable modules and it seems hopeless to classify *all* possibilities; therefore one cannot classify all the subgroups of $\mathrm{GL}(V)$ with $\dim V$ unbounded.

On a more sanguine note: since *irreducible* representations of H are classified by their highest weight, the *completely reducible* representations correspond to lists of such weights. If $p = 0$ then all representations are completely reducible and Weyl's dimension formula gives the dimensions of the irreducible factors, so listing semisimple subgroups of $G = \mathrm{GL}(V)$ reduces to straightforward combinatorics with roots systems and weights. In positive characteristic, understanding subgroups acting completely reducibly means understanding the dimensions of irreducible modules. This is a hard problem. The most recent progress on this problem is due to Williamson and his collaborators – for example, see [95] – but regrettably there is no clear description for the dimensions of the irreducible H-modules in all characteristics, unless the rank of H is small.

But now suppose that we bound the rank of G; more specifically take G to be simple of exceptional type, so that the rank of G is at most 8. Then there is a realistic prospect of understanding the poset of conjugacy classes of semisimple subgroups of G in all characteristics: any semisimple subgroup $H \subseteq G$ also has rank at most 8.

Main Problem *Let G be a simple algebraic group of exceptional type. Describe the poset of conjugacy classes of semisimple subgroups of G.*

Such a project naturally divides along lines prescribed by G-complete reducibility, which we explain in §5.4. With this in mind, our purpose is twofold:

(i) We introduce G-complete reducibility and the links it provides between representation theory, geometric invariant theory and group theory. We will be light on technical details, but aim to impart the flavour of the techniques and the most important results.

(ii) We discuss the current state of affairs in describing semisimple subgroups of the exceptional algebraic groups. We start with a historical overview and then collate the principal results from across the literature. We also correct some errors and omissions that have arisen in this study. Significantly, we update the table in [108]; see Table 5.3.

Prerequisite knowledge

This is an article about linear algebraic groups over algebraically closed fields and so a healthy knowledge of such groups would be appropriate.

Most of the results here do not require a scheme-theoretic background, and for these one of [49, 103, 17] would suffice. A particularly accessible overview of the theory surrounding parabolic subgroups, their Levi factors and associated combinatorics can be found in [78]. Our use of representation theory will not frequently stray far from the classification of irreducible modules by highest weight, which can be found in these same references. Occasionally a discussion of cohomology takes us into the world of [55]; though we will typically content ourselves with pointing out references to deeper material when appropriate. At points, knowledge of the theory of finite-dimensional complex Lie algebras would be helpful, as covered for example in [50, 39].

Notation

We will mostly introduce relevant notation as needed. Dynkin diagrams and conventions on roots will be as in [22]. Actions will always be on the left; thus conjugation will be written ${}^{g}h = ghg^{-1}$. In keeping with this, a group G which decomposes into a semidirect product of a normal subgroup N with a complement H will be written $N \rtimes H \cong G = NH$, since then $(n,h)(m,k) = (n^{h}m, hk)$ maps to $n^{h}mhk$ under this isomorphism.

As mentioned, our algebraic groups are all varieties over algebraically closed fields, and can be thought of as subgroups of an ambient group $\mathrm{GL}(V)$ for some finite-dimensional vector space V. The identity component of an algebraic group G is denoted G°, and G/G° is the (finite) component group.

Structure of the paper

The structure of the paper is as follows. In Part I, §§5.1–5.2 motivate the theory of G-complete reducibility as the natural generalisation of the representation-theoretic notion. This includes a sufficient overview of reductive algebraic groups to state the fundamental definition (Definition 5.3.1) and discuss the most useful tools for our applications. In §5.4 we describe a strategy for classifying subgroups of reductive groups, which arises naturally from the dichotomy between subgroups of G which are G-completely reducible, and those which are not.

In Part II we turn to the particular problem of understanding semisimple subgroups of exceptional simple algebraic groups. We begin with a brief historical overview of results in characteristic zero (essentially due to Dynkin) and their translation to positive characteristic (largely due to Seitz and his collaborators), before delving into the current state of the art, and the application of the strategy described in Part I. Finally, in §5.4 we discuss further ongoing research directions in the area.

Part I. G-complete reducibility

5.1 Complete reducibility

Almost the first result one encounters in group representation theory is Maschke's Theorem: Given a finite group H and a finite-dimensional $\mathbb{C}H$-module V, or equivalently a representation $H \to \mathrm{GL}(V)$, every H-submodule (i.e. H-stable subspace) $U \subseteq V$ admits an H-stable complement U', so that $V \cong U \oplus U'$ as H-modules. The proof proceeds by taking a vector-space direct sum $V = U_0 \oplus U$ and averaging the projection map $\phi : V \to V/U_0 = U$ over H to form the H-module homomorphism

$$V \to U, \qquad v \mapsto \frac{1}{|H|} \sum_{h \in H} h\phi(h^{-1}v)$$

whose kernel is then the required submodule U'. An identical proof works over any field k (in fact, over any commutative ring) in which the order $|H|$ is invertible, in particular Maschke's theorem holds whenever $\mathrm{char}\, k$ is sufficiently large relative to $|H|$. The analogous result (a consequence of the Peter–Weyl theorem) holds for unitary representations of connected compact topological groups, in particular for compact real Lie groups. Further still, given a complex semisimple Lie group, its Lie algebra is obtained by complexifying the (real) Lie algebra of a compact real Lie group – and the finite-dimensional representations of all these objects are essentially the same (this is Weyl's unitarian trick [119, §5][1]). For a direct approach to semisimple complex Lie algebras, one can also follow [39, C.15].

These are some examples of module categories which are *completely reducible* or *semisimple*. For a module of finite dimension over a field, repeated decomposition into submodules expresses it as a direct sum

[1] For an accessible overview in English, see [111]

of *irreducible* modules, i.e. non-zero modules having no proper, non-zero submodules. So when all modules are completely reducible, understanding the category amounts to understanding the irreducible modules. Complete reducibility is certainly not ubiquitous, however. A common example is the following representation of the group of integers under addition:

$$\mathbb{Z} \to \mathrm{GL}_2(\mathbb{C}); \quad n \mapsto \begin{pmatrix} 1 & n \\ 0 & 1 \end{pmatrix},$$

whose image stabilises a unique 1-dimensional subspace, spanned by $\binom{1}{0}$, which admits no complementary $\mathbb{Z}$-submodule. Performing a reduction modulo a prime p produces a cyclic group of order p acting on a module over a field of characteristic p, which again fails to be completely reducible.[2]

This formulation of complete reducibility places the focus on the acting group H. One could equally decide to fix the target of the representation and ask:

Given a finite-dimensional vector space V, for which subgroups H of $G = \mathrm{GL}(V)$ is V a completely reducible H-module?

This is now a question about the subgroup structure of G. It may also be more tractable, since H is either trivial or has a faithful module of dimension at most $\dim V$, which puts limitations on H. Moreover, this reformulation can be put in purely group-theoretic terms. Recall that a *parabolic subgroup* P of $\mathrm{GL}(V)$ is the stabiliser of a flag of subspaces

$$V = V_0 \supset V_1 \supset \cdots \supset V_r = \{0\}.$$

(See for example [78, Proposition 12.13].) One sees from this that the maximal (proper) parabolic subgroups are stabilisers of (proper, non-zero) subspaces, so a subgroup $H \subseteq \mathrm{GL}(V)$ acts irreducibly if and only if H is contained in no proper parabolic subgroup, and H is completely reducible if and only if: whenever H stabilises a flag of subspaces V_i as above, H stabilises an *opposite flag*

$$V = W_0 \supset W_1 \supset \cdots \supset W_r = \{0\}.$$

where $V = V_i \oplus W_{r-i}$ as H-modules for all i. In other words, whenever H is contained in a parabolic subgroup of $\mathrm{GL}(V)$, it is also contained in an *opposite* parabolic subgroup P^- of G. The subgroup of P acting

[2] Note that the condition of Maschke's theorem is violated, since the characteristic p of the field divides $|\mathbb{Z}/p| = p$.

trivially on each quotient V_i/V_{i+1} is upper unitriangular according to an appropriate basis, and is therefore a unipotent group. Indeed this turns out to be the largest normal connected unipotent subgroup of P, i.e. its *unipotent radical* $\mathscr{R}_u(P)$. The intersection $L := P \cap P^-$ turns out to be a reductive complement to $\mathscr{R}_u(P)$ in P, in other words a *Levi subgroup*. In $\mathrm{GL}(V)$, if P corresponds to a flag as above then Levi subgroups of P are stabilisers of vector-space direct-sum decompositions giving that flag as intermediate sums. That is, a decomposition $V = U_1 \oplus \cdots \oplus U_r$, where $V_i = U_{i+1} \oplus U_{i+2} \oplus \cdots \oplus U_r$ for each i, corresponds to the Levi factor $\mathrm{GL}(U_1) \times \ldots \times \mathrm{GL}(U_r)$. By change of basis, one sees that Levi subgroups of a fixed P are all conjugate (even by elements of $\mathscr{R}_u(P)$), so that a choice of basis in which the V_i and W_i are spanned by standard basis vectors leads to the following picture:

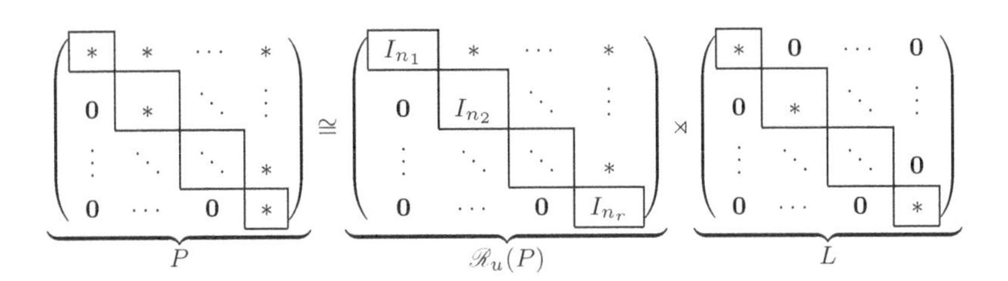

5.2 Reductive algebraic groups

The above discussion generalises at once to other groups with an appropriate notion of parabolic subgroup and Levi subgroup; in particular when G is a reductive algebraic group. The formal definition in this case is that a subgroup P of G is parabolic if G/P is a projective variety. However, a more useful characterisation for us can be given using the structure theory of reductive groups, which we now outline.

Amongst connected linear algebraic groups, the simple objects are by definition those which are non-abelian and have no non-trivial proper connected normal subgroups. Such groups are determined up to isogeny – i.e. a homomorphism with a finite kernel – by the algebraically closed field k and one of the Dynkin diagrams below, which are divided into those of *classical type* A_n $(n \geq 1)$, B_n $(n \geq 2)$, C_n $(n \geq 3)$, D_n $(n \geq 4)$, and those of *exceptional type* E_6, E_7, E_8, F_4 or G_2.

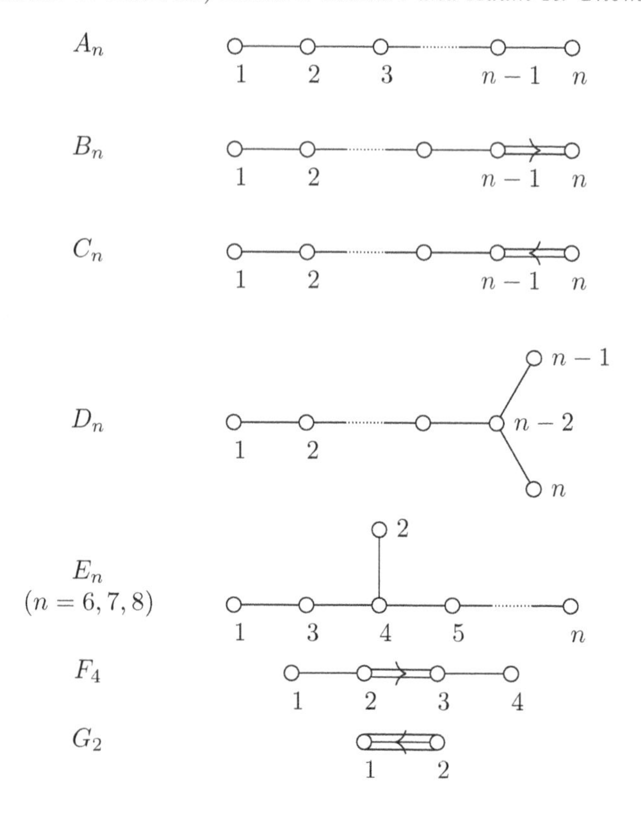

We also need to mention the additive group of the field k, often denoted $\mathbb{G}_a$, and the multiplicative group k^*, denoted $\mathbb{G}_m$. Since our groups are varieties over an algebraically closed field, it follows that groups with a subnormal series whose successive quotients are all $\mathbb{G}_a$ coincide with connected unipotent groups. Similarly, linear algebraic groups with a composition series whose quotients are isomorphic to $\mathbb{G}_m$ are precisely direct products $\mathbb{G}_m^r$ for some r, and are called *tori*. It is then a theorem that a a connected soluble linear algebraic group is the semidirect product of a connected unipotent group and a torus.

Tori play an important role in the structure theory of reductive groups. All maximal tori in a linear algebraic group are conjugate to one another, and their dimension is called the *rank* of the group. This is the number n of nodes in the Dynkin diagram when G is simple. For a maximal torus T of G, we let $W(G) = N_G(T)/T$ be the Weyl group of G with respect to T.

The simply-connected groups of type A_n–D_n are respectively SL_{n+1}, Spin_{2n+1}, Sp_{2n} and Spin_{2n}, and from these one obtains the others as quotients with finite kernels. For our purposes there is often no harm in working with SL_{n+1}, SO_n and Sp_{2n} which have more accessible descriptions in terms of the natural module and its quadratic or symplectic form.

A linear algebraic group G contains a unique maximal connected normal soluble subgroup $\mathscr{R}(G)$, the *radical* of G, containing the unique maximal connected normal unipotent subgroup $\mathscr{R}_u(G)$ of G, the *unipotent radical.* Then G is called:

- *reductive* if $\mathscr{R}_u(G)$ is trivial; this is equivalent to G° being an almost-direct product[3] of simple algebraic groups and a torus.
- *semisimple* if G is connected and $\mathscr{R}(G)$ is trivial; this is equivalent to G being an almost-direct product of simple algebraic groups.

For any linear algebraic group G, the quotient $G/\mathscr{R}_u(G)$ is reductive and if G is connected then $G/\mathscr{R}(G)$ is semisimple.

From now on let G be connected. We are interested in finite-dimensional *rational* representations of G, which can be identified with homomorphisms $G \to \mathrm{GL}(V)$ of algebraic groups. One easy case to describe is when $G = \mathbb{G}_m$. The irreducible representations of $\mathbb{G}_m$ are 1-dimensional; if $V = \langle v\rangle$ is one such, then $x \cdot v = x^r v$ for some $r \in \mathbb{Z}$ and so corresponds with a homomorphism $x \mapsto x^r$ of $\mathbb{G}_m$ to $\mathbb{G}_m$. Even better, $\mathbb{G}_m$ always acts completely reducibly, so that a representation up to isomorphism is identified with a list of integers. More generally if $G = T$ is a torus $T \cong \mathbb{G}_m \times \cdots \times \mathbb{G}_m$, then an irreducible representation V is still 1-dimensional, determined up to isomorphism by the action

$$(x_1, \ldots, x_s) \cdot v = x_1^{r_1} \cdots x_s^{r_s} v,$$

which identifies with an element $\lambda \in \mathrm{Hom}(T, \mathbb{G}_m) =: X(T)$; then λ is called a *weight* of T. (We often identify λ with the corresponding 1-dimensional representation.) Since T acts completely reducibly, a representation for T is simply a list of its weights.

Taking inspiration from the theory of Lie groups, one can construct a Lie algebra $\mathfrak{g} = \mathrm{Lie}(G)$ from G which affords a representation of G through an *adjoint action.* Since maximal tori in G are conjugate, the collection of weights of a maximal torus T on $\mathrm{Lie}(G)$ does not depend on the choice of T. The zero weight-space is $\mathrm{Lie}(T)$, and when G is reductive, the non-zero weights are called *roots* and form the *root system* which is denoted $\Phi = \Phi(G, T)$. Also by the conjugacy of maximal tori,

[3] commuting product of normal subgroups, with pairwise finite intersections

the isomorphism type of the Weyl group $W(G)$ is independent of T, and $W(G)$ acts naturally on $X(T)$ and $\Phi(G,T)$.

A reductive group G has a maximal connected soluble subgroup B called a *Borel subgroup* which by Borel's fixed point theorem is unique up to conjugacy in G. This decomposes as $B = US$, where $U = \mathscr{R}_u(B)$ is a maximal connected unipotent subgroup of G, and S is a maximal torus of G. When $S = T \subseteq B$, the set of roots $\Phi(B,T)$ contains exactly half the elements of $\Phi(G,T)$ and defines a positive system $\Phi^+ \subset \Phi$. Moreover, Φ has a *base of simple roots* Δ: a minimal subset of Φ^+ from which all elements of Φ^+ are obtained as non-negative integer sums. We call $|\Delta|$ the *semisimple rank* of G. One can define a scalar product (the Killing form) on the $\mathbb{R}$-linear span of the roots; the Dynkin diagram of G then encodes the resulting lengths and relative angles of the simple roots.

Within $X(T)$ we single out the *dominant weights*, $X(T)^+$, which are those having non-negative inner product with each positive root. If $\lambda \in X(T)^+$ then we identify λ with the corresponding 1-dimensional T-module. Composing with the map $B \to B/U \cong T$, we get a B-module λ. One can define an induced module $\mathrm{H}^0(\lambda) := \mathrm{Ind}_B^G(\lambda)$ in the category of rational G-modules. Since G/B is projective, this module has finite dimension, and can be viewed as a reduction modulo p of the irreducible module $L_{\mathbb{C}}(\lambda)$ for the complex Lie algebra $\mathfrak{g}_{\mathbb{C}}$, so the weights of $\mathrm{Ind}_B^G(\lambda)$ are given by Weyl's character formula. If we define $L(\lambda)$ to be the socle of $\mathrm{Ind}_B^G(\lambda)$ then it turns out to be simple, with λ as its highest weight. These turn out to be all the irreducible modules: [55, II.2.4].[4]

Parabolic subgroups P of G can now be characterised as follows: P is any subgroup containing a Borel subgroup. Fixing $B \subseteq P$ it turns out that $\Phi(P,T)$ is an enlargement of $\Phi(B,T)$ obtained by a suitable choice of simple roots $\Delta' \subseteq \Delta$ and taking the smallest additively-closed subset of $\Phi(G,T)$ containing $\Phi(B,T)$ and $-\Delta'$. If r is the semisimple rank of G, then there are 2^r non-conjugate parabolic subgroups containing a given Borel subgroup B; these correspond to the possible subsets of nodes in the Dynkin diagram. The remaining parabolic subgroups are all conjugate to one of these under the action of G.

Just as we saw in $\mathrm{GL}(V)$, a parabolic subgroup admits a *Levi decomposition* $P = \mathscr{R}_u(P) \rtimes L$, where L is a reductive group called a *Levi subgroup* of P; all such Levi subgroups are conjugate by elements of $\mathscr{R}_u(P)$. If one insists that $T \subseteq L$ then L is unique, and the Levi decom-

[4] Another reduction modulo p gives the Weyl module $V(\lambda)$, which has $L(\lambda)$ as its head.

position can be seen at the level of roots. If P corresponds to Δ' then the roots arising as sums of roots contained in $\Delta' \cup -\Delta'$ give those of $\Phi(L,T)$; their complement in $\Phi(P,T)$ are the roots in $\mathscr{R}_u(P)$.

Example 5.2.1 Let G be simple of type F_4. Elementary root system combinatorics (see e.g. [51, §2.10]) tell us that G has 48 roots (hence 24 positive roots). If we pick the two middle nodes of the Dynkin diagram of G (cf. page 197), the corresponding roots and their negatives generate a subsystem of type B_2, which has 8 roots. So in the corresponding parabolic subgroup $P = \mathscr{R}_u(P)L$ of G, the derived subgroup of L will be simple of type B_2, which has dimension 10 (8 roots, plus a 2-dimensional maximal torus). The connected centre of L will be a 2-dimensional torus since L contains a maximal torus of G, which has rank 4. Finally, counting positive roots in G and L, the unipotent radical $\mathscr{R}_u(P)$ has dimension $24 - 4 = 20$.

In fact, much of the structure of $\mathscr{R}_u(P)$ as an L-group can also be quickly deduced from the root system. We will return to this in §5.4.3.

The following result [18, Theorem 2.5] is fundamental in our study and will henceforth be called the *Borel–Tits theorem.*

Theorem 5.2.2 (Borel–Tits) *Let G be a connected reductive algebraic group, and let X be a unipotent subgroup of G. Then there exists a (canonically-defined) parabolic subgroup P of G such that $X \subseteq \mathscr{R}_u(P)$.*

This implies in particular that a maximal subgroup of a reductive group G is either reductive or parabolic. And a maximal connected subgroup of a semisimple group G is either semisimple or parabolic.

5.3 G-complete reducibility

At last, we come to the central definition.

Definition 5.3.1 ([101, p. 19], [102, §3.2.1]) Let G be a connected reductive algebraic group over the algebraically closed field k.

A subgroup H of G is called *G-completely reducible* (G-cr) if, whenever H is contained in a parabolic subgroup P of G, there exists a Levi subgroup L of P with $H \subseteq L$. Similarly, H is called *G-irreducible* (G-irr) if H is contained in no proper parabolic subgroup of G; it is *G-reducible* if it is not G-irr. Lastly H is *G-indecomposable* if H is in no proper Levi subgroup of G and *G-decomposable* otherwise.

Remark 5.3.2

(i) An abstract subgroup H of G and its Zariski closure are contained in precisely the same closed subgroups of G, in particular, the same parabolic subgroups and Levi subgroups thereof. Therefore H is G-cr if and only if its Zariski closure is, so it does no harm to work only with closed subgroups.

(ii) For disconnected groups G, one can define *R-parabolic* and *R-Levi* subgroups of G in terms of limits of certain morphisms. These coincide with parabolic and Levi subgroups when G is connected, hence G-complete reducibility and many results here generalise to disconnected groups. A full discussion is beyond the scope of this article; we direct the reader to [11, §6] for a thorough overview.

Many general results in the representation theory of groups can be viewed as cases of statements about G-complete reducibility, when G is specialised to $\mathrm{GL}(V)$. We now collect some of these.

5.3.1 Characteristic criteria for complete reducibility

Jantzen proved in [54] that if H is connected reductive and V is an H-module with $\dim V \leq p$ then V is completely reducible; this bound was improved by McNinch in [80]. A more general statement is:

Theorem 5.3.3 *Let G be a connected reductive algebraic group over k of characteristic $p \geq 0$ and let H be a connected subgroup of G. If H is G-cr then H is reductive. Conversely, if H is reductive and either $p = 0$ or p is greater than the ranks of all simple factors of G, then H is G-cr.*

The implication 'G-cr $\Rightarrow$ reductive' follows from the Borel–Tits theorem: one can construct a parabolic subgroup P containing H such that $\mathscr{R}_u(H) \subseteq \mathscr{R}_u(P)$. Thus if $\mathscr{R}_u(H)$ is non-trivial then H is in no Levi subgroup of P.

The 'reductive $\Rightarrow$ G-cr' direction can be deduced from Jantzen's or McNinch's result if G is simple and classical. The exceptional case, tackled in [63], relies on a careful study of the 1-cohomology arising from the conjugation action of subgroups of parabolics on unipotent radicals. We discuss this technique in more detail in §5.3, where we explain how one can find non-G-cr subgroups, when they exist.

Given the result for G simple, the general case follows in short order, since parabolic subgroups and Levi subgroups in G are commuting products of the central torus $Z(G)^\circ$ with parabolic and Levi subgroups of the simple factors of G.

One of Serre's initial motivations for studying G-complete reducibility was in studying complete reducibility of representations. Specifically, the starting point for Serre's lectures [101] is the observation of Chevalley [26] that in characteristic 0, tensor products of completely reducible modules for an *arbitrary* group are again completely reducible. This fails in positive characteristic, however Serre observed [99] that it does hold when the characteristic is large relative to the dimension of the modules in question, and he then began asking questions of the form: If a tensor product (or symmetric power, or alternating power) of modules is completely reducible, must the initial module(s) also be completely reducible? Or conversely? The answer [100] depends naturally on certain congruence conditions on the characteristic. See §5.4.1 for more on this.

In a related vein, in [102, §5.2], for a reductive group G with a finite-dimensional G-module V, Serre defines an invariant $n(V)$ in terms of the weights of G on V; he then proves that if the characteristic is larger than $n(V)$ and the identity component of the kernel of $G \to \mathrm{GL}(V)$ is a torus, then $H \subseteq G$ is G-cr if and only if V is a completely reducible H-module [102, Theorem 5.4]. Thus in sufficiently large characteristic, G-complete reducibility can indeed be detected on the level of G-modules. If $V = \mathrm{Lie}(G)$ is the adjoint module then $n(V) = 2h_G - 2$, where h_G is the Coxeter number of G. Note that this bound is typically much larger than that given in Theorem 5.3.3.

5.3.2 Equivalence with strong reductivity

A major result in complete reducibility generalises the idea that a completely reducible module is the direct sum of a *unique* list of simple ones. In place of the simple summands of a module, the focus is on the Levi subgroup stabilising the decomposition.

Theorem 5.3.4 ([11, Corollary 3.5]) *For a subgroup H of a connected reductive algebraic group G, the following are equivalent.*

(i) H is G-cr;
(ii) H is $C_G(S)$-irr for some maximal torus S of $C_G(H)$;
(iii) for every parabolic subgroup P of G which is minimal with respect to containing H, the subgroup H is L-irr for some Levi subgroup L of P;

(iv) *there exists a parabolic subgroup P of G which is minimal with respect to containing H, such that H is L-irr for some Levi subgroup L of P.*

In the prototype setting $G = \mathrm{GL}(V)$, if $V = V_1 \oplus \cdots \oplus V_r$ is a decomposition of V into irreducible H-modules then the torus S in ii consists of all elements inducing scalars on each summands V_i, and $C_G(S)$ is the direct product of the subgroups $\mathrm{GL}(V_i)$.

Remark 5.3.5 Levi subgroups of G are precisely the centralisers of sub-tori of G, and moreover all maximal tori in any linear algebraic group are conjugates of one another. In particular all maximal tori of $C_G(H)$ are $C_G(H)$-conjugate, and it follows that their centralisers, which are those Levi subgroups of G that are minimal subject to containing H, are also conjugate to one another by elements of $C_G(H)$, and the ranks of their centres equal the rank of $C_G(H)$. For the same reason, if H is $C_G(S)$-irr for some maximal torus S of $C_G(H)$, it is in fact $C_G(S)$-irr for *all* such S.

Subgroups satisfying ii in Theorem 5.3.4 were termed *strongly reductive* by Richardson [94, Def. 16.1]. Richardson studied strongly reductive subgroups from a geometric viewpoint, centred around the following result.

Theorem 5.3.6 ([94, §16]) *Suppose that H is the Zariski closure of a finitely-generated subgroup $\langle h_1, \ldots, h_n \rangle$ of G. Then H is G-cr (resp. G-irr) if and only if the orbit $G \cdot (h_1, \ldots, h_n)$ is Zariski closed in G^n (resp. a* stable point *of G^n).*

Remark 5.3.7

(i) This theorem is a lynchpin of the results in [11] and subsequent work. It relies on some geometry and geometric invariant theory which we omit; we direct the reader to [94] and [11].
(ii) In geometric invariant theory, a *stable point* of a G-variety is a point whose G-orbit is closed and whose stabiliser is a finite extension of the kernel of the action. Thus when G acts on G^n by simultaneous conjugation, a stable point is one whose orbit is closed and whose centraliser is a finite extension of the centre $Z(G)$.
(iii) One can drop the hypothesis that H is the closure of a finitely generated subgroup – one need only pick a sufficiently large n-tuple so that H and the elements of the n-tuple generate the same associative subalgebra of $\mathrm{End}(V)$ for some faithful representation

$G \to \mathrm{GL}(V)$; this is always possible for dimension reasons. Such an n-tuple is called a *generic tuple* [10, Def. 2.5].

A highlight of what can be proved through this approach is the following theorem.

Theorem 5.3.8 ([11]) *Let G be a connected reductive algebraic group and H be a G-cr subgroup of G.*

(i) If $N \triangleleft H$ then N is G-cr.
(ii) The subgroups $N_G(H)$ and $C_G(H)$ are G-cr. More generally, if K is a subgroup of G with

$$HC_G(H)^\circ \subseteq K \subseteq N_G(H),$$

then K is G-cr.

In particular, a subgroup $X \subseteq G$ is G-cr if and only if $N_G(X)$ is G-cr.

Part i generalises Clifford's theorem in representation theory: a completely reducible module for a group remains completely reducible upon restriction to a normal subgroup. Part ii gives a converse to this, and inspires the following question:

If H is a commuting product $H = AB$, where A and B are G-cr, must H also be G-cr?

By ii the answer is yes if $A = C_G(B)$ and $B = C_G(A)$. The answer is also positive when the characteristic is large enough:

Theorem 5.3.9 ([12, Theorem 1.3]) *Let G be a connected reductive algebraic group in characteristic $p \geq 0$ and let A and B be commuting G-cr subgroups of G such that either $p = 0$; or $p > 3$; or $p = 3$ and G has no simple factors of exceptional type. Then AB is also G-cr.*

The proof in [12] involves some case-by-case arguments depending on the Lie type of G, although a uniform argument is also possible if one is willing to relax the bound on the characteristic [83, Proposition. 40].

Remark 5.3.10 To date, the authors are not aware of a reductive group G and a pair of commuting G-cr subgroups A and B (connected or otherwise) whose product AB is non-G-cr when $p = 3$. It is therefore plausible that the above theorem holds whenever $p \neq 2$. Such examples do however exist when $p = 2$, cf. [12, Ex. 5.3].

One powerful feature of G-complete reducibility is that one can allow the group G to change under various constructions. For instance, in [11] it is shown that if $G = G_1 \times G_2$ is a direct product of reductive groups then $H \subseteq G$ is G-cr or G-irr if and only if both images under projection to a factor G_i are G_i-cr (resp. G_i-irr). Similarly, taking a quotient by a normal subgroup N sends G-cr subgroups to (G/N)-cr subgroups, and the converse also holds if N° is a torus.

The following result is useful in what follows. Recall that a linear algebraic group S is called *linearly reductive* if all its rational representations are completely reducible. In characteristic 0 this is equivalent to S being reductive, whereas in positive characteristic this means that S° is a torus and the order of the finite group $|S/S^\circ|$ is coprime to p.

Theorem 5.3.11 ([11, Lemma 2.6, Corollary 3.21]) *Let S be a linearly reductive subgroup of a connected reductive algebraic group G. Then S is G-cr, and if $H = C_G(S)^\circ$ then a subgroup of H is H-cr if and only if it is G-cr.*

In the particular case that S is a torus, $C_G(S)$ is a Levi subgroup of G and the result in this case was first proved in [102, Proposition 3.2]; this forms part of the proof of Theorem 5.3.4. The following corollary justifies our focus on semisimple subgroups:

Corollary 5.3.12 *Let G be connected reductive, H a subgroup of G and S a subgroup of $C_G(H)$. Then H is G-cr if and only if HS is G-cr. In particular, a connected reductive subgroup H of G is G-cr if and only if its (semisimple) derived subgroup $\mathscr{D}(H)$ is G-cr.*

Proof In characteristic 0, the subgroup H is G-cr if and only if it is reductive (Theorem 5.3.3), which is the case if and only if HS is reductive.

In positive characteristic, using Theorem 5.3.11 we can replace G with $C_G(S)^\circ$ so that S is central in G. Since S° is a torus, by the above discussion the subgroups H and HS are each G-cr if and only if HS/S is (G/S)-cr. □

5.3.3 Separability, reductive pairs

A subgroup H of G is called *separable* if the Lie algebra $\mathrm{Lie}(C_G(H))$ coincides with the fixed-point space $C_{\mathrm{Lie}(G)}(H)$; the latter always contains the former but can in general be larger.[5] If $\mathbf{h} \in G^n$ is a topological generating tuple (or generic tuple) for H, and G acts on G^n by simultaneous conjugation, the statement is also equivalent to saying the orbit map $G \to G \cdot \mathbf{h}$ is a separable morphism [17, Proposition 6.7], whence the terminology. In $\mathrm{GL}(V)$, all closed subgroups are separable, cf. [45, Lemma 3.5]. For a general reductive group G, non-separable subgroups are always a low-characteristic phenomenon. Indeed, the main result of [45] shows that the statement "all closed subgroups are separable" holds if and only if the characteristic is 0 or *pretty good* for G. The latter is a very mild condition - see [45, Definition 2.11] for the precise definition. For instance if G is simple of exceptional type then a prime p is pretty good for G unless $p = 2$ or 3, or $G = E_8$ and $p = 5$.

Next, a pair (G, H) of reductive groups with $H \subseteq G$ is called a *reductive pair* [92] if $\mathrm{Lie}(H)$ is an H-module direct summand of $\mathrm{Lie}(G)$.[6] Again, (G, H) is always a reductive pair if the underlying characteristic is large relative to G (for instance $\mathrm{Lie}(G)$ is a completely reducible module for all reductive subgroups in sufficiently large characteristic).

The application of these two concepts to complete reducibility is now as follows.

Theorem 5.3.13 ([11, Theorem 3.35, Corollary 3.36]) *Let (G, M) be a reductive pair and let H be a separable subgroup of G which is contained in M. If H is G-cr then H is M-cr.*

In particular, if $(\mathrm{GL}(V), M)$ is a reductive pair and $H \subseteq M$ acts completely reducibly on V, then H is M-cr.

The proof in *op. cit.* is geometric, following [92]. If H corresponds to $(h_1, \ldots, h_n) \in G^n$, that is, if this is a generic tuple for H or if H is the closure of $\langle h_1, \ldots, h_n \rangle$, then under the given hypotheses it is shown that the G-orbit $\mathcal{O} = G \cdot \mathbf{h}$ under simultaneous conjugacy splits into finitely many Zariski-closed M-orbits in $\mathcal{O} \cap M^n$. So if $\mathcal{O}$ is closed in G^n, then the M-orbits on $\mathcal{O} \cap M^n$ are closed in M^n, which in turn implies that H is M-cr.

5.3.4 G-complete reducibility in classical groups

In this section, we write $G = \mathrm{Cl}(V)$ to mean that G is one of the groups $\mathrm{SL}(V)$, $\mathrm{Sp}(V)$ or $\mathrm{SO}(V)$, where V carries either the zero form or a non-

[5] In scheme-theoretic language, this is equivalent to the centraliser $C_G(H)$ being a smooth subgroup scheme of G.

[6] While [92] introduces this only for connected groups, the concept and results apply equally well for disconnected subgroups.

degenerate alternating or quadratic form, respectively. Then parabolic subgroups of G are the stabilisers of flags of subspaces where the relevant form vanishes, i.e. *totally isotropic* and *totally singular* subspaces (respectively) when the form is non-degenerate, cf. [78, Proposition 12.13]. When the form is non-degenerate, two parabolic subgroups corresponding to flags $(V_i)_{i=1,\ldots,r}$ and $(W_i)_{i=1,\ldots,s}$ are opposite if $r = s$ and V is an orthogonal direct sum $V_i \perp W_i^\perp$ for each i, where $W_i^\perp$ is the annihilator of W_i relative to the alternating or quadratic form on V. Thus sufficient understanding of the action of H on V tells us whether or not H is G-cr.

When $G = \mathrm{SL}(V)$ we have seen that a subgroup of G is G-cr if and only if it is completely reducible on V. By Theorem 5.3.11 this also holds for classical groups $G = \mathrm{Cl}(V)$ in characteristic not 2, since G is then the centraliser in $\mathrm{SL}(V)$ of an involutory outer automorphism.[7] In this case, classifying all G-cr semisimple subgroups of $G = \mathrm{Cl}(V)$ amounts to understanding the dimensions and Frobenius–Schur indicators of all irreducible modules of dimension at most $\dim V$, for all semisimple groups.

The condition $\mathrm{char}\, k = p \neq 2$ is necessary to make these assertions. If $p \neq 2$ then recall that a quadratic form q gives rise to a symmetric bilinear form B on V via

$$B(v, w) := \frac{1}{2}\left(q(v + w) - q(v) - q(w)\right),$$

and q can be recovered from B via $q(x) = B(x, x)$. If $p = 2$ then $B(v, w) := q(v + w) - q(v) - q(w)$ defines a symmetric bilinear form, but $B(x, x) = 0$ so q can no longer be recovered. If B is non-degenerate and $B(v, w) \neq 0$ then B is also non-degenerate on $\langle v, w\rangle^\perp$ and V must have even dimension. Thus the bilinear form on the natural module for SO_{2n+1} has a 1-dimensional radical, spanned by a non-singular vector for q. For convenience, define the *natural irreducible module* for $\mathrm{Cl}(V)$ to be the largest non-trivial irreducible quotient of V; this is V itself unless $p = 2$ and $G = \mathrm{SO}(V) \cong \mathrm{SO}_{2n+1}$, in which case it is the $2n$-dimensional quotient by the radical of the bilinear form.

The following example is attributed to M. Liebeck in [11, Example 3.45].

Example 5.3.14 Let $\mathrm{char}\, k = 2$ and let H be a group preserving a symplectic (resp. orthogonal) form on an irreducible module W. Let $V = W \perp W$ be an orthogonal direct sum. Then V has a unique non-

[7] As an alternative proof, when $p \neq 2$ one can show that $(\mathrm{GL}(V), G)$ is a reductive pair; and since every subgroup is separable in $\mathrm{GL}(V)$, Theorem 5.3.13 tells us that every $\mathrm{GL}(V)$-cr subgroup of G is G-cr.

zero totally isotropic (resp. totally singular) H-submodule. Thus H is $\mathrm{GL}(V)$-cr but not $\mathrm{Cl}(V)$-cr.

Proof By Schur's lemma, each proper non-zero H-submodule of V is the image of a diagonal embedding $W \to V$ by $w \mapsto (aw, bw)$ for $[a : b] \in \mathbb{P}^1(k)$. If B is an H-invariant non-degenerate alternating form on W then $B((aw, bw), (au, bu)) = (a^2 + b^2)B(w, u) = (a + b)^2 B(w, u)$. This is zero if and only if $a = b$. The orthogonal case is similar. □

5.3.5 G-irreducibility in classical groups

By the characterisation of parabolic subgroups above, a subgroup of $G = \mathrm{Cl}(V)$ is G-irr if and only if it preserves no proper non-zero totally isotropic (or totally singular) subspace. In more detail:

Proposition 5.3.15 *Let G be a simple algebraic group of classical type with V the natural irreducible G-module, and let H be a subgroup of G. Then H is G-irr if and only if one of the following holds:*

(i) G has type A_n and V is an irreducible H-module;
(ii) G has type B_n, C_n or D_n and $V = V_1 \perp \ldots \perp V_k$ as H-modules, where the V_i are non-degenerate, irreducible and pairwise inequivalent;
(iii) $p = 2$, G has type D_n and H fixes a non-singular vector $v \in V$, such that H is G_v-irr in the point stabiliser G_v and does not lie in a subgroup of G_v of type D_{n-1}.

(See [75, Proposition 3.1] for a full proof.)

Remark 5.3.16

(a) In case iii, the bilinear form preserved by $G \cong \mathrm{SO}_{2n}$ is alternating when $\operatorname{char} k = 2$, hence every 1-space is isotropic. Now G_v preserves the space $\langle v \rangle^{\perp}$ of codimension 1 in V, and as v is nonsingular, q is non-degenerate on this space, so $G_v \cong \mathrm{SO}_{2n-1}$ is simple of type B_{n-1}. For more details, see for instance [57, Proposition 4.1.7].
(b) This proposition can be applied to any simple group of type A–D, such as PSp_{2n} or the half-spin groups of type D_n. For if G is simple and classical then G is related to some $\mathrm{Cl}(V)$ by a quotient or an extension by a finite central subgroup – or if G is a half-spin group, then one of each[8]. If Z is such a finite central subgroup then as G is

[8] In characteristic 2 or if G has type A_n with $p \mid n + 1$, one must work with central

connected and reductive, Z is contained in all maximal tori, hence in all parabolic subgroups. It follows that a subgroup H of G is G-irr if and only if its preimage HZ or quotient HZ/Z is G-irr. Furthermore if $\operatorname{char} k = 2$ then taking the quotient by the 1-dimensional radical of the bilinear form induces an *exceptional isogeny* $\psi : \mathrm{SO}_{2n+1} \to \mathrm{Sp}_{2n}$ (more details on p. 213). In this case, ψ is bijective and it follows that $H \subseteq \mathrm{SO}_{2n+1}$ is SO_{2n+1}-cr if and only if $\psi(H)$ is Sp_{2n}-cr.

5.4 A strategy for classifying semisimple subgroups

We remind the reader that we wish to tackle the following:

Main Problem *Let G be a simple algebraic group of exceptional type. Describe the poset of conjugacy classes of semisimple subgroups of G.*

Just as in representation theory, where one may begin by studying irreducible modules for a given object (immediately yielding the completely reducible modules) and then considering extensions, one can stratify the search for subgroups of G, beginning with G-cr subgroups and building up from these to non-G-cr subgroups.

Suppose that we know the maximal connected subgroups of simple groups up to a certain rank (such as 8), and let G be one of these simple groups. Let $H \subseteq G$ be semisimple and let M be a maximal connected subgroup of G containing H. If M is reductive then $M = \mathscr{D}(M) \cdot Z(M)$ and since H is perfect, we have $H \subseteq \mathscr{D}(M)$. Since $\mathscr{D}(M)$ is a central product of simple groups of smaller dimension, we hope to know H by induction on the dimension. The problem is that one may have $H \subset M$ where M is not reductive; this means $M = P$ is a parabolic subgroup by the Borel–Tits theorem. So we would also like to know the semisimple subgroups of $P = QL$. Since L is reductive of the same rank as G we may again assume that we know its semisimple subgroups. But it remains to find those semisimple subgroups of P which are not conjugate to subgroups of L; in other words, the non-G-cr subgroups. Of course, Theorem 5.3.3 tells us that such subgroups do not exist if $\operatorname{char} k$ is 0 or large enough relative to the root system of G.

subgroup *schemes*, since the possibilities for G can be isomorphic as abstract groups. However, this does not impact our discussion here.

5.4.1 G-cr subgroups

To start, we need to know the G-conjugacy classes of G-cr semisimple subgroups of G. Theorem 5.3.4 reduces our task to finding those which are L-irr, as L varies over all Levi subgroups of G. Let $\{L_1, \ldots, L_s\}$ be a complete list of representatives of the G-conjugacy classes of Levi subgroups. Then we can find all conjugacy classes of semisimple G-cr subgroups at least once, by listing the L_i-conjugacy classes of L_i-irr subgroups. The full analogue of the Jordan–Hölder theorem given below says that this will give each G-class of G-cr subgroups exactly once. It can be proven in various ways (cf. [25, Propositions 2.8.2, 2.8.3]) but the cleanest is via geometric invariant theory [13, Theorem 5.8].

Lemma 5.4.1 *Let H be a subgroup of the connected reductive group G, let P be minimal amongst parabolic subgroups of G containing H and let $\pi : P \to P/\mathscr{R}_u(P) = L$ be the natural projection to a Levi subgroup L. Then $\pi(H)$ is L-irr, and the G-conjugacy classes of L and $\pi(H)$ are uniquely determined by the G-conjugacy class of H.*

Moreover, H is G-cr if and only if H and $\pi(H)$ are $\mathscr{R}_u(P)$-conjugate.

As L contains a maximal torus of G, the group $N_G(L)$ is an extension of L by a finite group: the part of the Weyl group of G which stabilises the root system of L. Thus $N_G(L)$ may induce non-trivial conjugacy between some simple factors of L. Such conjugacy is easy to describe. cf. [78, Corollary 12.11]. In light of this, the key issue is to find the $\mathscr{D}(L)$-classes of $\mathscr{D}(L)$-irr semisimple subgroups for each G-class of Levi subgroup L; this includes the case that $L = G$.

Definition 5.4.2 For a connected reductive group G, let $\mathrm{ConIrr}(G)$ denote the poset of G-classes of connected G-irr subgroups of G under inclusion.

Suppose that $[H] \in \mathrm{ConIrr}(G)$. It follows from the Borel–Tits theorem that H is reductive. Since it cannot centralise a non-central torus of G, we get:

Lemma 5.4.3 ([69, Lemma 2.1], [11, Corollary 3.18]) *Suppose G is semisimple and let H be a G-irr connected subgroup. Then H is semisimple and $C_G(H)$ is finite and linearly reductive.*

Remark 5.4.4 Work of the third author and Liebeck in [70] classifies those finite subgroups which can occur as centralisers of semisimple G-irr subgroups for any simple algebraic group G.

As H is not contained in any proper parabolic subgroup of G, it must be contained in some semisimple maximal connected subgroup M of G. Moreover H is M-irr, since any proper parabolic subgroup of M is contained in a proper parabolic subgroup of G, again by the Borel–Tits theorem. When M is simple of classical type, one may determine $\mathrm{ConIrr}(M)$ at once by use of Proposition 5.3.15. If instead M is simple of exceptional type then induction on $\dim G$ yields $\mathrm{ConIrr}(M)$. For general semisimple M, knowledge of $\mathrm{ConIrr}(M_i)$ for each simple factor M_i of M yields $\mathrm{ConIrr}(M)$ (more on this shortly). Now, it can happen that an M-irr subgroup H lies in a proper parabolic subgroup of G. We call $[H] \in \mathrm{ConIrr}(M)$ a *candidate* and aim to decide which candidates are actually G-irr. Also, a G-irr subgroup can be contained in more than one semisimple maximal connected subgroup M. To detect this, we want to know when two candidates $H_1 \subseteq M_1$ and $H_2 \subseteq M_2$ are in the same G-class.

Returning to the issue of finding $\mathrm{ConIrr}(M)$ when M is not simple, the parabolic subgroups of M are the products of parabolic subgroups of its factors, so any M-irr subgroup needs to project to an M_i-irr subgroup of each simple factor M_i [115, Lemma 3.6]. As a partial converse, if for each i we have an M_i-irr subgroup H_i, then $H_1 \dots H_r \subseteq M$ is M-irr; however, these do not quite exhaust all the M-irr subgroups. To complete the list, one must also discuss *diagonal subgroups*: Whenever M has one or more simple factors of a given type, let $\hat{H}$ be the simply-connected simple group of this type. Then $\hat{H}$ admits a homomorphism to M with non-trivial projection to each simple factor of the of the appropriate type. By definition, a diagonal subgroup is the commuting product of images of such homomorphisms; this will be M-irr precisely when it has non-trivial projection in every simple factor of M.

Example 5.4.5 Let $M = M_1 M_2$ where the factors are simple of the same type. Then M has a diagonal subgroup H with simply-connected cover $\hat{H}$. Then the composed maps $\hat{H} \to H \to M_i$ are isogenies. It follows that up to M-conjugacy, these compositions of powers of Frobenius maps F (or their square roots in some very special cases) with automorphisms which induce a symmetry of the Dynkin diagram of $\hat{H}$. For instance, if M has type $A_1 A_1$ then, since the A_1 Dynkin diagram has trivial symmetry group, H corresponds to a pair of non-negative integers (r, s) and the map $\hat{H} \to H$ is $x \mapsto (F^r(x), F^s(x))$. Since $H \cong F(H)$ we may assume $rs = 0$. For brevity, we use the notation $A_1 \hookrightarrow A_1 A_1$

via $(1^{[r]}, 1^{[s]})$ for these diagonal subgroups. See [117, Chapters 2,11] for further discussion and notation.

To recap, our recipe is now to iterate through the maximal connected subgroups M of G, collecting (semisimple) candidates H. We throw away all those candidates which fall into a proper parabolic subgroup of G, and then determine the poset $\mathrm{ConIrr}(G)$ by identifying conjugacy amongst the remaining candidates.

The passage from G-irr semisimple subgroups to all G-cr semisimple subgroups is now easy using Lemma 5.4.1. Suppose that H is $\mathscr{D}(L)$-irr. By the above remarks we can write down the $\mathscr{D}(L)$-classes of $\mathscr{D}(L)$-irr subgroups; then from the lemma we need only establish conjugacy amongst those classes by examining the action of the stabiliser of L in the Weyl group of G.

Remark 5.4.6 A related representation-theoretic question is to classify triples (G, H, V) where $H \subseteq G$ and V is an irreducible G-module which remains irreducible as an H-module. This property is strictly stronger than G-irreducibility and has its own extensive literature, cf. [23] and the references therein.

5.4.2 Non-G-cr subgroups

We now turn our attention to non-G-cr semisimple subgroups H (recalling again that G is connected, reductive). Then $H \subseteq P$ for some proper parabolic subgroup $P = QL$ with $Q = \mathscr{R}_u(P)$ and we may assume P is minimal subject to containing H. Let $\bar{H}$ denote the image of H in L under the projection $\pi : P \to L$. Then by Lemma 5.4.1 $\bar{H}$ is L-irr (hence G-cr) and is not $\mathscr{R}_u(P)$-conjugate to H. We may assume from the previous section that we know $\bar{H}$ up to conjugacy. To make further progress, we use non-abelian cohomology, whose techniques are similar to those employed in Galois cohomology.

Firstly, note that either $\pi : H \to \bar{H}$ is an isomorphism of algebraic groups or a very special situation occurs, namely, $p = 2$ and H has a simple factor SO_{2n+1} with image Sp_{2n} in $\bar{H}$, so that the differential $d\pi : H \to \bar{H}$ has a non-zero kernel.[9]

Suppose for now that this special situation does not hold, so that H and $\bar{H}$ are isomorphic as algebraic groups and H is a complement to

[9] On the level of schemes, H intersects Q non-trivially, giving rise to a non-zero scheme-theoretic kernel. See [19] for the theory surrounding this map, or one of [88, Lemma 2.2], [118], [37] for more concrete treatments.

Q in the semidirect product $\bar{H}$. Any element of H can thus be written uniquely as $\gamma(h)h$ with $h \in \bar{H}$, for some map $\gamma : \bar{H} \to Q$ which is a morphism of varieties. The definition of the semidirect product implies that γ satisfies a 1-cocycle condition; namely:

$$\gamma(gh) = \gamma(g)(g \cdot \gamma(h)).$$

If H' is another complement to Q, corresponding to a map γ', then H is Q-conjugate to H' if and only if γ is related to γ' via a coboundary; in other words the Q-conjugacy classes of complements are given by classes $[\gamma] \in \mathrm{H}^1(\bar{H}, Q)$. We leave the description of the precise relationship of γ and γ' to [106, §2], but note that since Q is typically non-abelian, the set $\mathrm{H}^1(\bar{H}, Q)$ does not admit the structure of a group – rather, it is only a pointed set, having a distinguished element corresponding to the class of the trivial cocycle. In contrast, when Q has the structure of an $\bar{H}$-module – i.e. Q is a vector space on which $\bar{H}$ acts linearly – then both Q and $\mathrm{H}^1(\bar{H}, Q)$ are naturally k-vector spaces. In the latter case it can be shown that $\mathrm{H}^1(\bar{H}, Q)$ is isomorphic to the first right-derived functor (applied to Q) of the fixed point functor $\mathrm{H}^0(\bar{H}, ?)$ in the category of rational G-modules. For more on this last point, see [55, I.4].

Example 5.4.7 By way of illustration, we list the semisimple subgroups of $G = \mathrm{SL}_3$. There are four parabolic subgroups of G up to conjugacy, respectively stabilising flags with submodule dimensions (3), (2, 1), (1, 2) and (1, 1, 1). The first is G itself, the last is a Borel subgroup (whose only reductive subgroups are tori) and the other two have G-conjugate Levi subgroups GL_2, one of which can be described as the image of the embedding

$$\mathrm{GL}_2 \to \mathrm{SL}_3, \quad A \mapsto \left(\begin{array}{c|c} A & 0 \\ \hline 0 & \det(A)^{-1} \end{array}\right).$$

Of course G is G-irr; and if L is a Levi subgroup isomorphic to GL_2 then $\mathscr{D}(L) \cong \mathrm{SL}_2$, which has no proper semisimple subgroups. If $p \neq 2$, then there is one further G-irr subgroup PGL_2, embedded via the irreducible adjoint action on its Lie algebra. These are all the G-cr semisimple subgroups.

Suppose that $H \subseteq G$ is semisimple and non-G-cr. By rank considerations, $\bar{H}$ is isomorphic to SL_2, lying in a parabolic subgroup $P = QL$ with $L \cong \mathrm{GL}_2$. It is easy to check that the conjugation action of $\bar{H}$ on Q gives it the structure of the natural module $L(1)$. This means $\mathrm{H}^1(\bar{H}, Q) = 0$, which rules out the existence of non-G-cr subgroups unless $p = 2$.

If $p = 2$, however, the action of SL_2 on its Lie algebra and its dual are not completely reducible: Both are indecomposable with two composition factors. The first is isomorphic to the Weyl module $V(2)$ which has $L(2)$ in the head and $L(0)$ in its socle; we denote this $L(2)/L(0)$. The second is upside down: $\mathrm{H}^0(2) \cong L(0)/L(2)$. This yields two non-conjugate, non-G-cr subgroups PGL_2, one in each of the two standard parabolic subgroups of G.[10]

Concluding, the subgroups above – G itself, the two derived Levi subgroups SL_2, a G-irr subgroup PGL_2 (when $p \neq 2$) and two non-G-cr subgroups PGL_2 (when $p = 2$) – are now all the non-trivial semisimple subgroups of G.

5.4.3 Abelian and non-abelian cohomology

Let G be a connected reductive group acting on a G-module V. To mount a proper investigation of $\mathrm{H}^1(G, V)$, a scheme-theoretic treatment such as [55] is essential. This is not least because one can make use of the Lyndon–Hochschild–Serre spectral sequence

$$E_2^{ij} = \mathrm{H}^i(G/N, \mathrm{H}^j(N, V)) \Rightarrow \mathrm{H}^{i+j}(G, V)$$

for calculations, where N is a normal subgroup scheme of G. In this framework, N is allowed to be an *infinitesimal subgroup scheme*, the most important example being the Frobenius kernel $G_1 := \ker(G \to F(G))$ of G. At the level of points, $G_1 = \{1\}$, but $\mathrm{Lie}(G_1) = \mathrm{Lie}(G)$ has far more structure. We leave the interested reader to pursue this further, but give some references: The first general investigation of $\mathrm{H}^1(G, V)$ using the LHS spectral sequence applied to $G_1 \lhd G$ is probably that of Jantzen in [53], which connects $\mathrm{H}^1(G, L(\lambda))$ with the structure of the Weyl module $V(\lambda)$. Other relevant papers are too numerous to mention, but some highlights are [27], [15], [16], [86].

On the understanding that $\mathrm{H}^1(\bar{H}, V)$ has been well-studied for V an $\bar{H}$-module, let us return to the calculation of $\mathrm{H}^1(\bar{H}, Q)$, where $P = QL$ is a parabolic subgroup of a reductive algebraic group G. The fact that Q is connected, smooth and unipotent means that it admits a filtration

$$Q = Q_0 \supseteq Q_1 \supseteq \cdots \supseteq Q_n = 1$$

for some n, such that $Q_i \lhd Q$ and the subquotients Q_i/Q_{i+1} admit the structure of $\bar{H}$-modules. The statement for general connected unipotent

[10] In fact, this is the 'special situation' mentioned earlier, since $H = \mathrm{SL}_2$ is abstractly isomorphic to its image PGL_2 in $G = \mathrm{SL}_3 = \mathrm{SL}(\mathrm{Lie}(\mathrm{SL}_2))$.

groups Q can be found in [106, Theorem 3.3.5] and [82, Theorem C], but one can be more explicit here since Q has a filtration by subgroups Q_i generated by root subgroups of G. Following [8], let P and L be a standard parabolic and Levi subgroup, corresponding to a subset I of the simple roots Δ (which can be identified with nodes of the Dynkin diagram of G). Then P is generated by a maximal torus and root subgroups U_α where α runs through positive roots, as well as negative roots in I. Expressing each root α uniquely as

$$\alpha = \left(\sum_{\alpha_i \in I} c_i \alpha_i \right) + \left(\sum_{\alpha_j \in \Delta \setminus I} d_j \alpha_j \right),$$

the roots in P are those with $\sum d_i \geq 0$; the roots occurring in L are those with $\sum d_i = 0$, and those in Q have $\sum d_i > 0$. The quantity $\sum d_j$ is called the *level* of the root, and $\sum d_j \alpha_j$ is called its *shape*. For each $i > 0$, denote by Q_i the subgroup generated by root subgroups of level i; then the Chevalley commutator relations imply that each Q_i is normal in Q, and in fact Q_i/Q_{i+1} is central in Q/Q_{i+1}. Furthermore, from knowledge of the root system of G, say by reference to [22], one can write down explicitly the representations Q_i/Q_{i+1} as Weyl modules $V(\lambda)$ for the Levi subgroup L.

Example 5.4.8 Recall that the Dynkin diagram of G_2 is (node 1 ⇛ node 2). The nodes represent the two simple roots which we denote by lists of coefficients, with $\alpha_1 = 10$ and $\alpha_2 = 01$. The remaining positive roots, in order of height, are 11, 21, 31, 32. Let P be the standard parabolic containing the negative of α_2, i.e. -01. Then a Levi factor L of P has roots ± 01, and $\mathscr{R}_u(P)$ has roots of three levels $\{11, 10\}$, $\{21\}$ and $\{31, 32\}$, which one checks induce modules $L(1)$, $L(0)$, $L(1)$ for the Levi subgroup $\mathscr{D}(L) \cong \mathrm{SL}_2$.

Once the modules Q_i/Q_{i+1} and cohomology groups $\mathrm{H}^1(\bar{H}, Q_i/Q_{i+1})$ are understood, one can take the direct sum $\mathbb{V} := \bigoplus \mathrm{H}^1(\bar{H}, Q_i/Q_{i+1})$ and use this to approximate $\mathrm{H}^1(\bar{H}, Q)$. In fact, one can define a partial map $\mathbb{V} \to \mathrm{H}^1(\bar{H}, Q)$, which turns out to be surjective, using a lifting process we now describe. Given any short exact sequence of $\bar{H}$-groups

$$1 \to R \to Q \to S \to 1$$

with R contained the centre of Q, there is an exact sequence of $\bar{H}$-sets[11]

$$\begin{aligned} 1 \to &\mathrm{H}^0(\bar{H}, R) \to \mathrm{H}^0(\bar{H}, Q) \to \mathrm{H}^0(\bar{H}, S) \\ &\xrightarrow{\delta_{\bar{H}}} \mathrm{H}^1(\bar{H}, R) \to \mathrm{H}^1(\bar{H}, Q) \to \mathrm{H}^1(\bar{H}, S) \xrightarrow{\Delta_{\bar{H}}} \mathrm{H}^2(\bar{H}, R). \end{aligned} \tag{5.1}$$

Now, taking $R = Q_i/Q_{i+1}$ and $S = Q/Q_i$ for each i, one can use these 'long' exact sequences to lift elements of $\mathrm{H}^1(\bar{H}, Q/Q_i)$ to elements of $\mathrm{H}^1(\bar{H}, Q/Q_{i+1})$, eventually reaching $\mathrm{H}^1(\bar{H}, Q)$ itself, as long as we understand two issues:

(i) When is $\mathrm{H}^1(\bar{H}, Q/Q_i) \to \mathrm{H}^1(\bar{H}, Q)$ not injective?
(ii) When is $\mathrm{H}^1(\bar{H}, Q/Q_i) \to \mathrm{H}^1(\bar{H}, Q/Q_{i+1})$ not defined?

Question i asks whether $\delta_{\bar{H}}$ is non-zero. This happens precisely when cocycle classes in $\mathrm{H}^1(\bar{H}, R)$ fuse inside $\mathrm{H}^1(\bar{H}, Q)$ due to conjugacy induced by the fixed points $S^{\bar{H}}$. Question ii asks whether $\Delta_{\bar{H}}$ is non-zero. If so then cocycles in $\mathrm{H}^1(\bar{H}, S)$ are obstructed from lifting to cocycles in $\mathrm{H}^1(\bar{H}, Q)$. In particular, this only happens when $\mathrm{H}^2(\bar{H}, R) \neq 0$.

In the end, this lifting process allows us to calculate $\mathrm{H}^1(\bar{H}, Q)$ completely. The matter is easy if we can show the maps $\delta_{\bar{H}}$ and $\Delta_{\bar{H}}$ to be zero. However, this is often not the case and one must resort to explicit computations with cocycles; this is the approach taken in [107].

Part II. Subgroup structure of exceptional algebraic groups

5.1 Maximal subgroups

Work on classifying maximal sub-objects of Lie type objects dates back to Sophus Lie [61]. Taking inspiration from Galois's work on univariate polynomials, *op. cit.* develops 'continuous transformation groups' – now Lie groups – with a view to classifying differential equations in terms of symmetries amongst their solutions. One builds up group actions from primitive actions, corresponding to maximal subgroups, motivating Lie

[11] Here, an exact sequence of pointed sets means only that the image of each map is the preimage of the distinguished element under the next.

to describe such subgroups. The same problem for finite groups was not to be posed until a paper of Aschbacher and Scott [2] rather later, and Lie concentrated on connected subgroups of connected Lie groups. Here, the exp and log make this equivalent to finding maximal subalgebras $\mathfrak{m}$ of real Lie algebras $\mathfrak{g}$, and Lie solved the problem when $\dim \mathfrak{g} \leq 3$. Otherwise the question lay dormant for another fifty years.

Using the Killing–Cartan–Weyl classification of finite-dimensional complex simple Lie algebras, E. Dynkin solved Lie's problem over $\mathbb{C}$ [38]. We give a quick example – stolen from Seitz's excellent tribute in *op. cit.* – to illustrate his results. As is well-known, the complex 3-dimensional Lie algebra $\mathfrak{sl}_2$ has a unique irreducible representation of each degree up to equivalence. This amounts to an embedding of $\mathfrak{sl}_2$ into $\mathfrak{so}_{2n-1}$ or $\mathfrak{sp}_{2n}$, where $2n-1$ or $2n$ respectively is the degree. Dynkin showed that for $n \geq 2$, the image of each of these embeddings is a maximal subalgebra, with precisely one exception: when $n = 7$ and the exceptional Lie algebra of type G_2 has a self-dual 7-dimensional module, it occurs as a (maximal) subalgebra of $\mathfrak{so}_7$ and in turn contains the irreducible $\mathfrak{sl}_2$ as a maximal subalgebra. There is a remarkably short list of such situations. Dynkin in effect classified the maximal subalgebras of the classical Lie algebras $\mathfrak{sl}_{n+1}$, $\mathfrak{so}_n$ and $\mathfrak{sp}_{2n}$ by classifying non-maximal ones which nevertheless act irreducibly on the natural modules for those algebras. A key ingredient in Dynkin's work was detailed information on the representations of these Lie algebras, developed by Weyl and others, in terms of the weights for their Cartan subalgebras.

Dealing with the Lie algebras of exceptional type required Dynkin to adopt a more exhaustive approach. He first showed how to produce all the semisimple subalgebras of $\mathfrak{g} = \mathrm{Lie}(G)$ containing a given Cartan subalgebra $\mathfrak{h}$, so-called *regular* subalgebras. Since root spaces are 1-dimensional, it follows that such a subalgebra will be the sum of $\mathfrak{h}$ and the root spaces corresponding to a subset Φ' of the root system Φ of G. Dynkin showed that one can find all regular subalgebras by iteratively extending the Dynkin diagram (adding a node corresponding to the negative of the highest long root) and then deleting some nodes.

Example 5.1.1 Let Φ be an irreducible root system of type F_4 with roots labelled as in the following diagram.

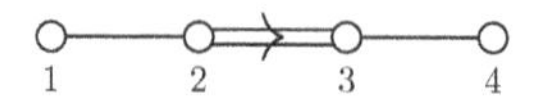

Then the highest long root is $\alpha_0 = 2\alpha_1 + 3\alpha_2 + 4\alpha_3 + 2\alpha_4$. The only simple root one can add to $-\alpha_0$ and still get a root is α_1. Therefore, the extended Dynkin diagram is:

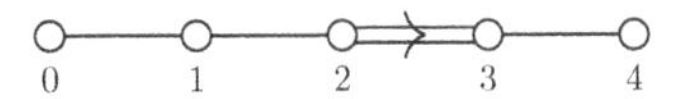

The maximal subalgebras of maximal rank correspond to deleting a node of this extended diagram corresponding to a simple root with prime coefficient in the expression of α_0: in our case, this is α_1, α_2 and α_4. Removing α_1 gives a Dynkin diagram of type A_1C_3, removing α_2 gives type A_2A_2 and removing α_4 gives type B_4.

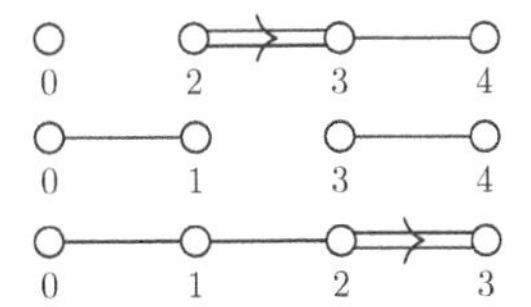

Non-semisimple subalgebras were described by a theorem of Morozov[12] and the maximal ones are the maximal parabolics. This leaves those maximal subalgebras which do not contain a Cartan subalgebra, so-called *S-subalgebras*. Dynkin tackled those of type $\mathfrak{sl}_2 = \langle e, f, h\rangle$ first, associating to each class of these under the adjoint action of G a Dynkin diagram with a label of 0, 1 or 2 above each node determining the conjugacy class of h. It turns out there is a unique conjugacy class of $\mathfrak{sl}_2$-subalgebras such that h is a *regular element*, i.e. the centraliser $\mathfrak{g}_h$ of h in $\mathfrak{g}$ is as small as possible, that is, $\mathfrak{g}_h$ is a Cartan subalgebra. The corresponding Dynkin diagram for this class of subalgebras has a 2 above each node and it is usually maximal. From there, if $\mathfrak{sl}_3$ is a subalgebra of $\mathfrak{g}$, then one can look to build it up from its own regular $\mathfrak{sl}_2$.

There are many reasons to extend this theory to positive characteristic, not least because algebraic groups and their points over finite fields give information about finite groups, for instance in furtherance of the Aschbacher–Scott programme. Over several important monographs of Seitz [97, 98], Liebeck–Seitz [67] and Testerman [113], Dynkin's classification is extended to describe the maximal subgroups of simple algebraic groups over algebraically closed fields of positive characteristic.

Unfortunately, there is no space to do anything else but state the main result in case when G is exceptional. In the following, conditions such as $p \geq 13$ also include the case $p = 0$. Note that $\tilde{H}$ denotes a subgroup of type H whose root groups are generated by short roots of G.

[12] the precursor of Borel–Tits' Theorem 5.2.2.

Theorem 5.1.2 ([67, Corollary 2], [35, Theorem. 1]) *Let G be a simple algebraic group of exceptional type in characteristic p and let M be maximal amongst connected subgroups of G. Then M is either parabolic or is G-conjugate to precisely one subgroup H in Table 5.1, where each H denotes one G-conjugacy class of subgroups.*

Table 5.1: The reductive maximal connected subgroups of exceptional algebraic groups.

G	H
G_2	A_2, $\tilde{A}_2$ $(p = 3)$, $A_1\tilde{A}_1$, A_1 $(p \geq 7)$
F_4	B_4, C_4 $(p = 2)$, A_1C_3 $(p \neq 2)$, A_1G_2 $(p \neq 2)$, $A_2\tilde{A}_2$, G_2 $(p = 7)$, A_1 $(p \geq 13)$
E_6	A_1A_5, A_2^3, F_4, C_4 $(p \neq 2)$, A_2G_2, G_2 $(p \neq 7; 2 \text{ classes})$, A_2 $(p \geq 5; 2 \text{ classes})$
E_7	A_1D_6, A_2A_5, A_7, G_2C_3, A_1F_4, A_1G_2 $(p \neq 2)$, A_2 $(p \geq 5)$, A_1A_1 $(p \geq 5)$, A_1 $(p \geq 17)$, A_1 $(p \geq 19)$
E_8	D_8, A_1E_7, A_2E_6, A_8, A_4^2, G_2F_4, F_4 $(p = 3)$, B_2 $(p \geq 5)$, A_1A_2 $(p \geq 5)$, A_1 $(p \geq 31)$, A_1 $(p \geq 29)$, A_1 $(p \geq 23)$

Remark 5.1.3

(i) As discussed in §5.2, the classes of parabolic subgroups are in bijection with subsets of the simple roots Δ, and the maximal ones correspond to subsets of size $|\Delta| - 1$.

(ii) In the caption of Table 5.1 we use the phrase *reductive maximal connected.* By this mean we reductive subgroups which are maximal amongst connected subgroups. Similarly, in the caption to the next table we say *reductive maximal positive-dimensional* to mean reductive subgroups which are maximal amongst positive-dimensional subgroups.

(iii) The subgroups of maximal rank can be enumerated using the Borel–de-Siebenthal algorithm. This is a more general version of Dynkin's procedure for regular subalgebras, and includes some extra cases where the Dynkin diagram has an edge of multiplicity $p = \operatorname{char} k$. For example if G is of type F_4, then there is a maximal regular subgroup of type C_4. See [78, §13.2] for a complete explanation.

(iv) When $G = E_6$ there are two classes of maximal subgroups of type

G_2 ($p \geq 7$) and A_2 ($p \geq 5$). The graph automorphism of G interchanges these two classes. See [114] for a proof of this and an explicit construction of the maximal subgroups.

(v) The maximal subgroup of type F_4 when $G = E_8$ and $p = 3$ was overlooked in [98] and subsequently missed in [67]. This was rectified by Craven and the second two authors; more information can be found in [35].

It is also natural to ask about non-connected maximal subgroups. For example, finite subgroups of E_8 remain unclassified. See §5.4.3 for a brief description of the latest developments. However one can successfully weaken 'connected' to 'positive-dimensional':

Theorem 5.1.4 ([67, Corollary 2]) *Let G be a simple algebraic group of exceptional type in characteristic p. Let M be a positive-dimensional maximal subgroup of G. Then M is either parabolic or G-conjugate to precisely one subgroup H as follows. Each isomorphism type of H denotes one G-conjugacy class of subgroups, and the notation T_i indicates an i-dimensional torus.*

Table 5.2: The reductive maximal positive-dimensional subgroups of exceptional algebraic groups.

G	H
G_2	$A_2.2$, $\tilde{A}_2.2$ ($p = 3$), $A_1\tilde{A}_1$, A_1 ($p \geq 7$)
F_4	B_4, $D_4.S_3$, C_4 ($p = 2$), $\tilde{D}_4.S_3$ ($p = 2$), A_1C_3 ($p \neq 2$), A_1G_2 ($p \neq 2$), $(A_2\tilde{A}_2).2$, G_2 ($p = 7$), A_1 ($p \geq 13$)
E_6	A_1A_5, $(A_2^3).S_3$, $(D_4T_2).S_3$, $T_6.W(E_6)$, F_4, C_4 ($p \neq 2$), A_2G_2, G_2 ($p \neq 7$), $A_2.2$ ($p \geq 5$)
E_7	A_1D_6, $(A_2A_5).2$, $A_7.2$, $(A_1^3D_4).S_3$, $(A_1^7).PSL_3(2)$, $(E_6T_1).2$, $T_7.W(E_7)$, G_2C_3, A_1F_4, $(2^2 \times D_4).S_3$, A_1G_2 ($p \neq 2$), $A_2.2$ ($p \geq 5$), A_1A_1 ($p \geq 5$), A_1 ($p \geq 17$), A_1 ($p \geq 19$)
E_8	D_8, A_1E_7, $(A_2E_6).2$, $A_8.2$, $(A_4^2).4$, $(D_4^2).(S_3 \times 2)$, $(A_2^4).(GL_2(3))$, $(A_1^8).AGL_3(2)$, $T_8.W(E_8)$, G_2F_4, $A_1(G_2^2).2$ ($p \neq 2$), F_4 ($p = 3$), B_2 ($p \geq 5$), A_1A_2 ($p \geq 5$), A_1 ($p \geq 31$), A_1 ($p \geq 29$), A_1 ($p \geq 23$), $A_1 \times S_5$ ($p \geq 7$)

Remark 5.1.5 The subgroup $A_2G_2 < E_6$ is a maximal subgroup and its presence above corrects a small mistake in [67, Table 1]. In *loc. cit.* it is claimed that $N_{E_6}(A_2G_2) = (A_2G_2).2$ with the outer involution acting

as a graph automorphism of the A_2 factor. This is not possible as the action of A_2G_2 on the 27-dimensional E_6-module V_{27} is not self-dual. Instead, it is the graph automorphism of E_6 which induces an outer involution on A_2G_2.

5.2 The connected G-irreducible subgroups

In light of §5.4.1 there are three things we need to determine ConIrr(G) (Definition 5.4.2) for G a simple exceptional algebraic group.

(i) Determine the semisimple maximal connected subgroups of G;
(ii) Decide whether a candidate subgroup[13] is G-irr;
(iii) Decide whether two G-irr candidate subgroups are G-conjugate.

Since G is a simple exceptional algebraic group, i is immediate from Theorem 5.1.2. We consider ii and iii in the next two sections.

5.2.1 Testing candidate subgroups

Let H be an M-irr connected subgroup of G, where M is maximal semisimple. We need to decide whether or not H is G-irr.

Proving that candidates are G-irr

If a candidate H is in fact contained in a parabolic subgroup $P = QL$ of G then we can consider the image $\pi(H)$ in a Levi subgroup. The action of H and $\pi(H)$ on G-modules may differ but their composition factors will always match.[14] Thus one way to prove that H is G-irr is to show that its composition factors on some G-module do not match those of any proper Levi subgroup of G. For instance, every proper Levi subgroup of G has a trivial composition factor on the adjoint module Lie(G), so a candidate is G-irr if it has no trivial composition factors on Lie(G).

Proving that candidates are not G-irr

Now let H be a candidate subgroup which we believe is not G-irr. If H is G-cr but not G-irr then H is contained in some proper Levi subgroup L of G by Lemma 5.4.1, and thus $C_G(H)$ will contain the non-trivial torus

[13] Recall that a candidate subgroup is an M-irr subgroup from a semisimple maximal connected subgroup M.

[14] see [115, Lemma 3.8] for the precise definition of *match*.

$Z(L)^{\circ}$. In this case, it is often easy to find a non-trivial torus commuting with H and thus conclude that H is not G-irr.

The most difficult cases are when a candidate H turns out to be non-G-cr (and thus not G-irr). Such cases are relatively rare: [117, Corollary 3] classifies the non-G-cr connected subgroups which are M-irr for every (and at least one) reductive maximal connected subgroup M in which they are contained. There are two main methods used in [115, 116]. Briefly, one either

(1) directly shows that H is contained in a parabolic subgroup P; or

(2) finds a non-G-cr subgroup $Z \subset P$ and show that Z is contained in M and conjugate to H.

One way to implement 1 is to exhibit a non-zero fixed point of H on the adjoint module $\mathrm{Lie}(G)$. By [98, Lemma 1.3], this places H in either a proper maximal-rank subgroup or a proper parabolic subgroup of G, and one can use representation theory to prove that H is not contained in a proper maximal-rank subgroup. Another way is to find a unipotent subgroup of G normalised by H, since the Borel–Tits theorem then places H in a proper parabolic subgroup. This is used in [115, Lemmas 7.9, 7.13], where calculations in Magma are used to construct an ad-nilpotent subalgebra $S \subset \mathrm{Lie}(G)$ stabilised by a 'large enough' finite subgroup $H(q) < H$, where $q = p^r$ for some $r > 0$. It then follows that H also stabilises S, and one checks that one can exponentiate S to yield a unipotent subgroup normalised by H.

Method 2 is implemented in [115, Lemmas 6.3, 7.4]. Here one starts with a candidate H contained in a semisimple maximal connected subgroup M of maximal rank. One constructs the relevant non-G-cr subgroup Z according to the recipe in §5.4.2, and then shows that $Z \subset M$. To establish the latter, one proves that any group acting with the same composition factors as Z on the adjoint module of G fixes a non-zero element of $\mathrm{Lie}(G)$. One then shows that Z does not fix any non-zero nilpotent element; thus it fixes a non-zero semisimple element and is contained in a maximal rank subgroup M', again by [98, Lemma 1.3]. This part is rather technical and requires the full classification of stabilisers of nilpotent elements and their structure, as found in [68]. It is however then possible to identify that $M' = M$ and show that Z is conjugate to H.

5.2.2 G-conjugacy

Once we know that two isomorphic candidates H_1 and H_2 are G-irr, we must check whether they are G-conjugate. One easy test is to check whether their composition factors on various G-modules agree. If so then it turns out that, with a single exception, the two candidates are in fact G-conjugate. The exception occurs when $G = E_8$, $p \neq 3$ and H_1 and H_2 are simple of type A_2, diagonally embedded in $A_2^2 \subset D_4^2 \subset D_8 \subset E_8$ via $(10, 10^{[r]})$ and $(10, 01^{[r]})$ respectively, with $r \neq 0$. For more detail see [117, Corollary 1] and its proof.

If H_1 and H_2 are not simple then it is usually straightforward to show they are conjugate when they have the same composition factors on $\mathrm{Lie}(G)$, by considering the centraliser of one of the simple factors.

Example 5.2.1 Let $G = E_8$, let M_1 be the maximal-rank subgroup of type A_1E_7 and let M_2 be the maximal-rank subgroup of type D_8. Take $H_1 = A_1^2D_6 \subset M_1$ and $H_2 = A_1^2D_6 \subset D_8 = M_2$. Then H_1 and H_2 are G-irr and G-conjugate. Indeed, taking Y to be one of the A_1 factors of H_2, we have $H_2 \subset YC_G(Y) = M_1$, by appealing to [62, p.333, Table 2]. As H_1 is the only subgroup of type $A_1^2D_6$ contained in M_1 up to conjugacy, it is conjugate to H_2.

When the candidates are simple this process can be slightly more involved. Often one of the candidates turns out to be the connected centraliser in M or G of an involution, or of an element of order 3.

Example 5.2.2 Again let $G = E_8$ and $p \neq 2$. Take $H = B_4 \subset A_8$, with H acting irreducibly on the natural 9-dimensional module for A_8. Then H is the centraliser in G of an involution t in the disconnected subgroup $A_8.2$. By calculating the trace of this involution on the adjoint module for G and using [65, Proposition 1.2], we find that $C_G(t) = D_8$ and hence $H \subset D_8$. Similar calculations are carried out in [63, pp. 56–68].

5.2.3 Main results

We now present results classifying the G-cr semisimple subgroups of exceptional algebraic groups. When p is large enough that all subgroups of a given type are G-cr, the simple G-cr subgroups were classified in [63, 60]. The G-irr subgroups of type A_1 were studied for G of exceptional type except E_8 in [1], and G-irr subgroups of $G = G_2, F_4$ are classified in [105], [107], respectively. The reductions in §5.4.1 now allow us to concentrate on semisimple, G-irr subgroups.

Theorem 5.2.3 *Let G be a simple algebraic group of exceptional type and let H be a G-irr connected subgroup of G. Then H is* $\mathrm{Aut}(G)$*-conjugate to exactly one subgroup in Tables [117, §11, Tables 1–5] and each subgroup in the tables is G-irr.*

This was proved in a sequence of papers [115, 116, 117]. The tables are lengthy and we do not reproduce them here. The last of these papers, specifically [117, §11, Tables 1A–5A], describes the poset structure of $\mathrm{ConIrr}(G)$ and gives a detailed explanation of the tables. The tables also provide the composition factors of each subgroup in $\mathrm{ConIrr}(G)$ on both the minimal and adjoint module.

Example 5.2.4 When $G = G_2$, the reductive maximal connected subgroups of G are:

$$A_1\tilde{A}_1, \quad A_2, \quad \tilde{A}_2 \ (p=3), \quad \text{and} \quad A_1 \ (p \geq 7),$$

where $\tilde{A}_1$ and $\tilde{A}_2$ denote subgroups whose roots are short roots of G. A group of type A_1 has no proper irreducible connected subgroups, so this requires no further consideration.

Take $M_1 = A_1\tilde{A}_1$. Since the factors have the same type, there are diagonal M-irr connected subgroups of type A_1. As in Example 5.4.5, these are determined by non-negative integers r, s with $rs = 0$ and we write $H^{r,s}$ for such a subgroup. So $\mathrm{ConIrr}(M_1) = \{H^{r,s} \mid rs = 0\}$. We now need to decide whether the members of $\mathrm{ConIrr}(M_1)$ are G-irr. If $(r,s) \neq (0,0)$ or $p > 3$ then $H^{r,s}$ acts on $\mathrm{Lie}(G)$ without trivial composition factors, hence it is G-irr. When $p = 3$, the composition factors of $H^{0,0}$ on $\mathrm{Lie}(G)$ do not match those of a Levi subgroup and so $H^{0,0}$ is also G-irr. When $p = 2$, however, $H^{0,0}$ is contained in an A_1-parabolic subgroup and is non-G-cr (appearing in Theorem 5.3.5). To see that $X := H^{0,0}$ is contained in a proper parabolic subgroup it suffices to demonstrate that X stabilises a 1-space on the irreducible 6-dimensional module $L_G(10)$, since by [71, Theorem B], G is transitive on 1-spaces of $L_G(10)$ and the stabiliser of such a 1-space is a long root parabolic subgroup. It remains to show that X is not conjugate to $\mathscr{D}(L)$ for some Levi subgroup L (up to conjugacy, these are the two simple factors of M). This can be done by calculating the action of X on $L_G(10)$, which is $T(2) + 2$ and then comparing it with the actions of the two subgroups $\mathscr{D}(L)$ on $L_G(10)$, which are $1^2 + 0^2$ and $1^2 + 2$, respectively.

Now let $M_2 = A_2$. Applying Proposition 5.3.15, we find a single candidate, H, which has type A_1 with $p \neq 2$, and this acts irreducibly on the adjoint 3-dimensional module $L(2)$. We must check that H is G-irr.

In fact, one can show H is conjugate to $H^{0,0}$ and thus G-irr. To see this, note that $N_G(M_2) = M_2\langle t\rangle$, with t an involution inducing a graph automorphism on M_2 [42, Table 4.3.1]. As $L(2)$ is self-dual, one concludes that H centralises t and hence $H \subset C_G(t) = M_1$. The only subgroups of type A_1 in M are its two simple factors and the subgroups $H^{r,s}$. Since $p \neq 2$, the composition factors of the action of these subgroups on $V_G(10)$ distinguish them and we conclude that H is conjugate to $H^{0,0}$.

The same method applies to $\tilde{A}_2$ when $p = 3$ and one finds a single candidate subgroup H of type A_1, which turns out to be G-irr and conjugate to $H^{1,0}$.

We present this classification in Figure 5.1, with a straight line depicting containment. This gives a small flavour of the additional information in [117].

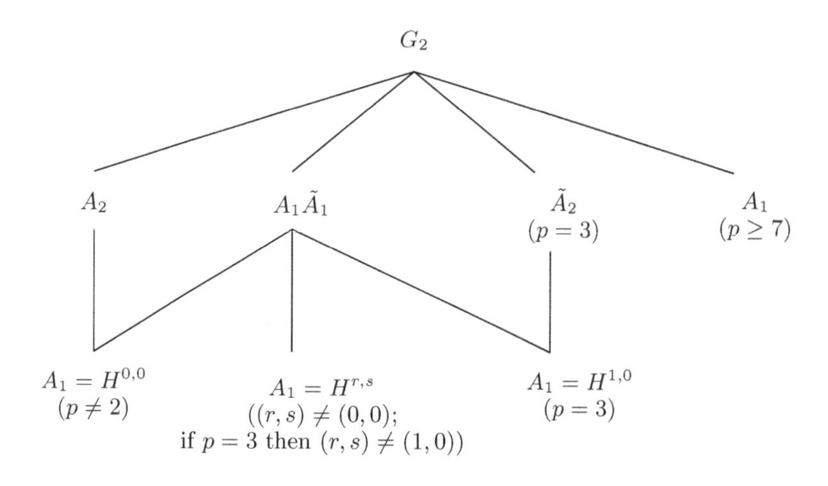

Figure 5.1 The poset of G_2-irr connected subgroups.

Remark 5.2.5 Further information about the G-reducible semisimple G-cr subgroups can be found in [74]. In particular, the semisimple G-reducible G-cr connected subgroups of $G = F_4$ are classified and written down explicitly in [75, §6.4], correcting [107, Corollary 5]. For each reducible G-cr subgroup H, the socle series of the action of H on $\text{Lie}(G)$ is computed together with the centraliser $C_G(H)^\circ$ (which is another G-cr subgroup by Theorem 5.3.8).

5.3 Non-G-completely reducible subgroups

The method in §5.4.2 has been applied to classify the non-G-cr semisimple subgroups of exceptional algebraic groups G in many cases. The following is a more precise version of [63, Theorem 1]. It generalises Theorem 5.3.3 for exceptional groups by ruling out certain non-G-cr subgroups of certain types even when $p \leq \mathrm{rank}(G)$.

Theorem 5.3.1 *Let G be a simple algebraic group of exceptional type in characteristic p. Let X be an irreducible root system and let $N(X,G)$ be the set of primes defined by Table 5.3, e.g. $N(B_2, E_8) = \{2,5\}$.*

If H is a connected reductive subgroup of G and $p \notin N(X,G)$ whenever H has a simple factor of type X, then H is G-cr. Conversely, whenever $p \in N(X,G)$, there exists a non-G-cr simple subgroup H of type X.

	$G = E_8$	E_7	E_6	F_4	G_2
$X = A_1$	≤ 7	≤ 7	≤ 5	≤ 3	2
A_2	≤ 3	≤ 3	≤ 3	3	
B_2	5 2	2	2	2	
G_2	7 3 2	7 2			
A_3	2	2			
B_3	2	2	2	2	
C_3	3				
B_4	2				
C_4, D_4	2	2			

Table 5.3 *Values of $N(X,G)$.*

Remark 5.3.2 This corrects [108, Theorem 1], which had claimed the existence of non-G-cr subgroups H of type G_2 when $p = 2$ and G is of type F_4 or E_6. We discuss this further in Example 5.3.4.

The result "$p \notin N(G,X) \Rightarrow$ all simple subgroups of type X are G-cr" is largely proven in [63]. The strategy is to show that for each parabolic subgroup $P = QL$ and each L-irr subgroup $\bar{X} < L$, the levels of Q restricted to $\bar{X}$ have trivial 1-cohomology. In *loc. cit.* and [107] this is accomplished by determining all modules V with $\mathrm{H}^1(\bar{X}, V) \neq 0$ such that $\dim V$ is small enough for V to potentially appear as an $\mathscr{D}(L)$-composition factor in the filtration of Q. For G of type E_8, the largest dimension of such a $\mathscr{D}(L)$-composition factor is 64, occurring when $\mathscr{D}(L)$ is of type D_7, which makes this a tractable problem. (See [108, Lemma 5.1] for more information on the upper bounds on $\dim V$.)

Example 5.3.3 We use results from §5.3.2 to exhibit a non-G-cr subgroup of type A_1 in $G = E_6$ when $p = 5$. Here, G has a Levi subgroup L with $\mathscr{D}(L)$ of type A_5. It follows from Theorem 5.3.11 that any non-L-cr subgroup is non-G-cr. So it suffices to find a non-L-cr subgroup of type A_1, which is equivalent to finding an indecomposable, reducible 6-dimensional module for SL_2. To that end, if $V \cong L(1)$ is the natural module for SL_2 then the p-th symmetric power $S^p(V)$ is indecomposable of dimension $p + 1$ with two composition factors, $L(p)$ and $L(0)$.

The following is a delicate case of the proof that every simple subgroup of type X is G-cr when $p \notin N(X, G)$; this corrects an error in [107].

Example 5.3.4 Let $G = F_4$ and $p = 2$. We show that every subgroup of type G_2 is G-cr, and note that a similar but easier argument applies for $G = E_6$. The only proper Levi subgroups with a subgroup of type G_2 are those with derived subgroup B_3 or C_3. Write $P = QL$ where $\mathscr{D}(L) = B_3$ and let H be the L-irr subgroup of type G_2. In [107, Lemma 4.4.3] it is claimed that $\mathrm{H}^1(G_2, Q)$ is 1-dimensional, which in turn relies on the claim that an element of $(Q/Q_2)^H$ lifts to an element of Q^H. However, this latter claim is false. To see this, note that $Q^H \subset C := C_G(H)^\circ$. Since H is G-cr, so is the subgroup C by Theorem 5.3.8 ii, hence this is reductive and moreover has rank 1 since $Z(L)$ is 1-dimensional, cf. Remark 5.3.5.

By the Borel–de-Siebenthal algorithm, $H \subset \mathscr{D}(L)$ centralises the subgroup $\tilde{A}_1$ of G. Thus C has type A_1, and has no 2-dimensional unipotent subgroup, so $\dim Q^H \leq 1$. By restricting the $\mathscr{D}(L)$-action on Q_2 to H, we find that $Q_2 \downarrow H \cong V(1,0) = L(1,0)/L(0,0)$, where $V(1,0)$ is the Weyl module of high weight $(1,0)$ with the trivial module $L(0,0)$ in its socle. Thus $Q_2^H \cong k$. As $Q_2^H \subseteq Q^H$, we have equality by comparing dimensions. Therefore, no non-trivial element of $(Q/Q_2)^H$ lifts to an element of Q^H. From the exact sequence (5.1), it then follows that the map $\mathrm{H}^1(H, Q_2) \to \mathrm{H}^1(H, Q)$, which is surjective since $\mathrm{H}^1(H, Q/Q_2) = 0$, is the zero map. Thus $\mathrm{H}^1(H, Q) = 0$, so H gives rise to no non-G-cr subgroups in P. Applying the graph morphism of G now allows one to conclude that the C_3-parabolic also contains no non-G-cr subgroups of type G_2.

Moving on, let us assume $p \in N(X, G)$. Then there exist non-G-cr semisimple subgroups with a factor of type X, and we wish to classify them all. The case $G = G_2$ was settled by the second author in [105]. Since the classification is unusually short, we present it here. Recall that

G_2 has semisimple maximal connected subgroups of type $A_1\tilde{A}_1$ and A_2 (see Theorem 5.1.2).

Theorem 5.3.5 *Let G be a simple algebraic group of type G_2 in characteristic p, and let H be a non-G-cr semisimple subgroup. Then $p = 2$, H has type A_1 and is G-conjugate to precisely one of Z_1 and Z_2 below.*

(i) $Z_1 \subset A_1\tilde{A}_1$ embedded diagonally via $x \mapsto (x, x)$;
(ii) $Z_2 \subset A_2$ embedded via $V(2) \cong L(2)/L(0)$.

This theorem exhibits an interesting feature: every non-G-cr subgroup has a proper reductive overgroup in G. This is not true in general, and we consider semisimple subgroups with no proper reductive overgroups further in §5.3.1.

For the next result, recall that a prime p is called *good for* G if G has type G_2, F_4, E_6, E_7 and $p > 3$, or if G has type E_8 and $p > 5$; otherwise p is called *bad* for G. The non-G-cr semisimple subgroups of F_4 were extensively studied by the second author in [107]; however the paper contains a number of errors. These are systemically dealt with in [40], from which we distil a headline result, presented together with the classification of non-G-cr semisimple subgroups in good characteristic due to the first and third authors [74].

Theorem 5.3.6 *Let G be a simple algebraic group of exceptional type in characteristic $p > 0$ and let H be a non-G-cr semisimple subgroup of G. Then one of the following holds:*

(i) $(G, p) = (G_2, 2)$ and H is a subgroup Z_1 or Z_2 from Theorem 5.3.5;
(ii) $(G, p) = (F_4, 3)$ and H has type A_2, A_1A_1 or ${}^\star A_1$;
(iii) $(G, p) = (F_4, 2)$ and H has type B_3, B_2, ${}^\star A_1B_2$, ${}^\star A_1A_2$ or ${}^\star A_1^n$ with $n \leq 3$;
(iv) $(G, p) = (E_6, 5)$ and H has type A_1^2 or A_1;
(v) $(G, p) = (E_7, 5)$ and H has type A_2A_1, A_1^2 or A_1;
(vi) $(G, p) = (E_7, 7)$ and H has type G_2 or A_1;
(vii) $(G, p) = (E_8, 7)$ and H has type A_1G_2, A_1^2 or A_1;
(viii) G has type E_6, E_7 or E_8 and p is bad for G.

Conversely, there is a non-G-cr semisimple subgroup of each type listed and infinitely many conjugacy classes for those marked *.

Case viii remains the subject of ongoing work of the first and third authors.

Remark 5.3.7 When p is good for G, a non-G-cr semisimple subgroup is G-conjugate to precisely one subgroup in [74, Tables 11–17] and conversely, each subgroup in those tables is non-G-cr. Furthermore, [74] also provides the connected centraliser of each subgroup and the action on minimal and adjoint modules.

5.3.1 Semisimple subgroups with no proper reductive overgroups

To describe the poset of connected reductive groups, one needs to describe the maximal elements. If H is one such, then either: H is maximal amongst all connected subgroups; or H has a non-trivial central torus S, so that $H = C_G(S)$ is a Levi subgroup; or H is semisimple and non-G-cr.

Example 5.3.8 Let $(G, p) = (E_7, 7)$. We exhibit a non-G-cr subgroup of type G_2 which is maximal amongst proper reductive subgroups of G. In fact, this is unique up to conjugacy (see [74, §6.1]). Let $P = QL$ be a parabolic subgroup of G, where the derived subgroup $\mathscr{D}(L)$ has type E_6. This has a maximal subgroup of type F_4 which itself has a maximal subgroup $\bar{H}$ of type G_2 when $p = 7$ (see Theorem 5.1.2 and [114, Theorem 1(c)]). The subgroup $\bar{H}$ turns out to be $\mathscr{D}(L)$-irr ([115, Theorem 1]). Now the unipotent radical Q is abelian, a 27-dimensional module for $\mathscr{D}(L)$, which implies that Q is isomorphic to either $L_{E_6}(\lambda_1)$ or its dual $L_{E_6}(\lambda_6)$. Moreover, $Q \downarrow \bar{H} = L(20) \oplus L(00)$, and $\mathrm{H}^1(\bar{H}, L(20))$ is 1-dimensional.[15] This implies that $\mathrm{H}^1(\bar{H}, Q) \cong k$. Furthermore, the torus $Z(L)^\circ$ acts by scalars on Q, inducing conjugacy amongst the non-zero elements of $\mathrm{H}^1(\bar{H}, Q)$ and it follows that there is a unique G-conjugacy class $[H]$ of non-G-cr subgroups of type G_2 complementing Q in $Q\bar{H}$.

In [74, §10] it is proved that if V is the 56-dimensional module for E_7 then $V \downarrow H = T(20) \oplus T(20)$ where $T(20) = L(00)/L(20)/L(00)$ is tilting, of high weight 20. It follows that any reductive overgroup of H acts on V either indecomposably or with two indecomposable summands of dimension $\dim T(20) = 28$. Inspecting the maximal subgroups of G and their actions on V, we see the only plausible maximal reductive connected overgroup has type A_7. But the only non-trivial 8-dimensional G_2 modules are Frobenius twists of $L(10) \oplus L(00)$, and so any G_2 subgroup of A_7 is contained in a Levi subgroup A_6. But these act on V with four

[15] This follows, for example, by a dimension-shifting argument [55, II.2.1(4)] with the induced module $\mathrm{H}^0(20) \cong L_{G_2}(00)/L_{G_2}(20)$.

indecomposable summands, which rules out A_7 as a reductive overgroup of H.

There are several more instances of non-G-cr semisimple subgroups with no proper connected reductive overgroups, which are thus maximal amongst connected reductive subgroups of G. The following partial result begins to extend Theorem 5.1.2 towards describing the classes of maximal connected reductive subgroups. It can be deduced by combining Theorem 5.1.2 and the main results of the references given for Theorem 5.3.6. Out of interest, we note here a geometric interpretation: Since reductive subgroups of G correspond to affine coset spaces [93], a subgroup H which is maximal amongst reductive subgroups corresponds to G/H being minimal (with respect to G-quotients) amongst non-trivial affine homogeneous G-spaces. Then H being maximal amongst *connected* reductive subgroups means G/H is minimal up to a *finite-sheeted* quotient.

Theorem 5.3.9 *Let G be a simple algebraic group of exceptional type in characteristic p, and let M be maximal amongst connected reductive subgroups of G. Then one of the following holds:*

(i) *M is maximal amongst connected subgroups and is G-conjugate to precisely one subgroup in Table 5.1;*
(ii) *M is a Levi subgroup, with $(G, \mathscr{D}(M)) = (E_6, D_5)$ or (E_7, E_6);*
(iii) *$(G, p) = (F_4, 3)$ and M is non-G-cr of type A_1;*
(iv) *$(G, p) = (F_4, 2)$ and M is non-G-cr of type B_2 or A_1A_1;*
(v) *$(G, p) = (E_7, 7)$ and M is non-G-cr of type G_2 (unique up to conjugacy);*
(vi) *G has type E_6, E_7 or E_8 and p is bad for G.*

Remark 5.3.10 Case vi will contain many conjugacy classes of subgroups. An interesting example is a class of subgroups of type A_2 in $G = E_8$ which act on the adjoint module with indecomposable summands of dimension 240 and 8, which already precludes its containment in a proper connected reductive subgroup.

5.4 Further directions and related problems

5.4.1 Hereditary subgroups

Recall from [102] and §5.3.1 that one of Serre's reasons to formalise G-complete reducibility was for the study of converse theorems. An archetypal question [102, §5.3, Remarque] is:

When does the exterior square of a module being semisimple imply that the original module is semisimple?

This can be translated into G-complete reducibility as follows. Let H be a semisimple algebraic group, V an H-module and suppose that $\wedge^2(V)$ is semisimple. The action of $\mathrm{SL}(V)$ on $\wedge^2(V)$ furnishes inclusions $\bar{H} \subseteq M \subseteq G := \mathrm{SL}(\wedge^2(V))$, where $\bar{H}$ and M are the images in G of the groups H and $\mathrm{SL}(V)$, respectively. Then the question above asks whether H being G-cr implies that H is M-cr.

Definition 5.4.1 Let M be a connected subgroup of a reductive group G. We define M to be *G-ascending hereditary* (G-ah) if, for all connected subgroups H of M, if H is M-cr then H is G-cr. And M is defined to be *G-descending hereditary* (G-dh) if, for all connected subgroups H of M, if H is G-cr then H is M-cr. We define M to be *G-hereditary* if it is both G-ah and G-dh.

Let G be a simple algebraic group with subgroups $H \subset M$. Theorem 5.3.11 says that centralisers of linearly reductive subgroups of G, such as Levi subgroups, are G-hereditary. Theorem 5.3.13 gives criteria for a subgroup M to be G-dh. We have also seen examples of non-hereditary subgroups. Theorem 5.3.5 shows that in $G = G_2$, the subgroup $H = Z_1$ is M-irreducible for $M = A_1\tilde{A}_1$, but that H is non-G-cr; so M is non-G-ah. The following example shows that non-G-dh subgroups also exist:

Example 5.4.2 Let $G = F_4$ and $p = 2$. Then G contains subgroups C_4 and B_4, which respectively contain subgroups $\tilde{D}_4$ and D_4, and these in turn respectively contain simple subgroups H_1 and H_2 of type G_2. The simple module $L_{C_4}(\lambda_1) \downarrow H_1 = T(10)$, and the Weyl module $V_{B_4}(\lambda_1) \downarrow H_2 = T(10) \oplus 00$. These subgroups are swapped by the exceptional isogeny of G. By Proposition 5.3.15, H_1 is non-C_4-cr and thus H_2 is non-B_4-cr. However, in Example 5.3.4 we saw that every subgroup of type G_2 is G-cr. So H_1 and H_2 are examples of G-cr subgroups that are non-M-cr in some reductive maximal subgroup M of G. Therefore, M is not G-dh.

Given the plethora of results (§5.3.1, [11] and elsewhere) which guarantee G-hereditary behaviour, it is natural to ask how often such behaviour fails. Ongoing work of the first and third authors seeks more precise results in this vein. For instance, it is very uncommon for a subgroup H in a reductive pair (G, H) (cf. p. 207) to be non-G-dh, in fact based on empirical evidence the first and third authors speculate:

Conjecture 5.4.3 *Let (G, H) be a reductive pair of algebraic groups in characteristic p. If H is non-G-dh then $p = 2$.*

5.4.2 Unipotent elements in exceptional algebraic groups

Much effort has been spent studying unipotent elements in algebraic groups. It is even non-trivial to show that there are finitely many unipotent classes in the exceptional groups. There is an extensive literature on this and the book [68] contains a comprehensive treatment. We mention those results most closely related to subgroup structure.

For many applications it is useful to understand how an embedding $H \to G$ fuses classes of unipotent elements. When H is a reductive maximal connected subgroup this has been completely determined by Lawther in [59] (supplemented by [35] for the newly-discovered maximal subgroup of type F_4 in E_8 when $p = 3$).

One can flip this question and ask: given a unipotent element, what are its overgroups? For example, *regular unipotent elements* are those whose centralisers have the smallest possible dimension (the rank of G), and these are all G-conjugate. Overgroups of regular unipotent elements have been heavily studied, first of all by Saxl–Seitz in [96], classifying the maximal positive-dimensional reductive subgroups containing a regular unipotent element. Extending this to all positive-dimensional reductive subgroups containing a regular unipotent element is difficult. Suppose that H is a subgroup of G containing a regular unipotent element. Testerman–Zalesski [112, Thm. 1.2] provided a full classification of the connected reductive subgroups containing a regular unipotent element. An important ingredient is to show that if H is connected then it is G-irr. If instead H is only assumed to be positive-dimensional then Malle–Testerman [79, Theorem 1] show that either H is G-irr or H° is a torus; the latter case does in fact occur. Indicative of the narrative in Part I, Bate–Martin–Röhrle in [14] were able to produce a uniform proof of [112, Theorem 1.2] and [79, Theorem 1], and further generalisations to disconnected groups, finite groups of Lie type and Lie algebras, without intricate case-by-case considerations. A key ingredient in their proof was the observation of Steinberg that regular unipotent elements normalise a unique Borel subgroup of G.

5.4.3 Finite subgroups and groups of Lie type

Historically, one of the main motivations for studying maximal and then reductive subgroups of exceptional algebraic groups was to deduce results for the exceptional finite groups of Lie type, see [66] for the work up to the early 2000s. A related problem is to understand finite subgroups of the exceptional algebraic groups, whose study cannot employ any of the techniques requiring connectedness of the subgroup. In attempting to use the strategy of §5.4 for a simple algebraic group G of exceptional type, the primary difficulty is in classifying G-irr subgroups, since ad-hoc methods are required in place of uniform statements about representations of reductive groups. A result of Borovik [21, Theorem 1] quickly reduces one to studying almost-simple finite subgroups, and the isomorphism types of finite simple subgroups have been enumerated by Cohen, Griess, Serre, Wales and others (e.g. [28, 21, 20]) over the complex numbers using character-theoretic methods, and by Liebeck and Seitz [65] in positive characteristic.

In positive characteristic, it is essential to understand *generic subgroups*, i.e. finite groups of Lie type $H(q)$ with embeddings $H(q) \to G$ which factor through an inclusion $H \to G$ of algebraic groups. The main result in this direction is due to Liebeck and Seitz [64], giving an explicit bound on q (usually $q > 9$) ensuring that *all* embeddings $H(q) \to G$ arise in this fashion.

Further progress on *non-generic* subgroups has been made by the first author [73] by comparing Brauer characters of finite simple groups with the Brauer traces of elements of exceptional algebraic groups G on low-dimensional modules, which limits the composition factors of simple groups acting on these modules. This carries sufficient information to rule out G-irr subgroups, for instance a subgroup fixing a vector on $\mathrm{Lie}(G)$ lies in the corresponding stabiliser, which is often parabolic. One can also rule out the existence of non-G-cr subgroups: For instance, if a subgroup is contained in a parabolic subgroup of some algebraic group G, then it normalises the unipotent radical, and the Lie algebra of this is a submodule of $\mathrm{Lie}(G)$. So if the subgroup has no composition factors on $\mathrm{Lie}(G)$ with non-zero cohomology, the subgroup cannot have non-zero cohomology in its action on the radical, so is G-cr. Craven [30] has extended these techniques to also bring in Jordan block sizes of unipotent elements of finite groups acting on the relevant modules, deriving still stronger conditions and ruling out further subgroup types.

For *maximal* subgroups of finite groups of Lie type, there has been considerable recent progress. The maximal subgroups of ${}^2B_2(q)$, ${}^2F_4(q)$, ${}^3D_4(q)$, ${}^2G_2(q)$ and $G_2(q)$ have been classified by Cooperstein, Kleidman, Malle, Suzuki [29, 56, 77]; leaving $F_4(q)$, $E_6^\epsilon(q)$, $E_7(q)$ and $E_8(q)$ to consider. The maximal subgroups of these groups have been completely classified for some very small q (e.g. $E_7(2)$ in [9]) and there has been considerable progress by Magaard [76] in the case $F_4(q)$ and Aschbacher [3, 4, 5, 6] in the case $E_6(q)$. The most recent work has led to a complete classification of the maximal subgroups of $F_4(q)$, $E_6(q)$ and ${}^2E_6(q)$ [31] and almost a complete classification for $E_7(q)$ [33]. This all builds on previous work [34, 32, 30]. The maximal subgroups of $E_8(q)$ are also a work in progress by Craven.

We round off our discussion here by mentioning one more way in which G-complete reducibility applies to finite groups of Lie type: Namely, through *optimality.* The details are somewhat technical (cf. [13, Def. 5.17]), but a key point is: A non-G-cr subgroup H is contained in a *canonical* parabolic subgroup of G, which is normalised by all automorphisms of G which normalise H. Applying this to a Frobenius endomorphism F of G, if H is a non-G-cr finite subgroup of some group of Lie type $G(q) = G^F$, we get a method of constructing subgroups in between H and $G(q)$, namely, F-fixed points of the corresponding canonical parabolic subgroup (cf. [72, Proposition 2.2] and [73, §4.1–4.2]).

5.4.4 Variations of complete reducibility

Serre's definition 5.3.1 admits generalisations in various directions. If G is equipped with a Frobenius endomorphism F and one considers only F-stable parabolic and Levi subgroups, one arrives at so called 'F-complete reducibility' [48]. In another direction, as mentioned in the introduction, G-complete reducibility generalises to disconnected reductive groups, if one is willing to work instead with R-parabolic and R-Levi subgroups, and many results in G-complete reducibility generalise at once (cf. [11, §6]). The resulting geometric invariant theory is in fact the natural setting in which to derive the most general results, only a handful of which have been mentioned in the present article. Since one is now working with collections of morphisms $\mathbb{G}_m \to G$, restricting these morphisms to land in a subgroup K yields yet another generalisation, 'G-complete reducibility with respect to K', and once again many natural results extend immediately [43, 7].

Ultimately, one can view complete reducibility as a property of the *spherical building* of G [102]. Here, opposite parabolics are opposite simplices, and non-G-cr subgroups correspond to contractible subcomplexes; omitting all details, we simply mention that recasting the above results in terms of the building allows one to unite the various generalisations above and derive yet stronger statements, e.g. [44], and even extend to Euclidean buildings, Kac–Moody groups and other settings, see for instance [36] and [24, §4.3].

5.4.5 Structure of the Lie algebra of exceptional algebraic groups

Through the exponential and logarithm maps, classifying maximal connected subgroups of a complex algebraic group G is equivalent to classifying maximal subalgebras of $\mathrm{Lie}(G)$ and indeed Dynkin's original work is set in this context. It was noticed by Chevalley that any complex finite-dimensional simple Lie algebra $\mathfrak{g}$ has a $\mathbb{Z}$-basis. This means there is an integral form $\mathfrak{g}_{\mathbb{Z}}$ from which one may build a Lie algebra $\mathfrak{g}_R := \mathfrak{g}_{\mathbb{Z}} \otimes_{\mathbb{Z}} R$ over any commutative ring R. In particular, we may take $R = k$ for k an algebraically closed field of characteristic $p > 0$. There is typically more than one $\mathbb{Z}$-form available, leading to non-isomorphic Lie algebras over k; Chevalley's recipe gives the simply-connected form, i.e. $\mathfrak{g}_k \cong \mathrm{Lie}(G)$, where G is the simply-connected algebraic group over k of the same type ([50, Chap. VII]).

It is natural to ask about maximal subalgebras of $\mathfrak{g} = \mathfrak{g}_k$ and more generally its subalgebra structure and their conjugacy under the adjoint action of G. Motivation for this question also arises from viewing G as a scheme. In that context, one gets a much wider collection of subgroups, due to the presence of non-smooth subgroup schemes of G. At the most extreme end, a subgroup H of G is *infinitesimal* if its only k-point is the identity element. The most natural non-trivial example of an infinitesimal subgroup is the first Frobenius kernel G_1 of G; one may view this as the functor from k-algebras to groups such such the A-points of G_1 applied to a k-algebra A is the group $G_1(A) = \{x \in G(A) \mid F(x) = 1\}$, where F is the (standard) Frobenius map on G.

Recall that a Lie algebra $\mathfrak{g}$ is *restricted* if it is equipped with a $[p]$-map $x \mapsto x^{[p]}$ which is p-semilinear in k and satisfies $\mathrm{ad}(x^{[p]})(y) = \mathrm{ad}(x)^p(y)$. A subalgebra $\mathfrak{h} \subseteq \mathfrak{g}$ is a p-subalgebra if it is closed under the $[p]$-map. There is an equivalence between G_1 and the Lie algebra $\mathfrak{g}$ in the following senses: $\mathrm{Lie}(G_1) = \mathfrak{g}$ is a restricted Lie algebra; any finite-dimensional restricted Lie algebra $\mathfrak{k}$ is $\mathrm{Lie}(K)$ for a unique connected height-one group scheme K; under this correspondence, any p-subalgebra $\mathfrak{h}$ of $\mathfrak{g}$ maps to a unique subgroup H of G_1; the finite-dimensional representation theory of G_1 is equivalent to the finite-dimensional restricted representation theory of $\mathfrak{g}$. For more on this, see the article of Brion from this volume.

Moving from the smooth subgroups of G to the subalgebras of $\mathfrak{g}$ introduces many new and difficult problems. First, the classification of simple Lie algebras in positive characteristic [91] is vastly more complicated than that of the algebraic groups, and is only complete when $p > 3$. Second, semisimple subalgebras are not the sums of simple Lie algebras, or even closely related to them [110, §3.3]. Third, it is not in general true that Theorem 5.2.2 has an analogue for $\mathfrak{g}$, and indeed maximal non-semisimple subalgebras of $\mathfrak{g}$ do not have to be parabolic – where a Lie subalgebra of $\mathfrak{g}$ is called *parabolic* if it is the Lie algebra of a parabolic subgroup. Indeed, [110, p. 149] describes (all) irreducible representations of the soluble Heisenberg Lie algebras, most of which have dimensions divisible by p.

Circumventing these problems in the case that G is classical is wide open. But at least when G is of exceptional type in good characteristic, these problems have been dealt with in [46], [89] and [90]. The analogue of the Liebeck–Seitz classification of maximal connected subgroups of G holds, with certain exceptions. For example, for $p \geq 3$ the first Witt algebra $W_1 := \mathrm{Der}(k[X]/X^p)$ is a simple Lie algebra of dimension p and appears between G and its regular $\mathfrak{sl}_2$ subalgebra whenever $p = h + 1$ where h is the Coxeter number of G. There are some maximal semisimple Lie algebras when G has type E_7 and $p = 5$ or 7, which have nothing to do with semisimple subgroups of G. We also point out that [90, Corollary 1.4] establishes an exact analogue of the Borel–Tits Theorem for exceptional Lie algebras in good characteristic.

One would like to consider the analogues for Lie algebras of the main problem addressed in this article. The following definition was given in [81] and developed in [10].

Definition 5.4.4 Let $\mathfrak{g} = \mathrm{Lie}(G)$ for G a reductive algebraic group. Then a subalgebra $\mathfrak{h}$ of $\mathfrak{g}$ is G-cr if whenever $\mathfrak{h}$ is in a parabolic subalgebra $\mathfrak{p} = \mathrm{Lie}(P)$ of G, then $\mathfrak{h}$ is in a Levi subalgebra $\mathfrak{l} = \mathrm{Lie}(L)$ of $\mathfrak{p}$, where $P = QL$ is a Levi decomposition of P.

An analogue of Theorem 5.3.3 for Lie algebras (building on work in [47]) is given in [109, Theorem 1.3]:

Theorem 5.4.5 *Let G be a connected reductive algebraic group in char-*

acteristic p with Lie algebra $\mathfrak{g}$. Suppose that $\mathfrak{h}$ is a semisimple subalgebra of $\mathfrak{g}$ and $p > h$. Then $\mathfrak{h}$ is G-cr.

In the case $\mathfrak{h} \cong \mathfrak{sl}_2$, the theorem interacts surprisingly closely with Kostant's uniqueness result about the embeddings of nilpotent elements into $\mathfrak{sl}_2$-subalgebras: it builds on the Jacobson–Morozov theorem [52, 85], which says that for any complex finite-dimensional semisimple Lie algebra $\mathfrak{g} = \text{Lie}(G)$, there is a surjective map

$$\{\text{conjugacy classes of } \mathfrak{sl}_2\text{-triples}\} \longrightarrow \{\text{nilpotent orbits in } \mathfrak{g}\}, \qquad (*)$$

where an $\mathfrak{sl}_2$-triple is a triple $(e, h, f) \in \mathfrak{g}^3$ satisfying $[h, e] = 2e$, $[h, f] = -2f$, $[e, f] = h$. The surjective map is induced by sending (e, h, f) to the nilpotent element e. So any such e can be embedded into some $\mathfrak{sl}_2$-triple. In [58], Kostant showed that this can be done uniquely up to conjugacy by the centraliser G_e of e; i.e. the map $(*)$ is actually a bijection. Much work has been done on extending this important result into characteristic $p > 0$. We mention some critical contributions. In [87], Pommerening showed that under the mild restriction that p is a good prime for G, one can always find an $\mathfrak{sl}_2$-subalgebra containing a given nilpotent element, but this may not be unique; in other words, the map $(*)$ is still surjective, but not necessarily injective. In [104] Springer and Steinberg prove that the uniqueness holds whenever $p \geq 4h - 1$ and in his book [25], Carter uses an argument due to Spaltenstein to reduce this bound to $p > 3h - 3$; both proofs make use of an exponentiation argument. One use of Theorem 5.4.5 is to prove the following.

Theorem 5.4.6 *Let G be a connected reductive group in characteristic $p > 2$ with Lie algebra $\mathfrak{g}$. Then $(*)$ is a bijection if and only if $p > h$.*

In fact, [109] also considers a map

$$\{\text{conjugacy classes of } \mathfrak{sl}_2\text{-subalgebras}\} \to \{\text{nilpotent orbits in } \mathfrak{g}\}, \quad (**)$$

and when a bijection exists, realises it in a natural way. The equivalence of bijections $(*)$ and $(**)$ is easily seen in large enough characteristics by exponentiation, but there are quite a few characteristics where there exists a bijection $(**)$, but not $(*)$.

Further progress on this theme has been made by Goodwin–Pengelly [41], characterising the subvarieties of nilpotent elements where bijections $(*)$ and $(**)$ hold.

Acknowledgment All three authors wish to thank Martin Liebeck for support and guidance from their PhD studies at Imperial College London through to the present day. The second author is supported by Leverhulme grant RPG-2021-080. The third author is supported by EPSRC grant EP/W000466/1.

References

[1] Amende, Bonnie. 2005. *G-irreducible subgroups of type A1*. ProQuest LLC, Ann Arbor, MI. Thesis (Ph.D.)–University of Oregon.

[2] Aschbacher, M., and Scott, L. 1985. Maximal subgroups of finite groups. *J. Algebra*, **92**(1), 44–80.

[3] Aschbacher, Michael. 1987. The 27-dimensional module for E_6. I. *Invent. Math.*, **89**(1), 159–195.

[4] Aschbacher, Michael. 1988. The 27-dimensional module for E_6. II. *J. London Math. Soc. (2)*, **37**(2), 275–293.

[5] Aschbacher, Michael. 1990a. The 27-dimensional module for E_6. III. *Trans. Amer. Math. Soc.*, **321**(1), 45–84.

[6] Aschbacher, Michael. 1990b. The 27-dimensional module for E_6. IV. *J. Algebra*, **131**(1), 23–39.

[7] Attenborough, Christopher, Bate, Michael, Gruchot, Maike, Litterick, Alastair, and Röhrle, Gerhard. 2020. On relative complete reducibility. *Q. J. Math.*, **71**(1), 321–334.

[8] Azad, H., Barry, M., and Seitz, G. 1990. On the structure of parabolic subgroups. *Comm. Algebra*, **18**(2), 551–562.

[9] Ballantyne, John, Bates, Chris, and Rowley, Peter. 2015. The maximal subgroups of $E_7(2)$. *LMS J. Comput. Math.*, **18**(1), 323–371.

[10] Bate, M., Martin, B., Röhrle, G., and Tange, R. 2011. Complete reducibility and conjugacy classes of tuples in algebraic groups and Lie algebras. *Math. Z.*, **269**(3-4), 809–832.

[11] Bate, Michael, Martin, Benjamin, and Röhrle, Gerhard. 2005. A geometric approach to complete reducibility. *Invent. Math.*, **161**(1), 177–218.

[12] Bate, Michael, Martin, Benjamin, and Röhrle, Gerhard. 2008. Complete reducibility and commuting subgroups. *J. Reine Angew. Math.*, **621**, 213–235.

[13] Bate, Michael, Martin, Benjamin, Röhrle, Gerhard, and Tange, Rudolf. 2013. Closed orbits and uniform S-instability in geometric invariant theory. *Trans. Amer. Math. Soc.*, **365**(7), 3643–3673.

[14] Bate, Michael, Martin, Benjamin, and Röhrle, Gerhard. 2022. Overgroups of regular unipotent elements in reductive groups. *Forum Math. Sigma*, **10**, Paper No. e13, 13.

[15] Bendel, C. P., Nakano, D. K., and Pillen, C. 2004. Extensions for Frobenius kernels. *J. Algebra*, **272**(2), 476–511.

[16] Bendel, Christopher P., Nakano, Daniel K., Parshall, Brian J., Pillen, Cornelius, Scott, Leonard L., and Stewart, David. 2015. Bounding cohomology for finite groups and Frobenius kernels. *Algebr. Represent. Theory*, **18**(3), 739–760.

[17] Borel, A. 1991. *Linear algebraic groups.* Second edn. Graduate Texts in Mathematics, vol. 126. New York: Springer-Verlag.

[18] Borel, A., and Tits, J. 1971. Éléments unipotents et sous-groupes paraboliques de groupes réductifs. I. *Invent. Math.*, **12**, 95–104.

[19] Borel, A., and Tits, J. 1973. Homomorphismes "abstraits" de groupes algébriques simples. *Ann. of Math. (2)*, **97**, 499–571.

[20] Borovik, A. V. 1989a. Jordan subgroups of simple algebraic groups. *Algebra i Logika*, **28**(2), 144–159, 244.

[21] Borovik, A. V. 1989b. The structure of finite subgroups of simple algebraic groups. *Algebra i Logika*, **28**(3), 249–279, 366.

[22] Bourbaki, N. 2005. *Lie groups and Lie algebras. Chapters 7–9.* Elements of Mathematics (Berlin). Berlin: Springer-Verlag.

[23] Burness, Timothy C., and Testerman, Donna M. 2019. Irreducible subgroups of simple algebraic groups – a survey. Pages 230–260 of: *Groups St Andrews 2017 in Birmingham.* London Math. Soc. Lecture Note Ser., vol. 455. Cambridge Univ. Press, Cambridge.

[24] Caprace, Pierre-Emmanuel. 2009. "Abstract" homomorphisms of split Kac-Moody groups. *Mem. Amer. Math. Soc.*, **198**(924), xvi+84.

[25] Carter, Roger W. 1993. *Finite groups of Lie type.* Wiley Classics Library. Chichester: John Wiley & Sons Ltd. Conjugacy classes and complex characters, Reprint of the 1985 original, A Wiley-Interscience Publication.

[26] Chevalley, Claude. 1951. *Théorie des groupes de Lie. Tome II. Groupes algébriques.* Actualités Sci. Ind. no. 1152. Hermann & Cie., Paris.

[27] Cline, E., Parshall, B., Scott, L., and van der Kallen, W. 1977. Rational and generic cohomology. *Invent. Math.*, **39**(2), 143–163.

[28] Cohen, Arjeh M., and Wales, David B. 1995. Finite simple subgroups of semisimple complex Lie groups – a survey. Pages 77–96 of: *Groups of Lie type and their geometries (Como, 1993).* London Math. Soc. Lecture Note Ser., vol. 207. Cambridge: Cambridge Univ. Press.

[29] Cooperstein, Bruce N. 1981. Maximal subgroups of $G_2(2^n)$. *J. Algebra*, **70**(1), 23–36.

[30] Craven, David A. 2017. Alternating subgroups of exceptional groups of Lie type. *Proc. Lond. Math. Soc. (3)*, **115**(3), 449–501.

[31] Craven, David A. 2021. The maximal subgroups of the exceptional groups $F_4(q), E_6(q)$ and ${}^2E_6(q)$ and related almost simple groups. *arXiv:2103.04869.*

[32] Craven, David A. 2022a. Maximal PSL_2 subgroups of exceptional groups of Lie type. *Mem. Amer. Math. Soc.*, **276**(1355), v+155.

[33] Craven, David A. 2022b. On the maximal subgroups of $E_7(q)$ and related almost simple groups. *arXiv:2201.07081.*

[34] Craven, David A. to appear. On medium-rank Lie primitive and maximal subgroups of exceptional groups of Lie type. *Mem. Amer. Math. Soc.*

[35] Craven, David A., Stewart, David I., and Thomas, Adam R. 2022. A new maximal subgroup of E_8 in characteristic 3. *Proc. Amer. Math. Soc.*, **150**(4), 1435–1448.

[36] Dawson, Denise Karin. 2011. *Complete reducibility in Euclidean twin buildings*. ProQuest LLC, Ann Arbor, MI. Thesis (Ph.D.)–Cornell University.

[37] Dowd, Michael F., and Sin, Peter. 1996. On representations of algebraic groups in characteristic two. *Comm. Algebra*, **24**(8), 2597–2686.

[38] Dynkin, E. B. 2000. *Selected papers of E. B. Dynkin with commentary*. American Mathematical Society, Providence, RI; International Press, Cambridge, MA. Edited by A. A. Yushkevich, G. M. Seitz and A. L. Onishchik.

[39] Fulton, William, and Harris, Joe. 1991. *Representation theory, A first course*. Graduate Texts in Mathematics, vol. 129. New York: Springer-Verlag.

[40] Ganeshalingam, Vanthana, and Thomas, Adam R. On the non-completely reducible subgroups of F_4. In preparation.

[41] Goodwin, Simon M., and Pengelly, Rachel. On $\mathfrak{sl}_2$-triples for classical algebraic groups in positive characteristic. *Transform. Groups*, to appear.

[42] Gorenstein, Daniel, Lyons, Richard, and Solomon, Ronald. 1998. *The classification of the finite simple groups. Number 3. Part I. Chapter A*. Mathematical Surveys and Monographs, vol. 40. American Mathematical Society, Providence, RI.

[43] Gruchot, Maike, Litterick, Alastair, and Röhrle, Gerhard. 2020. Relative complete reducibility and normalized subgroups. *Forum Math. Sigma*, **8**, Paper No. e30, 32.

[44] Gruchot, Maike, Litterick, Alastair, and Röhrle, Gerhard. 2022. Complete reducibility: variations on a theme of Serre. *Manuscripta Math.*, **168**(3-4), 439–451.

[45] Herpel, Sebastian. 2013. On the smoothness of centralizers in reductive groups. *Trans. Amer. Math. Soc.*, **365**(7), 3753–3774.

[46] Herpel, Sebastian, and Stewart, David I. 2016a. Maximal subalgebras of Cartan type in the exceptional Lie algebras. *Selecta Math.*, **22**(2), 765–799.

[47] Herpel, Sebastian, and Stewart, David I. 2016b. On the smoothness of normalisers, the subalgebra structure of modular Lie algebras, and the cohomology of small representations. *Doc. Math.*, **21**, 1–37.

[48] Herpel, Sebastian, Röhrle, Gerhard, and Gold, Daniel. 2011. Complete reducibility and Steinberg endomorphisms. *C. R. Math. Acad. Sci. Paris*, **349**(5-6), 243–246.

[49] Humphreys, James E. 1975. *Linear algebraic groups*. New York: Springer-Verlag. Graduate Texts in Mathematics, No. 21.

[50] Humphreys, James E. 1978. *Introduction to Lie algebras and representation theory*. Graduate Texts in Mathematics, vol. 9. Springer-Verlag, New York-Berlin. Second printing, revised.

[51] Humphreys, James E. 1990. *Reflection groups and Coxeter groups*. Cambridge Studies in Advanced Mathematics, vol. 29. Cambridge University Press, Cambridge.

[52] Jacobson, Nathan. 1951. Completely reducible Lie algebras of linear transformations. *Proc. Amer. Math. Soc.*, **2**, 105–113.

[53] Jantzen, J. C. 1991. First cohomology groups for classical Lie algebras. Pages 289–315 of: *Representation theory of finite groups and finite-dimensional algebras (Bielefeld, 1991)*. Progr. Math., vol. 95. Basel: Birkhäuser.

[54] Jantzen, J. C. 1997. Low-dimensional representations of reductive groups are semisimple. Pages 255–266 of: *Algebraic groups and Lie groups*. Austral. Math. Soc. Lect. Ser., vol. 9. Cambridge Univ. Press, Cambridge.

[55] Jantzen, J. C. 2003. *Representations of algebraic groups*. Second edn. Mathematical Surveys and Monographs, vol. 107. Providence, RI: American Mathematical Society.

[56] Kleidman, Peter. 1988. The maximal subgroups of the Steinberg triality groups ${}^3D_4(q)$ and of their automorphism groups. *J. Algebra*, **115**(1), 182–199.

[57] Kleidman, Peter, and Liebeck, Martin. 1990. *The subgroup structure of the finite classical groups*. London Mathematical Society Lecture Note Series, vol. 129. Cambridge: Cambridge University Press.

[58] Kostant, Bertram. 1959. The principal three-dimensional subgroup and the Betti numbers of a complex simple Lie group. *Amer. J. Math.*, **81**, 973–1032.

[59] Lawther, R. 2009. Unipotent classes in maximal subgroups of exceptional algebraic groups. *J. Algebra*, **322**(1), 270–293.

[60] Lawther, R., and Testerman, D. M. 1999. A_1 Subgroups of Exceptional Algebraic Groups. *Mem. Amer. Math. Soc.*, **141**(674).

[61] Lie, Sophus. 1880. Theorie der Transformationsgruppen I. *Math. Ann.*, **16**(4), 441–528.

[62] Liebeck, Martin W., and Seitz, Gary M. 1994. Subgroups generated by root elements in groups of Lie type. *Ann. of Math. (2)*, **139**(2), 293–361.

[63] Liebeck, Martin W., and Seitz, Gary M. 1996. Reductive subgroups of exceptional algebraic groups. *Mem. Amer. Math. Soc.*, **121**(580), vi+111.

[64] Liebeck, Martin W., and Seitz, Gary M. 1998. On the subgroup structure of exceptional groups of Lie type. *Trans. Amer. Math. Soc.*, **350**(9), 3409–3482.

[65] Liebeck, Martin W., and Seitz, Gary M. 1999. On finite subgroups of exceptional algebraic groups. *J. Reine Angew. Math.*, **515**, 25–72.

[66] Liebeck, Martin W., and Seitz, Gary M. 2003. A survey of maximal subgroups of exceptional groups of Lie type. Pages 139–146 of: *Groups, combinatorics & geometry (Durham, 2001)*. World Sci. Publ., River Edge, NJ.

[67] Liebeck, Martin W., and Seitz, Gary M. 2004. The maximal subgroups of positive dimension in exceptional algebraic groups. *Mem. Amer. Math. Soc.*, **169**(802), vi+227.

[68] Liebeck, Martin W., and Seitz, Gary M. 2012. *Unipotent and nilpotent classes in simple algebraic groups and Lie algebras*. Mathematical Surveys and Monographs, vol. 180. American Mathematical Society, Providence, RI.

[69] Liebeck, Martin W., and Testerman, Donna M. 2004. Irreducible subgroups of algebraic groups. *Q. J. Math.*, **55**(1), 47–55.

[70] Liebeck, Martin W., and Thomas, Adam R. 2017. Finite subgroups of simple algebraic groups with irreducible centralizers. *J. Group Theory*, **20**(5), 841–870.

[71] Liebeck, Martin W., Saxl, Jan, and Seitz, Gary M. 1996. Factorizations of simple algebraic groups. *Trans. Amer. Math. Soc.*, **348**(2), 799–822.

[72] Liebeck, Martin W., Martin, Benjamin M. S., and Shalev, Aner. 2005. On conjugacy classes of maximal subgroups of finite simple groups, and a related zeta function. *Duke Math. J.*, **128**(3), 541–557.

[73] Litterick, Alastair J. 2018. On non-generic finite subgroups of exceptional algebraic groups. *Mem. Amer. Math. Soc.*, **253**(1207), v+156.

[74] Litterick, Alastair J., and Thomas, Adam R. 2018a. Complete reducibility in good characteristic. *Trans. Amer. Math. Soc.*, **370**(8), 5279–5340.

[75] Litterick, Alastair J., and Thomas, Adam R. 2018b. Reducible subgroups of exceptional algebraic groups. *Journal of Pure and Applied Algebra*, 2489–2529.

[76] Magaard, Kay. 1990. *The maximal subgroups of the Chevalley groups F(,4)(F) where F is a finite or algebraically closed field of characteristic not equal to 2,3.* ProQuest LLC, Ann Arbor, MI. Thesis (Ph.D.)–California Institute of Technology.

[77] Malle, Gunter. 1991. The maximal subgroups of ${}^2F_4(q^2)$. *J. Algebra*, **139**(1), 52–69.

[78] Malle, Gunter, and Testerman, Donna. 2011. *Linear algebraic groups and finite groups of Lie type.* Cambridge Studies in Advanced Mathematics, vol. 133. Cambridge: Cambridge University Press.

[79] Malle, Gunter, and Testerman, Donna. 2021. Overgroups of regular unipotent elements in simple algebraic groups. *Trans. Amer. Math. Soc. Ser. B*, **8**, 788–822.

[80] McNinch, George J. 1998. Dimensional criteria for semisimplicity of representations. *Proc. London Math. Soc. (3)*, **76**(1), 95–149.

[81] McNinch, George J. 2007. Completely reducible Lie subalgebras. *Transformation Groups*, **12**(1), 127–135.

[82] McNinch, George J. 2014. Linearity for actions on vector groups. *J. Algebra*, **397**, 666–688.

[83] McNinch, George J., and Testerman, Donna M. 2007. Completely reducible SL(2)-homomorphisms. *Trans. Amer. Math. Soc.*, **359**(9), 4489–4510.

[84] Milne, J. S. 2017. *Algebraic groups. The theory of group schemes of finite type over a field.* Vol. 170. Cambridge: Cambridge University Press.

[85] Morozov, V. V. 1942. On a nilpotent element in a semi-simple Lie algebra. *C. R. (Doklady) Acad. Sci. URSS (N.S.)*, **36**, 83–86.

[86] Parker, Alison E. 2007. Higher extensions between modules for SL_2. *Adv. Math.*, **209**(1), 381–405.

[87] Pommerening, Klaus. 1980. Über die unipotenten Klassen reduktiver Gruppen. II. *J. Algebra*, **65**(2), 373–398.

[88] Prasad, Gopal, and Yu, Jiu-Kang. 2006. On quasi-reductive group schemes. *J. Algebraic Geom.*, **15**(3), 507–549. With an appendix by Brian Conrad.

[89] Premet, Alexander. 2017. A modular analogue of Morozov's theorem on maximal subalgebras of simple Lie algebras. *Adv. Math.*, **311**, 833–884.

[90] Premet, Alexander, and Stewart, David I. 2019. Classification of the maximal subalgebras of exceptional Lie algebras over fields of good characteristic. *J. Amer. Math. Soc.*, **32**(4), 965–1008.

[91] Premet, Alexander, and Strade, Helmut. 2006. Classification of finite dimensional simple Lie algebras in prime characteristics. Pages 185–214 of: *Representations of algebraic groups, quantum groups, and Lie algebras*. Contemp. Math., vol. 413. Amer. Math. Soc., Providence, RI.

[92] Richardson, R. W. 1967. Conjugacy classes in Lie algebras and algebraic groups. *Ann. of Math. (2)*, **86**, 1–15.

[93] Richardson, R. W. 1977. Affine coset spaces of reductive algebraic groups. *Bull. London Math. Soc.*, **9**(1), 38–41.

[94] Richardson, R. W. 1988. Conjugacy classes of n-tuples in Lie algebras and algebraic groups. *Duke Math. J.*, **57**(1), 1–35.

[95] Riche, Simon, and Williamson, Geordie. 2021. A simple character formula. *Ann. H. Lebesgue*, **4**, 503–535.

[96] Saxl, Jan, and Seitz, Gary M. 1997. Subgroups of algebraic groups containing regular unipotent elements. *J. London Math. Soc. (2)*, **55**(2), 370–386.

[97] Seitz, Gary M. 1987. The maximal subgroups of classical algebraic groups. *Mem. Amer. Math. Soc.*, **67**(365), iv+286.

[98] Seitz, Gary M. 1991. Maximal subgroups of exceptional algebraic groups. *Mem. Amer. Math. Soc.*, **90**(441), iv+197.

[99] Serre, Jean-Pierre. 1994. Sur la semi-simplicité des produits tensoriels de représentations de groupes. *Invent. Math.*, **116**(1-3), 513–530.

[100] Serre, Jean-Pierre. 1997. Semisimplicity and tensor products of group representations: converse theorems. *J. Algebra*, **194**(2), 496–520. With an appendix by Walter Feit.

[101] Serre, Jean-Pierre. 1998. *Morsund lectures, University of Oregon.*

[102] Serre, Jean-Pierre. 2005. Complète réductibilité. *Astérisque*, Exp. No. 932, viii, 195–217. Séminaire Bourbaki. Vol. 2003/2004.

[103] Springer, T. A. 1998. *Linear algebraic groups*. Second edn. Progress in Mathematics, vol. 9. Boston, MA: Birkhäuser Boston Inc.

[104] Springer, T. A., and Steinberg, R. 1970. Conjugacy classes. Pages 167–266 of: *Seminar on Algebraic Groups and Related Finite Groups (The Institute for Advanced Study, Princeton, N.J., 1968/69)*. Lecture Notes in Mathematics, Vol. 131. Springer, Berlin.

[105] Stewart, David I. 2010. The reductive subgroups of G_2. *J. Group Theory*, **13**(1), 117–130.

[106] Stewart, David I. 2013a. On unipotent algebraic G-groups and 1-cohomology. *Trans. Amer. Math. Soc.*, **365**(12), 6343–6365.

[107] Stewart, David I. 2013b. The reductive subgroups of F_4. *Mem. Amer. Math. Soc.*, **223**(1049), vi+88.

[108] Stewart, David I. 2014. Non-G-completely reducible subgroups of the exceptional algebraic groups. *Int. Math. Res. Not. IMRN*, 6053–6078.

[109] Stewart, David I., and Thomas, Adam R. 2018. The Jacobson-Morozov theorem and complete reducibility of Lie subalgebras. *Proc. Lond. Math. Soc. (3)*, **116**(1), 68–100.

[110] Strade, H. 2004. *Simple Lie algebras over fields of positive characteristic. I.* de Gruyter Expositions in Mathematics, vol. 38. Berlin: Walter de Gruyter & Co. Structure theory.

[111] Sury, B. 2016. Hermann Weyl and representation theory. *Resonance*, **21**, 1073–1091.

[112] Testerman, Donna, and Zalesski, Alexandre. 2013. Irreducibility in algebraic groups and regular unipotent elements. *Proc. Amer. Math. Soc.*, **141**(1), 13–28.

[113] Testerman, Donna M. 1988. Irreducible subgroups of exceptional algebraic groups. *Mem. Amer. Math. Soc.*, **75**(390), iv+190.

[114] Testerman, Donna M. 1992. The construction of the maximal A_1's in the exceptional algebraic groups. *Proc. Amer. Math. Soc.*, **116**(3), 635–644.

[115] Thomas, Adam R. 2015. Simple irreducible subgroups of exceptional algebraic groups. *J. Algebra*, **423**, 190–238.

[116] Thomas, Adam R. 2016. Irreducible A_1 subgroups of exceptional algebraic groups. *J. Algebra*, **447**, 240–296.

[117] Thomas, Adam R. 2020. The irreducible subgroups of exceptional algebraic groups. *Mem. Amer. Math. Soc.*, **268**(1307), v+191.

[118] Vasiu, A. 2005. Normal, unipotent subgroup schemes of reductive groups. *C. R. Math. Acad. Sci. Paris*, **341**(2), 79–84.

[119] Weyl, H. 1925. Theorie der Darstellung kontinuierlicher halb-einfacher Gruppen durch lineare Transformationen. I. *Math. Z.*, **23**(1), 271–309.

6

Axial algebras of Jordan and Monster type

Justin McInroy[a] and Sergey Shpectorov[b]

Abstract

Axial algebras are a class of non-associative commutative algebras whose properties are defined in terms of a fusion law. When this fusion law is graded, the algebra has a naturally associated group of automorphisms and thus axial algebras are inherently related to group theory. Examples include most Jordan algebras and the Griess algebra for the Monster sporadic simple group.

In this survey, we introduce axial algebras, discuss their structural properties and then concentrate on two specific classes: algebras of Jordan and Monster type, which are rich in examples related to simple groups.

6.1 Introduction

Axial algebras are a relatively recent class of non-associative algebras which have a strong natural link to group theory and to physics. The area has grown significantly since its inception in 2015 and so it is a good time for an article to survey all the recent developments and interesting open problems.

Many interesting classes of algebras, such as associative algebras, Jordan algebras and Lie algebras, are defined by asserting that all the el-

[a] Department of Mathematics, University of Chester, Parkgate Rd, Chester, CH1 4BJ, UK.
J.McInroy@chester.ac.uk
[b] School of Mathematics, University of Birmingham, Edgbaston, Birmingham, B15 2TT, UK.
S.Shpectorov@bham.ac.uk

ements should satisfy prescribed polynomial identities. However, axial algebras are not defined in this way. Instead, they are defined as being generated by certain idempotents called axes which obey a fusion law. Obeying an arbitrary generic fusion law here is probably too weak to lead to any interesting theory and the examples that could occur would be wild. However, specific tight fusion laws, similar to the ones considered in this survey, are both natural enough to allow a rich variety of interesting examples and strong enough to allow them to be classified.

Historically, examples of fusion laws have been known for quite a while. The Pierce decomposition for associative algebras goes back to the 19th century and later in the 1940s, Pierce decompositions were generalised to Jordan algebras by Albert. These are in fact nothing other than fusion laws which we will see later in Section 6.2.1. However, the fusion language really came to the fore in the 1990s following the discovery of the Monster and the Griess algebra in the 1970s and 80s. Recall that the Monster is the largest of the 26 sporadic simple groups and it was conjectured to exist by Fischer and Griess in 1973 and shown to exist by Griess in 1982 [29] as the group of automorphisms of the Griess algebra. Research around the Moonshine conjecture led to the introduction of Vertex Operator Algebras (VOAs) by Borcherds [3] and in particular to the Moonshine VOA $V^\natural$ (the Griess algebra is the weight 2 part of $V^\natural$). In 1996, Miyamoto [56] introduced Ising vectors for an OZ-type VOA and he proved that every Ising vector u leads to an involution τ_u of the VOA. This is a consequence of the fact that the fusion rules for the representations of Virasoro algebra generated by u are C_2-graded. This idea of associating an involution with an element of the algebra using the fusion law is the start of the axial theory. (We note that Miyamoto's work built on earlier contributions of Norton and Conway [11], who investigated in detail the properties of the Griess algebra. In particular, they knew of the correspondence between 2A involutions in the Monster and idempotents called 2A-axes in the Griess algebra, though without a fusion law.)

Miyamoto began a study of VOAs generated by two Ising vectors and this was completed by Sakuma [66]. Analysing his proof, Ivanov realised that the calculation is done entirely in the weight 2-component of the VOA. He took the properties of this subalgebra used in Sakuma's proof as axioms for a new class of real commutative non-associative algebras, which he called Majorana algebras [38]. This was the first class where one of the key axioms was a fusion law. Similarly to the VOA case, special elements in the Majorana algebra, called *Majorana axes*, lead to

Majorana involutions in the automorphism group of the algebra. Initially, the focus was on trying to determine the Majorana algebras for small, particularly simple, groups. Whilst doing this, it became clear that the fusion law and its grading are the main tools, whilst other axioms are much less important.

Axial algebras were introduced by Hall, Rehren and Shpectorov in [31, 32] to further generalise Majorana algebras, removing the unnecessary axioms and allowing arbitrary fusion laws and arbitrary fields. Historically, most people in this area are group theorists. This is because of the natural connection to groups tying into the long-standing task of finding a unified theory for simple groups, whereby all of them, including the sporadic groups, could be treated within the same setup. The class of axial algebras of Monster type, that we discuss below, includes Jordan algebras realising classical groups and G_2, Matsuo algebras realising 3-transposition groups, and of course the Griess algebra and its subalgebras realising the Monster and many other sporadic groups from the Happy Family. Whilst it might be presumptuous to suppose that axial theory can provide the ultimate unified setup, still we can reasonably expect to find new ties between different kinds of simple groups.

Axial algebra theory has grown significantly since 2015 and we cannot hope to cover everything in this survey. We are going to introduce axial algebras to the reader and then focus on the structure theory and on two specific classes of axial algebras, those of Jordan and Monster type.

Due to length considerations, we are forced to leave aside some very interesting material. This includes the link to VOAs and physics, but also recently discovered connections to analysis and other areas of mathematics. In particular, Tkachev [71] found that algebras arising in the global geometry and regularity theory of non-linear PDEs are axial algebras for suitable compact fusion laws, and similarly Fox [18] shows axial properties of the algebra of curvature tensors. Another recent development that we do not cover is the theory of decomposition algebras [13] due to De Medts, Peacock, Shpectorov and Van Couwenberghe, which generalises the axial setup even further and provides a categorical point of view. Rowen and Segev in [63] suggest a partial generalisation to the non-commutative case. Even though Majorana algebras are squarely within the class of axial algebras of Monster type, the motivation and approach of Majorana theory is slightly different and most importantly Ivanov himself published two recent surveys on this topic [36, 37]. Finally, it is very likely we missed some other interesting work and if so, we would like to apologise for this.

The work of the second author was partially supported by the Mathematical Center in Akademgorodok under agreement No. 075-15-2019-1675 with the Ministry of Science and Higher Education of the Russian Federation.

6.2 Background

We begin by introducing fusion laws and axial algebras and giving some examples. We then explain how we get a naturally associated group of automorphisms called the Miyamoto group.

Notation: For an algebra A and a set of elements $X \subseteq A$, we will write $\langle X \rangle$ for the subspace of A spanned by X and $\langle\langle X \rangle\rangle$ for the subalgebra generated by X.

6.2.1 Fusion laws

The starting point is the following concept.

Definition 6.2.1 A *fusion law* is a pair $\mathcal{F} := (\mathcal{F}, \star)$, where $\mathcal{F}$ is a set and $\star$ is a map,

$$\star \colon \mathcal{F} \times \mathcal{F} \to 2^{\mathcal{F}},$$

where $2^{\mathcal{F}}$ is our notation for the set of all subsets of $\mathcal{F}$.

The fusion laws we are most interested in have a small set $\mathcal{F}$, and it will be convenient to present them in a table similar to a group multiplication table. For example, in Table 6.1, we show the three fusion laws: the associative law $\mathcal{A}$, the Jordan type law $\mathcal{J}(\eta)$ and the (generalised) Monster type law $\mathcal{M}(\alpha, \beta)$. For simplicity, we drop the set signs and simply list the elements of the set $\lambda \star \mu$ in the (λ, μ) cell. In particular, the empty cell represents the empty set.

Note that these examples are all *symmetric* fusion laws, i.e. $\lambda \star \mu = \mu \star \lambda$, for all $\lambda, \mu \in \mathcal{F}$. From now on, we will consider only symmetric fusion laws.

6.2.2 Axial algebras

Let A be a commutative algebra over a field $\mathbb{F}$. The adjoint map ad_a of an element $a \in A$ is the endomorphism of the algebra which takes x to

$\star$	1	0
1	1	
0		0

$\star$	1	0	η
1	1		η
0		0	η
η	η	η	$1, 0$

$\star$	1	0	α	β
1	1		α	β
0		0	α	β
α	α	α	$1, 0$	β
β	β	β	β	$1, 0, \alpha$

Table 6.1 *The* $\mathcal{A}$, $\mathcal{J}(\eta)$, $\mathcal{M}(\alpha, \beta)$ *fusion laws.*

ax. For $\lambda \in \mathbb{F}$,

$$A_\lambda(a) = \{x \in A : ax = \lambda x\}$$

is the corresponding eigenspace. Note that $A_\lambda(a) = 0$ if λ is not an eigenvalue of the adjoint map ad_a. We further extend the eigenspace notation to sets $S \subseteq \mathbb{F}$ by defining $A_S(a) := \bigoplus_{\lambda \in S} A_\lambda(a)$.

We use fusion laws to impose partial control on the multiplication in an algebra.

Definition 6.2.2 Let A be a commutative, non-associative algebra over a field $\mathbb{F}$ and $(\mathcal{F}, \star)$ be a fusion law with $\mathcal{F} \subseteq \mathbb{F}$. An element a is said to be an $\mathcal{F}$-*axis* (or just an axis if $\mathcal{F}$ is clear from context) if

(1) a is an idempotent, that is $a^2 = a$;

(2) the adjoint map ad_a is semisimple with all eigenvalues contained in $\mathcal{F}$, that is,

$$A = A_{\mathcal{F}}(a);$$

(3) the above decomposition obeys the fusion law $\mathcal{F}$. That is, for all $\lambda, \mu \in \mathcal{F}$,

$$A_\lambda(a)A_\mu(a) \subseteq A_{\lambda\star\mu}(a).$$

From the above definition, as a is an idempotent, 1 is an eigenvalue of ad_a. So we always assume that $1 \in \mathcal{F}$.

Definition 6.2.3 An axis $a \in A$ is said to be *primitive* if $A_1(a) = \mathbb{F}a$.

Definition 6.2.4 A *primitive* $\mathcal{F}$-*axial algebra* is a pair $A = (A, X)$, where A is a commutative, non-associative algebra over the field $\mathbb{F}$ and X is a set of primitive $\mathcal{F}$-axes which generate A.

We will (mostly) just consider primitive axes and axial algebras and so for simplicity we will typically just speak of axes and axial algebras assuming primitivity.

It is also important to note that X is just *a* set of axes. We never assume that X is the set of all idempotents that have the property of axes. In fact, the problem of finding all such idempotents in a known algebra A is a very difficult problem. This question is related to finding the full automorphism group of an axial algebra, which we discuss later in Section 6.7.2.

We use the fusion law as a property which defines a class of axial algebras. For example, $\mathcal{J}(\eta)$- axial algebras constitute the class of axial algebras of Jordan type η and $\mathcal{M}(\alpha, \beta)$-axial algebras constitute the class of algebras of Monster type (α, β). The following example was one of the motivations for defining axial algebras.

Example 6.2.5 The Monster sporadic simple group M was constructed as the automorphism group of a $196{,}884$-dimensional real algebra V, known as the Griess algebra. The algebra V together with the set of so-called 2A-axes in it is an axial algebra of Monster type $(\frac{1}{4}, \frac{1}{32})$.

Recall that a Jordan algebra is a commutative non-associative algebra A satisfying the *Jordan identity*

$$(xy)x^2 = x(yx^2)$$

for all $x, y \in A$ (see for example [48]). The class of Jordan algebras also gives us examples of axial behaviour, namely every idempotent (primitive or not) in a Jordan algebra is a $\mathcal{J}(\frac{1}{2})$-axis. We should be careful here, because not every Jordan algebra is generated by non-zero idempotents. However, if a Jordan algebra A is generated by primitive idempotents, then it is an axial algebra of Jordan type $\frac{1}{2}$. In particular, all simple Jordan algebras over an algebraically closed field are examples.

One example which we will see again later is the spin factor Jordan algebra, which can be defined as follows.

Example 6.2.6 Let V be a vector space over a field $\mathbb{F}$ of characteristic not 2 and $b\colon V \times V \to \mathbb{F}$ be a symmetric bilinear form. Let $A = S(b) = \mathbb{F}\mathbf{1} \oplus V$ and define multiplication on A by

$$\mathbf{1}^2 = \mathbf{1}, \qquad \mathbf{1}u = u$$
$$uv = \tfrac{1}{2}b(u, v)\mathbf{1}$$

for all $u, v \in V$. This algebra is known as the *spin factor* algebra, or the Jordan algebra of *Clifford type*. The (primitive) axes have the form $\frac{1}{2}(1+u)$, where $b(u,u)=2$, and so the spin factor is an axial algebra of Jordan type $\frac{1}{2}$ if and only if V is spanned by vectors u with $b(u,u)=2$.

Here is another interesting example of algebras of Jordan type that we will see again later.

Example 6.2.7 Let (G, D) be a group of 3-transpositions. That is, D is a normal set of involutions (i.e. elements of order 2) in G such that $G = \langle D \rangle$ and $|cd| \leq 3$ for all $c, d \in D$. For example, we can take $G = S_n$ and D the set of all transpositions in G.

Suppose $\mathbb{F}$ is a field of characteristic not 2 and $\eta \in \mathbb{F} - \{0, 1\}$. The *Matsuo algebra*[1] $A := M_\eta(G, D)$ has basis D and the product of $a, b \in D$ given by

$$a \circ b = \begin{cases} a & \text{if } b = a, \\ 0 & \text{if } |ab| = 2, \\ \frac{\eta}{2}(a+b-c) & \text{if } |ab| = 3, \text{ where } c = a^b = b^a, \end{cases}$$

where we use $\circ$ for the algebra product to distinguish it from the multiplication in G. It is easy to see that for $a \in D$, $A_1(a) = \langle a \rangle$, $A_0(a) = \langle b : |ab| = 2 \rangle \oplus \langle b + c - \eta a : |ab| = 3 \rangle$ and $A_\eta(a) = \langle b - c : |ab| = 3 \rangle$, where as above $c = a^b = b^a$. By counting dimensions, we can see that ad_a is semisimple. A short calculation shows that the eigenspaces obey the $\mathcal{J}(\eta)$ fusion law and so every $a \in D$ is a (primitive) $\mathcal{J}(\eta)$-axis. Since the set of all $a \in D$ generate the algebra A, it is an axial algebra of type $\mathcal{J}(\eta)$.

Let us finish this subsection with a brief discussion of what additional assumptions we can make about the fusion laws due to our axial algebra definitions.

First of all, we have already mentioned that we always assume that $1 \in \mathcal{F}$, because axes are idempotents. Furthermore, our assumption that the fusion law be symmetric is natural because of commutativity of axial algebra. If we were to consider non-commutative generalisations, we would also have to allow non-symmetric fusion laws. Finally, primitivity also has some implications. Namely, for a primitive axis $a \in A$, we have

[1] In [32], Matsuo algebras are defined more broadly for an arbitrary triple system. However, they are only algebras of Jordan type when the triple system comes from a group of 3-transpositions.

that $A_1(a) = \langle a \rangle$ and, because of this,

$$A_1(a)A_\lambda(a) = aA_\lambda(a) = \begin{cases} A_\lambda(a) & \text{if } \lambda \neq 0; \\ 0 & \text{if } \lambda = 0. \end{cases}$$

So, for primitive axial algebras, we always assume that $1 \star \lambda = \{\lambda\}$ if $\lambda \neq 0$ and $1 \star 0 = \emptyset$.

6.2.3 Miyamoto group

The important feature of axial algebras is that they are closely related to groups. One can see this in the examples we presented earlier. The Griess algebra is clearly related to the Monster group, which is its automorphism group. The Matsuo algebras are clearly related to 3-transposition groups. Also, the Jordan algebras are very symmetric allowing classical and even some exceptional groups of Lie type as their automorphism groups. This is not an accident because the fusion laws of these axial algebras (see Table 6.1) are graded by the group C_2.

The exact meaning of this is as follows. For the law $\mathcal{F} = \mathcal{J}(\eta) = \{1, 0, \eta\}$, let the plus part of it be $\mathcal{F}_+ = \{1, 0\}$ and the minus part be $\mathcal{F}_- = \{\eta\}$. Then we have from the fusion law that $\mathcal{F}_+ \star \mathcal{F}_+ \subseteq \mathcal{F}_+$, $\mathcal{F}_- \star \mathcal{F}_- \subseteq \mathcal{F}_+$, and $\mathcal{F}_+ \star \mathcal{F}_- \subseteq \mathcal{F}_-$. In an algebra A of Jordan type η, this results in a C_2-grading for every axis a: setting $A_+ := A_{\mathcal{F}_+}(a)$ and $A_- := A_{\mathcal{F}_-}(a)$, we have that

$$A_+A_+ \subseteq A_+, A_-A_- \subseteq A_+, \text{ and } A_+A_- \subseteq A_-.$$

It is well known that such a C_2-grading of the algebra leads to an automorphism τ_a of A. This τ_a acts as identity on A_+ and as minus identity on A_-. (Then, $\tau_a^2 = 1$ and $\tau_a = 1$ only if $A_- = 0$, which is an exceptional situation.) We stress that we get a separate involution for each axis a, and thus, since A typically contains many axes, this leads to a substantial group of automorphisms.

For example, in the case of Matsuo algebras, this construction allows us to recover the 3-transposition group the algebra was constructed from, albeit we only recover the group up to its centre. In the case of Jordan algebras, we recover in this way a class of involutions in the corresponding group of Lie type.

The same grading trick works for the fusion law $\mathcal{F} = \mathcal{M}(\alpha, \beta)$, where we split $\mathcal{F}$ as follows: $\mathcal{F}_+ = \{1, 0, \alpha\}$ and $\mathcal{F}_- = \{\beta\}$. Again, this is a C_2-grading of the fusion law and it results in C_2-gradings and corresponding involutions for all axial algebras of Monster type. In the Griess algebra, these involutions are the 2A involutions in the Monster and thus we recover the Monster M in a natural way from its algebra.

Let us now give the formal definitions. Initially, for example, in [31, 32], the focus was specifically on the case of a C_2-grading, as above, but a general abelian grading group T can be handled in pretty much the same way, in terms of partitions. For example, such a definition was used in [66]. (A more categorical approach to gradings will be discussed later.)

Definition 6.2.8 Suppose $\mathcal{F}$ is a fusion law and T is an abelian group. A T-grading of $\mathcal{F}$ is a partition $\{\mathcal{F}_t : t \in T\}$ (with some parts $\mathcal{F}_t$ possibly empty) such that $\mathcal{F}_s \star \mathcal{F}_t \subseteq \mathcal{F}_{st}$ for all $s, t \in T$.

Given a T-grading of $\mathcal{F}$ and an axis a in an $\mathcal{F}$-axial algebra A, it is immediate that

$$A = \bigoplus_{t \in T} A_{\mathcal{F}_t}(a)$$

is a T-grading of the algebra A. It now follows that, for each linear character χ of T, we can define an automorphism of A as follows. Let $\tau_a(\chi)$ be the linear map $A \to A$, which acts on $A_t(a) := A_{\mathcal{F}_t}(a)$ by multiplying with the scalar $\chi(t)$.

The automorphism $\tau_a(\chi)$ is called a *Miyamoto automorphism.* It is immediate that the map $\chi \mapsto \tau_a(\chi)$ is a homomorphism from the group T^* of linear characters of T to $\mathrm{Aut}(A)$. The image of this map, $T_a \leq \mathrm{Aut}(A)$ is called the *axial subgroup.*

Definition 6.2.9 For a T-graded fusion law $\mathcal{F}$ and an $\mathcal{F}$-axial algebra A with the set of generating axes X, the group

$$\mathrm{Miy}(A, X) = \langle T_a : a \in X \rangle = \langle \tau_a(\chi) : a \in X, \chi \in T^* \rangle \leq \mathrm{Aut}(A)$$

is called the *Miyamoto group* of A. We will also use $\mathrm{Miy}(A)$ and $\mathrm{Miy}(X)$, depending on what our focus is.

In the case of $T = C_2$, when the characteristic of the ground field $\mathbb{F}$ is not 2, the group T^* contains a unique non-trivial character χ and so in this case we simplify $\tau_a(\chi)$ to just τ_a and call it the Miyamoto involution, as above.

The paper [13] introduced the category of fusion laws allowing us to separate the fusion law from the ground field $\mathbb{F}$. In this more modern language, a grading is simply a morphism from the fusion law $\mathcal{F}$ to the group fusion law T. Let us provide the exact definitions.

The category **Fus** has as objects all fusion laws, as given in Definition 6.2.1, and morphisms are as follows.

Definition 6.2.10 Let $(\mathcal{F}_1, \star_1)$ and $(\mathcal{F}_2, \star_2)$ be fusion laws. A *morphism* from $(\mathcal{F}_1, \star_1)$ to $(\mathcal{F}_2, \star_2)$ is a map $f\colon \mathcal{F}_1 \to \mathcal{F}_2$ such that

$$f(\lambda \star_1 \mu) \subseteq f(\lambda) \star_2 f(\mu)$$

for all $\lambda, \mu \in \mathcal{F}_1$.

The category of groups is a full subcategory of **Fus** via the following construction.

Definition 6.2.11 For a group T, we define the *group fusion law* $(T, \star)$, where $s \star t = \{st\}$ for all $s, t \in T$.

It is easy to see that morphisms of group fusion laws are the same as group homomorphisms, so here we indeed have a full subcategory of **Fus**.

In this categorical language, a T-grading of a fusion law $\mathcal{F}$ becomes simply a morphism from $\mathcal{F}$ to the group fusion law of T. (The parts $\mathcal{F}_t$ are then the fibres of the morphism.) In this sense, we do not even need to assume that T is abelian. However, since our algebras are commutative and Miyamoto automorphisms come from linear characters, we cannot gainfully utilise such a generalisation.

Note that for a T-graded fusion law $\mathcal{F}$, we can always unnecessarily enlarge the group T and still have a grading. We say the grading $f\colon \mathcal{F} \to T$ is *adequate* if $f(\mathcal{F})$ generates T. Clearly, we can restrict ourselves to only adequate gradings. A grading $f\colon \mathcal{F} \to T$ is a *finest* grading if every other grading of $\mathcal{F}$ factors uniquely through f. Every fusion law admits a unique finest grading and this is adequate [13].

Notice that if $g \in \mathrm{Aut}(A)$ and a is an axis, then a^g also has all the properties required for an axis and it is primitive if and only if a is primitive. So we can add it to our set X of generating axes. If we do this with $g \in \mathrm{Miy}(X)$, this gives us the concept of the *closure* $\bar{X} = X^{\mathrm{Miy}(X)} = \{x^g : g \in \mathrm{Miy}(X), x \in X\}$ of the set X of generators.

Lemma 6.2.12 *The Miyamoto group* $\mathrm{Miy}(\bar{X}) = \mathrm{Miy}(X)$. *In particular,* $\bar{\bar{X}} = \bar{X}$.

So our notation $\mathrm{Miy}(A)$ is unambigous even though we allow an extension of the generating set X to its closure as above. Furthermore, in what follows, we will often assume that X is *closed*, i.e. that $X = \bar{X}$. This has the advantage that we have an action of $\mathrm{Miy}(A)$ on the set X of generating axes. This action is necessarily faithful as the axes generate the algebra and so any automorphism fixing all the axes acts trivially on the algebra.

6.2.4 Frobenius form

Often algebras admit bilinear forms that associate with the algebra product. For example, the Killing form on a Lie algebra has this property.

Definition 6.2.13 Suppose A is an axial algebra. A non-zero bilinear form $(\cdot,\cdot)$ on A is called a *Frobenius* form if

$$(u, vw) = (uv, w)$$

for all $u, v, w \in A$.

This terminology came from the theory of Frobenius algebras which are associative algebras admitting a non-degenerate form satisfying the above associativity condition. Note that we do not assume that a Frobenius form is necessarily non-degenerate. In fact, as we will see later, the radical of the Frobenius form plays an important role in the structure theory of axial algebras.

Also note that in some areas of mathematics, where algebras with axial properties arise, algebras admitting a Frobenius form are called metrizable.

Often a Frobenius form would have additional nice properties. For example, the Griess algebra admits a Frobenius form which is positive definite and each axis has norm 1. These properties became some of the axioms of Majorana algebras. Similarly, by [34], every Jordan type algebra admits a Frobenius form with respect to which every primitive idempotent has norm 1.

Let us mention a couple of general facts about Frobenius forms on axial algebras.

Proposition 6.2.14 *(1) Every Frobenius form is symmetric.*

(2) For an axis $a \in A$, the eigenspaces $A_\lambda(a)$ and $A_\mu(a)$, where $\lambda \neq \mu$, are orthogonal with respect to every Frobenius form.

As of today, we actually do not know any examples of axial algebras which do not admit a Frobenius form. It is hard to imagine that no such algebras exist.

6.3 Structure theory

The structure theory of axial algebras was developed in [45] and the results in this section are from there unless otherwise referenced.

6.3.1 Radical

Theorem 6.3.1 *Suppose that (A, X) is a primitive axial algebra. Then there exists a unique maximal ideal disjoint from the set X of generating axes.*

Definition 6.3.2 The unique maximal ideal from Theorem 6.3.1 is called the *radical* of the axial algebra A.

Our notation for the radical is $R(A, X)$. Note that the radical a priori depends on the set of generating axes. However, in most cases, this set X is assumed, and then we just write $R(A)$.

The above definition is implicit, which is an inconvenience. When the algebra admits a Frobenius form, this can be remedied.

Theorem 6.3.3 *Suppose that A admits a Frobenius form such that $(a, a) \neq 0$ for all $a \in X$. Then $R(A, X)$ coincides with the radical $A^{\perp}$ of the Frobenius form.*

Note that if $(a, a) = 0$ for some $a \in X$ then, as a is primitive, it follows from Proposition 6.2.14 that $a \perp A$, that is, a is contained in the radical of the form. So the condition in the theorem is a necessary one.

Clearly, the above theorem gives us a practical and easy way to compute the radical by means of linear algebra. When the form is non-degenerate, the radical is trivial. In particular, this is the case for Majorana algebras where the Frobenius form is required to be positive definite. Note also that the condition on axes from the above theorem becomes mute for a positive definite form, so in a Majorana algebra every non-zero ideal contains a generating primitive axis.

In fact, this observation can be generalised.

Theorem 6.3.4 *If (A, X) is a primitive axial algebra admitting a non-degenerate Frobenius form then $R(A, X) = 0$, i.e. every non-zero ideal contains a generating primitive axis.*

Indeed, as we already pointed out, if $(a, a) = 0$ for a primitive axis $a \in A$, then $a \in A^{\perp}$. Since $A^{\perp} = 0$, we have that the assumptions of Theorem 6.3.3 are satisfied and so $R(A, X) = A^{\perp} = 0$.

6.3.2 Connectivity

In the previous subsection, we looked at ideals which do not contain any axes, so now we consider ideals which do contain axes. We begin with some observations which apply to any ideal and in fact are used in the proofs of the results from the previous subsection.

Since an ideal I is invariant under ad_a for all axes a and ad_a is semisimple, then ad_a is also semisimple acting on I and we have

$$I = \bigoplus_{\lambda \in \mathcal{F}} I_\lambda(a),$$

where $I_\lambda(a) := A_\lambda(a) \cap I$. In particular, this means that I is invariant under every Miyamoto automorphism $\tau_a(\chi)$ and hence every ideal I is invariant under the Miyamoto group $\mathrm{Miy}(A)$. This will be generalised later.

Suppose that a is an axis and $v \in A$. We can decompose v with respect to the eigenspace decomposition given by ad_a:

$$v = \sum_{\lambda \in \mathcal{F}} v_\lambda,$$

where $v_\lambda \in A_\lambda(a)$. In particular, if a is a primitive axis, then $v_1 = \varphi_a(v)a$ for some $\varphi_a(v) \in \mathbb{F}$. We call $\varphi_a(v)$ the *projection* of v onto a.

In particular, if v is in an ideal I, then $v_\lambda \in I$ for all $\lambda \in \mathcal{F}$. If, furthermore, a is primitive, then we have the following dichotomy: either the projection $\varphi_a(v) = 0$ and so $v_1 = 0$, or the projection $\varphi_a(v) \neq 0$ and then $0 \neq v_1 = \varphi_a(v)a \in I$, meaning $a \in I$. We will now apply this dichotomy with $v = b$ an axis in A to get the following definition.

Definition 6.3.5 The *projection graph* Γ is a directed graph whose vertices are the axes X and there is a directed edge $b \to a$ if $\varphi_a(b) \neq 0$.

This immediately gives us the following lemma.

Lemma 6.3.6 *If an axis $a \in I$ and there exists a directed path from a to b in the projection graph Γ, then $b \in I$.*

In other words, if a is in I, then all the successors of a in the projection graph are also in I. If Γ is strongly connected, i.e. every vertex is a successor to every other vertex, then A has no proper ideals containing an axis.

Assuming the set X of axes is closed under the action of the Miyamoto group, then we can form the *orbit projection graph* $\bar{\Gamma} := \Gamma/\mathrm{Miy}(A)$ as the quotient of the projection graph Γ. We then have the analogous result in Lemma 6.3.6 for the orbit projection graph $\bar{\Gamma}$.

If our algebra has a Frobenius form, then the picture is further simplified.

Lemma 6.3.7 *Let A be a primitive axial algebra with a Frobenius form and suppose that $(a, a) \neq 0 \neq (b, b)$, for axes a and b. Then the following are equivalent.*

(1) There is a directed edge $a \to b$.
(2) There is a directed edge $a \leftarrow b$.
(3) $(a, b) \neq 0$.

So in particular, if the Frobenius form is non-zero on all the axes, i.e. $(a, a) \neq 0$ for all axes $a \in X$, then we can consider the projection and orbit projection graphs as undirected graphs.

One further observation about ideals which contain axes is the following. If $a \in I$, then it is easy to see that $A_\lambda(a) \subseteq I$ for all $\lambda \neq 0$.

6.3.3 Associativity

Although axial algebras usually have elements $x, y, z \in A$ where $(xy)z \neq x(yz)$, they can be associative algebras (so "non-associative" for us means "not necessarily associative"). In fact, it is easy to characterise the associative axial algebras. Recall the fusion law $\mathcal{A}$ from Table 6.1.

Theorem 6.3.8 ([32, Corollary 2.9]) *Let A be a (primitive) $\mathcal{A}$-axial algebra generated by a set of axes X. Then A is associative and $A = \mathbb{F}X$. Conversely, if $A = \langle\langle X \rangle\rangle$ is an associative axial algebra, then every axis $a \in X$ satisfies the fusion law $\mathcal{A}$.*

This justifies us calling $\mathcal{A}$ the associative fusion law.

Even if our axial algebra is not associative, if the fusion law has the so-called Seress property, then the axes have some measure of associativity.

Definition 6.3.9 A fusion law $\mathcal{F}$ is *Seress* if $0 \in \mathcal{F}$ and for all $\lambda \in \mathcal{F}$, $0 \star \lambda \subseteq \{\lambda\}$.

Since axes are idempotents and $\lambda \in \mathcal{F}$ are eigenvalues for the adjoint ad_a, we must have $1 \in \mathcal{F}$ and $1 \star \lambda \subseteq \{\lambda\}$. So for a Seress fusion law, we have $1 \star 0 \subseteq \{0\} \cap \{1\} = \emptyset$. Also, for a Seress fusion law $0 \star 0 \subseteq \{0\}$ and so $A_0(a)$ is a subalgebra of A for any axis $a \in X$. Looking at Table 6.1 of examples of fusion laws, we can see that they are all Seress.

Lemma 6.3.10 (Seress Lemma[2]) *Let A be an axial algebra with a Seress fusion law and $a \in X$. Then for all $y \in A_1(a) \oplus A_0(a)$ and $x \in A$, we have*

$$a(xy) = (ax)y.$$

6.3.4 Sum decompositions

The subject of axes and ideals naturally leads onto direct sums and more generally sum decompositions of axial algebras.

Definition 6.3.11 An algebra A has a *sum decomposition* if it is generated by a set $\{A_i : i \in I\}$ of pairwise annihilating subalgebras, i.e. $A_i A_j = 0$ for all $i \neq j$. We write $A = \square_{i \in I} A_i$.

Note that we immediately have that $A = \sum_{i \in I} A_i$ as a vector space if A has a sum decomposition. If A has a sum decomposition and this sum is direct, then we write $A = \boxplus_{i \in I} A_i$.

Recall that the *annihilator ideal* is $\mathrm{Ann}(A) := \{x \in A : xA = 0\}$. For an axial algebra, as axes are idempotents, $\mathrm{Ann}(A)$ does not contain any axes and so $\mathrm{Ann}(A) \subseteq R(A)$.

Proposition 6.3.12 *Let $A = \square_{i \in I} A_i$ be an axial algebra.*

(1) $\mathrm{Ann}(A) = \square_{i \in I} \mathrm{Ann}(A_i)$.
(2) If $\mathrm{Ann}(A) = 0$*, then* $A = \boxplus_{i \in I} A_i$.

So for the sum not to be direct, at least one of the A_i must have a non-trivial annihilator. In fact, it can be shown that at least two of the A_i must have a non-trivial annihilator.

Just like for (not necessarily commutative) algebras we define the *commutator* $[x, y] := xy - yx$ to measure commutativity, for a non-associative algebra A we define the *associator* $(x, y, z) := (xy)z - x(yz)$ (see, for example, [67]). The *centre* $Z(A)$ is the subalgebra

$$Z(A) := \{a \in A : [a, A] = 0, (a, A, A) = (A, a, A) = (A, A, a) = 0\}$$

In fact, it is easy to see that if the algebra is commutative (as axial algebras are), then $(a, A, A) = 0$ is equivalent to $(A, A, a) = 0$ and either of these imply $(A, a, A) = 0$.

[2] This was first proved by Seress for axial algebras of Monster type $(\frac{1}{4}, \frac{1}{32})$, but Hall, Rehren and Shpectorov [31, Proposition 3.9] noticed that the same proof held in general.

Proposition 6.3.13 (Turner[3], 2022) *Let A be an axial algebra.*

(1) For every axis a, $Z(A) \subseteq A_1(a) \oplus A_0(a)$.
(2) If A is primitive, then $\mathrm{Ann}(A) = Z(A) \cap R(A, X)$.

Suppose that we have an axial algebra A which has a sum decomposition $A = \square_{i \in I} A_i$. So far, we have not put any restriction on the A_i – these subalgebras might not be axial subalgebras. That is, they might not be axial algebras in their own right.

Theorem 6.3.14 *Suppose that $A = \square_{i \in I} A_i$ is an axial algebra generated by a set of axes X. Let $X_i := X \cap A_i$ and define $B_i := \langle\langle X_i \rangle\rangle$. Then the X_i partition the set X and $A = \square_{i \in I} B_i$.*

So whenever we have a sum decomposition of an axial algebra, we in fact have a sum decomposition of A with respect to axial subalgebras of A. Note, however, that we may have $A_i \neq B_i$.

It is clear that if we have two different sum decompositions of an axial algebra into axial subalgebras, we can take a refinement of the two simply by taking a refinement of the two partitions of the set X. This raises the question of what the finest sum decomposition of an axial algebra is. In order to tackle this, we introduce the following graph.

Definition 6.3.15 The *non-annihilating* graph $\Delta(X)$ of an axial algebra generated by a set of axes X has vertex set X and an edge $a \sim b$ between distinct vertices if $ab \neq 0$.

This was first introduced for axial algebras of Jordan type η in [33] and for arbitrary axial algebras in [45]. It is easy to see that the projection graph is always a subgraph of the non-annihilating graph. In fact, in the important case of algebras of Monster type $(\frac{1}{4}, \frac{1}{32})$, the two graphs coincide provided the Frobenius form is non-zero on every axis, i.e. $(a, a) \neq 0$ for all $a \in X$.

If A has a sum decomposition $\square_{i \in I} A_i$, then by definition $A_i A_j = 0$ and hence $X_i X_j = 0$. So each X_i must be a union of connected components in the non-annihilating graph $\Delta(X)$. This naturally gives us the following conjecture.

Conjecture 6.3.16 *The finest sum decomposition of an axial algebra A arises when the X_i are the connected components of the non-annihilating graph $\Delta(X)$.*

[3] Turner is currently a PhD student at the University of Birmingham and this result is part of his MRes (Qual) project.

In other words, let X_i be the connected components of the non-annihilating graph and set $B_i := \langle\langle X_i \rangle\rangle$. It is clear that the B_i generate A as A is generated by the set $X = \bigcup_{i \in I} X_i$. So to prove the conjecture, we need to show that $B_i B_j = 0$ for all $i \neq j$ and hence $A = \Box_{i \in I} B_i$.

This conjecture holds for axial algebras of Jordan type η [33, Theorem A]. Before we say what progress has been made for this conjecture, note that if A has a non-trivial sum decomposition with axes in different parts, then $0 \in \mathcal{F}$. Likewise, if $\Delta(X)$ is not connected, then $0 \in \mathcal{F}$.

Suppose that the fusion law is graded and so the axial algebra has a non-trivial Miyamoto group. This group must behave well with respect to the finest sum decomposition.

Theorem 6.3.17 *Suppose A is a T-graded axial algebra, with $0 \in \mathcal{F}_{1_T}$. Let X_i, $i \in I$, be the connected components of $\Delta(X)$ and set $B_i := \langle\langle X_i \rangle\rangle$. Then,* $\mathrm{Miy}(A)$ *is a central product of the* $\mathrm{Miy}(B_i)$.

Before stating our partial result for Conjecture 6.3.16, we must first introduce quasi-ideals which generalise the notion of ideals.

Definition 6.3.18 Let A be an axial algebra with set of generating axes X. A *quasi-ideal* is a subspace $I \subseteq A$ such that $aI \subseteq I$ for all $a \in X$.

Definition 6.3.19 The *spine* of an axial algebra A is the quasi-ideal $Q(A, X)$ generated by the axes X. We say A is *slender* if $Q(A, X) = A$.

We can now state our partial result for the conjecture.

Theorem 6.3.20 *Suppose A is an axial algebra with a Seress fusion law. Let X_i be the connected components of $\Delta(X)$ and set $B_i := \langle\langle X_i \rangle\rangle$. If all but possibly one of the B_i are slender, then $A = \Box_{i \in I} B_i$ and so the conjecture holds.*

In fact, we know of very few examples of axial algebras which are not slender.

6.4 Axial algebras of Jordan type

An axial algebra is of *Jordan type* η, or just of Jordan type, if it has the Jordan fusion law $\mathcal{J}(\eta)$. This class was introduced in 2015 in [31] and is one of the most widely studied. We saw in Examples 6.2.6 and 6.2.7 several examples of axial algebras of Jordan type. One of these was related to 3-transposition groups.

6.4.1 3-transposition groups

Recall that a 3-transposition group is a pair (G, D), where G is a group generated by a normal set of involutions D such that $|cd| \leq 3$ for all $c, d \in D$. The *diagram* on a set $X \subseteq D$ is the graph having X as a vertex set, where $a, b \in X$ are adjacent whenever $|ab| = 3$ (so the complement of the diagram is the commuting graph). We say that (G, D) is *connected* when the diagram on D is connected.[4] It is not hard to see that (G, D) is connected if and only if D is a single conjugacy class. A general 3-transposition group decomposes as a central product of connected 3-transposition groups corresponding to the conjugacy classes within D, so it is enough to classify the connected 3-transposition groups.

Fischer began this when he classified the connected irreducible 3-transposition groups. Here irreducibility is a technical condition that requires $[G, O_2(G)] = 1 = [G, O_3(G)]$. Fischer then proved the following.

Theorem 6.4.1 ([17]) *Let (G, D) be a connected irreducible 3-transposition group. Then, up to the centre, D is one of*

(1) the transposition class in S_n,
(2) the transvection class in $O_n^{\pm}(2)$,
(3) the transvection class in $Sp_{2n}(2)$,
(4) a reflection class in $O_n^{\pm}(3)$,
(5) the transvection class in $U_n(2)$,
(6) a unique class of involutions in one of: $\Omega_8^+(2){:}S_3$, $\Omega_8^+(3){:}S_3$, Fi_{22}, Fi_{23}, or Fi_{24}.

Cuypers and Hall [12] extended this by classifying the (connected) non-irreducible 3-transposition groups and they fall into twelve cases (a convenient description can be found in [35, Theorem 5.3]).

6.4.2 Matsuo algebras

Recall from Example 6.2.7 that given any 3-transposition group (G, D) and $\eta \in \mathbb{F} - \{0, 1\}$, the *Matsuo algebra* $A := M_\eta(G, D)$ has basis D and

[4] Cuypers and Hall use the term "3-transposition group" only in the connected case, reserving a "group of 3-transpositions" for the general case.

multiplication given by

$$a \circ b = \begin{cases} a & \text{if } b = a, \\ 0 & \text{if } |ab| = 2, \\ \frac{\eta}{2}(a+b-c) & \text{if } |ab| = 3, \text{ where } c = a^b = b^a, \end{cases}$$

for $a, b \in D$. This is an axial algebra of Jordan type η. Moreover it also has some nice properties. All Matsuo algebras have a Frobenius form which is given by

$$(a,b) = \begin{cases} 1 & \text{if } b = a, \\ 0 & \text{if } |ab| = 2, \\ \eta/2 & \text{if } |ab| = 3. \end{cases}$$

The Miyamoto group of a Matsuo algebra $M_\eta(G, D)$ is $G/Z(G)$. In particular, we have examples of axial algebras for each group in Theorem 6.4.1.

Every Matsuo algebra $M_\eta(G, D)$ is an axial algebra of Jordan type η and in fact, for $\eta \neq \frac{1}{2}$, we have the converse.

Theorem 6.4.2 ([32]) *Every axial algebra of Jordan type $\eta \neq \frac{1}{2}$ is isomorphic to a quotient of a Matsuo algebra.*

6.4.3 Classification

For $\eta = \frac{1}{2}$, Matsuo algebras are also examples of axial algebras of Jordan type $\frac{1}{2}$. However, it is well known that a Jordan algebra satisfies the Peirce decomposition with respect to every idempotent. This decomposition is nothing other than the statement that the idempotents are axes of Jordan type $\frac{1}{2}$. So every Jordan algebra generated by primitive idempotents (such as, say, a spin factor from Example 6.2.6) is an axial algebra of Jordan type $\frac{1}{2}$.

Conjecture 6.4.3 *Every axial algebra A of Jordan type η is either isomorphic to a quotient of a Matsuo algebra, or $\eta = \frac{1}{2}$ and A is a Jordan algebra.*

The conjecture is known to hold in some cases. The 2-generated case, i.e. where the algebra is generated by two primitive axes, has been classified in the same paper [32] where the class of Jordan type algebras was introduced. When $\eta \neq \frac{1}{2}$, we only have two Matsuo algebras. One of these is 2B $:= M_\eta(V_4, \{a, b\})$, where $V_4 = \langle a, b \rangle$, which is just isomorphic to $\mathbb{F} \oplus \mathbb{F}$ and the other is $3C(\eta) := M_\eta(S_3, D)$, where D is the class of involutions in S_3. This fact almost immediately implies Theorem 6.4.2 and hence completes the classification for $\eta \neq \frac{1}{2}$.

For $\eta = \frac{1}{2}$, there is an infinite 1-parameter family of 3-dimensional algebras and every 2-generated algebra is one of these, or a quotient. The 3-dimensional examples are, with one exception, isomorphic to the spin factor algebras $S(b)$, with b a bilinear form on a 2-dimensional vector space, as described in Example 6.2.6. The exceptional algebra does not have an identity and is baric; that is, it has an algebra homomorphism to $\mathbb{F}$ with respect to which both generating axes map to 1. Almost all of these 3-dimensional algebras are simple and so there are only quotients for some very restricted values of the parameters.

In addition to Theorem 6.4.2, the classification of the 2-generated case also gives the following.

Theorem 6.4.4 ([34]) *Every axial algebra of Jordan type η has a Frobenius form with respect to which every primitive axis is of length* 1.

The 3-generated case has also been classified by Gorshkov and Staroletov [28], namely every axial algebra of Jordan type $\frac{1}{2}$ is covered by a 9-dimensional Jordan algebra from a 4-parameter family. Generically, they are isomorphic to the 3×3 matrix algebra.

The case with 4, or more generators remains wide open and in the next Section we will discuss some recent ideas. Note that the 4-generated case is the first one where some Matsuo algebras are not Jordan.[5]

6.4.4 Solid subalgebras

Suppose A is an algebra of Jordan type $\frac{1}{2}$ and $B = \langle\langle a, b\rangle\rangle$ is a 2-generated subalgebra, where a and b are primitive axes in A.

Definition 6.4.5 We say that B is *solid* if every primitive idempotent in B is a primitive axis in A.

For example, if $B = \langle\langle a, b\rangle\rangle \cong 2B$, then it is trivially solid as a and b are the only primitive axes in B. This is the only case of a 2-generated Jordan type algebra with two primitive axes.

Every 2-generated subalgebra of a Jordan algebra is solid. The work-

[5] De Medts, Rowen and Segev show in a recent paper [15] that a 4-generated algebra of Jordan type $\frac{1}{2}$ has dimension at most 81.

shop 'Algebras of Jordan type and groups'[6], which took place in Birmingham in January 2022, was focused on the following recent result.

Recall that by Theorem 6.4.4, every algebra of Jordan type has a Frobenius form.

Theorem 6.4.6 (Gorshkov, Shpectorov, Staroletov, 2021+) *If* $\alpha := (a, b) \notin \{0, \frac{1}{4}, 1\}$, *then* B *is a solid subalgebra.*

Note that if $\alpha = 0$, then B is isomorphic to either 2B (which is trivially solid), or its 3-dimensional cover. If $\alpha = 1$, then B is baric and hence has a 2-dimensional radical. Finally, when $\alpha = \frac{1}{4}$, $B \cong 3\mathrm{C}(\frac{1}{2})$. If the characteristic is 0, then an argument using algebraic geometry shows that the 3-dimensional cover of 2B and the baric algebra are solid. This gives us the following.

Corollary 6.4.7 *In characteristic* 0, *every* 2*-generated subalgebra* B *of* A *either contains three primitive axes, or it is solid.*

Consequently, it follows (in the same way as Theorem 6.4.2 follows from the 2-generated case) that if A contains no solid subalgebras apart from 2B, then A is a quotient of a Matsuo algebra. It is tempting to conjecture that solid subalgebras identify the case where the algebra has to be Jordan. However, the following example shows that this is not always the case.

Example 6.4.8 (Gorshkov, Staroletov, 2022) The Matsuo algebra $M = M_{\frac{1}{2}}(3^3{:}S_4, D)$, contains exactly six solid non-2B subalgebras while all other 2-generated subalgebras contain two, or three primitive axes. Needless to say, this means that M is not a Jordan algebra[7].

So this idea seems to lead to a geometric point of view (at least in the characteristic 0 case) where the primitive axes are points and we have several different types of lines corresponding to 2-generated subalgebras: either non-solid with three points, or several types of solid lines.

6.5 2-generated axial algebras of Monster type

An axial algebra is of *Monster type* (α, β), or just of Monster type, if it has the $\mathcal{M}(\alpha, \beta)$ fusion law (see Table 6.1). This class is particularly

[6] This was supported by a Focused Research Workshop grant from the Heilbronn Institute.

[7] De Medts and Rehren determined in [14], with a correction by Yabe [74], all Matsuo algebras for $\eta = \frac{1}{2}$ that are Jordan.

interesting as it is a natural extension of the class of algebras of Jordan type. Indeed every algebra of Jordan type α, or β is at the same time an algebra of Monster type (α, β). In addition, it is also interesting as the prize example of the Griess algebra occurs here and so we are guaranteed to have algebras with many interesting groups.

We would like to classify all such axial algebras, however, as hinted at by the Griess algebra example, this is a much more difficult problem. As with axial algebras of Jordan type, we can begin by classifying the 2-generated algebras.

6.5.1 Sakuma's Theorem

As we have seen, the Griess algebra is a motivating example of an axial algebra and it has the fusion law $\mathcal{M}(\frac{1}{4}, \frac{1}{32})$. Its axes are in bijection with the involutions in the Monster of class 2A and so they became known as 2A-axes. It was Norton [11] who first studied the 2-generated subalgebras of the Griess algebra. He showed that the isomorphism class of the subalgebra generated by distinct 2A-axes a and b is fully determined by the conjugacy class of the element $\tau_a \tau_b$ in the Monster. There are eight such conjugacy classes of products, namely 2A, 2B, 3A, 3C, 4A, 4B, 5A, and 6A. So, we label the 2-generated subalgebras by these conjugacy classes.

The concept of a 2A-axis was generalised in the VOA setting to the concept of an Ising vector in a general OZ VOA. Miyamoto noticed in [56] that the fusion property of Ising vectors gave a τ automorphism of the VOA and he proposed to study VOAs generated by Ising vectors, starting with two. He himself finished one of the cases and the remaining cases were completed by Sakuma [66]. Amazingly, he showed that the weight 2 component algebra of such a VOA is isomorphic to one of the above eight subalgebras of the Griess algebra.

Ivanov introduced the class of Majorana algebras [38] by taking as axioms the properties he extracted from Sakuma's proof. In particular, this is the first time that the fusion law, namely $\mathcal{M}(\frac{1}{4}, \frac{1}{32})$, appeared as an axiom. One of the early successes was the proof of Sakuma's Theorem in this setting, namely it was shown in [39] that every 2-generated Majorana algebra is one of the same eight Norton-Sakuma algebras.

Axial algebras were introduced as a broad generalisation of Majorana algebras, obtained by generalising some and dropping other axioms. Accordingly, we have the task of classifying 2-generated algebras for broad classes of axial algebras and we call such results Sakuma-like theorems.

The first of these was for axial algebras of Monster type $(\frac{1}{4}, \frac{1}{32})$ that have a Frobenius form [31]. The ultimate form of a Sakuma theorem for the class of Monster type $(\frac{1}{4}, \frac{1}{32})$ was finally obtained later as a consequence of work in [21].

Theorem 6.5.1 *Let A be a 2-generated axial algebra of Monster type $(\frac{1}{4}, \frac{1}{32})$ over a field $\mathbb{F}$ of characteristic 0. Then, A is one of the* Norton-Sakuma algebras*:* 2A, 2B, 3A, 3C, 4A, 4B, 5A, *and* 6A.

In Table 6.2, we give the structure constants and Frobenius forms for the Norton-Sakuma algebras. For an algebra labelled nL, $X = \{a_0, a_1\}$ is the set of generating axes. The Miyamoto group $\mathrm{Miy}(X) = \langle \tau_{a_0}, \tau_{a_1} \rangle$ is a dihedral group and we let $\rho := \tau_{a_0}\tau_{a_1}$. Let $\bar{X}$ be the closure of X; this has size n. The axes $a_i \in \bar{X}$ are labelled

$$a_{\varepsilon+2k} := a_\varepsilon^{\rho^k}$$

where $\varepsilon = 0, 1$. It is clear from this, that every a_i is conjugate to either a_0, or a_1. In fact, if n is odd, then $a_0^{\mathrm{Miy}(X)} = a_1^{\mathrm{Miy}(X)}$ and so there is a single orbit of axes under the action of the Miyamoto group. If n is even, then $a_0^{\mathrm{Miy}(X)}$ and $a_1^{\mathrm{Miy}(X)}$ are disjoint orbits and they both have size $\frac{n}{2}$. The additional basis elements in Table 6.2 which occur for some algebras are indexed by a power ρ^k of ρ and this indicates that the basis element is fixed by ρ^k.

It is of course desirable to generalise Theorem 6.5.1 to positive characteristics. Looking at Table 6.2, we see that some small characteristics 2, 3 and 5 might cause problems. In addition, we should avoid 3, 7 and 31, if we want to keep the eigenvalues $1, 0, \frac{1}{4}, \frac{1}{32}$ distinct. We expect the following is true and can possibly be obtained by generalising the arguments in [21].

Conjecture 6.5.2 *If* $\mathrm{char}(\mathbb{F})$ *is sufficiently large, then every 2-generated axial algebra of Monster type $(\frac{1}{4}, \frac{1}{32})$ is isomorphic to one of the Norton-Sakuma algebras realised over $\mathbb{F}$.*

6.5.2 Generalised Sakuma's Theorem

Rehren in [61, 62] was the first person to attempt to generalise the above from $\mathcal{M}(\frac{1}{4}, \frac{1}{32})$ to an arbitrary Monster type (α, β). He discovered that there are special cases $\alpha = 2\beta$ and $\alpha = 4\beta$ and proved the following:

Theorem 6.5.3 *Suppose* $\mathrm{char}(\mathbb{F}) \neq 2$, *with* $1, 0, \alpha, \beta \in \mathbb{F}$ *distinct. If $\alpha \neq 2\beta, 4\beta$, then every 2-generated axial algebra A of Monster type (α, β)*

Table 6.2 *Norton-Sakuma algebras*

Type	Basis	Products & form
2A	a_0, a_1, a_ρ	$a_0 \cdot a_1 = \frac{1}{8}(a_0 + a_1 - a_\rho)$ $a_0 \cdot a_\rho = \frac{1}{8}(a_0 + a_\rho - a_1)$ $(a_0, a_1) = (a_0, a_\rho) = (a_1, a_\rho) = \frac{1}{8}$
2B	a_0, a_1	$a_0 \cdot a_1 = 0$ $(a_0, a_1) = 0$
3A	a_{-1}, a_0, a_1, u_ρ	$a_0 \cdot a_1 = \frac{1}{2^5}(2a_0 + 2a_1 + a_{-1}) - \frac{3^3 \cdot 5}{2^{11}} u_\rho$ $a_0 \cdot u_\rho = \frac{1}{3^2}(2a_0 - a_1 - a_{-1}) + \frac{5}{2^5} u_\rho$ $u_\rho \cdot u_\rho = u_\rho$, $(a_0, a_1) = \frac{13}{2^8}$ $(a_0, u_\rho) = \frac{1}{4}$, $(u_\rho, u_\rho) = \frac{2^3}{5}$
3C	a_{-1}, a_0, a_1	$a_0 \cdot a_1 = \frac{1}{2^6}(a_0 + a_1 - a_{-1})$ $(a_0, a_1) = \frac{1}{2^6}$
4A	a_{-1}, a_0, a_1, a_2 v_ρ	$a_0 \cdot a_1 = \frac{1}{2^6}(3a_0 + 3a_1 + a_{-1} + a_2 - 3v_\rho)$ $a_0 \cdot v_\rho = \frac{1}{2^4}(5a_0 - 2a_1 - a_2 - 2a_{-1} + 3v_\rho)$ $v_\rho \cdot v_\rho = v_\rho$, $a_0 \cdot a_2 = 0$ $(a_0, a_1) = \frac{1}{2^5}$, $(a_0, a_2) = 0$ $(a_0, v_\rho) = \frac{3}{2^3}$, $(v_\rho, v_\rho) = 2$
4B	a_{-1}, a_0, a_1, a_2 a_{ρ^2}	$a_0 \cdot a_1 = \frac{1}{2^6}(a_0 + a_1 - a_{-1} - a_2 + a_{\rho^2})$ $a_0 \cdot a_2 = \frac{1}{2^3}(a_0 + a_2 - a_{\rho^2})$ $(a_0, a_1) = \frac{1}{2^6}$, $(a_0, a_2) = (a_0, a_{\rho^2}) = \frac{1}{2^3}$
5A	a_{-2}, a_{-1}, a_0, a_1, a_2, w_ρ	$a_0 \cdot a_1 = \frac{1}{2^7}(3a_0 + 3a_1 - a_2 - a_{-1} - a_{-2}) + w_\rho$ $a_0 \cdot a_2 = \frac{1}{2^7}(3a_0 + 3a_2 - a_1 - a_{-1} - a_{-2}) - w_\rho$ $a_0 \cdot w_\rho = \frac{7}{2^{12}}(a_1 + a_{-1} - a_2 - a_{-2}) + \frac{7}{2^5} w_\rho$ $w_\rho \cdot w_\rho = \frac{5^2 \cdot 7}{2^{19}}(a_{-2} + a_{-1} + a_0 + a_1 + a_2)$ $(a_0, a_1) = \frac{3}{2^7}$, $(a_0, w_\rho) = 0$, $(w_\rho, w_\rho) = \frac{5^3 \cdot 7}{2^{19}}$
6A	a_{-2}, a_{-1}, a_0, a_1, a_2, a_3 a_{ρ^3}, u_{ρ^2}	$a_0 \cdot a_1 = \frac{1}{2^6}(a_0 + a_1 - a_{-2} - a_{-1} - a_2 - a_3 + a_{\rho^3}) + \frac{3^2 \cdot 5}{2^{11}} u_{\rho^2}$ $a_0 \cdot a_2 = \frac{1}{2^5}(2a_0 + 2a_2 + a_{-2}) - \frac{3^3 \cdot 5}{2^{11}} u_{\rho^2}$ $a_0 \cdot u_{\rho^2} = \frac{1}{3^2}(2a_0 - a_2 - a_{-2}) + \frac{5}{2^5} u_{\rho^2}$ $a_0 \cdot a_3 = \frac{1}{2^3}(a_0 + a_3 - a_{\rho^3})$, $a_{\rho^3} \cdot u_{\rho^2} = 0$ $(a_0, a_1) = \frac{5}{2^8}$, $(a_0, a_2) = \frac{13}{2^8}$ $(a_0, a_3) = \frac{1}{2^3}$, $(a_{\rho^3}, u_{\rho^2}) = 0$

has an explicit spanning set of size 8 *defined in terms of the two generating axes. In particular, the dimension of* A *is at most* 8.

Additionally, he generalised the Norton-Sakuma algebras to 1- and 2-parameter families of 2-generated algebras of Monster type (α, β). For three algebras, namely 2A, 2B and 3C, generalisations exist for all values of (α, β) (excluding where $1, 0, \alpha, \beta$ are not distinct). The algebra 3A generalises for all (α, β), except for $\alpha = \frac{1}{2}$. However, the remaining algebras generalise to families with fusion law $\mathcal{M}(\alpha, \beta)$, but only when α and β are tied by a simple polynomial relation.

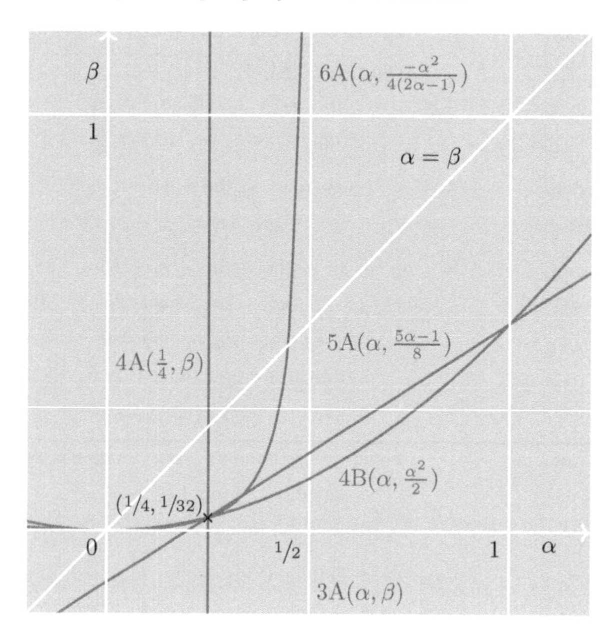

Figure 6.1 Rehren's generalised Norton-Sakuma algebras

As can be seen in Figure 6.1, the point $(\frac{1}{4}, \frac{1}{32})$ is the only point in the (α, β)-plane for which all eight examples exist. This underscores the exceptional nature of the Griess algebra and the Monster.

6.5.3 The generic case

Rehren's work was revisited and extended by Franchi, Mainardis and Shpectorov in [21, 22, 23]. In [21], they develop the general setup and derive several equations on the parameters involved. These equations are quite complex, however it was possible to solve them in some specific situations. For $(\alpha, \beta) = (\frac{1}{4}, \frac{1}{32})$, these equations yield Theorem 6.5.1. The other case considered is the *generic case*, that is where α and β are algebraically independent. So in other words, we can think of the ground field as $\mathbb{Q}(\alpha, \beta)$, where α and β are indeterminates. Any algebra arising over $\mathbb{Q}(\alpha, \beta)$ can be transferred into any characteristic and α and β can be specialised (nearly) arbitrarily. In other words, these are the algebras that always exist.

Theorem 6.5.4 *The generic 2-generated algebras of Monster type are* 2B, 3C(α), 3C(β) *and* 3A(α, β).

As a consequence of this, the Miyamoto group of a generic algebra with an arbitrary number of generators is always a 3-transposition group. It is an interesting problem to determine which 3-transposition groups actually arise here.

Additionally, Rehren's dimension bound was extended in the same paper to the exceptional cases $\alpha = 2\beta$ and $\alpha = 4\beta$. In the first case, the bound is also 8 but with a different spanning set. In the follow up paper [22], all the algebras in the $\alpha = 2\beta$ case were completely classified. The list includes the generic algebras specialised to $(2\beta, \beta)$, Joshi's algebras obtained by the double axis construction (see Section 6.6.2) as well as a handful of exceptional algebras, some of which exist only in a specific characteristic.

In the second case, $\alpha = 4\beta$, the situation is even more interesting. If $(\alpha, \beta) \neq (2, \frac{1}{2})$, then the dimension bound of 8 also holds, under a mild symmetry assumption (see Section 6.5.5). However, for $(\alpha, \beta) = (2, \frac{1}{2})$, there is an infinite dimensional example, called the Highwater algebra.

6.5.4 The Highwater algebra

The Highwater algebra $\mathcal{H}$ was discovered independently by Franchi, Mainardis and Shpectorov [23], and by Yabe [75], and it is an infinite dimensional algebra of Monster type $(2, \frac{1}{2})$. It has infinitely many axes a_i, one for each integer $i \in \mathbb{Z}$, and these can be completed to a basis of $\mathcal{H}$ by adding s_j, for $j \in \mathbb{N}$, which correspond to the 'distance' between the axes.

Definition 6.5.5 Let $\mathrm{char}(\mathbb{F}) \neq 2$. The *Highwater algebra* is the algebra $\mathcal{H}$ on $\bigoplus_{i\in\mathbb{Z}} \mathbb{F}a_i \oplus \bigoplus_{j\in\mathbb{N}} \mathbb{F}s_j$ with multiplication given by

$$\begin{aligned} a_i a_j &:= \tfrac{1}{2}(a_i + a_j) + s_{|i-j|} \\ a_i s_j &:= -\tfrac{3}{4}a_i + \tfrac{3}{8}(a_{i-j} + a_{i+j}) + \tfrac{3}{2}s_j \\ s_j s_k &:= \tfrac{3}{4}(s_j + s_k) - \tfrac{3}{8}(s_{|j-k|} + s_{j+k}) \end{aligned}$$

for all $i \in \mathbb{Z}$ and $j, k \in \mathbb{N}$, where we set $s_0 = 0$.

In characteristic 3, it requires infinitely many generators; otherwise it is 2-generated. The algebra is also baric, that is the map $\lambda\colon A \to \mathbb{F}$ defined by $\lambda(a_i) = 1$ and $\lambda(s_j) = 0$ is an algebra homomorphism. This immediately shows that $\mathcal{H}$ has a Frobenius form, given by $(x, y) = \lambda(x)\lambda(y)$, for all $x, y \in \mathcal{H}$.

For every axis $a_i \in \mathcal{H}$, the Miyamoto involution $\tau_i := \tau_{a_i}$ is given by $a_j \mapsto a_{2i-j}$ which is a reflection on the indices of the axis. It is then easy to see that $\mathrm{Miy}(\mathcal{H}) \cong D_\infty$. There is also an additional symmetry given by swapping a_0 and a_1, which we write as $\tau_{\frac{1}{2}}$ and is a reflection in $\frac{1}{2}$. It turns out that $\mathrm{Aut}(\mathcal{H}) = \langle \mathrm{Miy}(\mathcal{H}), \tau_{\frac{1}{2}} \rangle \cong D_\infty$.

In characteristic 5 only, the Highwater algebra admits a cover $\hat{\mathcal{H}}$ found by Franchi and Mainardis [19]. This is also a 2-generated axial algebra of Monster type $(2, \frac{1}{2})$. It is slightly more complicated than $\mathcal{H}$, with extra basis elements $p_{\bar{r},j}$, where $\bar{r} \in \{\bar{1}, \bar{2}\} \in \mathbb{Z}_3$ and $j \in 3\mathbb{N}$ (see [20, Definition 3.2]). These generate an ideal $J = \langle p_{\bar{r},j} : \bar{r} \in \{\bar{1}, \bar{2}\}, j \in 3\mathbb{N} \rangle$ and $\hat{\mathcal{H}}/J \cong \mathcal{H}$.

6.5.5 Symmetric 2-generated algebras of Monster type

One can note that all of the (generalised) Norton-Sakuma algebras have the property that there is an involutory automorphism which switches the two generating axes. We call such algebras *symmetric*. In a major breakthrough, Yabe [75] produced an almost complete classification of the symmetric 2-generated algebras of general Monster type (α, β). As part of this, he discovered some further infinite families of examples in addition to Rehren's generalised Norton-Sakuma algebras and Joshi's examples coming from the double axis construction (which we will see in Section 6.6.2) and he also independently discovered the Highwater algebra. The missing characteristic 5 case was completed by Franchi and Mainardis [19], who showed that the only additional examples are quotients of the cover $\hat{\mathcal{H}}$ of the Highwater algebra. Finally, the quotients of the Highwater algebra and its cover were classified by Franchi, Mainardis and McInroy in [20].

Theorem 6.5.6 ([75, 19, 20]) *A symmetric 2-generated axial algebra of Monster type (α, β) is a quotient of one of the following:*

(1) an axial algebra of Jordan type α, or β;
(2) an algebra in one of the following families:
(a) $3\mathrm{A}(\alpha, \beta)$, $4\mathrm{A}(\frac{1}{4}, \beta)$, $4\mathrm{B}(\alpha, \frac{\alpha^2}{2})$, $4\mathrm{J}(2\beta, \beta)$, $4\mathrm{Y}(\frac{1}{2}, \beta)$, $4\mathrm{Y}(\alpha, \frac{1-\alpha^2}{2})$, $5\mathrm{A}(\alpha, \frac{5\alpha-1}{8})$, $6\mathrm{A}(\alpha, \frac{-\alpha^2}{4(2\alpha-1)})$, $6\mathrm{J}(2\beta, \beta)$ *and* $6\mathrm{Y}(\frac{1}{2}, 2)$*;*
(b) $\mathrm{IY}_3(\alpha, \frac{1}{2}, \mu)$ *and* $\mathrm{IY}_5(\alpha, \frac{1}{2})$*;*
(3) the Highwater algebra $\mathcal{H}$, or its characteristic 5 *cover $\hat{\mathcal{H}}$.*

Note that our cases and notation are different to Yabe's[8]. Similarly to the (generalised) Norton-Sakuma algebras, we label an algebra $n\mathrm{L}$ to emphasis the size n of the closed set of axes. The two algebras which are labelled IY_d are of axial dimension d (the dimension of the subspace spanned by the axes) and generically have infinitely many axes, however they can have finitely many axes for some characteristics and values of their parameters. We use the same labels L as for the (generalised) Norton-Sakuma algebras, adding J for Joshi's examples and Y for the algebras introduced by Yabe. The parameters after this are (α, β) (and in the case of $\mathrm{IY}_3(\alpha, \frac{1}{2}, \mu)$ there is an additional parameter μ).

We list some properties of the algebras in Table 6.3, but we do not propose to give the structure constants for all them here. See [75] for the original definition of these algebras and [53] for new bases for some of them. However, we will comment on the permitted values (α, β) for these algebras as there are some errors in both Rehren's and Yabe's tables. It is clear from our definition of axial algebras, that we must have that $\{1, 0, \alpha, \beta\}$ are distinct, otherwise some eigenspaces would merge (however, see decomposition algebras [13] for a way round this). For all the algebras except $3\mathrm{A}(\alpha, \beta)$ and $6\mathrm{A}(\alpha, \frac{-\alpha^2}{4(2\alpha-1)})$, this is the only restriction. For these two, we additionally only need to exclude $\alpha = \frac{1}{2}$ (Rehren and Yabe excluded other values here unnecessarily). This information can be readily checked on the computer in many cases. McInroy has written a Magma package [51] with these algebras and quotients of the Highwater algebra.

Question 6.5.7 The value $\alpha = \frac{1}{2}$ is excluded for $3\mathrm{A}(\alpha, \beta)$ as some of

[8] Yabe's names for these algebras are: (a) $\mathrm{III}(\alpha, \beta, 0)$, $\mathrm{IV}_1(\frac{1}{4}, \beta)$, $\mathrm{IV}_2(\alpha, \frac{\alpha^2}{2})$, $\mathrm{IV}_1(\alpha, \frac{\alpha}{2})$, $\mathrm{IV}_2(\frac{1}{2}, \beta)$, $\mathrm{IV}_2(\alpha, \frac{1-\alpha^2}{2})$, $\mathrm{V}_1(\alpha, \frac{5\alpha-1}{8})$, $\mathrm{VI}_2(\alpha, \frac{-\alpha^2}{4(2\alpha-1)})$, $\mathrm{VI}_1(\alpha, \frac{\alpha}{2})$ and $\mathrm{IV}_3(\frac{1}{2}, 2)$ and (b) $\mathrm{III}(\alpha, \frac{1}{2}, \delta)$, where $\delta = -2\mu - 1$, and $\mathrm{V}_2(\alpha, \frac{1}{2})$. Here the Roman numeral indicates axial dimension.

Table 6.3 *The symmetric 2-generated axial algebras of Monster type*

Name	$\{1,0,\alpha,\beta\}$ coinciding	Additional	Dimension	Quotients	Dimension
$3\mathrm{A}(\alpha,\beta)$	$\alpha,\beta \neq 0,1,\ \alpha\neq\beta$	$\alpha\neq\frac{1}{2}$	4	$3\mathrm{A}(\alpha,\frac{1-3\alpha^2}{3\alpha-1})^\times$	3
$4\mathrm{A}(\frac{1}{4},\beta)$	$\mathrm{char}(\mathbb{F})\neq 3,\ \alpha\neq 0,1,\frac{1}{4}$		5	$4\mathrm{A}(\frac{1}{4},\frac{1}{2})^\times$	4
$4\mathrm{B}(\alpha,\frac{\alpha^2}{2})$	$\alpha\neq\{0,1,2,\pm\sqrt{2}\}$		5	$4\mathrm{B}(-1,\frac{1}{2})^\times$	4
$4\mathrm{J}(2\beta,\beta)$	$\beta\neq\{0,1,\frac{1}{2}\}$		5	$4\mathrm{J}(-\frac{1}{2},-\frac{1}{4})^\times$	4
$4\mathrm{Y}(\frac{1}{2},\beta)$	$\beta\neq\{0,1,\frac{1}{2}\}$		5		
$4\mathrm{Y}(\alpha,\frac{1-\alpha^2}{2})$	$\alpha\neq\{0,\pm1,\pm\sqrt{-1},$		5		
	$-1\pm\sqrt{2}\}$				
$5\mathrm{A}(\alpha,\frac{5\alpha-1}{8})$	$\alpha\neq\{0,1,-\frac{1}{3},\frac{1}{5},\frac{9}{5}\}$		6		
$6\mathrm{A}(\alpha,\frac{-\alpha^2}{4(2\alpha-1)})$	$\alpha\neq\{0,1,\frac{4}{9},$	$\alpha\neq\frac{1}{2}$	8	$6\mathrm{A}(\frac{2}{3},-\frac{1}{3})^\times$	7
	$-4\pm2\sqrt{5}\}$			$6\mathrm{A}(\frac{1\pm\sqrt{97}}{24},\frac{53\pm5\sqrt{97}}{192})^\times$	7
$6\mathrm{J}(2\beta,\beta)$	$\beta\neq\{0,1,\frac{1}{2}\}$		8	$6\mathrm{J}(-\frac{2}{7},-\frac{1}{7})^\times$	7
$6\mathrm{Y}(\frac{1}{2},2)$	$\mathrm{char}(\mathbb{F})\neq 3$		5	$6\mathrm{Y}(\frac{1}{2},2)^\times$	4
$\mathrm{IY}_3(\alpha,\frac{1}{2},\mu)$	$\alpha\neq 0,1,\frac{1}{2}$		4	$\mathrm{IY}_3(-1,\frac{1}{2},\mu)^\times$	3
				$\mathrm{IY}_3(\alpha,\frac{1}{2},1)^\times$	3
$\mathrm{IY}_5(\alpha,\frac{1}{2})$	$\alpha\neq 0,1,\frac{1}{2}$		6	$\mathrm{IY}_5(\alpha,\frac{1}{2})^\times$	5

the structure constants are not well-defined. Is there another description of this family which makes sense for $\alpha = \frac{1}{2}$? Note that for $6\mathrm{A}(\alpha, \frac{-\alpha^2}{4(2\alpha-1)})$, β tends to infinity as α tends to $\frac{1}{2}$, so this cannot be helped.

It is also evident from Table 6.3, that any symmetric 2-generated algebra with $\beta \neq \frac{1}{2}$ has at most 6 axes. It turns out that when $\beta = \frac{1}{2}$, the algebras $\mathrm{IY}_3(\alpha, \frac{1}{2}, \mu)$, $\mathrm{IY}_5(\alpha, \frac{1}{2})$, $\mathcal{H}$ and $\hat{\mathcal{H}}$ have generically infinitely many axes and they (or their quotients) can have any finite number of axes.

Problem 6.5.8 Explain the difference in the number of axes when $\beta = \frac{1}{2}$ and $\beta \neq \frac{1}{2}$.

Possibly linked to this, Krasnov and Tkachev [46] have found exceptional behaviour for the eigenvalue $\frac{1}{2}$. Roughly speaking if $\frac{1}{2}$ never appears as an eigenvalue of an idempotent, then the total number of idempotents is at most $2^{\dim(A)}$.

We note that every symmetric 2-generated algebra of Monster type in Theorem 6.5.6 has a Frobenius form.

Finally, in Figure 6.2, we give a picture similar to Figure 6.1 showing the intersection of the permitted values of (α, β) for the symmetric 2-generated algebras of Monster type. In this figure we again exclude 2A, 2B and 3C, which exist for all (α, β), and we additional exclude $3\mathrm{A}(\alpha, \beta)$ which exists for all (α, β) with $\alpha \neq \frac{1}{2}$. The Highwater algebra $\mathcal{H}$ appears in the picture in the correct place $(2, \frac{1}{2})$, however the algebra $6\mathrm{Y}(\frac{1}{2}, 2)$ is, unfortunately, out of the picture.

From Figure 6.2, we can now see some additional interesting points where several algebras exist at the same time.

Problem 6.5.9 Investigate the algebras and associated Miyamoto groups occurring at the intersection points in Figure 6.2.

6.5.6 Quotients of the Highwater algebra and its cover

In [20], Franchi, Mainardis and McInroy classify the ideals of the Highwater algebra and its cover. In fact, they introduce a cover $\hat{\mathcal{H}}$ in all characteristics, which is necessarily not of Monster type except in characteristic 5. They then use this to produce a unified classification for the ideals, which can then be specialised to the Highwater algebra, or its cover in characteristic 5.

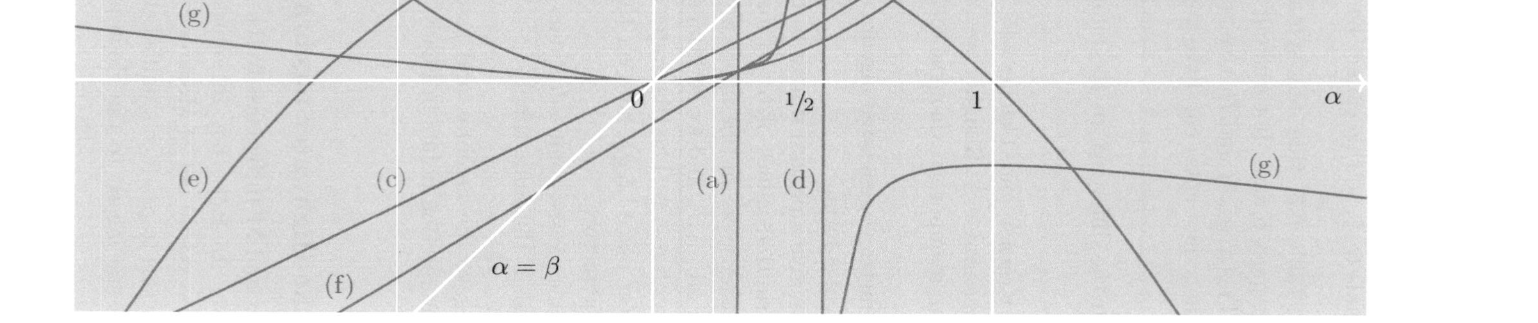

(a) $4A(\frac{1}{2}, \beta)$
(b) $4B(\alpha, \frac{\alpha}{2})$
(c) $4J(2\beta, \beta)$, $6J(2\beta, \beta)$
(d) $4Y(\frac{1}{2}, \beta)$
(e) $4Y(\alpha, \frac{1-\alpha^2}{2})$
(f) $5A(\alpha, \frac{5\alpha-1}{8})$
(g) $6A(\alpha, \frac{-\alpha^2}{4(2\alpha-1)})$
(h) $IY_3(\alpha, \frac{1}{2}, \mu), IY_5(\alpha, \frac{1}{2})$

Figure 6.2 *Symmetric 2-generated axial algebras of Monster type* (α, β)

Theorem 6.5.10 *Every ideal of the Highwater algebra $\mathcal{H}$ and its cover $\hat{\mathcal{H}}$ is invariant under the full automorphism group. In particular, every quotient is a symmetric 2-generated algebra of Monster type $(2, \frac{1}{2})$.*

For $\mathcal{H}$, they show that every ideal has finite codimension – so there are no infinite dimensional quotients of $\mathcal{H}$. In particular, we say that an ideal I has *axial codimension* D if the quotient $\mathcal{H}/I$ has axial dimension D. That is, if the images of the axes in quotient span a D-dimensional subspace. The results can now be stated very succinctly with the use of one definition.

Definition 6.5.11 A tuple $(\alpha_0, \dots, \alpha_D) \in \mathbb{F}^{D+1}$ is of *ideal type* if

(1) $\alpha_0 \neq 0 \neq \alpha_D$
(2) $\sum_{i=0}^{D} \alpha_i = 0$
(3) There exists $\varepsilon = \pm 1$, such that for all $i = 0, \dots, D$, $\alpha_i = \varepsilon \alpha_{D-i}$.

Theorem 6.5.12 ([20]) *For each $D \in \mathbb{N}$, there is a bijection between the ideal type tuples, up to scalars, and the minimal ideals of $\mathcal{H}$ of axial codimension D given by*

$$(\alpha_0, \dots, \alpha_D) \mapsto \left(\sum_{i=0}^{D} \alpha_i a_i \right) \trianglelefteq \mathcal{H}$$

In particular, it is clear that $(a_0 - a_D) \trianglelefteq \mathcal{H}$, for all $D \in \mathbb{N}$, and hence $\mathcal{H}$ has a quotient with D axes for every $D \in \mathbb{N}$. In addition, the above theorem also leads to an explicit basis for a general ideal and hence for the quotient.

The ideals for $\hat{\mathcal{H}}$ are only slightly more complicated. Recall that for $\hat{\mathcal{H}}$, there is a distinguished ideal J such that $\hat{\mathcal{H}}/J \cong \mathcal{H}$. Every ideal $I \trianglelefteq \hat{\mathcal{H}}$ has either finite codimension in $\hat{\mathcal{H}}$ and is described by the above theorem, or it is contained in the ideal J and has finite codimension in J. The paper [20] contains a theorem on the ideals contained in J and explicit bases, that is similar to Theorem 6.5.12.

6.5.7 Non-symmetric 2-generated axial algebras

The case of non-symmetric 2-generated algebras of Monster type (α, β) remains open. However, some examples are known which we will briefly discuss.

Example 6.5.13 One of the series of the algebras built by Joshi using the double axis construction (see Section 6.6.2) is called $Q_2(2\beta, \beta)$ and this is a 1-parameter family of 2-generated algebras of Monster type $(2\beta, \beta)$. It contains four axes in total, two of which are single axes and the other two are double axes. The algebra is non-symmetric as single and double axes have different lengths with respect to the unique Frobenius form.

Further non-symmetric examples come from a generalised Norton-Sakuma algebra, which was noticed by McInroy.

Example 6.5.14 For $\alpha = -1$, the algebra $A = 4\mathrm{B}(-1, \frac{1}{2})$ has a 2-dimensional radical $R \trianglelefteq A$. Furthermore, every 1-dimensional subspace of R is also an ideal. It can be seen that the automorphism σ switching the two generating axes a_0 and a_1 of A acts on R with eigenvalues 1 and -1. Hence every 1-dimensional space in R that is not one of the two eigenspaces is not invariant under σ and hence it gives a quotient of A that is not a symmetric algebra.

We call a 2-generated algebra $A = \langle\langle a, b \rangle\rangle$ *skew* if $a^{\mathrm{Miy}(A)}$ and $b^{\mathrm{Miy}(A)}$ have different lengths. The two examples above are not skew. It is clear that a skew algebra cannot be symmetric. This idea ties into the concept of an axet which we will discuss in Section 6.6.3 and we will see there Example 6.6.13 by Turner of a skew 2-generated algebra of Monster type.

Problem 6.5.15 Classify the non-symmetric 2-generated axial algebras of Monster type.

Needless to say, it is difficult to use the negative condition that the algebra is non-symmetric in a meaningful way. In the problem above, we mean that we wish to classify the 2-generated algebras without relying on the symmetry condition. There has been some work in this direction by Franchi, Mainardis and Shpectorov in [21] and they complete the classification in the case $\alpha = 2\beta$ in [22].

6.6 Larger algebras

Beyond the 2-generated case, very little is known about axial algebras of Monster type (α, β). In particular, we are far from knowing all the examples and hence not in a position to formulate a conjecture for the complete list. However, we do have some recent constructions of examples, both theoretical and computational, that demonstrate that there is a very rich variety of such algebras.

We note that all of the currently known examples of algebras of Monster type admit a Frobenius form. Hence we can at least formulate the following conjecture.

Conjecture 6.6.1 *Every axial algebra of Monster type (α, β) admits a Frobenius form.*

Let us now turn to some families of examples.

6.6.1 Split spin factor algebras

An interesting new family of algebras of Monster type arose when trying to understand the symmetric 2-generated algebras discovered by Yabe. In particular, M[c]Inroy and Shpectorov found in his algebra $\mathrm{III}(\alpha, \frac{1}{2}, \delta)$ (denoted in Theorem 6.5.6 as $\mathrm{IY}_3(\alpha, \frac{1}{2}, \mu)$) a different, more natural basis, and the way multiplication works in this basis readily generalises to any number of generators.

Definition 6.6.2 ([54]) Let $\mathbb{F}$ be a field of characteristic other than 2 and let E be a quadratic space over $\mathbb{F}$, i.e. a vector space over $\mathbb{F}$ endowed with a symmetric bilinear form $b\colon E \times E \to \mathbb{F}$. Choose $\alpha \in \mathbb{F} \setminus \{0, \frac{1}{2}, 1\}$. The *split spin factor algebra* is the commutative algebra $S(b, \alpha) = \mathbb{F}z_1 \oplus \mathbb{F}z_2 \oplus E$ with multiplication given by

$$z_1^2 = z_1, \quad z_2^2 = z_2, \quad z_1 z_2 = 0,$$
$$z_1 e = \alpha e, \quad z_2 e = (1-\alpha)e,$$
$$ef = -b(e,f)z,$$

where $z = (\alpha - 2)\alpha z_1 + (\alpha - 1)(\alpha + 1)z_2$ and $e, f \in E$.

It is clear that $\mathbf{1} = z_1 + z_2$ is the identity of the algebra. The above construction is very reminiscent of the construction of the spin factor algebra in Example 6.2.6. There one extends a quadratic space by a 1-dimensional algebra $1A \cong \mathbb{F}$ containing the identity; here we instead extend E by a 2-dimensional algebra $2B \cong \mathbb{F} \oplus \mathbb{F}$ in such a way that the generating idempotents split the identity and act on E asymmetrically by different scalars.

The nonzero idempotents in $A = S(b, \alpha)$ can be listed explicitly. In addition to the obvious idempotents z_1, z_2, and $\mathbf{1}$, there are only two families of idempotents:

(a) $\frac{1}{2}(e + \alpha z_1 + (\alpha + 1)z_2)$,
(b) $\frac{1}{2}(e + (2 - \alpha)z_1 + (1 - \alpha)z_2)$,

where $e \in E$ with $b(e,e) = 1$.

It turns out that z_1 and z_2 are primitive idempotents of Jordan type α and $1-\alpha$ respectively and, similarly, the idempotents from (a) (respectively, (b)) are primitive axes of Monster type $(\alpha, \frac{1}{2})$ (respectively, $(1-\alpha, \frac{1}{2})$). The following result now follows:

Theorem 6.6.3 *The split spin factor algebra $S(b, \alpha)$ is an algebra of Monster type $(\alpha, \frac{1}{2})$ (and $(1-\alpha, \frac{1}{2})$) if and only if $\alpha \neq -1$ (respectively, 2) and E is spanned by vectors e with $b(e,e) = 1$.*

For the first type we select z_1 and family (a) as the set of generating axes and, symmetrically, for the second type we select z_2 and family (b).

When $\alpha = 2$ (respectively, -1), A is baric with the non-generator idempotents being contained in the radical. Otherwise, A is simple as long as b is non-degenerate.

If $\dim E \geq 2$, the automorphism group of A coincides with the orthogonal group $O(E, b)$ of the quadratic space E. When $\dim E = 1$, the eigenvalue $\frac{1}{2}$ disappears and so A is a 3-dimensional algebra of Jordan type α, assuming that E is spanned by a vector e with $b(e,e) = 1$. Namely, A is isomorphic to $3\mathrm{C}(\alpha)$ in this case.

Finally, to relate this back to Yabe's algebras, let E be a 2-dimensional vector space with basis $\{e, f\}$ where $b(e,e) = 1 = b(f,f)$. Note that the bilinear form is then entirely determined by $\mu := b(e,f)$.

Theorem 6.6.4 *If $\alpha \neq -1$ and $\mu \neq 1$, then $S(b, \alpha) \cong \mathrm{IY}_3(\alpha, \frac{1}{2}, \mu)$.*

6.6.2 Double axis construction

The double axis construction was introduced by Joshi in [40, 41] and appeared in print in a much developed form in [26].

Let $M = M_\eta(G, D)$ be a Matsuo algebra, where $\eta \neq \frac{1}{2}$. We say that the axes $a, b \in D$ are *orthogonal* if $ab = 0$, or equivalently a and b commute as involutions in G. If a and b are orthogonal, then $x := a + b$ is an idempotent in M and we will call such idempotents *double axes.*

Theorem 6.6.5 ([40]) *Every double axis satisfies the fusion law $\mathcal{M}(2\eta, \eta)$.*

The double axis x is not primitive as both a and b are in the 1-eigenspace of ad_x. However, double axes may be primitive in some proper subalgebra. Joshi in [40] explored this idea and constructed all primitive algebras generated by two axes, where these may be single (note that every $d \in D$ trivially satisfies $\mathcal{M}(2\eta, \eta)$ and so we call these *single axes*), or double axes and at least one of these is a double axis. Joshi found three new infinite families of examples.

Theorem 6.6.6 ([40]) *A primitive 2-generated subalgebra of M, where at least one of the generating axes is a double axis, is one of $Q_2(2\eta, \eta)$, $4\mathrm{J}(2\eta, \eta)$, or $6\mathrm{J}(2\eta, \eta)$.*

Note that the first of these is a non-symmetric example mentioned in Example 6.5.13 and the other two are symmetric and can be seen in the statement of Theorem 6.5.6.

It is easy to see that the two generators of the subalgebra in Theorem 6.6.6 involve no more than four elements of D. This means that the entire calculation happens within a small ambient group of 3-transpositions. The cases are organised in terms of the diagram on the set of generators of this ambient group. It was observed that the cases leading to the examples in Theorem 6.6.6 are those where the diagram admits a automorphism of order 2 that extends to an automorphism of the ambient group. We call this a *flip*. This leads to the following general construction.

Let $\sigma \in \mathrm{Aut}(G)$ of order 2 which preserves D. Then the group $S = \langle \sigma \rangle$ acts on D and the orbits are classified into three types: orbits of length 1, orbits $\{x, y\}$ of length 2 which are orthogonal (i.e. $xy = 0$) and orbits of length 2 which are non-orthogonal (i.e. $xy \neq 0$). Taking orbit sums, we obtain single axes and double axes from the first two, and vectors called *extras* from the last. All these vectors lie in the fixed subalgebra M^σ and in fact they form a basis. The subalgebra of M^σ generated by all single and double axes above is called the *flip subalgebra*.

Theorem 6.6.7 ([26]) *Every flip subalgebra is a (primitive) axial algebra of Monster type $(2\eta, \eta)$.*

As can be seen from Theorem 6.4.1, there are many classes of groups of 3-transpositions and they admit many different flips. Hence flip subalgebras give a very large family of examples.

Joshi determined all flips and flip subalgebras for $G = S_n$ and $G = Sp_{2n}(2)$ [40, 41]. Alsaeedi [1, 2] completed the $2^{n-1}{:}S_n$ case. Shi [70] analysed the case of $O^{\pm}_{2n}(2)$ and finally, in as yet unpublished work, Hoffman, Rodrigues and Shpectorov finished the $U_n(2)$ case.

Question 6.6.8 Are there any examples of axial algebras of Monster type $(2\eta, \eta)$ which cannot be constructed as a flip subalgebra?

Such an example would require at least three generators.

6.6.3 Axets and shapes

While computing axial algebras for concrete groups, one has to deal with a closed set of generating axes together with the action of the Miyamoto group on it and the τ-map. Different configurations of 2-generated subalgebras correspond to different *shapes*, introduced by Ivanov in the context of Majorana algebras. In essence, shapes are the cases which one needs to consider. However, this leads to a slight technical problem: we want to talk about shapes which collapse, i.e. there is no algebra with that shape, so shapes should not be defined in terms of an ambient algebra.

Axets

The following concept, introduced by McInroy and Shpectorov in [53], frees the set of axes X from the algebra. Recall that the τ-map relates the Miyamoto automorphisms to each axis.

Definition 6.6.9 Let S be an abelian group. An S-*axet* (G, X, τ) is a G-set X together with a map $\tau\colon X \times S \to G$ (written $\tau_x(s) = \tau(x, s)$), such that

(1) $\tau_x(s) \in G_x$

(2) $\tau_x(ss') = \tau_x(s)\tau_x(s')$

(3) $\tau_{xg}(s) = \tau_x(s)^g$

for all $s, s' \in S$, $x \in X$ and $g \in G$.

We call elements of X *axes*. Clearly, this definition mimics how, in an axial algebra, the Miyamoto group $G = \mathrm{Miy}(A)$ acts on the closed set of axes X, linked via the τ-map. In particular, every axial algebra A contains an axet $X(A)$.

Note that, in this survey, we are mostly interested in C_2-graded algebras, so here we will take $S = C_2$ and talk about C_2-axets.

Even though an axet is just a combinatorial and group-theoretic object, it still carries much information about the axes in an axial algebra. For example, the Miyamoto group of an axet (G, X, τ) is defined in the obvious way as $\mathrm{Miy}(X) = \langle \mathrm{Im}(\tau) \rangle \trianglelefteq G$. We can also talk about subaxets and, in particular, the subaxet $\langle Y \rangle$ generated by a subset $Y \subseteq X$.

Example 6.6.10 Let $n \in \mathbb{N} \cup \{\infty\}$ and consider the natural action of $G = D_{2n}$ on the regular n-gon (or $\mathbb{Z}$ if $n = \infty$). Let X be the set of vertices and τ_x be the reflection in x, for each $x \in X$. We call $X(n) := (G, X, \tau)$ the *n-gonal C_2-axet.*

It is not too difficult to see that if n is odd, then there is one orbit of axes under $\mathrm{Miy}(X)$ and if n is even, or infinite, then there are two orbits of equal size. Moreover, if n is even, then opposite vertices in the n-gon give the same reflection and hence the same Miyamoto involution. This observation forms the basis of our next example.

Example 6.6.11 Consider the axet $X = X(4k)$. The Miyamoto group $\mathrm{Miy}(X) \cong D_{4k}$ has two orbits X_1 and X_2 on X, each of length $2k$. In particular, each vertex in the $4k$-gon is in the same orbit as its opposite vertex. We now 'fold' one of these orbits, X_1 say, by identifying opposite vertices in X_1. Let Y be the set corresponding to pairs of a vertex and its opposite vertex in X_1. Finally, let $X' = Y \cup X_2$ and let $\tau' \colon X' \to D_{4k}$ be the obvious adjustment of the map τ. Then $X'(3k) := (D_{4k}, X', \tau')$ is a C_2-axet of size $3k$ with two orbits of axes, one of size k and the other of size $2k$.

Theorem 6.6.12 ([53]) *Let $X = \langle a, b\rangle$ be a 2-generated C_2-axet with n axes, where $n \in \mathbb{N} \cup \{\infty\}$. Then $(\mathrm{Miy}(X), X, \tau)$ is isomorphic (up to the kernel of the action) to*

(1) $X(n)$, or
(2) $X'(3k)$, where $n = 3k$ and $k \in \mathbb{N}$.

Note that we have not used the fusion law, or any algebraic information here, and yet we can classify the possibilities for the axets of any 2-generated axial algebra with an arbitrary C_2-graded fusion law. Clearly, the second axet $X'(3k)$ cannot arise from a symmetric 2-generated algebra. In fact, all the examples of a 2-generated algebra we have seen here so far have the $X(n)$ axet. One can ask whether the *skew* axet $X'(3k)$ can arise from a 2-generated axial algebra.

The following observation by Turner gives a positive answer and hints at a possible class of very interesting algebras.

Example 6.6.13 Let $A = 3\mathrm{C}(\eta)$ with axes a, b, c. If $\eta \neq -1$, then it is not difficult to see that A has an identity $\mathbf{1} := \frac{1}{\eta+1}(a + b + c)$. Now, $a' := \mathbf{1} - a$ is an idempotent and it turns out that it is an axis of Jordan type $1 - \eta$. Consider A as an axial algebra of Monster type $(\alpha, 1 - \alpha)$ with the two generators a' and b. Then, $\tau_{a'}$ switches b and c while τ_b

and τ_c are trivial. Hence we see that $A = \langle\langle a', b \rangle\rangle$ is an axial algebra of Monster type $(\alpha, 1-\alpha)$ with a skew axet $X'(1+2)$.

However, this example is somewhat degenerate and no further examples are known.

Problem 6.6.14 Are there any other 2-generated axial algebras of Monster type (α, β) with a skew axet? If so, classify them.

McInroy and Shpectorov [52] also classified the *binary* axets, which are the axets where all orbits of the Miyamoto group have size 2. This classification is in terms of a digraph on the set of orbits.

Shapes

Let $\mathcal{F}$ be a fusion law. A *shape* on an axet (G, X, τ) is an assignment of a 2-generated $\mathcal{F}$-axial algebra A_Y to each 2-generated subaxet Y of X, so that Y is naturally identified with $X(A_Y)$. This assignment should satisfy the obvious consistency requirements. If we have inclusion between two subaxets $Y \subseteq Z$, then we should have an inclusion of $A_Y \subseteq A_Z$. Also, if Y and Z are two conjugate subaxets of X, where $Y^g = Z$, then we should have a suitable isomorphism $\theta_g \colon A_Y \to A_Z$. The technical details of this are given in [53], but to give the reader an idea: a shape is roughly an amalgam of small (2-generated) algebras organised on an axet.

Just like every axial algebra contains an axet, every axial algebra includes a shape on its axet. Thus shapes classify different axial algebras. Note that shapes rely on us knowing every 2-generated $\mathcal{F}$-axial algebra and this is why the investigation of a class of axial algebras starts by classifying the 2-generated examples.

As mentioned in the beginning, not every shape arises in an $\mathcal{F}$-axial algebra. In this case, we say that the shape *collapses*. This can be made more precise with the definition of a universal algebra [31, 62, 21]. In essence, for a fusion law $\mathcal{F}$ and a number of generators k, there exists a universal k-generated $\mathcal{F}$-axial algebra A, which is defined not over a field, but a certain commutative ring. The field versions of the algebra are obtained by factoring the coefficient ring by a prime ideal. Therefore, the spectrum of the coefficient ring gives us a variety of all k-generated $\mathcal{F}$-axial algebras. The shape is an invariant on this variety. For the fusion law $\mathcal{M}(\frac{1}{4}, \frac{1}{32})$, there are only eight 2-generated algebras, the Norton-Sakuma algebras, and so the shape is a discrete invariant and it is constant on every component.

Initial calculations with small groups and the Monster fusion law $\mathcal{M}(\frac{1}{4}, \frac{1}{32})$, would typically output a single algebra for each shape, with a small number of shapes collapsing. This suggested that the variety was perhaps 0-dimensional. However, Whybrow [73] gave an example of a 1-parameter family of 3-generated $\mathcal{M}(\frac{1}{4}, \frac{1}{32})$-axial algebras all with the same shape. This means that components of positive dimensions exist. This shifts the focus from individual algebras to components which are parametrised families of algebras.

While the initial calculations produced almost no collapsing shapes, subsequent systematic work of Khasraw [43] and Khasraw, McInroy and Shpectorov [44], seems to imply that the majority of the shapes for $\mathcal{M}(\frac{1}{4}, \frac{1}{32})$ collapse. In [53], for a family of shapes with the general $\mathcal{M}(\alpha, \beta)$ fusion law it is in fact shown that the non-collapsing shapes are a rare exception.

Problem 6.6.15 Find conditions to impose on shapes to eliminate the majority of collapsing shapes while preserving shapes leading to non-trivial algebras.

6.6.4 Algorithms

Seress developed an algorithm [68] and a GAP program for computing algebras using the 'resurrection' trick [39]. This program was very successful for small groups as long as the algebra was 2-closed (spanned by axes and products of two axes). Later the same algorithm was redeveloped and expanded by Pfeiffer and Whybrow [59], with the 2-closed restriction partly removed, and it is implemented in GAP [60]. This was used, for example, in the project of Mamontov, Staroletov and Whybrow [47] to classify minimal 3-generated axial algebras of Monster type $(\frac{1}{4}, \frac{1}{32})$.

An alternative approach based on the idea of expansions, which overcomes any m-closed restriction, was originally explored in GAP by Shpectorov and then properly developed by Rehren. The algorithm was developed much further for a general fusion law by McInroy and Shpectorov in [49] and a MAGMA implementation by McInroy is available in [50]. This program was able to find much larger algebras and tens of thousands of shapes have been analysed with several hundred non-trivial algebras found.

The main computational bottleneck is working with large-dimensional G-modules. This can be tackled in two ways. Firstly, by working with the modules decomposed into homogeneous components and preserving these when taking tensor and symmetric products; this means that taking submodules becomes a few much smaller linear algebra problems rather than a large G-module problem. Secondly, by doing the entire calculation over several finite fields and using a Chinese-remainder-theorem-like argument to recover the result for characteristic 0. Implemented either or both of these is an interesting computational problem and would allow us to explore much larger algebras.

Problem 6.6.16 Can these ideas lead to an explicit computed implementation of the Griess algebra?

6.7 Further areas of research

In this section, we briefly discuss some additional research in this area.

6.7.1 Algebras for simple groups

One of the motivations for group theorists to consider axial algebras is to have a unified approach to simple groups that arise from these algebras. Classical groups and G_2 are realised as automorphism groups of Jordan algebras, whereas the Monster and many of its subgroups and sections (subquotients) act on the Griess algebra and its subalgebras. So, within the class of axial algebras of Monster type, we have algebras for many simple groups. It would be very interesting to find axial realisations for the remaining simple groups.

Norton [58] proposed a class of non-associative algebras coming from 3-transposition groups, which are similar to Matsuo algebras. In [5], Cameron, Goethals and Seidel reinterpreted Norton's examples in terms of the association scheme coming from a permutation action of a group G. This class of algebras are now called Norton algebras. Note that Norton algebras are commutative non-associative algebras and they are typically realised from an irreducible component of the permutation module of G.

If one wants to have an identity in the algebra, then one can extend the Norton algebra using a cocycle (a G-invariant bilinear form). For example, Norton himself noticed that the Griess algebra can be obtained in this way. The Norton construction spurred interest in algebras for simple groups, in particular for sporadic groups. Algebras were constructed for $3.F_{24}'$ by Norton [58], A_6, A_7, $\Omega_6^-(3)$, $\Omega_7(3)$ and Fi_{24}' by Smith [69] and

Kitazume [42], J_3 [24] and ON [25] by Frohardt, and HN [65] and B [64] in positive characteristic by Ryba. However, this was before the axial language was introduced and so the axial structure for these algebras has not been investigated.

Problem 6.7.1 Explore the above algebras from the axial point of view and construct interesting axial algebras for other groups.

There has been some progress in this direction. Recently, Shumba in [57], investigated the axial structure of two such algebras of dimensions $1 + 77$ and $1 + 780$ for the sporadic groups HS and Suz, respectively. He found the fusion law for specific idempotents and showed that these algebras are axial algebras whose Miyamoto group recovers the original sporadic group. (Strictly speaking these are axial decomposition algebras [13] as to have a grading, we need to split one of the eigenspaces into two separate parts.) McInroy, Peacock and Van Couwenberghe have investigated the axial structure for J_3, which has nilpotent 'axes' and has a $C_3 \times C_3$-grading, and for Ly, which is suspected to have a C_3-grading.

In [72] and [16], De Medts and Van Couwenberghe gave a general procedure for constructing a commutative non-associative algebra for any simply-laced Chevalley group, by looking at a constituent of the symmetric square of the Lie algebra. They considered these from an axial point of view, calculating the fusion law in all cases and showing that they are axial decomposition algebras. They concentrated, in particular, on the example for E_8 of dimension $1 + 3875$. In this case, the algebra had been constructed previously from a different point of view by Garibaldi and Guralnick [27], who were interested in groups stabilising a polynomial on a vector space.

One deficiency in all the algebras in this section, is that the fusion laws tend not to be 'nice' for algebras constructed in this way. They typically have more eigenvalues than for the Jordan, or Monster fusion laws and some eigenspaces split. It should be noted, however, that they are typically constructed from a single irreducible module and we can instead consider algebras built in a similar way from a sum of several irreducibles. In fact, many of the known algebras of Monster type are far from being irreducible.

Question 6.7.2 Develop methods for constructing algebras on reducible modules. How can a prescribed fusion law be used to restrict possibilities for the product?

6.7.2 Automorphism groups and finding all axes

From the axial point of view, it is interesting to find all axes in an axial algebra. On the other hand, from the group-theoretic point of view, it is interesting to find the full automorphism group of an algebra. It turns out that these questions are intimately related for axial algebras. Indeed, the automorphism group has a natural permutation action on the full set of axes and conversely the Miyamoto group of the full set of axes is a normal subgroup in the automorphism group.

For simple Jordan algebras the full automorphism groups are the classical groups (and the group G_2) and it is known that the Monster is the full automorphism group of the Griess algebra and that the 2A-axes are the full set of $\mathcal{M}(\frac{1}{4}, \frac{1}{32})$-axes. Castillo-Ramirez found all the idempotents and calculated the automorphism groups for the Norton-Sakuma algebras [9] and also for two algebras, of dimensions 6 and 9, with the Miyamoto group S_4 in [6]. However, in general, the number of all idempotents in A is either infinite, or finite but growing exponentially, being generically $2^{\dim(A)}$. Currently there is a project underway with McInroy, Shpectorov and Shumba to find computationally the automorphism groups of the remaining algebras for S_4, which have dimension up to 25. The method involves reducing the problem from A to the smaller subalgebra $A_0(a)$ for an axis a.

Problem 6.7.3 Develop computational methods for finding axes and the full automorphism groups of axial algebras.

Combining such methods with group-theoretic approaches, we can potentially do much larger algebras – the Griess algebra was completed using group theory.

Another reason why finding the full automorphism group is important is that in the list of known examples of algebras, for example in [49, Tables 4 and 5] and [59, Table 4], the dimensions tend to repeat. This suggests that there may be several seemingly different algebras which are, in fact, the same.

Suppose that two axial algebras $A = (A, X)$ and $B = (B, Y)$ are isomorphic via an isomorphism $\varphi \colon A \to B$. Then in B, in addition to the closed set of axes (axet) Y, there is also the closed set of axes $\varphi(X)$. Either one set is contained in the other, which is typically easy to check, or neither of them is the full set of axes. In this case, B has a larger set of axes and perhaps a larger automorphism group. In most cases, there are not additional axes, however recently Alharbi[9] found three series of Matsuo algebras, with $\eta \neq \frac{1}{2}$, which contain additional axes and have a larger automorphism group (also being a 3-transposition group).

The above discussion of possibly isomorphic algebras A and B equally applies where the axes from X and Y obey different fusion laws. This situation does indeed happen. A simple example is $3\mathrm{C}(\alpha)$, $\alpha \neq -1$. In addition to the three axes a, b, c of Jordan type α, we have three axes $\mathbf{1} - a, \mathbf{1} - b, \mathbf{1} - c$ of Jordan type $1 - \alpha$. The following example noticed by McInroy extends this idea.

Example 6.7.4 Let $A = 4\mathrm{B}(\alpha, \frac{\alpha^2}{2})$ with axes $a_0, \dots, a_3$. It can be seen that $\langle\langle a_0, a_2 \rangle\rangle \cong 3\mathrm{C}(\alpha) \cong \langle\langle a_1, a_3 \rangle\rangle$. When $\alpha \neq -1$, let $\mathbf{1}_+$ be the identity in $\langle\langle a_0, a_2 \rangle\rangle$ and $\mathbf{1}_-$ be the identity in $\langle\langle a_1, a_3 \rangle\rangle$. Define $a_i' := \mathbf{1}_+ - a_i$, if $i = 0, 2$, and $a_i' := \mathbf{1}_- - a_i$, if $i = 1, 3$. Then the idempotents a_i' are in fact axes of Monster type $(1 - \alpha, \frac{\alpha(2-\alpha)}{2})$. Setting $\gamma = 1 - \alpha$, we check that $4\mathrm{B}(\alpha, \frac{\alpha^2}{2}) = \langle\langle a_0', a_1' \rangle\rangle \cong 4\mathrm{Y}(\gamma, \frac{1-\gamma^2}{2})$. So the two algebras $4\mathrm{B}(\alpha, \frac{\alpha^2}{2})$ and $4\mathrm{Y}(\gamma, \frac{1-\gamma^2}{2})$ are isomorphic as algebras, but have different fusion laws and disjoint sets of axes.

6.7.3 Code algebras

Code algebras were introduced by Castillo-Ramirez, McInroy and Rehren in [8]. They axiomatise some properties found in code VOAs in a similar way that axial algebras do for OZ-type VOAs. Code VOAs are an important class of VOAs, where a binary code controls the representation theory.

Definition 6.7.5 Let $C \subseteq \mathbb{F}_2^n$ be a binary linear code of length n and $\mathbb{F}$ a field, where $a, b, c \in \mathbb{F}$. Set $C^* := C - \{\mathbf{0}, \mathbf{1}\}$. The *code algebra* A_C is the algebra on $\bigoplus_{i=1}^n \mathbb{F} t_i \oplus \bigoplus_{\alpha \in C^*} e^\alpha$ with multiplication given by

$$t_i t_j = \delta_{i,j} t_i$$

$$t_i e^\alpha = \begin{cases} a\, e^\alpha & \text{if } \alpha_i = 1 \\ 0 & \text{if } \alpha_i = 0 \end{cases}$$

$$e^\alpha e^\beta = \begin{cases} b\, e^{\alpha+\beta} & \text{if } \alpha \neq \beta, \beta^c \\ c \sum_{i \in \mathrm{supp}(\alpha)} t_i & \text{if } \alpha = \beta \\ 0 & \text{if } \alpha = \beta^c \end{cases}$$

[9] Alharbi is currently a PhD student at the University of Birmingham and this is part of his PhD project.

The t_i are axes of Jordan type b, however they do not generate the algebra. In [8, 7] Castillo-Ramirez, McInroy and Rehren give a construction, called the s-map, which gives an idempotent $s(D, v)$ from a subcode $D \subseteq C$ and a vector $v \in \mathbb{F}_2^n$. The axial properties of this idempotent can be difficult to analyse, but this can be done for the smallest such subcodes $\langle \alpha \rangle$ and we call such idempotent *small idempotents.* The small idempotents whose fusion law is C_2-graded are classified in [7] and the resulting Miyamoto groups of the algebras are given in [10]. However, it turns out that these Miyamoto groups for small idempotents are all abelian.

The following example, shows that we do indeed get interesting groups for idempotents coming from the s-map construction for larger subcodes $D \subseteq C$.

Example 6.7.6 Let $C = H_8$ be the Hamming code of length 8. Then A_{H_8} is a 22-dimensional code algebra. We consider the s-map idempotents $s(C, v)$. Choosing $v \in \mathbb{F}_2^8 - C$ to have odd weight gives a set of eight mutually orthogonal axes of Jordan type b and choosing v to have even weight gives a different set of eight mutually orthogonal axes of Jordan type b. Both of these are different from the set $\{t_i : i = 1, \dots, 8\}$ of axes of Jordan type b. These 24 axes together generate the algebra and hence A_{H_8} is an axial algebra of Jordan type b and its Miyamoto group has shape $2^6{:}S_3$. Moreover, $\mathrm{Aut}(H_8)$ also acts on the algebra, so we see a group of automorphism of shape $2^6{:}(L_3(2) \times S_3)$.

For code VOAs, the example V_{H_8} coming from the Hamming code is very important. Miyamoto used V_{H_8} in a new construction of the Moonshine VOA $V^\natural$ and other VOAs [55]. Using the parameters $(a, b, c) = (\frac{1}{4}, \frac{1}{2}, 1)$ above, A_C is isomorphic to the weight 2 part of V_{H_8} and moreover, we see the full automorphism group $\mathrm{Aut}(V_{H_8}) = 2^6{:}(L_3(2) \times S_3)$ as a group of automorphisms of A_C.

References

[1] M. Alsaeedi, *A new series of axial algebras of Monster type $(2\eta, \eta)$ related to the extended symmetric groups*, PhD thesis, University of Birmingham, 2022.

[2] M. Alsaeedi, A new series of axial algebras of Monster type $(2\eta, \eta)$, *Arab. J. Math.* **10** (2021), no. 1, 11–20.

[3] R.E. Borcherds, Vertex algebras, Kac-Moody algebras, and the Monster, *Proc. Natl. Acad. Sci. USA*, **83** (1986), 3068–3071.

[4] F. Buekenhout, La géometrie des groupes de Fischer, preprint, Univ. Libre de Bruxelles.

[5] P.J. Cameron, J.-M. Goethals, J.J. Seidel, The Krein condition, spherical designs, Norton algebras and permutation groups, *Nederl. Akad. Wetensch. Indag. Math.* **40** (1978), no. 2, 196–206.

[6] A. Castillo-Ramirez, Associative subalgebras of low-dimensional Majorana algebras, *J. Algebra* **421** (2015), 119–135.

[7] A. Castillo-Ramirez and J. M^c^Inroy, Code algebras which are axial algebras and their $\mathbb{Z}_2$-grading, *Israel J. Math.* **233** (2019), 401–438.

[8] A. Castillo-Ramirez, J. McInroy and F. Rehren, Code algebras, axial algebras and VOAs, *J. Algebra* **518** (2019), 146–176.

[9] A. Castillo-Ramirez, Idempotents of the Norton-Sakuma algebras, *J. Group Theory* **16** (2013), no. 3, 419–444.

[10] A. Castillo-Ramirez and J. McInroy, Miyamoto groups of code algebras, *J. Pure Appl. Algebra* **225** (2021), no. 6, 19 pages.

[11] J.H. Conway, A simple construction for the Fischer-Griess monster group, *Invent. Math.* **79** (1985), no. 3, 513–540.

[12] H. Cuypers, J.I. Hall, *The 3-transposition groups with trivial center*, *J. Algebra* **178** (1995), 149–193.

[13] T. De Medts, S. F. Peacock, S. Shpectorov and M. Van Couwenberghe, Decomposition algebras and axial algebras, *J. Algebra* **556** (2020), 287–314.

[14] T. De Medts, F. Rehren, Jordan algebras and 3-transposition groups, *J. Algebra* **478** (2017), 318–340.

[15] T. De Medts, L. Rowen and S. Segev, Primitive 4-generated axial algebras of Jordan type, *Proc. Amer. Math. Soc.* **152** (2024), no. 2, 537–551.

[16] T. De Medts, M. Van Couwenberghe, Non-associative Frobenius algebras for simply laced Chevalley groups, *Trans. Amer. Math. Soc.* **374** (2021), no. 12, 8715–8774.

[17] B. Fischer, Finite groups generated by 3-transpositions, *Invent. Math.* **13** (1971), 232–246.

[18] D. Fox, The commutative nonassociative algebra of metric curvature tensors, *Forum Math. Sigma* **9** (2021), Paper No. e79, 48 pages.

[19] C. Franchi and M. Mainardis, Classifying 2-generated symmetric axial algebras of Monster type, *J. Algebra* **596** (2022), 200–218.

[20] C. Franchi, M. Mainardis and J. McInroy, Quotients of the Highwater algebra, *J. Algebra* **640** (2024), 432–476.

[21] C. Franchi, M. Mainardis and S. Shpectorov, 2-generated axial algebras of Monster type, *arXiv*:2101.10315, 22 pages, Jan 2021.

[22] C. Franchi, M. Mainardis and S. Shpectorov, 2-generated axial algebras of Monster type $(2\beta, \beta)$, *J. Algebra* **636** (2023), 123–170.

[23] C. Franchi, M. Mainardis and S. Shpectorov, An infinite-dimensional 2-generated primitive axial algebra of Monster type, *Ann. Mat. Pura Appl.* (2021), 1279–1293.

[24] D. Frohardt, A trilinear form for the third Janko group, *J. Algebra* **83** (1983), no. 2, 349–379.

[25] D. Frohardt, Toward the construction of an algebra for O'Nan's group, *Proceedings of the Rutgers group theory year,* 1983–1984 (*New Brunswick, N.J.,* 1983–1984), 107–110, Cambridge Univ. Press, Cambridge, 1985.

[26] A. Galt, V. Joshi, A. Mamontov, S. Shpectorov and A. Staroletov, Double axes and subalgebras of Monster type in Matsuo algebras, *Comm. Algebra* **49**, no. 10, 4208–4248.

[27] S. Garibaldi, R.M. Guralnick, Simple groups stabilizing polynomials, *Forum Math. Pi* **3** (2015), e3, 41 pages.

[28] I. Gorshkov, A. Staroletov, On primitive 3-generated axial algebras of Jordan type, *J. Algebra* **563** (2020), 74–99.

[29] R.L. Griess. The friendly giant. *Invent. Math.* **69** (1982), 1–102.

[30] J.I. Hall, *The general theory of 3-transposition groups*, Math. Proc. Camb. Phil. Soc. **114** (1993), 269–294.

[31] J.I. Hall, F. Rehren and S. Shpectorov, Universal axial algebras and a theorem of Sakuma, *J. Algebra* **421** (2015), 394–424.

[32] J.I. Hall, F. Rehren, S. Shpectorov, Primitive axial algebras of Jordan type, *J. Algebra* **437** (2015), 79–115.

[33] J.I. Hall, Y. Segev and S. Shpectorov, Miyamoto involutions in axial algebras of Jordan type half, *Israel J. Math.* **223** (2018), no. 1, 261–308.

[34] J.I. Hall, Y. Segev and S. Shpectorov, On primitive axial algebras of Jordan type, *Bull. Inst. Math. Acad. Sin. (N.S.)*, **13** (2018), no. 4, 397–409.

[35] J.I. Hall, S. Shpectorov, The spectra of finite 3-transposition groups. *Arab. J. Math.* **10** (2021), no. 3, 611–638.

[36] A.A. Ivanov, The future of Majorana theory, *Group theory and computation*, 107–118, Indian Stat. Inst. Ser., Springer, Singapore, 2018.

[37] A.A. Ivanov, The future of Majorana theory II, *Algebraic Combinatorics and the Monster Group*, London Mathematical Society Lecture Notes,*to appear.*

[38] A.A. Ivanov, *The Monster Group and Majorana Involutions*, Cambridge Tracts in Mathematics **176**, Cambridge University Press, 2009.

[39] A.A. Ivanov, D.V. Pasechnik, Á. Seress, S. Shpectorov, Majorana representations of the symmetric group of degree 4, *J. Algebra* **324** (2010), 2432–2463.

[40] V. Joshi, *Double Axial Algebras*, MRes Thesis, University of Birmingham, 2018.

[41] V. Joshi, *Axial algebras of Monster type* $(2\eta, \eta)$, PhD thesis, University of Birmingham, 2020.

[42] M. Kitazume, The Conway-Norton algebras for $\Omega^-(6,3)$, $\Omega(7,3)$, F'_{24}, and their full automorphism groups, *Invent. Math.* **88** (1987), no. 2, 277–318.

[43] S.M.S. Khasraw, *M-Axial Algebras Related to 4-Transposition Groups*, PhD thesis, University of Birmingham, 2015.

[44] S.M.S. Khasraw, J. McInroy, S. Shpectorov, Enumerating 3-generated axial algebras of Monster type, *J. Pure Appl. Algebra* **226** (2022), no. 2, 21 pages.

[45] S.M.S. Khasraw, J. McInroy, S. Shpectorov, On the structure of axial algebras, *Trans. Amer. Math. Soc.* **373** (2020), 2135–2156.

[46] Y. Krasnov, V.G. Tkachev, Variety of idempotents in nonassociative algebras, *Topics in Clifford analysis – special volume in honor of Wolfgang Sprößig*, Trends Math., Birkhäuser/Springer, 2019, 405–436.

[47] A. Mamontov, A. Staroletov and M. Whybrow, Minimal 3-generated Majorana algebras, *J. Algebra* **524** (2019), 367–394.

[48] K. McCrimmon, *A taste of Jordan algebras*, Universitext, Springer-Verlag, New York, 2004.

[49] J. McInroy and S. Shpectorov, An expansion algorithm for constructing axial algebras, *J. Algebra* **550** (2020), 379–409.

[50] J. McInroy, Partial axial algebras – a MAGMA package, `https://github.com/JustMaths/AxialAlgebras`.

[51] J. McInroy, Axial algebras of Monster type – a MAGMA package, `https://github.com/JustMaths/AxialMonsterType`.

[52] J. McInroy and S. Shpectorov, Binary axets and the related algebras, *in preparation.*

[53] J. McInroy and S. Shpectorov, From forbidden configurations to a classification of some axial algebras of Monster type, *J. Algebra* **627** (2023), 58–105.

[54] J. McInroy and S. Shpectorov, Split spin factor algebras, *J. Algebra* **595** (2022), 380–397.

[55] M. Miyamoto, A new construction of the Moonshine vertex operator algebra over the real number field, *Ann. of Math.* **159** (2004), no. 2, 535–596.

[56] M. Miyamoto, Griess algebras and conformal vectors in vertex operator algebras, *J. Algebra* **179** (1996), no. 2, 523–548.

[57] T.M. Mudziiri Shumba, *Axial algebras for sporadic simple groups HS and Suz*, PhD thesis, University of KwaZulu-Natal, 2020.

[58] S. Norton, *F and other simple groups*, PhD thesis, University of Cambridge, 1975.

[59] M. Pfeiffer and M. Whybrow, Constructing Majorana Representations, *arXiv*:1803.10723, 19 pages, Mar 2018.

[60] M. Pfeiffer and M. Whybrow, Majorana Algebras – a GAP package, `https://github.com/MWhybrow92/MajoranaAlgebras`.

[61] F. Rehren, *Axial algebras*, PhD thesis, University of Birmingham, 2015.

[62] F. Rehren, Generalised dihedral subalgebras from the Monster, *Trans. Amer. Math. Soc.* **369** (2017), 6953–6986.

[63] L. Rowen, Y. Segev, Axes in non-associative algebras, *Turkish J. Math.* **45** (2021), no. 6, 2366–2381.

[64] A.J.E. Ryba, A natural invariant algebra for the Baby Monster group, *J. Group Theory* **10** (2007), no. 1, 55–69.

[65] A.J.E. Ryba, A natural invariant algebra for the Harada-Norton group, *Math. Proc. Cambridge Philos. Soc.* **119** (1996), no. 4, 597–614.

[66] S. Sakuma, 6-transposition property of τ-involutions of vertex operator algebras, *Int. Math. Res. Not. IMRN* 2007, no. 9, 19 pages.

[67] R.D. Schafer, *An introduction to nonassociative algebras*, Pure and Applied Mathematics, Vol. 22 Academic Press, New York-London, 1966.

[68] Á. Seress, Construction of 2-closed M-representations, *ISSAC* 2012–*Proceedings of the 37th International Symposium on Symbolic and Algebraic Computation*, ACM, New York, 2012, 311–318.

[69] S.D. Smith, Nonassociative commutative algebras for triple covers of 3-transposition groups, *Michigan Math. J.* **24** (1977), no. 3, 273–287.

[70] Y. Shi, *Axial Algebras of Monster Type* $(2\eta, \eta)$ *for Orthogonal Groups over* $\mathbb{F}_2$, MRes thesis, University of Birmingham, 2020.

[71] V.G. Tkachev, *The universality of one half in commutative nonassociative algebras with identities*, *J. Algebra* **569** (2021), 466–510.

[72] M. Van Couwenberghe, *Decomposition algebras and axial algebras*, PhD thesis, Ghent University, 2020.

[73] M. Whybrow, An infinite family of axial algebras, *J. Algebra* **577** (2021), 1–31.

[74] T. Yabe, Jordan Matsuo algebras over fields of characteristic 3, *J. Algebra* **513** (2018), 91–98.

[75] T. Yabe, On the classification of 2-generated axial algebras of Majorana type, *J. Algebra* **619** (2023), 347–382.

7

An introduction to the local-to-global behaviour of groups acting on trees and the theory of local action diagrams

Colin D. Reid[a] and Simon M. Smith[b]

Abstract

The primary tool for analysing groups acting on trees is Bass–Serre Theory. It is comprised of two parts: a decomposition result, in which an action is decomposed via a graph of groups, and a construction result, in which graphs of groups are used to build examples of groups acting on trees. The usefulness of the latter for constructing new examples of 'large' (e.g. nondiscrete) groups acting on trees is severely limited. There is a pressing need for new examples of such groups as they play an important role in the theory of locally compact groups. An alternative 'local-to-global' approach to the study of groups acting on trees has recently emerged, inspired by a paper of Marc Burger and Shahar Mozes, based on groups that are 'universal' with respect to some specified 'local' action. In recent work, the authors of this survey article have developed a general theory of universal groups of local actions, that behaves, in many respects, like Bass–Serre Theory. We call this the theory of local action diagrams. The theory is powerful enough to completely describe all closed groups of automorphisms of trees that enjoy Tits' Independence Property (P).

This article is an introductory survey of the local-to-global behaviour of groups acting on trees and the theory of local action diagrams. The article contains many ideas for future research projects.

[a] The University of Newcastle, School of Mathematical and Physical Sciences, Callaghan, NSW 2308, Australia.
colin@reidit.net

[b] Charlotte Scott Research Centre for Algebra, University of Lincoln, Lincoln, UK.
sismith@lincoln.ac.uk

7.1 Introduction

Actions on trees have a significant role in the general theory of finite and infinite groups. In finite group theory such actions are, for example, a natural setting for questions about vertex-transitive groups G of automorphisms of connected finite graphs. For such a pair $G \leq \mathrm{Aut}(\Gamma)$ with G nontrivial, the universal cover of Γ is an infinite regular tree T, and there is a natural projection π of T to Γ. The fundamental[1] group $\Pi(\Gamma, v)$ of Γ at any vertex v can be identified with a subgroup of $\mathrm{Aut}(T)$ in such a way that Γ can be identified with the quotient graph $\Pi(\Gamma, v)\backslash T$ and the lift $\tilde{G}$ of G along π can be identified with the quotient $\tilde{G}/\Pi(\Gamma, v)$. For a thorough description of this correspondence, see [18] for example.

In this context, important open problems in finite groups can translate into natural questions about groups acting on trees. Consider, for example, the well-known Weiss conjecture ([30]) due to Richard Weiss. In the language of finite groups, it states that there exists a function $f : \mathbb{N} \to \mathbb{N}$ such that if G is vertex-transitive and locally primitive[2] on a graph with finite valency k, then all vertex stabilisers satisfy $|G_v| < f(k)$. Equivalently, in the language of groups acting on trees, it states that for each $k \in \mathbb{N}_{\geq 3}$ the automorphism group of the k-regular tree T_k contains only finitely many conjugacy classes of discrete, locally primitive and vertex transitive subgroups.

The role played by actions on trees in the general theory of infinite groups is, of course, well-known. Much of our understanding of groups acting on trees comes from the celebrated Bass–Serre Theory, described in detail in Jean-Pierre Serre's book [21]. We give an introductory overview to Bass–Serre Theory in Section 7.2.1. Serre's theory concerns a group G *acting on a tree* T; by this we mean that G acts on T as a group of automorphisms of T. We will denote such an action by the pair (T, G), and will often identify (T, G) with its image in the group $\mathrm{Aut}(T)$ of automorphisms of T. In Bass–Serre theory, the algebraic structure of a group G acting on a tree T is 'decomposed' into pieces with the decomposition described via a combinatorial structure called a *graph of groups*. This graph of groups associated to the action (T, G) is essentially a vertex- and edge-coloured graph, where each colour

[1] We use the notation $\Pi(\Gamma, v)$ for the fundamental group, rather than the more common $\pi(\Gamma, v)$, because for us π will be reserved for projection maps.

[2] For a group G acting on a graph Γ, the *neighbours* of a vertex v are the vertices in Γ at distance one from v. Each vertex stabiliser G_v induces a permutation group on the neighbours of v; if every such induced permutation group has some permutational property $\mathcal{P}$ (e.g. transitive, primitive), we say that G is *locally* $\mathcal{P}$.

is in fact a group. The groups in this palette are called the *vertex groups* and *edge groups* of the graph of groups. Importantly, this decomposition process can be reversed, where one starts with a graph of groups Γ and then constructs the universal cover $\tilde{T}$ (which is a tree) of Γ and the fundamental group Π of Γ. The group Π acts on $\tilde{T}$ in such a way that its associated graph of groups is again Γ. This 'decomposition' of (T, G) in Bass–Serre Theory hinges on the observation that if Γ is the associated graph of groups for (T, G), and $\tilde{T}$ and Π are respectively the universal cover and fundamental group of Γ, then the actions (T, G) and $(\tilde{T}, \Pi)$ can be identified.

Two algebraic constructions, the HNN extension and the amalgamated free product, play a foundational role in this theory: the graph of groups of an HNN extension is a single vertex with a loop, and the graph of groups of an amalgamated free product is a pair of vertices joined by an edge. Intuitively, edges and loops in the graph of groups of the action give information about how the action decomposes into HNN extensions and amalgamated free products of the vertex groups.

Naturally, there are areas in which the usefulness of Bass–Serre Theory is limited. One of the most significant limitations can be seen when attempting to construct a group acting on a tree with certain desired properties, by first writing down a graph of groups. We discuss this in Section 7.3.1. Essentially Bass–Serre Theory can be used to construct an action on a tree with a given graph of groups, but the action itself cannot, in general, be fully controlled. This limitation is the source of the intractability of several famous conjectures, including the aforementioned Weiss Conjecture and the related Goldschmidt–Sims Conjecture (see [4], [11]). This limitation is particularly problematic when trying to construct nondiscrete actions on trees, and recent developments in the theory of locally compact groups have made the need to address this limitation more acute.

In the theory of locally compact groups, the structure theory of compactly generated locally compact groups is known to depend on the class $\mathscr{S}$ of nondiscrete, compactly generated, locally compact topologically simple groups; see Pierre-Emmanuel Caprace and Nicolas Monod's paper [5]. For a survey of the class $\mathscr{S}$, placing it in a broad mathematical context, see Caprace's survey paper [6]. Groups acting on trees are one of the main sources of examples of groups in $\mathscr{S}$. In fact the very first examples of nonlinear, nondiscrete, locally compact simple groups were constructed using groups acting on trees, by Jacques Tits (answering a question of J. P. Serre) in [29]. Tits used an independence condition he called Property (P) for groups acting on trees, and showed that for any group G of automorphisms of a tree T, if G has Property (P), fixes (setwise) no nonempty proper subtree of T and fixes no end of T, then the subgroup G^+ of G generated by arc stabilisers is (abstractly) simple[3] (see Section 7.2.2). As an immediate consequence, for the infinite regular tree T_n of (finite or infinite) valency n we have that $\mathrm{Aut}(T_n)^+$ is simple for $n \geq 3$. This group is nondiscrete and when n is finite it is easily seen to be compactly generated and locally compact (under the permutation topology).

The limitations of Bass–Serre Theory when constructing nondiscrete groups acting on trees can be avoided using a variety of techniques inspired by a 2000 article [4] by Marc Burger and Shahar Mozes; these techniques might collectively be known as *local-to-global constructions of groups acting on trees*. In their paper, Burger and Mozes take a permutation group $F \leq S_n$ of finite degree n and build a group $\mathbf{U}(F)$ of automorphisms of T_n that has *local action* F (or is *locally-F*); that is, vertex stabilisers in $\mathbf{U}(F)$ induce F on the set of neighbours of the stabilised vertex. For example, $\mathbf{U}(S_n) = \mathrm{Aut}(T_n)$. These *Burger–Mozes* groups $\mathbf{U}(F)$ enjoy Tits' Independence Property (P), and when F is transitive $\mathbf{U}(F)$ is 'universal' among subgroups of $\mathrm{Aut}(T_n)$ that have local action F; that is, $\mathbf{U}(F)$ contains an $\mathrm{Aut}(T_n)$-conjugate of every subgroup of $\mathrm{Aut}(T_n)$ that is locally-F. When F is transitive and generated by its point stabilisers, $\mathbf{U}(F)$ contains a simple subgroup $\mathbf{U}(F)^+$ of index 2. Since 2000, the majority of constructions of compactly generated, simple, locally compact groups have used the ideas of Tits [29] and Burger–Mozes [4].

In [25], the second author generalised the Burger–Mozes construction to *biregular trees* $T_{m,n}$, where $m, n \geq 2$ are (possibly infinite) cardinals and $T_{m,n}$ is the infinite tree in which vertices in one part of its bipartition have valency m and those in the other part have valency n. Given permutation groups $F_1 \leq S_m$ and $F_2 \leq S_n$, this generalisation is a group $\mathbf{U}(F_1, F_2) \leq \mathrm{Aut}(T_{m,n})$, called *the box product* of F_1 and F_2. The group $\mathbf{U}(F_1, F_2)$ has local actions F_1 (at vertices of $T_{m,n}$ in one part of the bipartition) and F_2 (at vertices in the other part of the bipartition); we say any such action on $T_{m,n}$ is *locally-(F_1, F_2)*. For $m \geq 3$

[3] We will always think of $\mathrm{Aut}(T)$ as a topological group under the *permutation topology*, defined below. A topological group is *topologically simple* if it has no nontrivial proper closed normal subgroups; it is *abstractly simple* if it has no nontrivial proper normal subgroups. When we write simple, we will always mean abstractly simple.

and $F \leq S_m$ it can be shown that the Burger–Mozes group $\mathbf{U}(F)$ is isomorphic as a topological group (but not as a permutation group) to $\mathbf{U}(F, S_2)$. The properties of $\mathbf{U}(F_1, F_2)$ mirror those of the Burger–Mozes groups: it has Tits' Independence Property (P), and when F_1 and F_2 are transitive, $\mathbf{U}(F_1, F_2)$ is 'universal' among subgroups of $\mathrm{Aut}(T_{m,n})$ that are locally-(F_1, F_2), in that it contains an $\mathrm{Aut}(T_{m,n})$-conjugate of every locally-(F_1, F_2) subgroup of $\mathrm{Aut}(T_{m,n})$. When F_1 and F_2 are generated by their point stabilisers and at least one group is nontrivial, $\mathbf{U}(F_1, F_2)$ is simple if and only if F_1 or F_2 is transitive.

The Burger–Mozes groups are automatically totally disconnected, locally compact (henceforth, "t.d.l.c.") and compactly generated groups because they are defined only on locally-finite trees. Despite admitting actions on non-locally finite trees, $\mathbf{U}(F_1, F_2)$ can still be a compactly generated t.d.l.c. group under mild conditions on F_1 and F_2. Indeed, if F_1 and F_2 are closed and compactly generated, with compact nontrivial point stabilisers and finitely many orbits, and either F_1 or F_2 is transitive (e.g. F_1 is closed, nonregular, subdegree-finite and primitive and F_2 is finite, nonregular and primitive), then $\mathbf{U}(F_1, F_2)$ is a nondiscrete compactly generated t.d.l.c. group. By taking F_1 to be infinite permutation representations of various Tarski–Ol'shanskiĭ Monsters (see [19]), the second author used this box product construction to show that there are precisely $2^{\aleph_0}$ isomorphism types of nondiscrete, compactly generated simple locally compact groups, answering a well-known open question; the analogous result for discrete groups was proved in 1953 by Ruth Camm in [8].

In light of the Burger–Mozes and box product constructions, the authors of this article wished to create a unified way of seeing $\mathbf{U}(F)$ and $\mathbf{U}(F_1, F_2)$, within a framework permitting yet more local actions to be specified. What emerged from this endeavour was something far deeper: a general theory of 'universal' groups of local actions, that behaves, in many respects, like Bass–Serre Theory. This work, which we call *the theory of local action diagrams*, is fully described in our paper [20].

To better understand this theory, let us define the (P)-*closure* of an action (T, G) of a group G on a tree T as being the smallest closed subgroup of $\mathrm{Aut}(T)$ with Tits Independence Property (P) that contains (T, G); we denote it $G^{(\mathrm{P})}$. That is, $G^{(\mathrm{P})}$ is the smallest closed subgroup of $\mathrm{Aut}(T)$ with Tits Independence Property (P) that contains the subgroup of $\mathrm{Aut}(T)$ induced by the action of G on T. If $G = G^{(\mathrm{P})}$ we say that G is (P)-*closed*. In the theory, any group G acting on a tree T is 'decomposed' into its local actions with the decomposition described via a combinatorial structure called a *local action diagram*. This local action diagram is a graph decorated with sets and groups that codify the local actions of G. Importantly, this decomposition process can be reversed, where one starts with a local action diagram Δ and then constructs an arc-coloured tree $\mathbf{T}$ called the Δ-tree and a group $\mathbf{U}(\Delta)$ called the *universal group of* Δ. The group $\mathbf{U}(\Delta)$ acts on $\mathbf{T}$ in such a way that its local action diagram is again Δ, and moreover it exhibits various desirable global properties that are often impossible to verify for groups arising from graphs of groups via Bass–Serre Theory. This 'decomposition' of (T, G) into local actions hinges on the observation that if Δ is the associated local action diagram for (T, G) and $\mathbf{T}$ and $\mathbf{U}(\Delta)$ are the Δ-tree and the universal group of Δ, then the actions $(T, G^{(\mathrm{P})})$ and $(\mathbf{T}, \mathbf{U}(\Delta))$ can be identified. Thus, viewing $\mathbf{U}(\Delta)$ as a subgroup of $\mathrm{Aut}(T)$, we see that $\mathbf{U}(\Delta)$ is 'universal' with respect to the local actions of G; that is, $\mathbf{U}(\Delta)$ contains an $\mathrm{Aut}(T)$-conjugate of every action (T, H) whose local action diagram is also Δ.

The Burger–Mozes groups $\mathbf{U}(F)$ and the box product construction $\mathbf{U}(F_1, F_2)$ play a foundational role in this new theory: the local action diagram of $\mathbf{U}(F)$ is a (suitably decorated) graph consisting of a single vertex with a set of loops, each of which is its own reverse; the local action diagram of $\mathbf{U}(F_1, F_2)$ when F_1 and F_2 are transitive is a (suitably decorated) graph consisting of a pair of vertices joined by an edge and no loops (c.f. the graphs of groups of HNN extensions and amalgamated free products).

The theory of local action diagrams is powerful enough to give a complete description of closed actions on trees with Tits' Independence Property (P): these actions are precisely the universal groups of local action diagrams. From this one immediately obtains a robust classification of all actions (T, G) with Tits' Independence Property (P), by taking closures in the permutation topology. Note that (T, G) and its closure in the permutation topology have the same orbits on all finite (ordered and unordered) sets of vertices.

Working with local action diagrams is remarkably easy; unlike for graphs of groups there are no embeddability issues to contend with. Moreover, all properties of faithful closed actions with property (P) can be read directly from the local action diagram; many can be read easily from the diagram, including – surprisingly – geometric density, compact generation and simplicity.

The theory of local action diagrams is described in [20]. This note is intended to be an accessible introduction to the theory, placing it in its broad context and giving illustrative diagrams and examples, while omitting most proofs. It largely follows the contents of the second author's plenary talk at Groups St Andrews 2022 in Newcastle. Our intention with this note is to make the motivations and ideas in the theory accessible to non-specialists. To that end, we give an overview of Bass–Serre Theory, with enough depth so that the reader can understand its limitations, appreciate how the theory of local action diagrams mirrors the rich relationship between action and combinatorial description in Bass–Serre Theory, and to see how fundamentally different the two theories are. Following this we introduce the theory of local action diagrams, giving examples and some consequences, but largely omitting proofs (all proofs can be found in [20]). We conclude with some suggestions for future research projects.

Since this note is intended to be an introduction to local action diagrams, many topics from [20] have been omitted. The most significant omissions concern subgroups of universal groups of local action diagrams. We refer the interested reader to [20] for further details.

Notation and conventions

Unless otherwise stated we follow the definitions in our paper [20]. In particular, our graphs can have multiple distinct edges between two vertices, and each edge is comprised of two arcs (one in each direction). Loops are allowed, meaning that our graphs are graphs in the sense of Serre (except for us a loop a may or may not equal its reverse $\overline{a}$, which is not the case for Serre). Our graphs Γ have a (finite or infinite) vertex set $V = V\Gamma$, a (finite or infinite) arc set $A = A\Gamma$, an arc-reversal map $a \mapsto \overline{a}$ (also called an arc or edge *inversion*) together with an *origin* map $o : A \to V$ and a *terminal* map $t : A \to V$, so that $a \in A$ is an arc *from* $o(a)$ *to* $t(a)$. Edges are pairs $\{a, \overline{a}\}$ and are said to *contain* vertices $o(a)$ and $t(a)$, or to be *between* $o(a)$ and $t(a)$. We define *loops* to be arcs a such that $o(a) = t(a)$. The *valency* of a vertex v is $|o^{-1}(v)|$, sometimes denoted $|v|$; if this is finite for all vertices then Γ is *locally finite*. A *leaf* in Γ is a vertex with exactly one edge containing it, and that edge is not a loop. A graph is *simple* if it has no loops and there is at most one edge between any two vertices.

Our graphs are not simple so we must define paths with care. Given an interval $I \subseteq \mathbb{Z}$, let $\hat{I} = \{i \in I : i+1 \in I\}$; a *path* in Γ indexed by I is then a sequence of vertices $(v_i)_{i \in I}$ and edges $(\{a_i, \overline{a_i}\})_{i \in \hat{I}}$ such that

$\{a_i, \overline{a_i}\}$ is an edge in Γ between v_i and v_{i+1} for all $i \in \hat{I}$. For finite I the path has *length* $|\hat{I}|$. A path is *simple* if all its vertices v_i are distinct from one another. We can now define *directed paths* in the obvious way, as a sequence of vertices and arcs. For $n > 0$, if $I = \{0, \ldots, n\}$ and $v_0 = v_n$ and vertices $\{v_0, \ldots, v_{n-1}\}$ are distinct, then the path is called a *cycle* of length n. The *distance* between two vertices v, w, denoted $d(v, w)$, is the length of the shortest path between them if it exists, and is infinite otherwise. A graph is *connected* if there is a path between any two distinct vertices. For a vertex v the set of vertices whose distance from v is at most k is called a k-*ball* and is denoted $B_v(k)$. We will sometimes write $B(v)$ for the set $B_v(1)$ of *neighbours* of v. An *orientation* of Γ is a subset $O \subseteq A\Gamma$ such that for each $a \in A\Gamma$, either a or $\overline{a}$ is in O, but not both. For graphs that are not simple, the graph subtraction operation is not well behaved and so we avoid it except for the following situation: for $a \in A\Gamma$ the graph $\Gamma \setminus \{a\}$ is obtained from Γ by removing arcs $a, \overline{a}$. For a simple graph Γ with subgraph Λ we define graph subtraction $\Gamma \setminus \Lambda$ in the usual way: $\Gamma \setminus \Lambda$ is obtained from Γ by removing all vertices that lie in Λ and their incident edges.

A *tree* is a nonempty simple, connected graph that contains no cycles. In a simple graph Γ, a *ray* is a one-way infinite simple path and a *double ray* or *line* is a two-way infinite simple path. For us the *ends* (sometimes called *vertex-ends* for non-locally-finite graphs) of Γ are equivalence classes on the set of rays, where rays R_1, R_2 lie in the same end if and only if there exists a ray R in Γ containing infinitely many vertices of R_i for $i = 1, 2$. In a tree T, there is a unique shortest path between any two vertices v and w, denoted $[v, w]$ (or $[v, w)$, for example, if we wish to exclude w). For an arc or edge e in T the graph $T \setminus \{e\}$ has two connected components; these are called the *half-trees* associated with e.

Actions of a group G on a set X are from the left, with Gx denoting the orbit of $x \in X$ under the action of G. We denote the stabiliser of x by G_x, and for a subset $Y \subseteq X$ we write $G_{\{Y\}}$ (resp. $G_{(Y)}$) for the setwise (resp. pointwise) stabiliser of Y in G. The action is *transitive* on X if $X = Gx$ for some $x \in X$. The group of all permutations of X is denoted $\mathrm{Sym}(X)$. Subgroups of $\mathrm{Sym}(X)$ in which all orbits of point stabilisers are finite are called *subdegree-finite*. A group $G \leq \mathrm{Sym}(X)$ acts *freely* or *semiregularly* if the stabiliser G_x of any $x \in X$ is trivial; if G is transitive and semiregular we say it is *regular*. If G is transitive, then it is *primitive* if and only if the only G-invariant equivalence relations on X are the trivial relation (where equivalence classes are singletons) or the universal relation (where X is an entire equivalence class).

There is a natural topology on G that can be obtained from the action of G on X, called the *permutation topology*, in which a neighbourhood basis of the identity is taken to be the pointwise stabilisers of finite subsets of X. If we think of X as a discrete space with elements of G as maps from X to X, then the topology is equal to the topology of pointwise convergence and the compact-open topology. Permutational properties of G have topological ramifications. For example, G is totally disconnected if and only if the action on X is faithful, and if $G \leq \mathrm{Sym}(X)$ is closed and subdegree-finite then all stabilisers G_x are compact and open, so G is a totally disconnected and locally compact group (henceforth, t.d.l.c.) with compact open stabilisers. See [17] for a thorough guide to this topology. Topological statements concerning $\mathrm{Sym}(X)$ will always appertain to the permutation topology. Note that a topological group is *compactly generated* if there is a compact subset that abstractly generates the group.

A *graph homomorphism* $\theta : \Gamma \to \Gamma'$ is a pair of maps $\theta_V : V\Gamma \to V\Gamma'$ and $\theta_A : A\Gamma \to A\Gamma'$ that respect origin vertices and edge reversal; if θ_V and θ_A are both bijections we say θ is an *isomorphism*. A group G acting on Γ gives rise to a *quotient graph* $G\backslash\Gamma$ whose vertex (resp. arc) set is the set of G-orbits on V (resp. on A), and for an arc Ga in $G\backslash\Gamma$ we have $o(Ga) = Go(a)$ and $t(Ga) = Gt(a)$ and $\overline{Ga} = G\overline{a}$. The group of automorphisms of Γ is denoted $\mathrm{Aut}(\Gamma)$. When Γ is a simple graph, $\mathrm{Aut}(\Gamma)$ acts faithfully on V as those elements in $\mathrm{Sym}(V)$ that respect the arc relation in $V \times V$, and in this case we identify $\mathrm{Aut}(\Gamma)$ with the corresponding subgroup of $\mathrm{Sym}(V)$.

For a tree T (which recall is always simple) and a line L in T, a *translation* of L is an orientation-preserving automorphism of L that does not fix any point on the line. If B is a subtree of T and $G \leq \mathrm{Aut}(T)$, we say that G leaves B *invariant* if G fixes setwise the vertices of B. Throughout, countable means finite or countably infinite. We say that an action on a tree T is *geometrically dense* if the action does not leave invariant any nonempty proper subtree of T and does not fix any end of T.

7.2 Groups acting on trees

7.2.1 An overview of Bass–Serre Theory

Traditionally, Bass–Serre Theory concerns groups acting on trees *without inversion*, meaning that no element of the group acts as an edge inversion. This is not a significant restriction, since given an action with inversion one can subdivide the edges of the tree and thus obtain an action without inversion. We abide by this tradition here and restrict our attention to inversion-free actions. Let T be a tree and let G act on T without inversion. We closely follow Section 5 of Serre's book [21], so readers seeking a more complete description of the theory can consult this source. Our aim in this section is for readers to see that a local action diagram and its corresponding universal group are thoroughly dissimilar to a graph of groups and its universal cover, but nevertheless the beautiful correspondence in Bass–Serre Theory between the action and its description as a graph of groups is mirrored in our theory of local action diagrams. In Serre's work, a loop always admits an automorphism of order two which changes its orientation. We temporarily adopt this convention for Section 7.2.1.

The combinatorial structure at the heart of Bass–Serre Theory is called a *graph of groups*. This is a connected nonempty graph Γ, together with some groups that will be associated with the vertices and edges of Γ; this association can be thought of as the vertices and edges of Γ being coloured with a pallet of colours comprised of these groups. More precisely, for each vertex $v \in V\Gamma$ we have a group $\mathcal{G}_v$ (these are called the *vertex groups*) and for each arc $a \in A\Gamma$ we have a group $\mathcal{G}_a$ satisfying $\mathcal{G}_{\overline{a}} = \mathcal{G}_a$ (these are called the *edge groups*); we also have a monomorphism $\mathcal{G}_a \to \mathcal{G}_{t(a)}$ allowing us to view any edge group as a subgroup of the vertex groups of the vertices that comprise the edge. Let us denote this graph of groups as $(\Gamma, (\mathcal{G}_v), (\mathcal{G}_a))$.

This concise combinatorial structure admits two natural universal objects. The first is a tree called the *universal cover* of the graph of groups $(\Gamma, (\mathcal{G}_v), (\mathcal{G}_a))$; the second is the *fundamental group* of the graph of groups. Before formally defining these, let us first explore their significance. The first significant part of the Fundamental Theorem of Bass–Serre Theory (see Theorem 7.2.4 below) is essentially a decomposition of the action (T, G) in the language of graphs of groups:

($\circledast$) *There is a graph of groups associated to (T, G), and G can be identified with the fundamental group of this graph of groups.*

The second significant part of the theorem is essentially a method of constructing inversion-free actions of groups on trees using graphs of groups:

($\circledcirc$) *Given a graph of groups $(\Gamma, (\mathcal{G}_v), (\mathcal{G}_a))$, its fundamental group acts on its* *universal cover (which is a tree) without inversion, and the graph of groups associated with this action via ⊛ is again* $(\Gamma, (\mathcal{G}_v), (\mathcal{G}_a))$.

These two components of the theorem are the foundation of Bass–Serre Theory. For our purposes they are significant for two reasons. The constructive part (◎) is not usable in general for constructing new examples of nondiscrete actions of groups on trees because of what might be thought of as a 'chicken or the egg' dilemma (see Section 7.3.1). This limitation can be overcome via the local-to-global theory of groups acting on trees, and this local-to-global approach provided the motivation for our theory of local action diagrams. The second significance is that the rich correspondence between the action (T, G) and the combinatorial object (i.e. the graph of groups) is mirrored in our theory of local action diagrams, in which the combinatorial object is a local action diagram.

Let us now formally define the objects in Bass–Serre Theory, and then give a precise statement of the theorem.

The associated graph of groups (see [21, §5.4])

Suppose G acts on a tree T without inversion. We first describe its *associated graph of groups.* The underlying graph of the graph of groups is the quotient graph $\Gamma := G\backslash T$. We next use Γ to choose a subtree of T in a coherent way so that we can take the vertex and arc stabilisers of this subtree to be the vertex and edge groups of the graph of groups.

Choose an orientation E^+ of Γ, and for each $a \in A\Gamma$ set $e(a) = 0$ whenever $a \in E^+$ and set $e(a) = 1$ otherwise. Consider the subgraphs of Γ that are trees; these form an ordered set (ordered by inclusion) and by Zorn's Lemma this set has a maximal element M, called a *maximal tree of* Γ, which is easily seen to contain every vertex of Γ (see [21, §2 Proposition 11] for example). One can lift M to a subtree T' of T; that is, T' is isomorphic to M via the natural map π which takes each vertex $v \in VT'$ (resp. arc $a \in AT'$) to the vertex Gv (resp. arc Ga) in $M \subseteq G\backslash T$.

Next, we construct a map φ that takes arcs in $A\Gamma$ to arcs in AT with the property that $\varphi(\overline{a}) = \overline{\varphi(a)}$ for all $a \in A\Gamma$. The map will also be defined on the vertices of M. On M take φ to be $\pi^{-1} : M \to T'$. For $a \in E^+ \setminus AM$ there is an arc $b \in AM$ such that $o(a) = o(b)$, and we set $\varphi(a) = \varphi(b)$. In this way we have $o(\varphi(a)) = \varphi(o(a))$. Now $t(\varphi(a))$ and $\varphi(t(a))$ have the same image under π since both project to $t(b) \in V\Gamma$. In particular, $t(\varphi(a))$ and $\varphi(t(a))$ lie in the same G orbit, so we can choose $\gamma_a \in G$ such that $t(\varphi(a)) = \gamma_a\varphi(t(a))$. From this we obtain elements

$\gamma_a \in G$ for all arcs $a \in A\Gamma$ by specifying that $\gamma_{\overline{a}} = \gamma_a^{-1}$ and γ_a is the identity whenever $a \in AM$.

We can now define the vertex and edge groups of our associated graph of groups. For each vertex v of Γ recall that v is a vertex in M, so $\varphi(v)$ is defined and is equal to a vertex in T; take the vertex group $\mathcal{G}_v$ to be the vertex stabiliser $G_{\varphi(v)} \leq G$. Similarly, for an arc a of Γ recall that $\varphi(a)$ is an arc in T; take the edge group $\mathcal{G}_a$ to be the arc stabiliser $G_{\varphi(a)} \leq G$, with $\mathcal{G}_{\overline{a}} = \mathcal{G}_a$. Finally, we define the monomorphism $\mathcal{G}_a \to \mathcal{G}_{t(a)}$ as $g \mapsto \gamma_a^{e(a)-1} g \gamma_a^{1-e(a)}$.

Example 7.2.1 Let T be the biregular tree $T_{m,n}$ for finite distinct $m, n > 2$. Notice that the action $(T, \mathrm{Aut}(T))$ has one edge orbit and two vertex orbits, and these vertex orbits correspond to the natural bipartition of $T_{m,n}$ into vertices with valency m, and vertices with valency n. Thus, the graph of groups associated with $(T, \mathrm{Aut}(T))$ is a pair of vertices connected by a single edge. The vertex groups are the stabilisers $\mathrm{Aut}(T)_v$ and $\mathrm{Aut}(T)_w$ where v, w are adjacent vertices in VT with v having valency m in T and w having valency n, and the edge group is $\mathrm{Aut}(T)_{(v,w)}$.

The fundamental group of a graph of groups (see [21, §5.1]) Suppose we have a graph of groups $(\Gamma, (\mathcal{G}_v), (\mathcal{G}_a))$. For each arc $a \in A\Gamma$ we have a monomorphism $\mathcal{G}_a \to \mathcal{G}_{t(a)}$, and we denote the image of any $h \in \mathcal{G}_a$ under this monomorphism by h^a. We need extra generators for the fundamental group, one for each arc in Γ, so for all $a \in A\Gamma$ we find new letters g_a not contained in any of the vertex or edge groups. Our generating set Ω is then the union of $\{g_a : a \in A\Gamma\}$ and the vertex groups $\bigcup_{v \in V\Gamma} \mathcal{G}_v$. Let M be a maximal tree of Γ. Then we define the *fundamental group* $\Pi_1((\Gamma, (\mathcal{G}_v), (\mathcal{G}_a))$, abbreviated to Π_1, to be

$$\langle \Omega : g_{\overline{a}} = g_a^{-1},\ g_a h^a g_a^{-1} = h^{\overline{a}}\ (\forall a \in A\Gamma, h \in \mathcal{G}_a),\ g_b = 1\ (\forall b \in AM) \rangle.$$

One can show (see [21, §5 Proposition 20]) that the definition of Π_1 is independent of the choice for M.

Example 7.2.2 If $(\Gamma, (\mathcal{G}_v), (\mathcal{G}_a))$ is the graph of groups in Example 7.2.1, then the definition above gives Π_1 to be the amalgamated free product $\mathrm{Aut}(T)_v *_{\mathrm{Aut}(T)_{(v,w)}} \mathrm{Aut}(T)_w$.

Universal covering of a graph of groups (see [21, §5.3]) Now suppose we are given a graph of groups $(\Gamma, (\mathcal{G}_v), (\mathcal{G}_a))$, a maximal tree M of Γ and an orientation E^+ of Γ. Let Π_1 be the fundamental group of $(\Gamma, (\mathcal{G}_v), (\mathcal{G}_a))$. Define the map $e : A\Gamma \to \{0,1\}$ as in Section 7.2.1, with $e(a) = 0$ whenever $a \in E^+$ and $e(a) = 1$ otherwise. For each arc $a \in \Gamma$, let $\hat{a} \in \{a, \overline{a}\}$ be such that $\hat{a} \in E^+$, and let $\mathcal{G}_a^a$ be the image (in $\mathcal{G}_{t(a)}$) of the edge group $\mathcal{G}_a$ under the monomorphism $h \mapsto h^a$. Notice that $\hat{a} = \hat{\overline{a}}$.

The *universal cover* of $(\Gamma, (\mathcal{G}_v), (\mathcal{G}_a))$ is a graph $\tilde{T}$ whose vertex set is the disjoint union of left cosets,

$$V\tilde{T} := \bigsqcup_{v \in V\Gamma} \Pi_1/\mathcal{G}_v,$$

and whose arc set is the disjoint union,

$$\bigsqcup_{a \in A\Gamma} \Pi_1/\mathcal{G}_{\hat{\overline{a}}}^{\hat{\overline{a}}}.$$

Arc inversion in $\tilde{T}$, and the maps o and t, are defined as follows. For each arc $a \in A\Gamma$, let $\tilde{a}$ denote the trivial coset in $\Pi_1/\mathcal{G}_{\hat{\overline{a}}}^{\hat{\overline{a}}}$ corresponding to $\mathcal{G}_{\hat{\overline{a}}}^{\hat{\overline{a}}}$, and for each vertex $v \in V\Gamma$, let $\tilde{v}$ denote the trivial coset in $\Pi_1/\mathcal{G}_v$ corresponding to $\mathcal{G}_v$. Any $\tilde{T}$-arc lies in $\Pi_1/\mathcal{G}_{\hat{\overline{a}}}^{\hat{\overline{a}}}$ for some arc $a \in A\Gamma$. Each $\tilde{T}$-arc in $\Pi_1/\mathcal{G}_{\hat{\overline{a}}}^{\hat{\overline{a}}}$ is of the form $g\tilde{a}$, for some $g \in \Pi_1$. Recalling our elements $g_a \in \Pi_1$ from Section 7.2.1 and setting $v_o := o(a)$ and $v_t := t(a)$, we then take

$$\overline{g\tilde{a}} = g\tilde{\overline{a}}, \quad o(g\tilde{a}) = gg_a^{-e(a)}\widetilde{v_o}, \quad t(g\tilde{a}) = gg_a^{1-e(a)}\widetilde{v_t}.$$

One can then (see [21, §5.3]) check that: $\tilde{T}$ is a tree and Π_1 acts on the tree $\tilde{T}$ via left multiplication, and moreover this action is as automorphisms of $\tilde{T}$. Under this action the quotient graph $\Pi_1\backslash\tilde{T}$ is Γ, and for all vertices $v \in V\Gamma$ (resp. arcs $a \in A\Gamma$) we have that the stabiliser $\Pi_{1\tilde{v}}$ (resp. $\Pi_{1\tilde{a}}$) is equal to $\mathcal{G}_v$ (resp. $\mathcal{G}_{\hat{\overline{a}}}^{\hat{\overline{a}}} \cong \mathcal{G}_a$). For any arc a in the maximal tree M of Γ we have $g_a = 1$ and thus $\widetilde{o(y)} = o(\tilde{y})$ and $\widetilde{t(y)} = t(\tilde{y})$, so we have a lift of M to a subtree $\tilde{T}'$ of the tree $\tilde{T}$ via $v \mapsto \tilde{v}$ and $a \mapsto \tilde{a}$ for $v \in VM$ and $a \in AM$. Thus, the associated graph of groups for $(\tilde{T}, \Pi_1)$ is $(\Gamma, (\mathcal{G}_v), (\mathcal{G}_a))$.

Example 7.2.3 Let us continue Examples 7.2.1–7.2.2, resuming their notation. For the tree $T = T_{m,n}$ and the action $(T, \mathrm{Aut}(T))$, the graph of groups $(\Gamma, (\mathcal{G}_v), (\mathcal{G}_a))$ consists of two vertices connected by a single edge, with vertex groups $\mathrm{Aut}(T)_v$ and $\mathrm{Aut}(T)_w$, and edge group $\mathrm{Aut}(T)_{v,w}$, where v has valency m in T and w has valency n. Recall that the fundamental group of this graph of groups is $\Pi_1 = \mathrm{Aut}(T)_v *_{\mathrm{Aut}(T)_{(v,w)}} \mathrm{Aut}(T)_w$. Now $\mathrm{Aut}(T)_v$ is transitive on the edges in T that are incident to v, so the index of the edge group $\mathrm{Aut}(T)_{(v,w)}$ in the vertex group $\mathrm{Aut}(T)_v$ is m. Similarly the index of $\mathrm{Aut}(T)_{(v,w)}$ in $\mathrm{Aut}(T)_w$ is n. The universal cover $\tilde{T}$ of $(\Gamma, (\mathcal{G}_v), (\mathcal{G}_a))$ is a tree, and by definition we have $V\tilde{T} = (\Pi_1/\mathrm{Aut}(T)_v) \sqcup (\Pi_1/\mathrm{Aut}(T)_w)$ with the two sets in this disjoint union naturally bipartitioning T.

Fundamental Theorem of Bass–Serre Theory (see [21, §5.4]) We are now able to formally state the fundamental theorem.

Theorem 7.2.4 ([21, §5]) *Let G be a group and T be a tree.*

($\circledast$) *Suppose G acts on T without inversion. Let $(\Gamma, (\mathcal{G}_v), (\mathcal{G}_a))$ be its associated graph of groups, and choose a maximal tree M of Γ. Let Π_1 be the fundamental group of this graph of groups (with respect to M) and let $\tilde{T}$ be the universal covering (with respect to M). Then the map $\phi : \Pi_1 \to G$ defined by the inclusion $\mathcal{G}_v \leq G$ (for $v \in V\Gamma$) and $\phi(g_a) = \gamma_a$ (for $a \in A\Gamma$) is an isomorphism of groups. The map $\psi : \tilde{T} \to T$ given, for all $h \in \Pi_1$, by $\psi(h\tilde{v}) = \phi(h)\varphi(v)$ (for all $v \in V\Gamma$) and $\psi(h\tilde{a}) = \phi(h)\varphi(a)$ (for all $a \in A\Gamma$) is an isomorphism of graphs. Moreover, the isomorphism ψ is ϕ-equivariant, and so the actions $(\tilde{T}, \Pi_1)$ and (T, G) can be identified.*

($\circledcirc$) *Suppose $(\Gamma, (\mathcal{G}_v), (\mathcal{G}_a))$ is a graph of groups. Let Π_1 be its fundamental group and $\tilde{T}$ its universal cover. Then $(\tilde{T}, \Pi_1)$ is an inversion-free action on a tree whose associated graph of groups is $(\Gamma, (\mathcal{G}_v), (\mathcal{G}_a))$.*

Example 7.2.5 Let us continue Examples 7.2.1–7.2.3, resuming their notation. We have the tree $T = T_{m,n}$ and a graph of groups $(\Gamma, (\mathcal{G}_v), (\mathcal{G}_a))$ for the action $(T, \mathrm{Aut}(T))$. The universal cover of $(\Gamma, (\mathcal{G}_v), (\mathcal{G}_a))$ is $\tilde{T}$ and its fundamental group is $\mathrm{Aut}(T)_v *_{\mathrm{Aut}(T)_{(v,w)}} \mathrm{Aut}(T)_w$. By the Fundamental Theorem of Bass–Serre Theory, we have that the actions of $\mathrm{Aut}(T)_v *_{\mathrm{Aut}(T)_{(v,w)}} \mathrm{Aut}(T)_w$ on $\tilde{T}$ and of $\mathrm{Aut}(T)$ on T are permutationally isomorphic.

7.2.2 Jacques Tits' Independence Property (P) and simplicity

We describe [29, §4.2], where *Tits' Independence Property* (P) (also called *property* (P) or the *independence property*) is introduced. Suppose $G \leq \mathrm{Aut}(T)$, where T is a tree. If C is a (finite or infinite) nonempty simple path in T, then for each $v \in VT$ there is a unique vertex $\pi_C(v)$ in C which is closest to v. This gives a well-defined map $v \mapsto \pi_C(v)$. For each $w \in VC$, the set $\pi_C^{-1}(w)$ is the vertex set of a subtree of T. Each of these subtrees is invariant under the action of the pointwise stabiliser $G_{(C)}$ of C, and so we define $G^w_{(C)}$ to be the subgroup of $\mathrm{Sym}(\pi_C^{-1}(w))$ induced by $G_{(C)}$. We therefore have homomorphisms $\varphi_w : G_{(C)} \to G^w_{(C)}$ for each $w \in VC$ from which we obtain the homomorphism,

$$\varphi : G_{(C)} \to \prod_{w \in VC} G^w_{(C)}.$$

Now G has the *independence property for* C if the homomorphism φ is an isomorphism, and G has *Tits' Independence Property* (P) if it has the independence property for C for every possible simple path C. Intuitively, G has the independence property for a path C if $G_{(C)}$ can act independently on all of the subtrees 'hanging' from C.

Theorem 7.2.6 ([29, Théorème 4.5]) *Let T be a tree and suppose $G \leq \mathrm{Aut}(T)$ has property* (P). *Let G^+ be the group generated by the pointwise stabilisers in G of edges in T. If G is geometrically dense, then $G^+ \trianglelefteq G$ and G^+ is trivial or abstractly simple.*

7.3 Universal groups acting on trees

7.3.1 Limitations to Bass–Serre Theory

In a graph of groups, we specify embeddings of edge groups into vertex groups. In doing so, the vertex stabiliser is specified *as an abstract group*, while the action on the 1-ball is also specified. In many commonly encountered situations, this leads to complications that are intractable. We describe three of them here; they are related but different.

1. In the action on a tree resulting from a graph of groups, the vertex stabiliser can only be a quotient of the vertex group. So, we can't obtain 'large' (e.g. nondiscrete) vertex stabilisers this way, unless the vertex group we started with already had an interesting action on a tree.

2. Suppose we are given a graph of groups and in addition we are told that the action is in some sense interesting. Even though we have been given the vertex stabiliser, how it acts on the tree on a large scale is unintuitive and often abstruse. Even basic questions, like whether or not the action is faithful, can be difficult to answer. Restricting to locally finite trees does not sufficiently reduce the complexity: we are still in a sense required to understand the dynamics of sequences of virtual isomorphisms as we travel along all possible walks through the graph of groups. This is typically an insurmountable problem when the vertex

stabilisers are infinite, or when one is interested in understanding ever larger finite vertex stabilisers as in the Weiss conjecture.

3. The possible combinations of vertex groups are difficult to understand because they are not independent of one another: we need a common edge group for any pair of neighbouring vertices. If we wish to choose a collection of local actions independently of one another, there are very few ways to do this and all have significant limitations. For example, we could make all the vertex groups infinitely generated free groups, but then determining the action of the resulting vertex groups on the tree is hopelessly complicated, rendering the task of understanding the group's closure in $\mathrm{Aut}(T)$ as a topological group futile.

Readers who are familiar with Bass–Serre Theory will no doubt recognise these limitations. For readers unfamiliar with Bass–Serre Theory, we set an (intractable) exercise that highlights some of these issues.

Exercise Let T be the biregular tree $T_{7,5}$. Using only Bass–Serre Theory, attempt to construct a subgroup $G \leq \mathrm{Aut}(T)$ that has two vertex orbits on T, such that every vertex stabiliser G_v is infinite and does not induce S_7 or S_5 on the set of neighbours of the vertex v. The point of this exercise is not to complete the construction, it is to make an attempt and in doing so experience the aforementioned limitations 1–3.

As we shall see, the exercise has an easy solution using the theory of local action diagrams. However, even when we take this solution and (pretending for a moment that we do not know it is a solution) use Bass–Serre Theory to analyse its action, our point (2) above becomes apparent.

7.3.2 Burger–Mozes groups

Here we largely follow Burger and Mozes' paper [4, §3.2] with one significant difference: we do not insist that the trees be locally finite. Let $d > 2$ be some finite or infinite cardinal and let T be a regular tree of valency d. Fix some set X such that $|X| = d$, and let $F \leq \mathrm{Sym}(X)$. We say that $G \leq \mathrm{Aut}(T)$ is *locally-F* if for all $v \in VT$ the permutation group induced on the set $B(v)$ of neighbours of v by the vertex stabiliser G_v is permutationally isomorphic to F. A *legal colouring* is a map $\mathcal{L} : AT \to X$ that satisfies the following:

(i) For each vertex $v \in VT$, the restriction $\mathcal{L}\big|_{o^{-1}(v)} : o^{-1}(v) \to X$ is a bijection;

(ii) For all arcs $a \in AT$ we have $\mathcal{L}(a) = \mathcal{L}(\overline{a})$.

One can always construct a legal colouring of T. For $g \in \mathrm{Aut}(T)$ and $v \in VT$ we define the *$\mathcal{L}$-local action of g at v* to be,

$$\sigma_{\mathcal{L},v}(g) := \mathcal{L}\big|_{o^{-1}(gv)} g \mathcal{L}\big|_{o^{-1}(v)}^{-1}. \tag{7.1}$$

Notice that $\sigma_{\mathcal{L},v}(g) \in \mathrm{Sym}(X)$. One might ask if we can constrain these bijections, so that they all lie in some common subgroup of $\mathrm{Sym}(X)$. Such a restriction gives rise to the Burger–Mozes universal groups.

The *Burger–Mozes universal group of F* (with respect to $\mathcal{L}$) is the group,

$$\mathbf{U}_{\mathcal{L}}(F) := \{g \in \mathrm{Aut}(T) : \sigma_{\mathcal{L},v}(g) \in F \quad \forall v \in VT\}.$$

Two legal colourings $\mathcal{L}$ and $\mathcal{L}'$ give rise to universal groups $\mathbf{U}_{\mathcal{L}}(F)$ and $\mathbf{U}_{\mathcal{L}'}(F)$ that are conjugate in $\mathrm{Aut}(T)$; for this reason we replace $\mathbf{U}_{\mathcal{L}}(F)$ with $\mathbf{U}(F)$ and speak of *the* Burger–Mozes universal group $\mathbf{U}(F)$ of F. Further properties of $\mathbf{U}(F)$ described in [4] are as follows. Note that some of these properties are expanded upon in Section 7.3.3.

(i) $\mathbf{U}(F)$ is a vertex transitive and locally-F subgroup of $\mathrm{Aut}(T)$, and if F has finite degree then $\mathbf{U}(F)$ is closed;
(ii) $\mathbf{U}(F)$ enjoys Tits Independence Property (P), and consequently by Theorem 7.2.6 the subgroup $\mathbf{U}(F)^+$ is trivial or simple;
(iii) If F has finite degree then the index $|\mathbf{U}(F) : \mathbf{U}(F)^+|$ is finite if and only if $F \leq \mathrm{Sym}(X)$ is transitive and generated by point stabilisers; when this happens $\mathbf{U}(F)^+ = \mathbf{U}(F) \cap (\mathrm{Aut}(T))^+$ and therefore $|\mathbf{U}(F) : \mathbf{U}(F)^+| = 2$;
(iv) when F is transitive on X, the group $\mathbf{U}(F)$ has the following universal property: $\mathbf{U}(F)$ contains an $\mathrm{Aut}(T)$-conjugate of every locally-F subgroup of $\mathrm{Aut}(T)$.

Thus, in the situation when F is transitive, we have a natural way to describe $\mathbf{U}(F)$: it is the largest locally-F subgroup of $\mathrm{Aut}(T)$.

Burger and Mozes used $U(F)$ to build towards a hoped-for classification of closed 2-transitive actions on locally finite trees. Here we instead see them as a first step towards a local-to-global theory of groups acting on trees. When T is locally finite, $\mathbf{U}(F)$ is obviously a t.d.l.c. subgroup of $\mathrm{Aut}(T)$ with compact vertex stabilisers. In fact, the group can be t.d.l.c. in more general situations; this was shown in [25], where the Burger–Mozes universal groups are viewed as a special case of a more general construction called the *box product*. In this more general setting, various global topological properties are easy to characterise using local conditions (i.e. conditions on F). We describe these in section 7.3.3.

7.3.3 The box product construction

Here we largely follow the second author's paper [25], where the box product was introduced as a product for permutation groups; as a permutational product it is, in some sense, the dual of the wreath product in its product action. In the same paper, the box product was used to produce the first examples of $2^{\aleph_0}$ distinct isomorphism classes of compactly generated locally compact groups that are nondiscrete and topologically simple, answering a long standing open question. The box product arises naturally in the structure theory of subdegree-finite primitive permutation groups due to the second author ([26]).

Let $d_1, d_2 > 1$ be two finite or infinite cardinal numbers and let $T = T_{d_1,d_2}$ be the biregular tree with valencies d_1, d_2. Fix disjoint sets X_1, X_2 such that $|X_1| = d_1$ and $|X_2| = d_2$. Thus there is a natural bipartition of T as $VT = V_1 \sqcup V_2$ such that vertices in V_1 have valency $|X_1|$ and vertices in V_2 have valency $|X_2|$. Let $F_i \leq \mathrm{Sym}(X_i)$, for $i = 1, 2$ with at least one of F_1, F_2 being nontrivial. We say $G \leq \mathrm{Aut}(T)$ is *locally-(F_1, F_2)* if G preserves setwise the parts V_1 and V_2, and for all $v \in VT$ the permutation group induced on $B(v)$ by the vertex stabiliser G_v is permutationally isomorphic to F_i whenever $v \in V_i$, for $i = 1, 2$.

In this new context we define a *legal colouring* to be a map $\mathcal{L} : AT \to X$ that satisfies the following:

(i) For each vertex $v \in V_i$, the restriction $\mathcal{L}\big|_{o^{-1}(v)} : o^{-1}(v) \to X_i$ is a bijection, for $i = 1, 2$;
(ii) For all $v \in VT$ the restriction $\mathcal{L}\big|_{t^{-1}(v)}$ is constant.

One can always construct a legal colouring of T. Using the same $\mathcal{L}$-local action defined in Equation 7.1, we define the *topological box product of F_1 and F_2* to be the group

$$\mathbf{U}_{\mathcal{L}}(F_1, F_2) := \left\{ g \in \mathrm{Aut}(T)_{\{V_1\}} : \sigma_{\mathcal{L},v}(g) \in F_i \quad \forall v \in V_i, \text{ for } i = 1, 2 \right\}.$$

The *permutational box product of F_1 and F_2* is the subgroup of $\mathrm{Sym}(V_2)$ induced by the action of $\mathbf{U}_{\mathcal{L}}(F_1, F_2)$ on V_2, and is denoted $F_1 \boxtimes_{\mathcal{L}} F_2$.

As with the Burger–Mozes groups, two legal colourings $\mathcal{L}$ and $\mathcal{L}'$ give rise to groups $\mathbf{U}(F_1, F_2)_{\mathcal{L}}$ and $\mathbf{U}(F_1, F_2)_{\mathcal{L}'}$ that are conjugate in $\mathrm{Aut}(T)$ and so we speak of *the* box product and write $\mathbf{U}(F_1, F_2)$ and $F_1 \boxtimes F_2$.

The Burger–Mozes groups arise as special cases of the box product construction: for any permutation group $F \leq \mathrm{Sym}(X)$ where $|X| \geq 3$, the Burger–Mozes group $\mathbf{U}(F)$ is permutationally isomorphic to $S_2 \boxtimes F$, where S_2 here denotes the symmetric group of degree 2. To see why, write $d := |X|$ and note that $\mathbf{U}(F)$ as a group of automorphisms of the d-regular tree T_d induces a faithful action on $T_{2,d}$ because $T_{2,d}$ is the barycentric subdivision of T_d. Now $\mathbf{U}(S_2, F)$ also acts as a group of automorphisms on $T_{2,d}$ and one can easily verify that the actions of $\mathbf{U}(F)$ and $\mathbf{U}(S_2, F)$ induced on the set of d-valent vertices of $T_{2,d}$ is permutationally isomorphic. From this we have that $\mathbf{U}(F)$ is topologically isomorphic to $S_2 \boxtimes F$, $\mathbf{U}(F, S_2)$, $F \boxtimes S_2$, and $\mathbf{U}(S_2, F)$.

As a permutational product, the box product has the following striking similarity to the wreath product in its product action. Recall that $F_1 \mathrm{Wr}\, F_2$ (in its product action) is a primitive permutation group if and only if F_1 is primitive and nonregular and F_2 is transitive and finite. Special cases of this fact were, according to Peter M. Neumann, first proved by W. A. Manning in the early 20th Century. For over a century, no other product of permutation groups was known to preserve primitivity in this kind of generality. Compare this with the box product: $F_1 \boxtimes F_2$ is a primitive permutation group if and only if F_1 is primitive and nonregular and F_2 is transitive.

With the benefit of hindsight, we can see hints of this local and global primitivity equivalence for the box product construction going back to the 1970s. Indeed, groups of automorphisms of simple graphs with connectivity one (see [20, §7.2]) can be realised as faithful inversion-free actions on trees, and in 1977 H. A. Jung and M. E. Watkins in [12, Theorem 4.2] proved that the automorphism group of a simple graph Γ with connectivity one is (i) vertex primitive if and only if (ii) all its lobes are vertex primitive, have at least 3 vertices and are pairwise isomorphic. Now (i) implies that the induced faithful tree action is primitive on one part P of the bipartition of the tree, and (ii) implies that the local actions at vertices in $T \setminus P$ are primitive. Arguments by Rögnvaldur G. Möller in the 1994 paper [15] (which used Warren Dicks and M. J. Dunwoody's powerful theory of structure trees from [9]) can be used to show all infinite, subdegree-finite primitive permutation groups with more than one end have a locally finite orbital digraph with connectivity one. In 2010 Jung and Watkins' result was generalised to directed graphs by the second author in [23], and in this context the primitivity condition in (ii) becomes primitive but not regular. These observations inspired constructions in the second author's 2010 paper [24] which, again with hindsight, can be seen as precursors to the permutational box product construction. As noted in [25], the box product $F_1 \boxtimes F_2$ can be constructed for closed groups F_1, F_2 using refinements of the arguments in [24] (the arguments are based on countable relational structures).

We also see hints coming from the world of groups acting on locally finite trees: obviously, in Burger and Mozes' 2000 paper [4], but also in Pierre-Emmanuel Caprace and Tom De Medts' 2011 result [7, Theorem 3.9], which states the following. Let T be a locally finite tree and suppose $G \leq \mathrm{Aut}(T)$ is nondiscrete, noncompact, compactly generated, closed, t.d.l.c. and topologically simple with Tits' Independence Property (P). Then the following conditions are equivalent: (i) every proper open subgroup of G is compact; and (ii) G splits as an amalgamated free product $G \cong A *_C B$, where A and B are maximal compact open subgroups, $C = A \cap B$ and the A-action on A/C (resp. the B-action on B/C) is primitive and noncyclic.

For a group G satisfying Caprace and De Medts' result, condition (i) implies G has a natural permutation representation that is primitive, and (ii) implies that the action of G on its Bass–Serre tree is primitive on *both* parts of its bipartition and moreover the local action at vertices in both parts of the bipartition is primitive but not regular.

The following are further properties of the box product construction. Since $\mathbf{U}(F)$ is topologically isomorphic to $\mathbf{U}(F, S_2)$ and permutationally isomorphic to $S_2 \boxtimes F$, many of these properties expand the properties of Burger–Mozes groups given in Section 7.3.2.

(i) $\mathbf{U}(F_1, F_2)$ is a locally-(F_1, F_2) subgroup of $\mathrm{Aut}(T)$.
(ii) Any subset $Y \subseteq X_1 \cup X_2$ is an orbit of F_1 or F_2 if and only if $t(\mathcal{L}^{-1}(Y))$ is an orbit of $\mathbf{U}_{\mathcal{L}}(F_1, F_2)$. In particular, $F_1 \boxtimes F_2$ is transitive if and only if F_1 is transitive, and $\mathbf{U}(F_1, F_2)$ has precisely two vertex-orbits (V_1 and V_2) if and only if F_1 and F_2 are transitive.
(iii) If F_i is a closed subgroup of $\mathrm{Sym}(X_i)$ for $i = 1, 2$ then $\mathbf{U}(F_1, F_2)$ is a closed subgroup of $\mathrm{Aut}(T)$.
(iv) $\mathbf{U}(F_1, F_2)$ enjoys Tits Independence Property (P).
(v) If F_1, F_2 are generated by point stabilisers, then $\mathbf{U}(F_1, F_2)$ is simple if and only if F_1 or F_2 is transitive.
(vi) When F_1, F_2 are transitive, the group $\mathbf{U}(F_1, F_2)$ has the following universal property: $\mathbf{U}(F_1, F_2)$ contains an $\mathrm{Aut}(T)$-conjugate of every locally-(F_1, F_2) subgroup of $\mathrm{Aut}(T)$.
(vii) If F_1, F_2 are closed, then $\mathbf{U}(F_1, F_2)$ is locally compact (and hence t.d.l.c.) if and only if all point stabilisers in F_1 and F_2 are compact.

Moreover, for all $v \in V_2$ the stabiliser $\mathbf{U}(F_1, F_2)_v$ is compact if and only if F_2 is compact and every point stabiliser in F_1 is compact.

(viii) If F_1, F_2 are closed with compact point stabilisers, then all point stabilisers in $\mathbf{U}(F_1, F_2)$ are compactly generated if and only if F_1 and F_2 are compactly generated. Moreover, if F_1 and F_2 are compactly generated with finitely many orbits, and at least one of the groups is transitive, then $\mathbf{U}(F_1, F_2)$ is compactly generated.

(ix) $\mathbf{U}(F_1, F_2)$ is discrete if and only if F_1 and F_2 are semiregular.

Thus, in the situation when F_1, F_2 are transitive, we again have a natural way to describe $\mathbf{U}(F_1, F_2)$: it is the largest locally-(F_1, F_2) subgroup of $\mathrm{Aut}(T)$. Furthermore, we can under mild *local* conditions (that is, conditions on F_1, F_2) ensure that $\mathbf{U}(F_1, F_2)$ is nondiscrete, compactly generated, t.d.l.c. and abstractly simple. An important point here is that one can use discrete local groups F_1, F_2 with certain topological properties, and obtain a nondiscrete topological group $\mathbf{U}(F_1, F_2)$ that inherits those local topological properties (except of course, discreteness) as global topological properties, as well as being guaranteed other desirable global properties like Tits' Independence Property (P) and nondiscreteness.

Constructing $2^{\aleph_0}$ distinct isomorphism classes of compactly generated locally compact groups that are nondiscrete and topologically simple is now relatively straightforward. For example, for a large enough prime p (e.g. $p > 10^{75}$) let $\{\mathcal{O}_i\}_{i \in I}$ be a set of representatives from the $2^{\aleph_0}$ isomorphism classes of A. Yu. Ol'shanskiĭ's p-Tarski Monsters ([19]). Each $\mathcal{O}_i$ is infinite and simple and any nontrivial proper subgroup $H_i < \mathcal{O}_i$ has finite order p. The groups $\mathcal{O}_i$ can therefore be viewed as (faithful) groups of permutations of $\mathcal{O}_i / H_i$. The groups $\mathbf{U}_i := \mathbf{U}(\mathcal{O}_i, S_3)$ are nondiscrete, compactly generated, t.d.l.c. and simple. It can then be shown (with a little work) that the $\mathbf{U}_i$ are pairwise nonisomorphic.

7.3.4 Further generalisations

In this section we survey a selection of generalisations of the Burger–Mozes groups, the box product construction, and Tits' Independence Property (P). These generalisations play no part in the theory of local action diagrams, but they suggest natural generalisations to the theory that we will revisit in Section 7.8.

Adrien Le Boudec in [13] considered universal groups of the locally finite d-valent tree T_d, for $d \geq 3$, that have local action $F \leq S_d$ at all but finitely many vertices. (Earlier, the special case where F is the alternating group of degree d was examined in [1].) More precisely, for a Burger–Mozes legal colouring $\mathcal{L}$ of T_d and a group $F \leq S_d$, the *Le Boudec group* $G_{\mathcal{L}}(F)$ is the group,

$$\{g \in \mathrm{Aut}(T_d) : \sigma_{\mathcal{L},v}(g) \in F \quad \text{for all but finitely many } v \in VT_d\}.$$

For a given $g \in G_{\mathcal{L}}(F)$, the finitely many vertices v for which $\sigma_{\mathcal{L},v}(g) \notin F$ are called the *singularities of* g. Of course we have $\mathbf{U}_{\mathcal{L}}(F) \leq G_{\mathcal{L}}(F) \leq \mathrm{Aut}(T_d)$. There is a topology on $G_{\mathcal{L}}(F)$ such that the inclusion map of $\mathbf{U}_{\mathcal{L}}(F)$ into $G_{\mathcal{L}}(F)$ is continuous and open (in general this is not the topology inherited from $\mathrm{Aut}(T_d)$). Under this topology $G_{\mathcal{L}}(F)$ is a t.d.l.c. group, and $G_{\mathcal{L}}(F)$ is discrete if and only if F is a semiregular permutation group.

For permutation groups $F \leq F' \leq S_d$, Le Boudec defines a group $G_{\mathcal{L}}(F, F')$ now called the *restricted universal group*, with $G_{\mathcal{L}}(F, F') := G_{\mathcal{L}}(F) \cap \mathbf{U}_{\mathcal{L}}(F')$ under the topology induced from $G_{\mathcal{L}}(F)$. Note that, while the groups $G_{\mathcal{L}}(F)$ and $G_{\mathcal{L}}(F, F')$ are subgroups of $\mathrm{Aut}(T_d)$, in general they are not closed in $\mathrm{Aut}(T_d)$.

The Le Boudec groups and restricted universal groups are an important source of examples of compactly generated locally compact simple groups that do not admit lattices[4] Prior to Le Boudec's work, the only example of a lattice-free compactly generated locally compact simple group was Neretin's group $\mathrm{AAut}(T_d)$ of almost automorphisms (see [1]). In particular Le Boudec uses these constructions to show the very nice result that there exist compactly generated abstractly simple t.d.l.c. groups $H \leq G$ with H is cocompact in G such that G contains lattices but H does not.

Waltraud Lederle in [14] simultaneously generalises Neretin's group $\mathrm{AAut}(T_d)$ and the Burger–Mozes groups for the locally finite tree T_d. Some of the terms defined in [14] have subsequently become known by different names, and we follow this more recent naming convention here.

Let us first define Neretin's group $\mathrm{AAut}(T_d)$. Finite subtrees of T_d in which every vertex is either a leaf or of valency d are called *complete*. An *almost automorphism* of T_d is an isomorphism g of rooted forests $g : T_d \setminus C_0 \to T_d \setminus C_1$ where C_0, C_1 are complete finite subtrees of T_d. If $g : T_d \setminus C_0 \to T_d \setminus C_1$ and $g' : T_d \setminus C_0' \to T_d \setminus C_1'$ are almost automorphisms, we say they are equivalent if and only if they agree on

[4] A *lattice* Λ in a locally compact group G is a discrete subgroup such that the quotient space G/Λ admits a G-invariant finite measure; the lattice is *uniform* if the quotient space G/Λ is compact.

$T_d \setminus C$ for some complete finite subtree C satisfying $C_0 \cup C_0' \subseteq C$. Equivalent almost automorphisms induce the same homeomorphism of the set ∂T_d of ends of T_d; in other words they give rise to the same *spheromorphism* of ∂T_d. The spheromorphisms are thus the equivalence classes of almost automorphisms under this equivalence relation. To multiply two spheromorphisms $[g]$ and $[g']$, choose representatives $\overline{g} \in [g]$ and $\overline{g}' \in [g']$ that are defined on the same forest $T_d \setminus C$ (where C is a finite complete subtree), then take $[g][g']$ to be the equivalence class containing $\overline{g}\,\overline{g}'$. The set of spheromorphisms under this multiplication form a group $\mathrm{AAut}(T_d)$ called *Neretin's group.* It was shown in [1] that Neretin's group can be given a group topology by taking the conjugates of vertex stabilisers in $\mathrm{Aut}(T_d) \leq \mathrm{AAut}(T_d)$ as a subbasis of neighbourhoods of the identity (this makes sense because $\mathrm{AAut}(T_d)$ commensurates each stabiliser $\mathrm{Aut}(T_d)_v$). Under this topology, $\mathrm{AAut}(T_d)$ is a compactly generated t.d.l.c. group that contains $\mathrm{Aut}(T_d)$ as an open subgroup. Moreover, $\mathrm{AAut}(T_d)$ is simple and contains no lattice (see [1, Theorem 1]).

Given $G \leq \mathrm{Aut}(T_d)$, Waltraud Lederle's group $\mathcal{F}(G)$ consists of elements of $\mathrm{AAut}(T_d)$ that intuitively 'locally look like' G. When G is closed and has Tits' Independence Property (P) (that is, when G is (P)-closed), there is a unique group topology on $\mathcal{F}(G)$ such that the inclusion of G into $\mathcal{F}(G)$ is continuous and open. Since G is t.d.l.c. the group $\mathcal{F}(G)$ is also a t.d.l.c. group. The focus of [14] is when G is taken to be a Burger–Mozes group $\mathbf{U}(F)$.

To define $\mathcal{F}(G)$ precisely, we first define a *G-almost automorphism* of T_d to be an almost automorphism $h : T_d \setminus C \to T_d \setminus C'$ such that for every component $\mathcal{T}$ of the forest $T_d \setminus C$ there is some $g \in G$ such that g and h agree on $\mathcal{T}$. The elements of $\mathcal{F}(G)$ are the equivalence classes of all G-almost automorphisms and one can easily see that $\mathcal{F}(G) \leq \mathrm{AAut}(T_d)$.

In [14] it is shown that if $F \leq S_d$ then $\mathcal{F}(\mathbf{U}(F))$ is a compactly generated t.d.l.c. group and its commutator subgroup is open, abstractly simple and has finite index in $\mathcal{F}(\mathbf{U}(F))$. Moreover, if F is a Young subgroup with strictly fewer than d orbits, then this commutator subgroup is a nondiscrete compactly generated simple group without lattices.

Jens Bossaert and Tom De Medts in [3] generalise the box product construction by defining universal groups over right-angled buildings where the local actions are again arbitrary permutation groups. As with the box product, there is no requirement for these local actions to have finite degree. Bossaert and De Medts characterise when their universal groups are locally compact, abstractly simple or have primitive action on the residues of the building.

An important generalisation of Tits' Independence Property (P) is described by Christopher Banks, Murray Elder and George A. Willis in [2], where it is called *Property* (P_k). This definition is made only for locally finite trees. For groups $H \leq \mathrm{Aut}(T)$ satisfying Property (P_k) there is a natural subgroup denoted H^{+_k} that is trivial or simple whenever H neither fixes an end nor setwise stabilises a proper subtree of T (see [2, Theorem 7.3]). In the same paper they define something they call the k-closure of a group. In a different context this notion was independently described by Sam Shepherd in [22]. The term k-closure has a different and well-established meaning in permutation group theory due to Wielandt, and so we do not use the term; we instead call this notion the (P_k)-*closure* and define it in the following paragraph. In both [2] and [22] the notion of (P_k)-closure pertains only to locally finite trees where issues surrounding the closure or non-closure of local actions do not occur, because they are always closed. In fact, the definition of (P_k)-closure can be adjusted slightly so as to work in a natural way for trees that are not locally finite, and this latter notion was introduced in our paper [20]. Here we follow [20], and note that for locally finite trees the definitions coincide.

Let T be a tree that may or may not be locally finite, and let $G \leq \mathrm{Aut}(T)$. For $k \in \mathbb{N}$, the (P_k)-*closure* of G, denoted $G^{(\mathrm{P}_k)}$, is the group consisting of all $g \in \mathrm{Aut}(T)$ such that for all $v \in VT$, and all finite subsets $\Phi \subseteq B_v(k)$, there exists $g_\Phi \in G$ such that $gw = g_\Phi w$ for every vertex $w \in \Phi$. We say that the group G is (P_k)-*closed* if G is equal to its (P_k)-closure. Intuitively, we can think of the k-local action of G as being its action on k-balls $B_v(k)$, and the (P_k)-closure of G is then the largest subgroup of $\mathrm{Aut}(T)$ whose k-local action is equal to the closure of the k-local action of G. Note that all of the groups described thus far in this subsection that arise as closed groups of automorphisms of trees are (P_k)-closed for $k = 1$.

In [2] the following properties of $G^{(\mathrm{P}_k)}$ are established for $G \leq \mathrm{Aut}(T)$ where T is locally finite; see [20] for trivial adjustments to the arguments so they continue to work in the non-locally finite case. Again let T be a (not necessarily locally finite) tree, let $G \leq \mathrm{Aut}(T)$, and fix $k \in \mathbb{N}$.

(i) $G^{(\mathrm{P}_k)}$ is a closed subgroup of $\mathrm{Aut}(T)$.
(ii) $G^{(\mathrm{P}_r)} = (G^{(\mathrm{P}_k)})^{(\mathrm{P}_r)}$ whenever $r \leq k$. In particular, $G^{(\mathrm{P}_k)}$ is (P_k)-closed.

(iii) If G is closed then $G = G^{(\mathrm{P}_1)}$ if and only if G satisfies Tits' Independence Property (P). Consequently, to decide if G satisfies Tits' Independence Property (P) it is sufficient to check only that the definition of Property (P) holds for paths in T of length one (i.e. edges).

The (P_k)-closures of G as $k \to \infty$ give a series of approximations to the action of G on T. Moreover, the (P_k)-closures converge to the closure of the action of G in $\mathrm{Aut}(T)$. Intuitively then the notion of (P_k)-closure gives a tool for understanding all actions of groups on trees as 'limits' of 'universal groups' of 'local actions' on k-balls. In our theory of local action diagrams we make precise this idea for $k = 1$, creating an overarching theory for (P_k)-closed groups when $k = 1$.

Understanding (P_k)-closures for $k > 1$ presents significant challenges, but it is the obvious direction in which to generalise our theory of local action diagrams (see Section 7.8). Stephan Tornier in [28] has generalised the Burger–Mozes construction (for locally finite trees T_d) in a way that allows the local action on balls of a given radius $k \geq 1$ to be prescribed, where the Burger–Mozes groups $\mathbf{U}(F)$ then arise precisely when $k = 1$. Even in this setting, Tornier's work highlights how the global behaviour of the resulting 'universal' group is difficult to control – interactions between the prescribed local action on overlapping k-balls is the source of this complexity.

Let $\mathcal{L}$ be a legal colouring in the sense of Burger–Mozes (see Section 7.3.2) of the locally finite d-regular tree T_d. Fix a subtree of the labelled tree T_d arising as a ball of radius k around some vertex, and denote this by $B_{d,k}$. For each vertex $v \in VT_d$ recall that $B_v(k)$ is the k-ball around v, which is isomorphic (as a labelled tree) to $B_{d,k}$; denote this label-respecting isomorphism (which is unique) by $l_v^k : B_v(k) \to B_{d,k}$. The *$k$-local action* of $g \in \mathrm{Aut}(T_d)$ is then the graph isomorphism

$$\sigma_{k,v}(g) : B_{d,k} \to B_{d,k} \qquad \sigma_{k,v}(g) = l_{gv}^k \circ g \circ (l_v^k)^{-1}.$$

For any group $F \leq \mathrm{Aut}(B_{d,k})$ we define $\mathbf{U}_k(F)$ as follows,

$$\mathbf{U}_k(F) := \{g \in \mathrm{Aut}(T_d) : \sigma_{k,v}(g) \in F \quad \forall v \in VT_d\}.$$

In [28] it is shown that $\mathbf{U}_k(F)$ is a closed, vertex transitive and compactly generated subgroup of $\mathrm{Aut}(T_d)$, however $\mathbf{U}_k(F)$ can fail to have k-local action F. Despite this, the group retains a universal property: if $H \leq \mathrm{Aut}(T_d)$ is locally transitive and contains an involutive inversion (that is, an involution g that maps some $e \in AT_d$ to $\overline{e}$), and F is the k-local action of H, then $\mathbf{U}_k(F)$ contains an $\mathrm{Aut}(T_d)$-conjugate of H.

Precisely when $\mathbf{U}_k(F)$ has local action F is characterised by a condition on F that Tornier calls Condition (C); this condition can be thought of as a 'compatibility' condition for the local action with itself as it interacts with an involutive automorphism. Another condition on F called Condition (D) is sufficient for $\mathbf{U}_k(F)$ to be discrete, and when F satisfies Condition (C) then (D) precisely characterises whether or not $\mathbf{U}_k(F)$ is discrete.

Finally we mention Caprace and De Medts' paper [7] which contains an abundance of interesting results. The paper largely concerns compactly generated locally compact groups which act on locally finite trees and satisfy Tits' independence property (P). In other words, compactly generated locally compact (P)-closed actions on locally finite trees. The central theme of the paper is to consider the extent to which selected global properties of these groups (e.g. having every proper open subgroup compact) are determined by the local action of a stabiliser of a vertex on its neighbours. The theory of local action diagrams shows that *all* properties of (P)-closed actions on trees are determined by these local actions, because every (P)-closed action is the universal group of a local action diagram.

7.4 The theory of local action diagrams

In this section we give an intuitive introduction to the various ingredients of our theory, and we describe the relationship between them. Everything here can be found in [20].

7.4.1 Local action diagrams

At the heart of our theory is a combinatorial object we call a *local action diagram*. It plays the same role in our theory that a graph of groups plays in Bass–Serre Theory (see Section 7.2.1).

Definition 7.4.1 A *local action diagram* is comprised of three things:

(i) A nonempty connected graph Γ.
(ii) A nonempty set X_a of colours for each arc $a \in A\Gamma$, such that the colour sets of distinct arcs are disjoint. For each vertex $v \in V\Gamma$ let $X_v := \bigsqcup_{a \in o^{-1}(v)} X_a$.

(iii) A closed group $G(v) \leq \mathrm{Sym}(X_v)$ for each vertex $v \in V\Gamma$, such that the arc colour sets X_a are the orbits of $G(v)$ on X_v.

We denote such a local action diagram as $\Delta = (\Gamma, (X_a), (G(v)))$. We call each X_a the *colour set* of a, and each group $G(v)$ the *local action* at v.

Notice how easy it is to construct a local action diagram. One chooses any nonempty connected graph and some groups (that have the specified orbit structure) to play the role of local actions; there are no further compatibility conditions governing whether or not various local actions can be combined on a local action diagram. Nevertheless, we will see they provide a complete description of all (P)-closed groups of automorphisms of trees.

Two local action diagrams are isomorphic if the graphs are isomorphic and their local actions are permutationally isomorphic, and these two types of isomorphisms are compatible. More precisely, if we have two local action diagrams $\Delta = (\Gamma, (X_a), (G(v)))$ and $\Delta' = (\Gamma', (X'_a), (G'(v)))$, an isomorphism between them consists of two things:

(i) an isomorphism $\theta : \Gamma \to \Gamma'$ of graphs; and
(ii) bijections $\theta_v : X_v \to X'_{\theta(v)}$ for each vertex $v \in V\Gamma$, that restrict to a bijection $X_a \to X_{\theta(a)}$ for each $a \in o^{-1}(v)$, and such that $\theta_v G(v) \theta_v^{-1} = G'(\theta(v))$.

Example 7.4.2 In Figures 7.1 and 7.2 we see examples of local action diagrams. For now they are nothing more than combinatorial objects, but we will soon see that they give rise to universal groups acting on trees. The second local action diagram shows how one can embed interesting permutation groups or topological groups as local actions: the automorphism group of the 3-regular tree T_3 is a component of the local action of v, where $\mathrm{Aut}(T_3)$ is acting on the colour set (i.e. VT_3) of the arc a.

7.4.2 The associated local action diagram

As we described in Section 7.2.1, for every group of automorphisms of a tree we can associate a graph of groups. The analogous situation arises here: for every group of automorphisms of a tree we can associate a local action diagram.

Definition 7.4.3 Suppose G is a group of automorphisms of a tree T. The *associated local action diagram* is a local action diagram $(\Gamma, (X_a), (G(v)))$ with the following parameters.

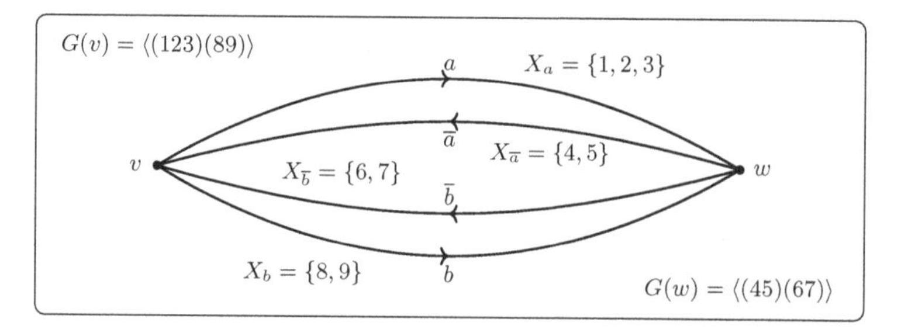

Figure 7.1 An example of a local action diagram.

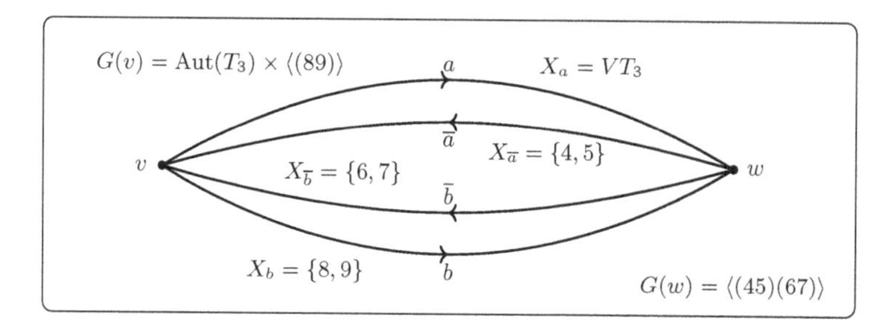

Figure 7.2 Interesting permutation groups can be embedded as local actions inside local action diagrams.

- For our connected graph we take Γ to be the quotient graph $G\backslash T$, with π denoting the natural quotient map.
- For our arc colours we proceed as follows. The vertices of Γ are orbits of G on VT, so for each vertex $v \in V\Gamma$ we choose a representative vertex v^* in VT such that $\pi(v^*) = v$, and write V^* for the set of all such representatives. Now the stabiliser G_{v^*} permutes the arcs in $o^{-1}(v^*)$, and so we take $X_v := o^{-1}(v^*)$. The arcs in Γ are the orbits of G on AT, so this set X_v breaks down into G_{v^*}-orbits as $X_a := \{b \in o^{-1}(v^*) : \pi(b) = a\}$ for each $a \in A\Gamma$ satisfying $o(a) = v$, and these sets X_a become our arc colours.
- For the local actions, take $G(v)$ to be the closure of the permutation group induced on X_v by the stabiliser G_{v^*}, and note that the orbits of $G(v)$ and G_{v^*} on X_v coincide.

There are many choices for the associated local action diagram, but they are all isomorphic as local action diagrams.

Example 7.4.4 Consider the diagram in Figure 7.3, which at the top

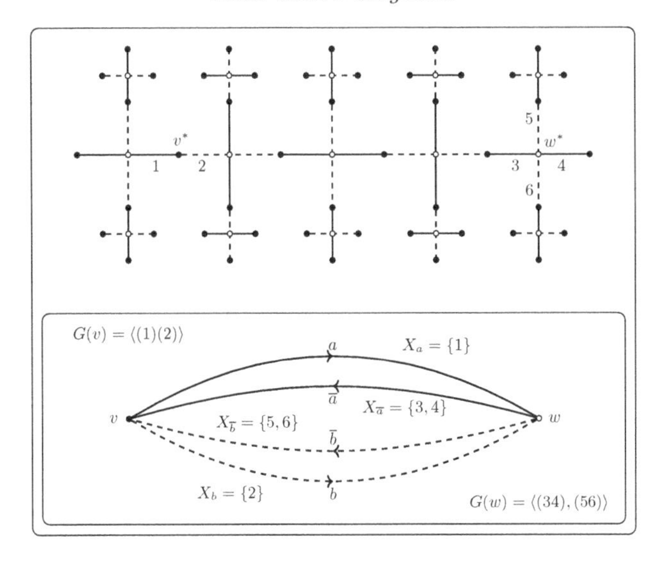

Figure 7.3 Part of $T := T_{2,4}$ with the arc and vertex orbits of $G \leq \mathrm{Aut}(T)$ indicated (top) and the associated local action diagram for (T, G) (bottom) described in Example 7.4.4.

shows part of the infinite $(2, 4)$-biregular tree $T := T_{2,4}$ (ignoring for now the decorations). Let G be the closed subgroup of $\mathrm{Aut}(T)$ with two orbits on vertices (indicated using black vertices and white vertices), and two orbits on edges (solid edges and dashed edges).

Thus the local action diagram has two vertices, say v, w, and two edges (each consisting of an arc in each direction), say $\{a, \overline{a}\}$ and $\{b, \overline{b}\}$. We pick vertex orbit representatives v^* and w^* in VT. The T-arcs leaving these vertices will become the arc colours in our local action diagram, so we name them as indicated: the two arcs leaving v^* are named 1, 2 and the four arcs leaving w^* are named 3 to 6. Now $o^{-1}(v^*) = \{1, 2\}$ and G_{v^*} induces the trivial subgroup of S_2 on this set. Thus, we take $G(v) = \langle(1)(2)\rangle$. Meanwhile, $o^{-1}(w^*) = \{3, 4, 5, 6\}$ and G_{w^*} induces $\langle(34), (56)\rangle \leq \mathrm{Sym}(\{3, 4, 5, 6\})$ on $o^{-1}(w^*)$, so we take $G(w) = \langle(34), (56)\rangle$.

From these observations we obtain the local action diagram for (T, G), shown at the bottom of Figure 7.3.

Example 7.4.5 Let $d \geq 3$ be a finite or infinite cardinal and let $F \leq S_d$. Consider the Burger–Mozes group $\mathbf{U}(F) \leq \mathrm{Aut}(T_d)$. Fix an arc $a \in AT_d$ and notice that the condition $\mathcal{L}(a) = \mathcal{L}(\overline{a})$ ensures that there is an element $g \in \mathrm{Aut}(T_d)$ inverting a (that is, mapping a to $\overline{a}$ and

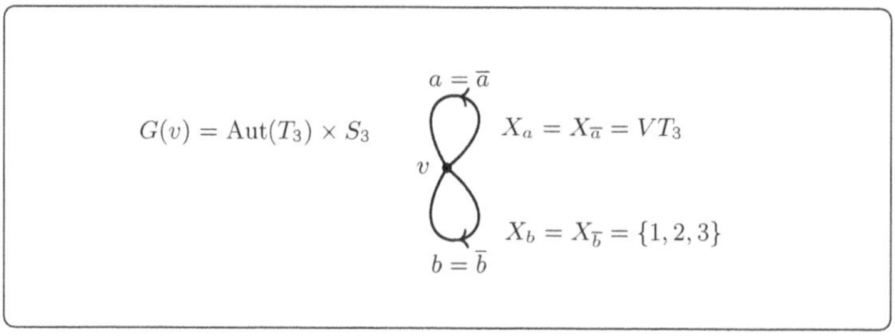

$G(v) = \mathrm{Aut}(T_3) \times S_3$ $a = \overline{a}$ $X_a = X_{\overline{a}} = VT_3$ v $X_b = X_{\overline{b}} = \{1, 2, 3\}$ $b = \overline{b}$

Figure 7.4 The associated local action diagram for Example 7.4.5.

vice versa) such that $\mathcal{L}(b) = \mathcal{L}(gb)$ for all $b \in AT_d$. For this element g we have that $\sigma_{\mathcal{L},v}(g)$ is trivial and therefore lies in F, for all $v \in VT_d$. Thus, $g \in \mathbf{U}(F)$. We have seen then that for any arc $a \in AT_d$ we have that a and $\overline{a}$ lie in the same orbit of $\mathbf{U}(F)$. Moreover, in Section 7.3.2 we saw that $\mathbf{U}(F)$ is vertex transitive. Hence the associated local action diagram for $\mathbf{U}(F)$ consists of a single vertex with some loops, each of which is its own reverse. Each loop corresponds to an orbit of F.

For example, taking F to be the group $\mathrm{Aut}(T_3) \times S_3$ acting via the product action on $VT_3 \times \{1, 2, 3\}$, we have that F has two orbits and therefore $\mathbf{U}(\mathrm{Aut}(T_3) \times S_3)$ has the associated local action diagram shown in Figure 7.4.

Example 7.4.6 Let $d_1, d_2 > 1$ be finite or infinite cardinals, let X_1, X_2 be sets of cardinality d_1, d_2 respectively, and let $F_1 \leq \mathrm{Sym}(X_1)$ and $F_2 \leq \mathrm{Sym}(X_2)$. Consider the box product group $\mathbf{U}(F_1, F_2) \leq \mathrm{Aut}(T)$, where T is the biregular tree T_{d_1,d_2}.

In this general setting the local action diagram of $\mathbf{U}(F_1, F_2)$ can be significantly more complicated than those of the Burger–Mozes groups; despite this it is still tractable. Indeed, in [25, Lemma 22] it is shown that the quotient graph $\mathbf{U}(F_1, F_2)\backslash T$ is the complete bipartite graph K_{n_1,n_2}, where n_i is the number of orbits of F_i on X_i for $i = 1, 2$. In particular, between any two vertices in K_{n_1,n_2} there is a single edge consisting of two arcs, one in each direction.

Identifying $\mathbf{U}(F_1, F_2)\backslash T$ and K_{n_1,n_2} we can now form the associated local action diagram. Let V_1, V_2 be the two parts corresponding to the bipartition of K_{n_1,n_2}, with each part V_i consisting of vertices with valency n_i for $i = 1, 2$. Vertices in V_i have local action F_i. Each arc a from V_1 to V_2 represents an orbit Ω of F_1 and so we set $X_a := \Omega$. Similarly,

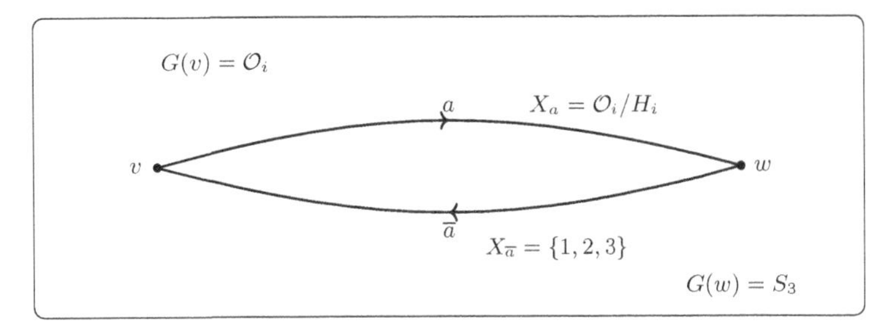

Figure 7.5 Local action diagrams for the groups described in Example 7.4.6.

each arc b from V_2 to V_1 represents and orbit Ω' of F_2, and so we set $X_b := \Omega'$. We have a local action diagram.

In the special case where F_1 and F_2 are transitive, the local action diagram associated to $\mathbf{U}(F_1, F_2)$ is then two vertices connected by a single edge which consists of a pair of arcs, one in each direction. For example, the $2^{\aleph_0}$ pairwise nonisomorphic simple nondiscrete compactly generated t.d.l.c. groups that were constructed in the proof of [25, Theorem 38] were of the form $\mathbf{U}(\mathcal{O}_i, S_3)$, where each group $\mathcal{O}_i$ is viewed as a permutation group $\mathcal{O}_i \leq \mathrm{Sym}(\mathcal{O}_i/H_i)$ as in Section 7.3.3. The local action diagrams of these groups are shown in Figure 7.5.

7.4.3 Building a Δ-tree

In Section 7.2.1 we described how every graph of groups in Bass–Serre Theory gives rise to a universal cover that is a tree. An analogous situation arises here: every local action diagram gives rise to an arc-coloured tree we call a *Δ-tree.* Intuitively this arc-coloured tree is built from taking 'coloured walks' around the local action diagram. Shortly, we will construct a natural universal group from the automorphism group of this arc-coloured tree.

Definition 7.4.7 Let $\Delta = (\Gamma, (X_a), (G(v))$ be a local action diagram. A *Δ-tree* is comprised of three things:

(i) a tree T;
(ii) a surjective graph homomorphism $\pi : T \to \Gamma$; and
(iii) a colouring map $\mathcal{L} : AT \to \bigsqcup_{a \in A\Gamma} X_a$, such that for every vertex $v \in VT$, and every arc a in $o^{-1}(\pi(v))$, the map $\mathcal{L}$ restricts to a bijection $\mathcal{L}_{v,a}$ from $\{b \in o^{-1}(v) : \pi(b) = a\}$ to X_a.

We denote such a Δ-tree as $\mathbf{T} = (T, \mathcal{L}, \pi)$.

In [20, Lemma 3.5] it is shown that for any local action diagram Δ there exists a Δ-tree, and moreover, any two Δ-trees $(T, \pi, \mathcal{L})$ and $(T', \pi', \mathcal{L}')$ are isomorphic in the following sense: there is a graph isomorphism $\alpha : T \to T'$ such that $\pi' \circ \alpha = \pi$.

The construction of a Δ-tree is intuitive, but it requires careful notation to describe formally. We choose a root vertex $v_0 \in V\Gamma$. Then, for $v \in V\Gamma$ and $c \in X_v$, we define the *type* $p(c)$ of the colour c to be the arc $a \in A\Gamma$ for which $c \in X_a$. A finite sequence of colours $(c_1, c_2, \dots, c_n)$ such that $o(p(c_{i+1})) = t(p(c_i))$ for all $1 \leq i < n$ is called a *coloured path*, and we think of our vertices in VT as being labelled by these coloured paths from the origin v_0.

Intuitively, we build T inductively by taking all coloured paths around Γ starting from v_0, but at each 'step' we will have a choice to make about the reverse colour for our step. We can choose arbitrarily, subject to two constraints: the colour must be of the correct type, and once we have chosen a reverse colour we cannot pick it for the coloured arc corresponding to our next step. To make this precise, we need more notation. If a vertex v has label $(c_1, c_2, \dots, c_n)$ then we write the length of v as $\ell(v) = n$, and we say any vertex with label $(c_1, c_2, \dots, c_m)$ for $m \leq n$ is a *prefix* of v. If v is a prefix of w and $\ell(v) = \ell(w) - 1$, then we write $v \ll w$. The *reverse label* of v is $\overline{v} = (d_1, d_2, \dots, d_n)$, where each d_i is a colour such that (i) $p(d_i) = \overline{p(c_i)}$, and (ii) whenever v is a prefix of w, the label $\overline{v}$ is the corresponding prefix of $\overline{w}$.

We build the vertex set VT of T inductively, starting at a base vertex $()$. If we have already defined a vertex $v = (c_1, c_2, \dots, c_n)$ with $\overline{v} = (d_1, d_2, \dots, d_n)$, then we create new vertices $v_{+c_{n+1}} = (c_1, \dots, c_n, c_{n+1})$ by letting c_{n+1} range over all colours satisfying $o(p(c_{n+1})) = t(p(c_n))$ and $c_{n+1} \neq d_n$. For each of these new vertices, choose d_{n+1} arbitrarily from $X_{\overline{p(c_{n+1})}}$ and let $\overline{v_{+c_{n+1}}} = (d_1, d_2, \dots, d_{n+1})$.

Now we have constructed the vertices of T, creating the remaining tree structure requires little effort. We define $AT_+ := \{(v, w) : v \ll w\}$ and $AT_- := \{(w, v) : (v, w) \in AT_+\}$ and $AT := AT_- \sqcup AT_+$, with functions o, t and edge reversal defined in the obvious way. It is clear that from this we obtain a tree. Furthermore, T is naturally arc-coloured by the following colouring map $\mathcal{L}$: for all $(v, w) \in AT_+$, take $\mathcal{L}(v, w)$ to be the last entry of w, and take $\mathcal{L}(w, v)$ to be the last entry of $\overline{w}$.

To define $\pi : T \to \Gamma$ we define π of the base vertex $()$ to be our fixed root vertex $v_0 \in V\Gamma$, and then set $\pi(v)$ to be the vertex $t(p(c_n)) \in$

$V\Gamma$, where $v \in VT$ has label $(c_1, \ldots, c_n)$. For arcs in $a \in AT$, we set $\pi(a) = p(\mathcal{L}(a))$. It is now routine to verify that π is a surjective graph homomorphism, and that $\mathcal{L}$ restricts to a bijection from $\{b \in o^{-1}(v) : \pi(b) = a\}$ to X_a. Whence, we have our Δ-tree $\mathbf{T} = (T, \mathcal{L}, \pi)$.

Remark 7.4.8 Given a tree action (T, G) and its associated local action diagram Δ, we can equip T with a natural colouring map $\mathcal{L}$ and projection map π so that $(T, \mathcal{L}, \pi)$ is a Δ-tree. Indeed, the associated local action diagram already has an appropriate map π, so we need only specify $\mathcal{L}$. Using the notation of Definition 7.4.3 we choose, for each vertex $v \in VT$, an element $g_v \in G$ such that $g_v v \in V^*$. Then g_v induces a bijection from $o^{-1}(v)$ to X_v, because X_v was defined to be $o^{-1}(v^*)$ for $v^* := g_v v \in V^*$. We then set $\mathcal{L}(b) := g_v b$ for all $b \in o^{-1}(v)$. It transpires that this colouring map satisfies the requirements of Definition 7.4.7, ensuring that $(T, \mathcal{L}, \pi)$ is indeed a Δ-tree.

7.4.4 The universal group of a local action diagram

In Section 7.2.1 we saw that every graph of groups in Bass–Serre Theory gives rise to a fundamental group that acts naturally on its universal cover. An analogous situation arises here: every local action diagram Δ gives rise to a universal group that acts naturally on a Δ-tree.

Definition 7.4.9 Let $\Delta = (\Gamma, (X_a), (G(v)))$ be a local action diagram with Δ-tree $\mathbf{T} = (T, \mathcal{L}, \pi)$. An *automorphism* of $\mathbf{T}$ is a graph automorphism ϕ of the tree T such that $\pi \circ \phi = \pi$. The group of all automorphisms of $\mathbf{T}$ is $\mathrm{Aut}_\pi(T)$. For $g \in \mathrm{Aut}_\pi(T)$ and $v \in VT$ we define the $\mathcal{L}$-local action of g at v as in Equation 7.1: $\sigma_{\mathcal{L},v}(g) := \mathcal{L}|_{o^{-1}(gv)} g \mathcal{L}|_{o^{-1}(v)}^{-1}$. Again we note that $\sigma_{\mathcal{L},v}(g) \in \mathrm{Sym}(X_{\pi(v)})$ for all $v \in VT$.

The *universal group* of Δ and $\mathbf{T}$ is the group consisting of all $\mathbf{T}$-automorphisms whose local action at any $v \in VT$ always lies in the corresponding local action $G(\pi(v))$ of the local action diagram. Formally,

$$\mathbf{U}_{\mathbf{T}}(\Delta) := \{g \in \mathrm{Aut}_\pi(T) : \sigma_{\mathcal{L},v}(g) \in G(\pi(v)) \quad \text{for all } v \in VT\}.$$

It transpires (see [20, Theorem 3.12]) that for a fixed Δ, different Δ-trees give rise to the same universal group; that is, if $\mathbf{T}, \mathbf{T}'$ are Δ-trees with underlying trees T, T' respectively, then there is a graph isomorphism $\phi : T \to T'$ such that $\phi \mathbf{U}_{\mathbf{T}}(\Delta) \phi^{-1} = \mathbf{U}_{\mathbf{T}'}(\Delta)$. For this reason we typically omit the subscripts and speak of *the* universal group $\mathbf{U}(\Delta)$ of a local action diagram Δ.

7.4.5 The correspondence theorem

Recall that the (P)-closure of an action (T, G) of a group G on a tree T is the smallest closed subgroup of $\mathrm{Aut}(T)$ with Tits Independence Property (P) that contains the action (T, G). We denote this by $G^{(\mathrm{P})}$, and if $G = G^{(\mathrm{P})}$ then we say that G is (P)-closed.

In Section 7.2.1 we described the Fundamental Theorem of Bass–Serre Theory, a correspondence theorem linking a tree action (T, G) with the action of the fundamental group of its graph of groups on its universal cover. There is an analogous correspondence theorem for local action diagrams, linking the (P)-closure of a tree action (T, G) with the action of the universal group of its local action diagram Δ on its Δ-tree.

Moreover, in the Fundamental Theorem of Bass–Serre Theory we have that the fundamental group of a graph of groups $\mathbf{\Gamma}$ in its action on the universal cover of $\mathbf{\Gamma}$ (which is a tree) has $\mathbf{\Gamma}$ as its associated graph of groups. An analogous statement holds for local action diagrams: the universal group of a local action diagram Δ in its action on a Δ-tree has an associated local action diagram that is isomorphic to Δ.

Theorem 7.4.10 ([20, Theorems 3.9 & 3.10]) *Let G be a group and T be a tree.*

($\circledast$) *Suppose G acts on T with associated local action diagram Δ, universal group $\mathbf{U}(\Delta)$ and Δ-tree $\mathbf{T}$. Then the actions $(T, G^{(\mathrm{P})})$ and $(\mathbf{T}, \mathbf{U}(\Delta))$ can be identified.*

($\circledcirc$) *Suppose Δ is a local action diagram. Let $\mathbf{T}$ be a Δ-tree and $\mathbf{U}(\Delta)$ its universal group. Then $\mathbf{U}(\Delta)$ is a (P)-closed group of automorphisms of the Δ-tree $\mathbf{T}$ and the associated local action diagram of $(\mathbf{T}, \mathbf{U}(\Delta))$ is isomorphic to Δ.*

The first statement ($\circledast$) follows from the following observations. We have seen in Remark 7.4.8 that we can equip T with a colouring map $\mathcal{L}$ so that T becomes a Δ-tree $\mathbf{T}' = (T, \mathcal{L}, \pi)$. Using these we can construct the universal group $\mathbf{U}_{\mathbf{T}'}(\Delta) \leq \mathrm{Aut}(T)$. In [20, Theorem 3.10] we show that in fact $\mathbf{U}_{\mathbf{T}'}(\Delta) = G^{(\mathrm{P})}$. As noted previously, the possibly different Δ-trees $\mathbf{T}'$ and $\mathbf{T}$ give rise to permutationally isomorphic universal groups, and via this relationship we can identify the actions of $(T, G^{(\mathrm{P})})$ and $(\mathbf{T}, \mathbf{U}(\Delta))$. The second statement ($\circledcirc$) is [20, Theorem 3.9].

From this correspondence theorem, we obtain the following universal property for $\mathbf{U}(\Delta)$, which holds because for any group $G \leq \mathrm{Aut}(T)$ with local action diagram Δ we have $G \leq G^{(\mathrm{P})} = \mathbf{U}(\Delta)$. This property clarifies the nature of the universal properties for the Burger–Mozes and box product groups, which both needed local transitivity to hold.

Corollary 7.4.11 *Suppose Δ is a local action diagram, and form a Δ-tree $\mathbf{T} = (T, \mathcal{L}, \pi)$. Then the universal group $\mathbf{U}(\Delta) \leq \mathrm{Aut}(T)$ contains a permutationally isomorphic copy of every group $G \leq \mathrm{Aut}(T)$ whose associated local action diagram is Δ.*

7.5 A classification of groups with Tits' Independence Property (P)

7.5.1 The classification

As we have already mentioned, Tits' Independence Property (P) plays an important role in the theory of infinite groups, because it gives a general tool for constructing infinite nonlinear simple groups and because of its obvious importance to the subject of groups acting on trees. Our theory of local action diagrams gives a complete and highly usable description of all closed actions that have Tits' Independence Property (P). From this one immediately obtains a classification of all groups that have Tits' Independence Property (P).

In Theorem 7.4.10 and the subsequent commentary, if we have a tree action (T, G) where G is closed with Tits' Independence Property (P), then $G = G^{(\mathrm{P})}$ and therefore (T, G) is equal to the universal group $\mathbf{U}_{\mathbf{T}'}(\Delta) \leq \mathrm{Aut}(T)$, where Δ is the local action diagram of (T, G) and $\mathbf{T}'$ is the Δ-tree obtained from T as described in Remark 7.4.8. On the other hand, if we have the universal group $\mathbf{U}(\Delta)$ of a local action diagram then it is a group of automorphisms of the underlying tree T of a Δ-tree, and the action $(T, \mathbf{U}(\Delta))$ is closed with Tits' Independence Property (P). Thus, we have the following description of closed groups with (P).

Closed groups of automorphisms of trees with Tits' independence property (P) *are precisely the universal groups of local action diagrams.*

In our paper [20], we state the correspondence between (P)-closed actions and local action diagrams as follows.

Theorem 7.5.1 ([20, Theorem 3.3]) *There is a natural one-to-one correspondence between isomorphism classes of* (P)*-closed actions on trees and isomorphism classes of local action diagrams.*

This correspondence is easy for us to state explicitly. For a local action diagram Δ, we have a corresponding pair $(T, \mathbf{U}(\Delta))$, where T is the underlying tree of a Δ-tree. As previously discussed, different Δ-trees give rise to permutationally isomorphic universal groups. Hence the pair $(T, \mathbf{U}(\Delta))$ is unique up to isomorphisms. Moreover, by construction we see that two isomorphic local action diagrams Δ and Δ' will produce isomorphic actions $(T, \mathbf{U}(\Delta))$ and $(T', \mathbf{U}(\Delta'))$. We have shown that there is a well-defined map θ from isomorphism class of actions (T, G), where T is a tree and G is a (P)-closed group of automorphisms of T, to isomorphism classes of local action diagrams. Our correspondence theorem (Theorem 7.4.10) shows that θ is a bijection. Thus θ is our claimed natural one-to-one correspondence.

Recall that a permutation group $G \leq \mathrm{Sym}(\Omega)$ and its closure in $\mathrm{Sym}(\Omega)$ have the same orbits on all n-tuples of Ω, for all $n \in \mathbb{N}$. Our complete description of all (P)-closed actions on trees immediately yields a useful classification of all (not just closed) actions (T, G), where T is a tree and G has Tits' Independence Property (P), whereby such actions (T, G) are classified according to the isomorphism type of their associated local action diagram. In such a classification, every such action (T, G) lies in precisely one class and the classification gives a complete description of the closure of G. Thus, we now have a deep understanding of all groups with Tits' Independence Property (P).

7.5.2 Some consequences

There are many consequences to our description of all (P)-closed actions on trees. In this section we describe two (see [20] for more).

The first is that any (P)-closed action on a tree can be described completely by drawing a local action diagram. This means that *all* properties of the action (e.g. whether it is simple, geometrically dense, etc) can be read directly from the local action diagram. We explore this further in Section 7.6.

The second is that for natural numbers d, n there are only finitely many conjugacy classes of (P)-closed actions (T, G) such that T is locally finite of bounded valency d and G has at most n vertex orbits. Indeed, any such action arises as the universal group of a local action diagram $\Delta = (\Gamma, (X_a), (G(v)))$ where Γ has at most n vertices and all groups $G(v)$ are finite permutation groups of degree at most d. In a sense, our theory reduces the study of such groups to questions about finite graphs and finite permutation groups. In particular, for sensible choices for d, n it would be possible to enumerate and describe (with the help of a computer) all such groups, for example by constructing all (finitely many) possible local action diagrams and then determining which are isomorphic.

In [20, §7.1] we consider a special case of this: (P)-closed actions (T_d, G) where T_d is the d-regular tree and G is vertex-transitive. By our classification theorem we know that up to conjugacy, there are only finitely many such actions. To determine them, we define an *orbit pairing* for $H \leq S_d$ to be a permutation of the set $H\backslash[d]$ of H-orbits whose square is the identity, where $[d]$ here denotes the set $\{1, \ldots, d\}$. We consider pairs (H, r), where $H \leq S_d$ and r is an orbit pairing for H, and say two pairs (H_1, r_1) and (H_2, r_2) are equivalent whenever there exists $g \in S_d$ such that $gH_1g^{-1} = H_2$ and the map $g' : H_1\backslash[d] \to H_2\backslash[d]$ induced by g satisfies $g'r_1 = r_2g'$. The (finitely many) $\mathrm{Aut}(T_d)$-conjugacy classes for (P)-closed vertex transitive actions (T_d, G) are in one-to-one correspondence with the set of equivalence classes of pairs (H, r).

Since each G is vertex transitive, its associated local action diagram $\Delta = (\Gamma, (X_a), (G(v)))$ has only one vertex v_0, together with some loops that may or may not be their own reverse. For each pair (H, r), the group H is the group $G(v_0)$; the arcs of Δ correspond to orbits of H; and the edges of Δ correspond to orbits of r on $H\backslash[d]$. Recall that the vertices, arcs and edges of Δ correspond to, respectively, the vertex-, arc- and edge-orbits of $\mathbf{U}(\Delta)$ on T. Those orbits of H that are fixed by r correspond to the arc-orbits of $\mathbf{U}(\Delta)$ on arcs that are reversed by some element in $\mathbf{U}(\Delta)$. The equivalence classes of these pairs (H, r) give rise to all isomorphism classes of local action diagrams of (P)-closed vertex transitive actions (T_d, G), and we can enumerate these pairs for reasonable values of d.

In [20, §7] we use this method to classify all such actions for $0 \leq d \leq 5$. The appendix to [20] is written by Stephan Tornier and contains a GAP ([10]) implementation that can perform this classification for values of d greater than 5. Even for $d = 3$ we find examples of such actions that do not arise as Burger–Mozes groups; for larger values of d the GAP implementation shows that the conjugacy classes of vertex transitive (P)-closed actions on T_d that do not arise as Burger–Mozes groups grows rapidly.

7.6 Reading simplicity from a local action diagram

Recall Tits' result, Theorem 7.2.6: If $G \leq \mathrm{Aut}(T)$ has property (P) then the subgroup G^+ generated by arc stabilisers is trivial or simple whenever the action of G is geometrically dense (i.e. G leaves invariant no nonempty proper subtree of T and G fixes no end of T). Closed groups $G \leq \mathrm{Aut}(T)$ with property (P) are completely described by their local action diagrams, so it is not surprising that the simplicity of G^+ can be read directly from the local action diagram. What is surprising, however, is how easily this can be done. It happens that the invariant subtrees and fixed ends of a tree action (T, G) correspond to combinatorial features of the local action diagram that we call *strongly confluent partial orientations* (or scopos). Again we note that T does not need to be locally finite.

Definition 7.6.1 A *strongly confluent partial orientation* (or *scopo*) in a local action diagram $\Delta = (\Gamma, (X_a), (G(v)))$ is a subset O of $A\Gamma$ satisfying:

(i) For all $a \in O$ we have $\overline{a} \notin O$ and $|X_a| = 1$;
(ii) For all $a \in O$ we have that $t^{-1}(o(a)) \setminus \{\overline{a}\} \subseteq O$.

The empty set is always a scopo; if the empty scopo is the only scopo of Δ then Δ is said to be *irreducible.*

In [20, Theorem 1.4] it is noted that the invariant subtrees and fixed ends of (T, G) correspond to scopos of the local action diagram Δ. Under this natural correspondence, the empty scopo (which always exists) corresponds to T (which is always invariant under G). Thus (T, G) being geometrically dense is equivalent to Δ being irreducible.

We can completely characterise all types of scopos that correspond to invariant subtrees and fixed ends of faithful actions (T, G) with Property (P). There are four types of these scopos, and we call them a *stray leaf*, a *focal cycle*, a *horocyclic end* and a *stray half-tree*. Before describing these types of scopos, we note the following.

Theorem 7.6.2 ([20, Corollary 1.5]) *If $G \leq \mathrm{Aut}(T)$ has Tits' Property* (P) *and local action diagram Δ, then the following are equivalent:*

(i) G is geometrically dense;
(ii) Δ is not a focal cycle and has no stray half-trees, no horocyclic ends and no stray leaves.

In particular, if ii holds then G^+ is abstractly simple or trivial.

In this theorem we have a complete characterisation in the local action diagram of when Tits' Theorem can be applied to an action with Tits' Property (P).

Definition 7.6.3 Let $\Delta = (\Gamma, (X_a), (G(v)))$ be a local action diagram. The following are the aforementioned scopos that correspond to fixed ends and proper invariant subtrees.

- If Γ is a finite cycle with a cyclic orientation O, such that for all $a \in O$ we have $|X_a| = 1$, then we say that Δ is a *focal cycle.*
- A *stray leaf* of Δ is a leaf v of Γ such that $|X_v| = 1$ (or equivalently such that $G(v)$ is trivial).
- A *horocyclic end* of Δ occurs only when Γ is a tree. It is an end ξ of the tree Γ such that all the arcs $a \in A\Gamma$ that are directed towards ξ satisfy $|X_a| = 1$.
- If $\Gamma \setminus \{a, \overline{a}\}$ is not connected and Γ_a is the connected component containing $t(a)$, then Γ_a is a *stray half-tree* of Δ whenever Γ_a is a tree that contains no leaves of Γ, and moreover within Γ_a all arcs b orientated towards $t(a)$ satisfy $|X_b| = 1$.

This characterisation of scopos allows us to quickly determine whether or not Δ is irreducible in the frequently encountered case where Γ is a finite graph that is not a cycle: Δ is irreducible if and only if Δ has no stray leaves. Thus, in this situation, if Δ has no stray leaves then G^+ is simple or trivial.

Determining whether or not G^+ is trivial is easy to detect in the local action diagram.

Proposition 7.6.4 ([20, Lemma 5.11]) *If T is a tree and $G \leq \mathrm{Aut}(T)$ with associated local action diagram $\Delta = (\Gamma, (X_v), (G(v)))$, then G^+ is trivial if and only if $G(v)$ acts freely (i.e. semiregularly) on X_v for all $v \in V\Gamma$.*

Our discussion so far concerned the simplicity of G^+. However, in various naturally occurring situations (see for example property v of the box product construction in Section 7.3.3) it is the case that G itself is simple. Regarding this, we have an almost complete characterisation of simplicity for faithful (P)-closed actions. It is almost complete because we must exclude some degenerate cases and we must ensure that the action results in a closed induced action on any invariant subtree.

Definition 7.6.5 A group $G \leq \mathrm{Aut}(T)$ is *strongly closed* if for every G-invariant subtree T' of T, the induced action of G on T' is closed.

Being strongly closed is easily achieved. For example, in [20, Corollary 6.4], we show that a locally compact (P)-closed subgroup of $\mathrm{Aut}(T)$ that acts with translation is always strongly closed.

Theorem 7.6.6 ([20, Theorem 1.8]) *If (T, G) is a faithful (P)-closed and strongly closed action on a tree T, then the following are equivalent:*

(i) G is nondiscrete, abstractly simple, and acts with translation.
(ii) There exists an infinite G-invariant subtree T' of T (not necessarily proper) on which G acts faithfully. Furthermore, if the associated local action diagram of (T', G) is $\Delta = (\Gamma, (X_a), (G(v)))$, then Γ is a tree; Δ is irreducible; all the groups $G(v)$ are closed and generated by point stabilisers; and at least one of the groups $G(v)$ is nontrivial.

7.7 Topological properties of universal groups

In this section we survey a selection of results in [20] concerning various topological properties of (P)-closed subgroups of $\mathrm{Aut}(T)$. All statements are with respect to the permutation topology. Of course the topological properties of (P)-closed subgroups of $\mathrm{Aut}(T)$ when T is locally finite are well-understood. The novelty in our results is that (i) we make no assumptions about local finiteness, and (ii) our results are typically concerned with deducing 'global' topological statements from 'local' properties that can be found in the local action diagram.

To characterise local compactness and compact generation of 'nondegenerate' (P)-closed groups via their local action diagrams, we first need to define a combinatorial feature of local action diagrams called a *cotree*. Let Γ be a connected graph. Directed paths $(v_0, \dots, v_n)$ in Γ are called *nonbacktracking* if $n = 0, 1$ or for $n \geq 2$ we have that $v_i \neq v_{i+2}$ for all $0 \leq i \leq n-2$. Given an induced subgraph Γ' of Γ, a finite directed nonbacktracking path $(v_0, \dots, v_n)$ in Γ such that $v_n \in V\Gamma'$ and $v_i \notin V\Gamma'$ for $i < n$, is called a *projecting path* from $v_0 \in V\Gamma$ to Γ'. We say that Γ' is a *cotree of* Γ if it is nonempty and for all $v \in V\Gamma \setminus V\Gamma'$ there is precisely one projecting path $(v_0, \dots, v_n)$ from v to Γ' (in particular this means that multiple arcs in Γ from any v_i to v_{i+1} are not permitted). Note that if Γ is a connected graph that is not a tree then there is always a (unique) smallest cotree of Γ and cotrees are connected induced subgraphs that contain this smallest cotree. Given a cotree Γ' of Γ, there is a scopo $O_{\Gamma'}$

associated with Γ', consisting of all arcs a satisfying (i) $o(a) \notin V\Gamma'$ and (ii) a lies on the projecting path from $o(a)$ to Γ.

By excluding some degenerate cases, we can characterise local compactness of (P)-closed actions via the local action diagram as follows.

Proposition 7.7.1 ([20, Proposition 6.3]) *Suppose T is a tree and $G \leq \mathrm{Aut}(T)$. Let $\Delta = (\Gamma, (X_a), (G(v)))$ be the local action diagram for (T, G) and let Γ' be the unique smallest cotree of Δ. Suppose further that there is a unique minimal G-invariant subtree T' of T that has at least 3 vertices. Then the following are equivalent.*

(i) $G^{(\mathrm{P})}$ is locally compact.
(ii) For all $a \in AT'$, the arc stabiliser $(G^{(\mathrm{P})})_a$ is compact.
(iii) For all $a \in A\Gamma$ such that $\overline{a} \notin O_{\Gamma'}$, and for all $x \in X_a$, the orbits of the stabiliser $(G(o(a)))_x$ in its action on X_v are finite.

By excluding some degenerate cases, we can characterise the compact generation of (P)-closed actions via their local action diagrams as follows.

Proposition 7.7.2 ([20, Proposition 6.5]) *Let T be a tree and suppose $G \leq \mathrm{Aut}(T)$ is closed with all vertex-orbits having unbounded diameter. Let $\Delta = (\Gamma, (X_a), (G(v)))$ be the local action diagram of (T, G). If some arc stabiliser in G is compact, then G and $G^{(\mathrm{P})}$ are locally compact, and the following are equivalent.*

(i) G is compactly generated;
(ii) $G^{(\mathrm{P})}$ is compactly generated;
(iii) there is a unique smallest G-invariant subtree T' such that G has finitely many orbits on $VT' \sqcup AT'$ and G_v is compactly generated for each $v \in VT'$;
(iv) there is a unique smallest cotree Γ' of Δ such that Γ' is finite and $G(v)$ is compactly generated for each $v \in V\Gamma'$.

Recall the class $\mathscr{S}$ of nondiscrete, topologically simple, compactly generated, locally compact groups. Let $\mathscr{S}_{td}$ be the class of totally disconnected groups in $\mathscr{S}$. For constructing groups in the class $\mathscr{S}_{td}$, we are typically interested in the situation where the universal group is compactly generated, locally compact, and acts geometrically densely on its associated tree (allowing us to deduce simplicity via Tits' Theorem). For this situation we have the following.

Theorem 7.7.3 ([20, Theorem 1.9]) *Suppose $\Delta = (\Gamma, (X_a), (G(v)))$ is a local action diagram. Then $\mathbf{U}(\Delta)$ is compactly generated, locally compact, and acts geometrically densely on its associated tree if and only if Δ is irreducible, Γ is finite, and all groups $G(v)$ are subdegree-finite and compactly generated.*

The following result is particularly useful, since it allows us to construct local action diagrams that immediately yield groups in $\mathscr{S}_{td}$. Again for faithful (P)-closed actions it is an almost perfect characterisation of membership of the class $\mathscr{S}_{td}$.

Theorem 7.7.4 ([20, Corollary 1.10]) *If (T, G) is a faithful (P)-closed and strongly closed action on a tree T, then the following are equivalent:*

(i) $G \in \mathscr{S}_{td}$ and G fixes no vertex of T.
(ii) There exists a unique smallest G-invariant subtree T' of T (not necessarily proper) on which G acts faithfully. Furthermore, if $\Delta = (\Gamma, (X_a), (G(v)))$ is the associated local action diagram of (T', G), then Γ is a finite tree; all of the groups $G(v)$ are closed, compactly generated, subdegree-finite and generated by point stabilisers; and for every leaf v of Γ the group $G(v)$ is nontrivial.

To conclude this section we give a theorem that establishes an entirely different universal property of the groups $\mathbf{U}(\Delta)$ within the class $\mathscr{S}_{td}$.

Theorem 7.7.5 ([20, Theorem 1.14]) *Let $G_1, \ldots, G_n$ be a finite list of nontrivial compactly generated t.d.l.c. groups, such that for each G_i there is a compact open subgroup U_i such that $G_i = \langle gU_ig^{-1} : g \in G_i \rangle$ and $\bigcap_{g \in G_i} gU_ig^{-1}$ is trivial. For example, we can take $G_i \in \mathscr{S}_{td}$ and U_i to be any compact open subgroup. Then there exists $\mathbf{U}(\Delta) \in \mathscr{S}_{td}$ acting continuously on a countable tree T, vertex stabilisers $O_1, \ldots, O_n$ of $\mathbf{U}(\Delta)$ and compact normal subgroups K_i of O_i, such that $O_i \cong K_i \rtimes G_i$ for $1 \leq i \leq n$.*

7.8 Project ideas and open problems

There are a number of directions in which our theory of local action diagrams can be generalised, and many areas where it can be applied; we describe some of them here.

Project 1 *The application of local action diagrams to understand the automorphism groups of various types of infinitely ended graphs.*

A connected, locally finite graph Γ with infinitely many thin ends (i.e. ends that do not contain infinitely many disjoint rays) and only countably many thick ends (i.e. ends that contain infinitely many disjoint rays) is known by work of Carsten Thomassen and Wolfgang Woess (see [27]) to resemble a tree in some precise way. Thomassen and Woess' work relies heavily on Dicks and Dunwoody's theory of structure trees (see [9]), and it is via this theory that we can see the 'tree-like' nature of Γ. See [16] for an 'accessible' introduction to these ideas.

Now the automorphism group $\mathrm{Aut}(\Gamma)$ will act on the structure tree T of Γ, and this action can be studied via local action diagrams. We give an example of this in [20, §7.2], where we find a complete description of all automorphism groups of simple, nontrivial, vertex-transitive graphs with vertex connectivity one: they are precisely the universal groups of a certain type of local action diagram.

The automorphism groups of many other classes of graphs could be understood in this way.

Project 2 *Generalise the ideas behind local action diagrams to better understand* (P_k)*-closures of groups acting on trees.*

This project idea appears in our paper ([20, §8 Question 2]). Such a project would be a significant undertaking, given the complexities of Tornier's generalisation of the Burger–Mozes groups (see Section 7.3.4). Nevertheless, it is our opinion that a usable theory could be developed, perhaps using a modified version of the local action action diagram built around k-arcs rather than arcs.

Project 3 *Find further examples of locally determined global properties of* (T, G).

This is [20, §8 Question 5]. A long term research theme could be built around continuing to find global properties of $G \leq \mathrm{Aut}(T)$ that are perfectly characterised by the associated local action diagram. Such properties could be found by looking for global properties of G that are characterised by properties of $G^{(\mathrm{P})}$ in its action on T. In [20] we call such properties *locally determined global properties of* (T, G). As we have seen, geometrically dense actions are a locally determined global property. An important measure of success here will be that the locally determined global properties should be expressed in terms of features of the local action diagram.

Project 4 *Write software to search local action diagrams for known* *features that characterise the locally determined global properties of actions (T, G).*

As we saw in 7.5.2, for given natural numbers d, n there are only finitely many isomorphism classes of local action diagrams for actions (T, G) where G has at most n vertex orbits on T, and T is locally finite with every valency bounded by d. Constructing all possible local action diagrams for a given pair (d, n) is then possible for reasonable choices of (d, n), since these are finite graphs decorated with finite groups and finite colour sets. Algorithms can be created to search these constructions for known locally determined global properties. In this way, we could obtain lists of $\mathrm{Aut}(T_d)$-conjugacy classes of, for example, vertex transitive actions on T_d that have some locally determined global property. As new properties come to light via Project 3, more software solutions will be needed.

In [20, §8 Question 2] we give a somewhat related project, asking for the asymptotics of the number N_d of conjugacy classes of vertex-transitive (P)-closed subgroups of $\mathrm{Aut}(T_d)$ as a function of d.

Project 5 *Create analogies of the constructions in Section 7.3.4, for universal groups of local action diagrams.*

In Section 7.3.4 we outlined some generalisations of either the Burger–Mozes groups or the box product construction. For each of these generalisations there are natural ways to modify their definitions so that they can be applied to local action diagrams. For example, inspired by Le Boudec groups one could allow finitely many (or boundedly many) singularities for local actions in local action diagrams, and the local action at these singularities could be restricted to give an analogy to Le Boudec's restricted universal groups.

For all of these generalisations, it would be interesting to understand what permutational and topological properties the resulting 'modified universal' groups possess.

Furthur projects As this is an introductory note, we have omitted several aspects of the theory (e.g. local subaction diagrams as a tool to better understand subgroups of universal groups of local action diagrams). There are several interesting questions arising from these omitted topics. We refer the interested reader to [20, §8], where further open questions and project ideas are given.

References

[1] Uri Bader, Pierre-Emmanuel Caprace, Tsachik Gelander and Shahar Mozes, Simple groups without lattices. *Bull. Lond. Math. Soc.* 44 (2012), 55–67.

[2] Christopher Banks, Murray Elder and George A. Willis, Simple groups of automorphisms of trees determined by their actions on finite subtrees. *J. Group Theory* 18 (2015), 235–261.

[3] Jens Bossaert and Tom De Medts, Topological and algebraic properties of universal groups for right-angled buildings. *Forum Math.* 33 (2021), 867–888.

[4] Marc Burger and Shahar Mozes, Groups acting on trees: from local to global structure. *Publ. Math. IHÉS* 92 (2000), 113–150.

[5] Pierre-Emmanuel Caprace and Nicolas Monod, Decomposing locally compact groups into simple pieces. *Math. Proc. Camb. Phil. Soc.* 150 (2011), 97–128.

[6] Pierre-Emmanuel Caprace, Non-discrete simple locally compact groups. In: *European congress of mathematics. Proceedings of the 7th ECM (7ECM) congress, Berlin, Germany, July 18–22, 2016,* (European Mathematical Society (EMS), Zürich, 2018), pp. 333–354.

[7] Pierre-Emmanuel Caprace and Tom De Medts, Simple locally compact groups acting on trees and their germs of automorphisms. *Transformation Groups*, 16 (2011) 375–411.

[8] Ruth Camm, Simple free products. *J. London Math. Soc.* 28 (1953), 66–76.

[9] Warren Dicks and M. J. Dunwoody, *Groups acting on graphs.* Cambridge Studies in Advanced Mathematics (Cambridge University Press, 1989).

[10] The GAP Group, GAP – Groups, Algorithms, and Programming, Version 4.8.7 (2017), `http://www.gap-system.org`

[11] David M. Goldschmidt, Automorphisms of trivalent graphs. *Ann. Math.* 111 (1980), 377–406.

[12] H. A. Jung and M. E. Watkins, On the structure of infinite vertex-transitive graphs. *Discrete Math.* 18 (1977), 45–53.

[13] Adrien Le Boudec, Groups acting on trees with almost prescribed local action. *Commentarii Mathematici Helvetici* 91 (2016), 253–293.

[14] Waltraud Lederle, Coloured Neretin groups. *Groups Geom. Dyn.* 13 (2019), 467–510.

[15] Rögnvaldur G. Möller, Primitivity and ends of graphs. *Combinatorica* 14 (1994), 477–484.

[16] Rögnvaldur G. Möller, Groups acting on locally finite graphs—a survey of the infinitely ended case. In: *Groups '93 Galway/St. Andrews. Proceedings of the international conference, Galway, Ireland, August 1–14, 1993. Volume 2.* Lond. Math. Soc. Lect. Note Ser. 212, (Cambridge University Press, 1995), pp. 426–456.

[17] Rögnvaldur G. Möller, Structure theory of totally disconnected locally compact groups via graphs and permutations. *Canad. J. Math.* 54 (2002), 795–827.

[18] Primož Potočnik and Pablo Spiga, Lifting a prescribed group of automorphisms of graphs. *Proc. Am. Math. Soc.* 147 (2019), 3787–3796.
[19] A. Yu. Ol'shanskiĭ, *Geometry of defining relations in groups.* Mathematics and its Applications (Soviet Series) 70 (Kluwer Acad. Publ., Dordrecht, 1991).
[20] Colin D. Reid and Simon M. Smith (with an appendix by Stephan Tornier), Groups acting on trees with Tits' independence property (P). *Preprint,* (2022) arXiv:2002.11766v2.
[21] Jean-Pierre Serre, *Trees.* Springer Monogr. in Math. (Springer, Berlin, 2003).
[22] Sam Shepherd (with appendix by Giles Gardam and Daniel J. Woodhouse), Two Generalisations of Leighton's Theorem. *Preprint,* (2019) to appear in *Groups, Geometry, and Dynamics.* arXiv:1908.00830.
[23] Simon M. Smith, Infinite primitive directed graphs. *J. Algebr. Comb.* 31 (2010), 131–141.
[24] Simon M. Smith, Subdegree growth rates of infinite primitive permutation groups. *J. Lond. Math. Soc.* 82 (2010), 526–548.
[25] Simon M. Smith, A product for permutation groups and topological groups, *Duke Math. J.* 166 (2017), 2965–2999.
[26] Simon M. Smith, The structure of primitive permutation groups with finite suborbits and t.d.l.c. groups admitting a compact open subgroup that is maximal. *Preprint,* (2019) arXiv:1910.13624v2.
[27] Carsten Thomassen and Wolfgang Woess, Vertex-transitive graphs and accessibility. *J. Comb. Theory, Ser. B* 58 (1993), 248–268.
[28] Stephan Tornier, Groups acting on trees with prescribed local action. *Preprint,* (2020) arXiv:2002.09876v3.
[29] Jacques Tits, Sur le groupe des automorphismes d'un arbre. In: *Essays on topology and related topics (Mémoires dédiés à Georges de Rham),* (Springer, New York, 1970), pp. 188–211.
[30] Richard Weiss, s-Transitive graphs, *Algebraic methods in graph theory* **25** (1978), 827–847.

8

Finite groups and the class-size prime graph revisited

Víctor Sotomayor[a]

Abstract

We report on recent progress concerning the relationship that exists between the algebraic structure of a finite group and certain features of its class-size prime graph.

8.1 Introduction

Hereafter, all groups considered are finite. In recent decades, the influence on the structure of a finite group G of the arithmetical properties of its set $cs(G)$ of conjugacy class sizes has been a widely investigated area. In order to better understand the arithmetical structure of that set of positive integers, the use of the prime graph built on the class sizes has been gaining an increasing interest. In general, the prime graph $\Delta(X)$ built on a set of positive integers X is the (simple undirected) graph whose vertex set consists of the prime divisors of the numbers in X, and the edge set contains pairs $\{p, q\}$ of vertices such that the product pq divides some number in X. In particular, when $X = cs(G)$, we briefly write $\Delta(G)$ for the class-size prime graph, $V(G)$ for the vertex set, and $E(G)$ for the edge set.

The relationship that exists between graph-theoretical properties of $\Delta(G)$ and the structure of G itself has attracted the interest of many authors. In this setting, two main questions naturally arise: what can be said about the structure of G if some information on $\Delta(G)$ is known,

[a] Instituto Universitario de Matemática Pura y Aplicada (IUMPA-UPV), Universitat Politècnica de València, Camino de Vera s/n, 46022 Valencia, Spain.
vsotomayor@mat.upv.es

and which graphs can occur as $\Delta(G)$ for some group G? We usually refer to the expository paper [14] due to Camina and Camina for an overview of results on this framework. Besides, the survey [35] of Lewis, which is mainly devoted to summarise results regarding graphs associated with character degrees, has also a section focused on graphs built on conjugacy class sizes. Nevertheless, the aforementioned papers were publised in 2011 and 2008, respectively, and numerous interesting results have appeared since then. Thus, our aim in this note is to bring them together in one place, providing references to the literature for their proofs. For the sake of completeness, the corresponding results that were already gathered in [14] or [35] will be again presented here.

The structure of the paper is mostly based on the two above mentioned questions. More concretely, in Section 8.2 we focus on how graph-theoretical properties of $\Delta(G)$ are reflected and influenced by the algebraic structure of G. On the other hand, in Section 8.3 we pay attention to the graphs that may appear as $\Delta(G)$ for some group G. In the last section, we review variations of these results, which consider the class-size prime graph built on some smaller subsets of elements of the group, as *p-regular* elements, *p-singular* elements, *real* elements, *vanishing* elements, elements lying in a normal subgroup, and elements that belong to the factors of a factorised group.

Notation: For a finite group G, let $x^G = \{g^{-1}xg \ : \ g \in G\}$ be the conjugacy class of an element $x \in G$, and let $|x^G|$ be its size. Recall that, in virtue of the orbit-stabiliser theorem, $|x^G| = |G : \mathbf{C}_G(x)|$. As previously said, $cs(G) = \{|x^G| \ : \ x \in G\}$. By $\mathrm{Irr}(G)$ we mean the set of all irreducible complex characters of the group G. The sets of all Sylow p-subgroups and Hall π-subgroups of G are $\mathrm{Syl}_p(G)$ and $\mathrm{Hall}_\pi(G)$, respectively, where p is a prime number and π is a set of primes. The set of all prime divisors of the order of G is $\pi(G)$. The semidirect product of A and B, where A is normalised by B, is written by $A \rtimes B$. Finally, we use CFSG to denote the classification of finite simple groups. The remaining notation and terminology is standard within Finite Group Theory.

8.2 The interplay between $\Delta(G)$ and the structure of G

Perhaps the earliest result in this strand of research is the following one due to Sylow in 1872: if $cs(G)$ only contains powers of a fixed prime p, then the centre of G is non-trivial. This fact, that elementarily follows from the class equation, can actually be rephrased in terms of the class-size prime graph:

Lemma 8.2.1 ([41], rephrased) *Let G be a group such that $\Delta(G)$ is an isolated vertex. Then $\mathbf{Z}(G)$ is non-trivial.*

In fact, Baer in 1953 proved that all the class sizes of prime power order elements of a group are powers of a fixed prime p if and only if G has a central p-complement (cf. [3, Proposition 1]). Thus, we have the next corollary.

Corollary 8.2.2 *Let G be a group. Then $\Delta(G)$ is an isolated vertex p if and only if $G = P \times H$ with $P \in \mathrm{Syl}_p(G)$ and $H \leqslant \mathbf{Z}(G)$.*

For the "dual" condition on the class sizes, i.e. a fixed prime p does not divide any class size of a group G, it is not difficult to prove that this happens if and only if G has a central Sylow p-subgroup. In other words, we have a characterisation of the absence of a vertex in the class-size prime graph:

Lemma 8.2.3 *Let G be a group, $p \in \pi(G)$ and $P \in \mathrm{Syl}_p(G)$. Then p does not lie in $V(G)$ if and only if $P \leqslant \mathbf{Z}(G)$.*

In this spirit, the absence of an edge in $\Delta(G)$ has also a strong impact on the group structure. This fact was studied in 1953 by Itô, but without any mention of a graph. More concretely, he proved in [34, Proposition 5.1] that whenever $p \neq q$ are primes that divide two distinct conjugacy class sizes of a group G, and pq does not divide any class size of G, then such a group has either a normal p-complement or a normal q-complement. It also follows from Itô's proof that the Sylow p-subgroups (or q-subgroups) are abelian, although he did not explicitly stated it. Indeed, Dolfi obtained in [20, Theorem 13] that both the Sylow p-subgroups and Sylow q-subgroups are also abelian if we assume the solubility of the group. However, Casolo and Dolfi moved beyond that, obtaining statement (ii) of the following theorem.

Theorem 8.2.4 *Let G be a finite group. Suppose that $p \neq q$ are non-adjacent vertices of $\Delta(G)$. Then the following conclusions hold:*

(a) ([34, Proposition 5.1]). *G has, up to interchanging p and q, a normal p-complement and abelian Sylow p-subgroups.*
(b) ([16, Theorem B]). *G is $\{p, q\}$-soluble of $\{p, q\}$-length at most 1, and both the Sylow p-subgroups and Sylow q-subgroups of G are abelian.*

It is worth mentioning that the proof of the $\{p, q\}$-solubility of G uses the CFSG via the next theorem.

Theorem 8.2.5 ([16, Theorem 9]) *Let G be a finite group with $\mathbf{F}(G) = 1$. Then $\Delta(G)$ is complete.*

In particular, simple groups have complete class-size prime graph.

As it can be perceived, non-adjacency between vertices of $\Delta(G)$ highly restricts the group structure. Indeed, the most extreme case was analysed in [10] by Bertram, Herzog and Mann, where they characterised the structure of those groups G whose *class-size common divisor graph* $\Gamma(G)$ is disconnected; the vertices of $\Gamma(G)$ are the non-central conjugacy classes of G, and two classes are adjacent if their sizes are not coprime. Certainly, $\Gamma(G)$ is connected if and only if $\Delta(G)$ is connected, so these authors particularly provided a characterisation of those groups whose class-size prime graph is disconnected. Later on, this fact was independently proved by Dolfi in [20, Theorem 4] using different tools:

Theorem 8.2.6 ([10, Theorem 2]) *Let G be a finite group. Then $\Delta(G)$ is disconnected if and only if up to abelian direct factors, $G = A \rtimes B$ with A and B abelian, coprime Hall subgroups, and $G/\mathbf{Z}(G)$ is a Frobenius group.*

In this last situation, $\Delta(G)$ turns out to be the union of two complete connected components (also called *cliques*), which are the sets of prime divisors of the Frobenius kernel and a Frobenius complement of $G/\mathbf{Z}(G)$, respectively.

In parallel to this research, Chillag and Herzog examined in [19] a more restrictive situation, namely groups all of whose class sizes are prime powers. That result, which can be viewed as a consequence of the above one, can be restated in terms of the class-size prime graph as follows.

Corollary 8.2.7 ([19, Theorem 2 and Corollary 2.2]) *Let G be a finite group. Then $\Delta(G)$ has no edges if and only if up to abelian direct factors:*

(a) G is a non-abelian p-group, for some prime p; or
(b) $G = Q \rtimes P$ with P a p-group, Q a q-group (p, q primes), both P and Q abelian, and $G/\mathbf{Z}(G)$ a Frobenius group.

Interestingly, that feature of $\Delta(G)$ in the disconnected case of having two complete induced subgraphs is not sparse. In fact, Alfandary proved in [2] that if G is soluble and it does not contain a certain subgroup of an affine semi-linear group, then $\Delta(G)$ also possesses two complete induced subgraphs. Further, this situation turns out to hold in full generality for every finite group (see Corollary 8.3.4).

As it will be shown in the next section (see Theorem 8.3.1), the diameter of $\Delta(G)$ in the connected case is at most 3. Casolo and Dolfi in 1996 characterised those groups for which the diameter of $\Delta(G)$ attains this maximal bound:

Theorem 8.2.8 ([15, Theorem 8]) *Let G be a group such that $\Delta(G)$ is connected. Then the diameter of $\Delta(G)$ is equal to 3 if and only if $G = A \rtimes B$ where A and B are abelian, coprime Hall subgroups, and there exist $p \in \pi(B)$, $q \in \pi(A)$, $B_p \in \mathrm{Syl}_p(B)$, and $A_q \in \mathrm{Syl}_q(A)$ such that:*

(a) $1 \neq \mathbf{C}_A(B_p) \neq A$, $1 \neq \mathbf{C}_B(A_q) \neq B$;
(b) $\mathbf{C}_A(y) \leqslant \mathbf{C}_A(B_p)$ *for all* $y \in B \setminus \mathbf{Z}(G)$;
(c) $\mathbf{C}_B(x) \leqslant \mathbf{C}_B(A_q)$ *for all* $x \in A \setminus \mathbf{Z}(G)$.

In particular, in [15, Proposition 7] it is also showed that if G is non-soluble, then the diameter of $\Delta(G)$ is at most 2. This bound is best possible, as Casolo and Dolfi noticed with the following example: take the direct product of an alternating group on 5 letters and a metacyclic Frobenius group of order coprime to 30.

The next research on this topic focused on refining the previously mentioned Itô's result (Theorem 8.2.4 (a)). That theorem explores the structure of G focusing on a prime p and a single prime non-adjacent to it. In this spirit, Casolo and Dolfi considered in [15, Lemma 5] a global, rather than local, perspective; that is, they analysed the structure of a soluble group G via the non-adjacency of p and a set of vertices π distinct from p, by proving that the Hall π-subgroups of G are abelian. Moreover, in the joint work [18] of these authors with Pacifici and Sanus, the solubility hypothesis was removed:

Theorem 8.2.9 ([18, Theorem C]) *Let G be a group, p a vertex of $\Delta(G)$, and π a subset of vertices that are non-adjacent to p and different from p. Then G is π-soluble, and the following conclusions hold.*

(a) G has abelian Hall π-subgroups and π-length 1.
(b) The vertices in π are pairwise adjacent.

In that paper, and still in the spirit of a global perspective, the authors also studied how the set of incomplete vertices (i.e. vertices that are not adjacent to at least one other vertex) of $\Delta(G)$ influences the group structure:

Theorem 8.2.10 ([18, Theorem A]) *Let G be a group, and let π_0 be the set of incomplete vertices of $\Delta(G)$. Then G admits a Hall π_0-subgroup H, and the following conclusions hold.*

(a) H is metabelian.
(b) $H \cap G'' \leqslant \mathbf{F}(G)$.

Certainly, statement (b) is equivalent to saying that the Hall π_0-subgroups of G'' are nilpotent and subnormal in G'', and therefore $|G''/\mathbf{F}(G'')|$ is not divisible by any prime in π_0. This fact and Burnside's $p^\alpha q^\beta$-theorem lead to the next result.

Corollary 8.2.11 ([18, Corollary B]) *Let G be a group. Then the following conclusions hold.*

(a) All the primes within $\pi(G''/\mathbf{F}(G''))$ are complete vertices of $\Delta(G)$.
(b) If $\Delta(G)$ has at most two complete vertices, then G is soluble.

Note that the class-size prime graph of an alternating group on 5 letters has three complete vertices, and clearly that group is non-soluble.

In an earlier paper, Casolo *et al.* addressed a more restrictive situation, namely $\Delta(G)$ has at most one complete vertex. Under this assumption, they proved in [17, Theorem 2.4] not only the solubility of G, but also that $G'\mathbf{F}(G)/\mathbf{F}(G)$ is nilpotent, so G is nilpotent-by-nilpotent-by-abelian. Further, if the prime 2 is not a complete vertex of $\Delta(G)$, then G is in fact nilpotent-by-metabelian, and they showed that this description is, in some sense, best possible (see [17, Remark 2.5 and Example 2.6]).

In the most extreme case when no vertex of $\Delta(G)$ is complete, by Theorem 8.2.10 we get that G is metabelian, and the following stronger conclusion was obtained by the previous authors:

Theorem 8.2.12 ([17, Theorem C]) *Let G be a group, and assume that no vertex of $\Delta(G)$ is complete. Then, up to abelian direct factors, $G = K \rtimes H$ where K and H are abelian, coprime Hall subgroups. Moreover, $K = G'$, $K \cap \mathbf{Z}(G) = 1$, and both $\pi(K)$ and $\pi(H)$ induce complete subgraphs in $\Delta(G)$.*

Graphs that are non-complete and regular clearly satisfy the above property, and in that paper a complete characterisation of groups with such class-size prime graphs is attained. Before presenting it, we recall that the *join* of two graphs Δ_1 and Δ_2, with disjoint vertex sets V_1 and

V_2, is the graph $\Delta_1 * \Delta_2$ whose vertex set is $V_1 \cup V_2$, and two vertices are adjacent whenever either one of them lies in V_1 and the other one in V_2, or they are vertices of the same Δ_i and they are adjacent in Δ_i for some $i \in \{1, 2\}$. Besides, in that paper, the concept of a *$\mathcal{D}$-group* is used for groups $G = A \rtimes B$ where A and B are abelian, coprime Hall subgroups, $\mathbf{Z}(G) \leqslant B$, and $G/\mathbf{Z}(G)$ is a Frobenius group.

Theorem 8.2.13 ([17, Theorem D]) *Let G be a group, and assume that $\Delta(G)$ is a non-complete regular graph of degree d with n vertices. Then, up to abelian direct factors, $G = G_1 \times \cdots \times G_{n/2m}$, where $m = (n-1)-d$, and G_i are pairwise coprime $\mathcal{D}$-groups satisfying that the orders of both the Frobenius kernel and the Frobenius complements of $G_i/\mathbf{Z}(G_i)$ have m prime divisors.*

Conversely, for a group G of this kind, $\Delta(G)$ is the join of $n/2m$ copies of a graph having two complete connected components of m vertices each.

As a direct consequence, we deduce that $\Delta(G)$ is a square with vertex set $V(G) = \{p, q, r, s\}$ and edge set $E(G) = \{\{p, q\}, \{p, r\}, \{q, s\}, \{r, s\}\}$ if and only if $G = A \times G_1 \times G_2$ where G_1 and G_2 are $\mathcal{D}$-groups with $\pi(G_1) = \{p, s\}$ and $\pi(G_2) = \{q, r\}$, and A is abelian.

A natural generalisation of this graph arises when one replaces each vertex by a set of vertices. With this in mind, a (simple undirected) graph is said to be a *block square* if its vertex set can be partitioned into four disjoint, non-empty subsets $\pi_1, \pi_2, \pi_3, \pi_4$ such that no prime in π_1 is adjacent to any prime in π_3, no prime in π_2 is adjacent to any prime in π_4, and there exist vertices in both π_1 and π_3 that are adjacent to vertices in π_2 and in π_4.

The result below, which I proved in [40], provides a complete characterisation of those groups whose class-size prime graph is a block square.

Theorem 8.2.14 ([40, Theorem A]) *Let G be a finite group. Then $\Delta(G)$ is a block square if and only if, up to an abelian direct factor, $G = A \times B$ where A and B are $\mathcal{D}$-groups of coprime orders.*

We close this section with recent research on how non-adjacency features of $\Delta(G)$ control the group structure. Let Δ be a graph with n connected components. A vertex v of Δ is said to be a *cut-vertex* if the number of connected components of the graph $\Delta - v$, obtained by removing the vertex v and all edges incident to v from Δ, is larger than n.

Dolfi *et al.* carried out in [27, Theorem 3.3] a complete, and somewhat technical, characterisation of the structure of a group G such that $\Delta(G)$

has a cut vertex. In particular, such a group G turns out to be soluble with Fitting height at most 3, and its Sylow p-subgroups are abelian for every prime p distinct from the cut vertex (cf. [27, Theorem A (a)]). These authors also characterised in [27, Theorem C] groups whose class-size prime graph has two cut vertices (which is actually the maximal number of cut vertices that can occur, as it will be shown in the next section). As an application, a classification of the finite groups G such that $\Delta(G)$ has no cycle as an induced subgraph is achieved (cf. [27, Corollary 3.4]).

8.3 Graphs that may occur as $\Delta(G)$

In this section, we put focus on which graphs Δ might satisfy that $\Delta(G) = \Delta$ for some finite group G. Recall that, in virtue of Theorem 8.2.6, the number of connected components of $\Delta(G)$ is less than or equal to 2 for every group G, having both components diameter 1. It is natural, then, to examine whether the diameter of $\Delta(G)$ can be upper bounded in the connected case. This problem was addressed by Alfandary in [1], where he proved the result below.

Theorem 8.3.1 ([1, Corollary 2.5]) *Let G be a group. If $\Delta(G)$ is connected, then its diameter is at most* 3.

Dolfi proved the same conclusion in [20, Theorem 17], and he provided an example which shows that this bound is best possible.

Roughly speaking, the class-size prime graph is very rich in edges. This property can be deduced from the next theorem, whose proof involves the CFSG via Theorem 8.2.5.

Theorem 8.3.2 ([21, Theorem A]) *Let G be a finite group. If $p, q, r \in V(G)$ are pairwise distinct, then two among them are adjacent in $\Delta(G)$.*

It is worthwhile to point out that this fact was firstly proved for soluble groups by Dolfi itself in [20]. From the above result, it follows again the bound of 3 for the diameter of $\Delta(G)$ in the connected case, and also the feature of $\Delta(G)$ in the non-connected case of being the union of two complete subgraphs.

At this point, it is convenient to recall the definition of the complement graph $\overline{\Delta(G)}$ of the class-size prime graph: this graph has the same vertex set $V(G)$, but two vertices are adjacent in $\overline{\Delta(G)}$ if and only if they are not adjacent in $\Delta(G)$. With this in mind, Theorem 8.3.2 can be expressed as follows: the graph $\overline{\Delta(G)}$ does not contain any cycle of length 3, for any group G. Now it is natural to wonder whether there exist cycles of larger lengths within $\overline{\Delta(G)}$; and indeed any Frobenius group G with abelian kernel and complement, both with orders divisible by two different primes, provides a cycle of length 4 in $\overline{\Delta(G)}$. However, Dolfi *et al.* showed in [26] that it is not possible to find cycles of length 5, and in general of any odd length:

Theorem 8.3.3 ([26, Theorem A]) *For any group G, the graph $\overline{\Delta(G)}$ does not contain any cycle of odd length.*

A first immediate consequence is that the class-size prime graph of a group cannot contain any pentagon as an induced subgraph. Further, the assertion of the above theorem is equivalent to the fact that $\overline{\Delta(G)}$ is always a bipartite graph, and so the following result directly follows.

Corollary 8.3.4 ([26, Corollaries B and C]) *For every finite group G, the vertex set of $\Delta(G)$ can be partitioned in two subsets of pairwise adjacent vertices. In particular, if $\omega(G)$ is the maximum size of a clique in $\Delta(G)$, then it holds $|V(G)| \leq 2\omega(G)$.*

Some results that have appeared in the previous section lead to characterisations of which non-complete regular graphs and which block square graphs may occur as $\Delta(G)$ for some group G. For instance, the following two results are consequences of Theorems 8.2.13 and 8.2.14, respectively.

Theorem 8.3.5 ([17, Corollary E]) *Let Δ be a non-complete regular graph. Then there exists a finite group G such that $\Delta(G) = \Delta$ if and only if Δ is the join of k copies of a graph that have two complete connected components of m vertices each, for some positive integers k and m.*

Theorem 8.3.6 ([40, Corollary B]) *Let Δ be a block square graph. Then there exists a finite group G such that $\Delta(G) = \Delta$ if and only if π_i is a clique for each $1 \leq i \leq 4$, and all the primes in $\pi_1 \cup \pi_4$ are adjacent to all the primes in $\pi_2 \cup \pi_3$.*

In particular, it cannot occur a *house graph* (i.e. a square graph in which two adjacent vertices are adjacent to an additional vertex) as $\Delta(G)$ for any group G.

Regarding graphs that possess a cut vertex, Dolfi *et al.* obtained the next conclusions.

Theorem 8.3.7 ([27, Theorem A]) *Let G be a group such that $\Delta(G)$ has a cut vertex r. Then the following conclusions hold:*

(a) *$\Delta(G) - r$ is a graph with two connected components, that are both complete graphs.*

(b) *If r is a complete vertex of $\Delta(G)$, then it is the unique complete vertex and the unique cut vertex of $\Delta(G)$. If r is non-complete, then $\Delta(G)$ is a graph of diameter 3, and it can have at most two cut vertices.*

Hence, $\Delta(G)$ can have at most two cut vertices for any group G, and a characterisation of the associated class-size prime graphs was also achieved in [27].

Theorem 8.3.8 ([27, Theorem D]) *Let Δ be a graph having a cut vertex. Then there exists a finite group G such that $\Delta(G) = \Delta$ if and only if Δ is connected and its vertex set can be partitioned in two subsets of pairwise adjacent vertices.*

8.4 Variations on the class-size prime graph

Over the last years, several authors have investigated whether the entire data contained in $cs(G)$ is required for studying the structure of G. In particular, some smaller subsets of elements of the group G have been considered, and hence it naturally arised the definition and the study of the prime graphs built on the associated conjugacy class sizes. It is worthwhile to remark that, in general, these graphs are not induced subgraphs of $\Delta(G)$.

p-regular and p-singular conjugacy classes

In 1972, Camina generalised Lemma 8.2.3 by considering only *p-regular* elements for a fixed prime p, i.e. elements of order not divisible by p. He proved that p does not divide any conjugacy class size of each p-regular element of a group G if and only if G has a Sylow p-subgroup as a direct factor. Let $cs_{p'}(G)$ be the set of conjugacy class sizes of p-regular elements of a group G. Then this result can be rephrased in terms of $\Delta(cs_{p'}(G))$:

Lemma 8.4.1 ([13, Lemma 1], rephrased) *Let G be a group, p a prime, and $P \in \mathrm{Syl}_p(G)$. Then p is not a vertex of $\Delta(cs_{p'}(G))$ if and only if $G = P \times H$ with $H \in \mathrm{Hall}_{p'}(G)$.*

Lu and Zhang studied in [36] the number of connected components of $\Delta(cs_{p'}(G))$, but assuming the p-solubility of the group G. Beltrán and Felipe attained best possible bounds for the diameter in [4] and [5].

Theorem 8.4.2 *Let G be a p-soluble group, for a fixed prime p. Then:*

(a) ([36, Theorem 1]). *The number of connected components of $\Delta(cs_{p'}(G))$ is at most 2.*
(b) ([4, Theorem 4] and [5, Theorem A]). *If $\Delta(cs_{p'}(G))$ is connected, then its diameter is at most 3; and if it is disconnected, then both connected components are complete subgraphs.*

Observe that these bounds cannot be directly deduced from the fact that $\Delta(cs_{p'}(G))$ is a subgraph of $\Delta(G)$.

In [6], Beltrán and Felipe extended this study by analysing the structure of a p-soluble group G with disconnected $\Delta(cs_{p'}(G))$. The authors did not achieved a complete characterisation, but they provided some necessary conditions, as well as some sufficient ones.

A characterisation of p-soluble groups such that all p-regular conjugacy classes have prime power size was addressed in [7] by the same authors:

Theorem 8.4.3 ([7, Theorem D], rephrased) *Let G be a p-soluble group, for a fixed prime p. Then $\Delta(cs_{p'}(G))$ has no edges (i.e. is an empty graph) if and only if one of the following conclusions hold:*

(a) G has abelian p-complements. This occurs if and only if $cs_{p'}(G)$ only contains powers of p.
(b) G is nilpotent with abelian Sylow r-subgroups for all primes $r \notin \{p, q\}$, for some prime $q \neq p$. This happens if and only if $cs_{p'}(G)$ only contains powers of q.
(c) $G = P \times H$ with $P \in \mathrm{Syl}_p(G)$ and $H \in \mathrm{Hall}_{p'}(G)$. Furthermore, $H = A \rtimes B$ with both A and B abelian, $H/\mathbf{Z}(H)$ is a Frobenius group, and the class sizes of H are powers of either q or r, for some primes q and r distinct from p. This happens if and only if $cs_{p'}(G)$ consists of powers of q and r.

In particular, $\Delta(cs_{p'}(G))$ is an isolated vertex if and only if either (a) or (b) above holds. These authors extended the previous necessary condition in [8, Theorem A], by removing the p-solubility assumption, and considering prime power class lengths of π-elements, for a set of primes π.

Alternatively, Qian and Wang focused on the class sizes of *p-singular* *elements*, which are elements whose order is divisible by p. Let $cs_p(G)$ denote the set of class sizes of such elements in a group G. Note that if $p \in \pi(\mathbf{Z}(G))$, then $cs(G) = cs_p(G)$. In [38] they proved the next results.

Theorem 8.4.4 *Let G be a group, p a prime, and $P \in \mathrm{Syl}_p(G)$.*

(a) ([38, Theorem B], rephrased). *p is not a vertex of $\Delta(cs_p(G))$ if and only if P is abelian, $P \cap P^g = 1$ for each $g \in G \smallsetminus \mathbf{N}_G(P)$, and $\mathbf{N}_G(P)/\mathbf{C}_G(P)$ acts Frobeniusly on P whenever $\mathbf{C}_G(P) < \mathbf{N}_G(P)$.*

(b) ([38, Theorem C]). *If $p \notin \pi(\mathbf{Z}(G))$, then $\Delta(cs_p(G))$ is connected with diameter at most 3.*

Real conjugacy classes

Various authors have investigated the sizes of the so-called *real conjugacy classes* of G, i.e. conjugacy classes x^G satisfying that $\chi(x)$ is a real number for each $\chi \in \mathrm{Irr}(G)$. This is analogous to saying that $x^g = x^{-1}$ for some $g \in G$, so $x^G = (x^{-1})^G$. Let us denote by $cs_{\mathbb{R}}(G)$ the set of sizes of the real conjugacy classes of G.

In 2009, Navarro, Sanus and Tiep studied the situation where all the real class sizes of G are 2-powers, which means that the set of vertices of $\Delta(cs_{\mathbb{R}}(G))$ is just the prime 2. In virtue of [37, Theorem C (a)], this happens if and only if G has a normal 2-complement K and all the real elements of G lie in $\mathbf{C}_G(K)$. One year before, Dolfi, Navarro and Tiep characterised when $cs_{\mathbb{R}}(G)$ satisfies the "opposite" situation, namely all real class sizes are odd:

Theorem 8.4.5 ([22, Theorem 6.1], rephrased) *Let G be a group and let $P \in \mathrm{Syl}_2(G)$. Then 2 is not a vertex of $\Delta(cs_{\mathbb{R}}(G))$ if and only if P is normal in G and each real element of P lies in $\mathbf{Z}(P)$.*

For odd primes, however, the approach is quite more difficult and it depends heavily on the CFSG, which contrasts with Lemma 8.2.3. Combining results of [32], [33] and [42], the next theorem follows. Recall that $\mathbf{O}^{p'}(G)$ is the smaller normal subgroup of G such that the corresponding quotient group has order not divisible by p.

Theorem 8.4.6 ([43, Lemma 2.7], rephrased) *Let G be a group and p be an odd prime. If $p = 3$, assume in addition that G has no composition factor isomorphic to $SL_3(2)$. If p is not a vertex of $\Delta(cs_{\mathbb{R}}(G))$, then G is p-soluble and $\mathbf{O}^{p'}(G)$ is soluble. Furthermore, $\mathbf{O}^{2'}(G)$ has a normal Sylow p-subgroup P and $P' \leqslant \mathbf{Z}(\mathbf{O}^{2'}(G))$.*

It is also proved in [22, Theorem 6.2] that the number of connected components of $\Delta(cs_{\mathbb{R}}(G))$ is at most 2. In the extreme case when $\Delta(cs_{\mathbb{R}}(G))$ is disconnected, Tong-Viet obtained the following result.

Theorem 8.4.7 ([43, Theorems A and B]) *Let G be a group such that $\Delta(cs_{\mathbb{R}}(G))$ is disconnected. Then G is soluble, 2 is a vertex of $\Delta(cs_{\mathbb{R}}(G))$, and one of the following conclusions holds.*

(a) G has a normal Sylow 2-subgroup.
(b) $\Delta(cs_{\mathbb{R}}(\mathbf{O}^{2'}(G)))$ is disconnected and $cs_{\mathbb{R}}(\mathbf{O}^{2'}(G))$ are either odd numbers or powers of 2.

Actually, in the above statement (b), both connected components are complete and one of them is just the prime 2 (see [43, Theorem 3.5]). Concerning statement (a), Tong-Viet conjectured that it cannot occur; in other words, $\Delta(cs_{\mathbb{R}}(G))$ is connected when G has a normal Sylow 2-subgroup. In fact, he proved that conjecture whenever the real elements of the Sylow 2-subgroup P of G are central in P (see [43, Lemma 3.6]).

Regarding the solubility assertion of Theorem 8.4.7, the same conclusion was achieved in [23, Theorem 3.1] by assuming a more restrictive hypothesis, namely all real class sizes of G are either odd or powers of 2. Thus, every group whose real classes all have prime power size is soluble, and its structure seems to be characterised in a current preprint due to Bonazzi (see [12]). On the other hand, the groups whose graph $\Delta(cs_{\mathbb{R}}(G))$ consists of only one odd vertex have not yet been characterised.

Vanishing conjugacy classes

In parallel to the aforementioned research on real class sizes, some authors started to filter the class sizes of G by the set $\mathrm{Irr}(G)$ in a different manner. Concretely, an element $g \in G$ is said to be *vanishing in G* if $\chi(g) = 0$ for some $\chi \in \mathrm{Irr}(G)$, and so g^G is called a *vanishing conjugacy class*. This definition is motivated by a celebrated theorem due to Burnside, which asserts that a non-linear irreducible character always vanishes on some conjugacy class; so it is natural to analyse the columns of the character table of a group that verify that property. We refer the reader to the expository paper [25] for a collection of results and conjectures on this topic.

Let $cs_v(G)$ be the set of vanishing class sizes of G. An immediate consequence of the mentioned Burnside's result is the next one: if $\Delta(cs_v(G))$ is a null graph (i.e. it has no vertices), then G is abelian. Note that the converse also holds, so the commutativity of G can be characterised in terms of $\Delta(cs_v(G))$.

In [24, Theorem A], the authors studied (via the CFSG) the situation when a given prime p does not divide any vanishing conjugacy class size of a group, which has the following transcription in terms of $\Delta(cs_v(G))$.

Theorem 8.4.8 ([24, Theorem A], rephrased) *Let G be a group, and p a prime. If p is not a vertex of $\Delta(cs_v(G))$, then G has a normal p-complement and abelian Sylow p-subgroups.*

It is significant to mention that the same assertion remains true if we consider only the class sizes of elements that are zeros of some (ordinary) irreducible character of G lying in the *principal p-block*, i.e. characters $\chi \in \mathrm{Irr}(G)$ such that $\sum_{g \in G\ p\text{-regular}} \chi(g) \neq 0$ (cf. [25, Theorem 6.3]).

Certainly, the vertex set of $\Delta(cs_v(G))$ can be smaller than that of $\Delta(G)$: it is enough to consider a symmetric group on 3 letters. Surprisingly enough, if G possesses a non-abelian minimal normal subgroup, then Bianchi *et al.* proved that these two vertex sets turn out to be equal:

Theorem 8.4.9 ([11, Proposition]) *Let G be a group that has a non-abelian minimal normal subgroup. Then the vertex sets of $\Delta(G)$ and $\Delta(cs_v(G))$ are equal.*

In the spirit of [16], and under the assumption that G has a non-abelian minimal normal subgroup, these authors also obtained the next "vanishing versions" of Theorems 8.2.4 (b) and Theorem 8.2.5:

Theorem 8.4.10 ([11, Theorem A]) *Let G be a group having a non-abelian minimal normal subgroup. If $\{p, q\}$ are vertices of $\Delta(cs_v(G))$ that are non-adjacent, then G is $\{p, q\}$-soluble.*

Theorem 8.4.11 ([11, Theorem B]) *Let G be a group with $\mathbf{F}(G) = 1$. Then $\Delta(cs_v(G))$ is a complete graph with vertex set $\pi(G)$.*

An immediate p-solubility criterion follows from Theorems 8.4.8 and 8.4.10: if p is not a complete vertex of $\Delta(cs_v(G))$, and G has a non-abelian minimal normal subgroup, then G is p-soluble (cf. [11, Corollary]). Besides, as noticed in [11], the assumption about the existence of a non-abelian minimal normal subgroup in Theorem 8.4.10 is fundamental.

All the previous results might suggest that, under the assumption that G has a non-abelian minimal normal subgroup, the graphs $\Delta(G)$ and $\Delta(cs_v(G))$ actually coincide, which is an open question.

In 2019, Felipe, Martínez-Pastor and myself analysed the influence of class sizes of vanishing elements of prime power order on the structure of certain factorised groups. In particular, from [31, Corollary 4] it can be obtained some information about groups satisfying that the vanishing class sizes are all prime powers, i.e. groups G such that $\Delta(cs_v(G))$ is an empty graph.

Corollary 8.4.12 *Let G be a group. If $\Delta(cs_v(G))$ has no edges, then $G/\mathbf{F}(G)$ is abelian.*

This result has also been proved independently by Robati and Hafezieh-Balaman in [39, Corollary 2.3]. In that paper, it is also addressed the particular situation in which the lenghts of the vanishing conjugacy classes of a group are all powers of a fixed prime p; in other words, it is addressed the feature of $\Delta(cs_v(G))$ of being a single vertex p (see [39, Theorem 1.1]).

Recently, the authors of [28] have further filtered the set of class sizes of G by considering only the *SM-vanishing conjugacy classes*, i.e. conjugacy classes that are zeros of the *strongly monolithic* characters of G. It is worth mentioning that some of the above situations also hold for the corresponding prime graph built on this last subset of class sizes of G (cf. [28]).

Normal subgroups

Beltrán, Felipe and Melchor considered in [9] the set $cs_G(N)$ of conjugacy class sizes in G of elements of a normal subgroup N of G, and they introduced the associated prime graph $\Delta(cs_G(N))$. In that paper, they obtained the next result.

Theorem 8.4.13 ([9, Theorems C and D]) *Let N be a normal subgroup of a group G. Then the number of connected components of $\Delta(cs_G(N))$ is at most 2. Moreover, if $\Delta(cs_G(N))$ is connected, then its diameter is at most 3; and if $\Delta(cs_G(N))$ is disconnected, then each connected component is a complete graph.*

Additionally, in the disconnected case, the authors also obtained strong restrictions on the structure of the normal subgroup N:

Theorem 8.4.14 ([9, Theorem E]) *Let N be a normal subgroup of a* *group G. If $\Delta(cs_G(N))$ is disconnected, then either $N = P \times A$ with P a p-group and $A \leqslant \mathbf{Z}(G)$, or $N = K \rtimes H$ where both K and H are abelian subgroups and $N/\mathbf{Z}(N)$ a Frobenius group.*

Note that in the previous result it appears a structure case for N that differs with the general case of $N = G$ in Theorem 8.2.6.

Products of groups

Another variation arises when one consider a factorised group and the conjugacy class sizes of only the elements that lie in the factors. More concretely, if $G = AB$ is the product of two subgroups A and B, then we can define $cs_G(A \cup B)$ as the set of class sizes of the elements that belong to $A \cup B$. Felipe *et al.* proved in [29, Corollary 1] that, in a factorised group $G = AB$, all the prime power order elements in $A \cup B$ have class sizes not divisible by a fixed prime p if and only if G has a central Sylow p-subgroup. Therefore, the next consequence directly follows.

Corollary 8.4.15 *Let the group $G = AB$ be the product of subgroups A and B, and let p be a prime. Then p is not a vertex of $\Delta(cs_G(A \cup B))$ if and only if G has a central Sylow p-subgroup.*

The proof of [29, Corollary 1] uses the CFSG, in contrast to the elementary proof of Lemma 8.2.3.

In [30], Felipe, Martínez-Pastor and myself introduced the concept of *Baer factorisations* $G = AB$, that is, products of groups such that the class sizes in G of the prime power order elements in $A \cup B$ are also prime powers. This is a generalisation of the so-called *Baer groups* (see [3]). In [30, Corollary 5], we characterised the structure of G in the particular situation when those class sizes are precisely powers of a fixed prime p. In particular, this holds when $cs_G(A \cup B)$ only contains powers of p, and so the next result follows.

Corollary 8.4.16 *Let the group $G = AB$ be the product of subgroups A and B. Then $\Delta(cs_G(A \cup B))$ is an isolated vertex p if and only if G has a central p-complement.*

More generally, we obtained some properties that are verified by Baer factorisations, and thus we can derive the same conclusions for a factorised group $G = AB$ such that $\Delta(cs_G(A \cup B))$ has no edges. For instance, we proved that $G/\mathbf{F}(G)$ is abelian, and both A and B are Baer groups (cf. [30]). Indeed, using [30, Proposition D], it can be verified that if $\Delta(cs_G(A \cup B))$ has no edges, then both $\Delta(A)$ and $\Delta(B)$ have no edges (although in general they are not subgraphs of $\Delta(G)$), and therefore by Corollary 8.2.7 we can control the structure of both subgroups.

As demonstrated in the previous subsections, if some information on $cs(G)$ is not considered, then it is not always possible to get analogous results to the ordinary case. As an example, Itô's result (Theorem 8.2.4 (a)) states that if $\{p, q\}$ are non-adjacent vertices of $\Delta(G)$, then G has either a normal p-complement or a normal q-complement. It is natural to wonder whether an analogous statement can be obtained when we consider this property on $\Delta(cs_G(A \cup B))$. Nevertheless, if $G = A \times B$ is the direct product of a symmetric group A on 3 letters and a dihedral group B of order 10, then $cs_G(A \cup B) = \{2, 3, 5\}$; so 3 and 5 are non-adjacent vertices of $\Delta(cs_G(A \cup B))$, and G has neither a normal 3-complement nor a normal 5-complement. But G has $\{3, 5\}$-length 1, and both the Sylow 3-subgroup and 5-subgroup are abelian, so it is an open question whether the claims in Theorem 8.2.4 (b) hold for factorised groups too.

Regarding the number of connected components of $\Delta(cs_G(A \cup B))$, following the techniques in [9] it is not difficult to prove that it can be at most 4. But the structure of A and B (and possibly that of G) in the disconnected case, and the diameter of $\Delta(cs_G(A \cup B))$ in the connected case are problems not addressed at the moment.

Acknowledgements. This research is supported by Proyecto PGC2018-096872-B-I00 (MCIU / AEI / FEDER, UE), and by Proyectos AICO/2020/298 and CIAICO/2021/163, Generalitat Valenciana (Spain). Additionally, the author would like to thank S. Dolfi, E. Pacifici and L. Sanus for introducing him to the research on the class-size prime graph.

References

[1] G. Alfandary, On graphs related to conjugacy classes of groups, *Israel J. Math.* **86** (1994), 211–220.

[2] G. Alfandary, A graph related to conjugacy classes of solvable groups, *J. Algebra* **176** (1995), 528–533.

[3] R. Baer, Group elements of prime power index, *Trans. Amer. Math. Soc.* **75** (1953), 20–47.

[4] A. Beltrán and M.J. Felipe, On the diameter of a p-regular conjugacy class graph of finite groups, *Comm. Algebra* **30** (2002), 5861–5873.

[5] A. Beltrán and M.J. Felipe, On the diameter of a p-regular conjugacy class graph of finite groups II, *Comm. Algebra* **31** (2003), 4393–4403.

[6] A. Beltrán and M.J. Felipe, Finite groups with a disconnected p-regular conjugacy class graph, *Comm. Algebra* **32** (2004), 3503–3516.

[7] A. Beltrán and M.J. Felipe, Certain relations between p-regular class sizes and the p-structure of p-solvable groups, *J. Aust. Math. Soc.* **77** (2004), 387–400.

[8] A. Beltrán and M.J. Felipe, Prime powers as conjugacy class lengths of π-elements, *Bull. Austral. Math. Soc.* **69** (2004), 317–325.

[9] A. Beltrán, M.J. Felipe and C. Melchor, Graphs associated to conjugacy classes of normal subgroups in finite groups, *J. Algebra* **443** (2015), 335–348.

[10] E.A. Bertram, M. Herzog and A. Mann, On a graph related to conjugacy classes of groups, *Bull. London Math. Soc.* **22** (1990), 569–575.

[11] M. Bianchi, J. Brough, R.D. Camina and E. Pacifici, On vanishing class sizes in finite groups, *J. Algebra*, **489** (2017), 446–459.

[12] L. Bonazzi, Finite groups whose real classes have prime-power size, *preprint*, (2022). https://doi.org/10.48550/arXiv.2108.06304

[13] A.R. Camina, Arithmetical conditions on the conjugacy class numbers of a finite group, *J. London Math. Soc.* **2** (1972), 127–132.

[14] A.R. Camina and R.D. Camina, The influence of conjugacy class sizes on the structure of finite groups: a survey, *Asian-Eur. J. Math.* **4** (2011), 559–588.

[15] C. Casolo and S. Dolfi, The diameter of a conjugacy class graph of finite groups, *Bull. London Math. Soc.* **28** (1996), 141–148.

[16] C. Casolo and S. Dolfi, Products of primes in conjugacy class sizes and irreducible character degrees, *Israel J. Math.* **174** (2009), 403–418.

[17] C. Casolo, S. Dolfi, E. Pacifici and L. Sanus, Groups whose prime graph on conjugacy class sizes has few complete vertices, *J. Algebra* **364** (2012), 1–12.

[18] C. Casolo, S. Dolfi, E. Pacifici and L. Sanus, Incomplete vertices in the prime graph on conjugacy class sizes of finite groups, *J. Algebra* **376** (2013), 46–57.

[19] D. Chillag and M. Herzog, On the length of the conjugacy classes of finite groups *J. Algebra* **131** (1990), 110–125.

[20] S. Dolfi, Arithmetical conditions on the length of the conjugacy-classes of a finite group, *J. Algebra* **174** (1995), 753–771.

[21] S. Dolfi, On independent sets in the class graph of a finite group, *J. Algebra* **303** (2006), 216–224.

[22] S. Dolfi, G. Navarro and P.H. Tiep, Primes dividing the degrees of the real characters, *Math. Z.* **259** (2008), 755–774.

[23] S. Dolfi, E. Pacifici and L. Sanus, Finite groups with real conjugacy classes of prime size, *Israel J. Math.* **175** (2010), 179–189.

[24] S. Dolfi, E. Pacifici and L. Sanus, Groups whose vanishing class sizes are not divisible by a given prime, *Arch. Math.* **94** (2010), 311–317.

[25] S. Dolfi, E. Pacifici and L. Sanus, *On zeros of characters of finite groups*, Group Theory and Computation, Indian Statistical Institute Series, Springer, Singapore (2018).

[26] S. Dolfi, E. Pacifici, L. Sanus and V. Sotomayor, The prime graph on class sizes of a finite group has a bipartite complement, *J. Algebra* **542** (2020), 35–42.

[27] S. Dolfi, E. Pacifici, L. Sanus and V. Sotomayor, Groups whose prime graph on class sizes has a cut vertex, *Israel J. Math.* **244** (2021), 775–805.

[28] T. Erkoç, S.B. Güngör and G. Akar, SM-vanishing conjugacy classes of finite groups, *J. Algebra Appl.* DOI: 10.1142/S0219498825500471.

[29] M.J. Felipe, L. Kazarin, A. Martínez-Pastor and V. Sotomayor, On products of groups and indices not divisible by a given prime, *Monatsh. Math.* **193** (2020), 811–827.

[30] M.J. Felipe, A. Martínez-Pastor and V.M. Ortiz-Sotomayor, Prime power indices in factorised groups, *Mediterr. J. Math.* **14** (2017), 225.

[31] M.J. Felipe, A. Martínez-Pastor and V.M. Ortiz-Sotomayor, Zeros of irreducible characters in factorised groups, *Ann. Mat. Pura Appl.* **198** (2019), 129–142.

[32] R.M. Guralnick, G. Navarro and P.H. Tiep, Real class sizes and real character degrees, *Math. Proc. Camb. Phil. Soc.* **150** (2011), 47–71.

[33] I.M. Isaacs and G. Navarro, Groups whose real irreducible characters have degrees coprime to p, *J. Algebra* **356** (2012), 195–206.

[34] N. Itô, On finite groups with given conjugate types, I, *Nagoya Math. J.* **6** (1953), 17–28.

[35] M.L. Lewis, An overview of graphs associated with character degrees and conjugacy class sizes in finite groups, *Rocky Mt. J. Math.* **38** (2008), 175–211.

[36] Z. Lu and J. Zhang, On the diameter of a graph related to p-regular conjugacy classes of finite groups, *J. Algebra* **231** (2000), 705–712.

[37] G. Navarro, L. Sanus and P.H. Tiep, Real characters and degrees, *Israel J. Math.* **171** (2009), 157–173.

[38] G. Qian and Y. Wang, On class size of p-singular elements in finite groups, *Comm. Algebra* **37** (2009), 1172–1181.

[39] S.M. Robati and R. Hafezieh-Balaman, Groups whose vanishing class sizes are p-powers, *Comm. Algebra*, (2023). https://doi.org/10.1080/00927872.2023.2175841

[40] V. Sotomayor, Finite groups whose prime graph on class sizes is a block square, *Comm. Algebra* **50** (2022), 3995-3999.

[41] M.L. Sylow, Théorèmes sur les groupes de substitutions, *Math. Ann.* **5** (1872), 584–594.

[42] P.H. Tiep, Real ordinary characters and real Brauer characters, *Trans. Amer. Math. Soc.* **367** (2015), 1273–1312.

[43] H.P. Tong-Viet, Real class sizes, *Israel J. Math.* **228** (2018), 753–769.

9
Character bounds for finite simple groups and applications

Pham Huu Tiep[a]

Abstract

Given the current knowledge of complex representations of finite quasi-simple groups, obtaining good upper bounds for their characters values still remains a difficult problem, a satisfactory solution of which would have significant implications in a number of applications. We will report on recent results that produce such character bounds, and discuss some such applications, in and outside of group theory.

9.1 Introduction

Let G be a finite group, $\Phi : G \to \mathrm{GL}_n(\mathbb{C})$ a faithful irreducible representation, and let

$$\chi : G \to \mathbb{C}, \ \chi(g) = \mathrm{Tr}(\Phi(g)),$$

denote the corresponding character. Since $\chi(g)$ is a sum of n roots of unity, it is clear that

$$|\chi(g)| \leq \chi(1), \tag{9.1}$$

with equality attained if and only if $g \in \mathbf{Z}(G)$.

The main theme of this survey paper is *how much better can we do than the obvious bound* (9.1). If we can, it will be helpful for a number of applications. Many, but not all, applications of this kind rely on the following classic result of F. G. Frobenius, in which $\mathrm{Irr}(G)$ denotes the set of irreducible complex characters of any finite group G.

a Department of Mathematics, Rutgers University, Hill Center – Busch Campus, 110 Frelinghuysen Road, Piscataway, NJ 08854-8019, USA.
tiep@math.rutgers.edu

Lemma 9.1.1 (Frobenius character formula) *Let G be a finite group and $g \in G$ a fixed element.*

(a) For any $k \geq 2$ (not necessarily distinct) conjugacy classes $\mathbf{C}_i = g_i^G$ of G, the number of ways to write $g = x_1 x_2 \dots x_k$ with $x_i \in \mathbf{C}_i$ is

$$\frac{|\mathbf{C}_1||\mathbf{C}_2|\dots|\mathbf{C}_k|}{|G|} \sum_{\chi \in \mathrm{Irr}(G)} \frac{\chi(g_1)\dots\chi(g_k)\chi(g^{-1})}{\chi(1)^{k-1}}.$$

(b) The number of ways to write $g = [x, y]$, $x, y \in G$ is

$$|G| \sum_{\chi \in \mathrm{Irr}(G)} \frac{\chi(g)}{\chi(1)}.$$

Proof of part (a) Let $\mathbf{C}_1, \dots, \mathbf{C}_r$ denote the distinct conjugate classes of G. Then the following elements of the group algebra $\mathbb{C}G$

$$\bar{\mathbf{C}}_i := \sum_{x \in \mathbf{C}_i} x, \ 1 \leq i \leq r$$

form a basis of $\mathbf{Z}(\mathbb{C}G)$. Suppose $\chi \in \mathrm{Irr}(G)$ is afforded by a representation

$$\Phi : G \to \mathrm{GL}_n(\mathbb{C}).$$

Then $\Phi(\bar{\mathbf{C}}_i)$ is a scalar matrix. Taking the trace, we obtain

$$\Phi(\bar{\mathbf{C}}_i) = \frac{\chi(g_i)|\mathbf{C}_i|}{\chi(1)} I_n.$$

Writing $\prod_{i=1}^k \bar{\mathbf{C}}_i = \sum_{j=1}^r a_j \bar{\mathbf{C}}_j$ for suitable $a_j \in \mathbb{C}$, we have

$$\sum_{j=1}^r a_j \frac{\chi(g_j)|\mathbf{C}_j|}{\chi(1)} I_n = \Phi\left(\prod_{i=1}^k \bar{\mathbf{C}}_i\right) = \frac{\prod_{i=1}^k \chi(g_i)|\mathbf{C}_i|}{\chi(1)^k} I_n,$$

and so

$$\sum_{j=1}^r a_j \chi(g_j)|\mathbf{C}_j| = \frac{\prod_{i=1}^k \chi(g_i)|\mathbf{C}_i|}{\chi(1)^{k-1}}. \tag{9.2}$$

Let us assume that $g \in \mathbf{C}_s$. Multiplying both sides of (9.2) by $\chi(g^{-1})$ and summing over $\chi \in \mathrm{Irr}(G)$, we get

$$\sum_{\chi \in \mathrm{Irr}(G)} \frac{\chi(g^{-1}) \prod_i \chi(g_i)|\mathbf{C}_i|}{\chi(1)^{k-1}} = \sum_{j=1}^r a_j |\mathbf{C}_j| \sum_{\chi \in \mathrm{Irr}(G)} \chi(g_j)\chi(g^{-1}) = |G| a_s.$$

Since a_s is the number in question, the statement follows. □

A finite group G is called *quasisimple* if G is perfect and $G/\mathbf{Z}(G)$ is simple, and *almost quasisimple* if $S \lhd G/\mathbf{Z}(G) \leq \mathrm{Aut}(S)$ for a non-abelian finite simple group S.

The main problem considered in this paper is

Problem 9.1.2 Let G be a finite almost quasisimple group and let $g \in G \smallsetminus \mathbf{Z}(G)$.

(a) Find an explicit, and as small as possible, constant $0 < \gamma = \gamma(g) < 1$, such that the following upper bound for character ratios

$$\frac{|\chi(g)|}{\chi(1)} \leq \gamma$$

holds for all $\chi \in \mathrm{Irr}(G)$ with $\chi(1) > 1$.

(b) Find an explicit, and as small as possible, constant $0 < \boldsymbol{\alpha} = \boldsymbol{\alpha}(g) < 1$, such that

$$|\chi(g)| \leq \chi(1)^{\boldsymbol{\alpha}}$$

for all $\chi \in \mathrm{Irr}(G)$.

The class of almost quasisimple groups includes well-known groups such as the symmetric group S_n, the alternating group A_n, as well as many finite groups of Lie type, e.g. the special linear group $\mathrm{SL}_n(q)$, the general linear group $\mathrm{GL}_n(q)$, and $E_8(q)$. In general, by a *finite group of Lie type over* $\mathbb{F}_q$ we mean any group G of the form

$$G = \mathcal{G}^F := \{g \in G \mid F(g) = g\},$$

where $\mathcal{G}$ is a connected reductive algebraic group over a field of characteristic p, $F : \mathcal{G} \to \mathcal{G}$ is a *Steinberg endomorphism* (i.e. a morphism F of $\mathcal{G}$ such that some power F^d is a standard Frobenius morphism $x \mapsto x^{p^a}$; in this context we have $q = p^{a/d}$). If $\mathcal{G}$ is simple and simply connected, then, aside from a couple of small examples, $\mathcal{G}^F$ is quasisimple.

Example 9.1.3 Let $\mathcal{G} := \mathrm{GL}_n(\overline{\mathbb{F}_p})$. For the Steinberg endomorphism

$$F_q : X := (x_{ij}) \mapsto (x_{ij}^q)$$

we have $\mathcal{G}^{F_q} = \mathrm{GL}_n(\mathbb{F}_q) =: \mathrm{GL}_n(q)$, whereas for

$$F_q' : X \to {}^t(F_q(X))^{-1},$$

we have $(F_q')^2 = F_{q^2}$ and $\mathcal{G}^{F_q'} = \mathrm{GU}_n(\mathbb{F}_{q^2}) =: \mathrm{GU}_n(q)$, the general unitary group.

The finite classical groups $G = \mathrm{Cl}_n(q) = \mathrm{Cl}(V)$, where $V = \mathbb{F}_q^n$, and $\mathrm{Cl} = \mathrm{GL}, \mathrm{SL}, \mathrm{GU}, \mathrm{SU}, \mathrm{Sp}, \mathrm{SO}, \mathrm{Spin}$, etc. and *exceptional groups of Lie type*, like $G_2(q)$, ${}^2B_2(q), \ldots, E_8(q)$, can all be obtained this way.

9.2 Results on Problem 9.1.2(A)

In a series of papers in 1991–95, see e.g. [19, 20], David Gluck obtained a universal upper bound on character ratios, giving the first substantial result on Problem 9.1.2(A):

Theorem 9.2.1 (Gluck) *There is an absolute constant $C > 0$ such that if G is a finite quasisimple group of Lie type over $\mathbb{F}_q$, $g \in G \smallsetminus \mathbf{Z}(G)$, and $1_G \neq \chi \in \mathrm{Irr}(G)$, then*

$$\frac{|\chi(g)|}{\chi(1)} \leq \begin{cases} \sqrt{C/q}, & g \text{ unipotent}, \\ C/q, & \text{otherwise}. \end{cases}$$

Outline of proof. Suppose that g is not contained in any (proper) parabolic subgroup of G. Then $|\mathbf{C}_G(g)|$ is tiny, and, by the orthogonality relations, $|\chi(g)| \leq \sqrt{|\mathbf{C}_G(g)|}$, proving the bound.

Suppose now that g is contained in a (proper) parabolic subgroup $P = U \rtimes L$ of G. Then we decompose

$$\chi|_P = \chi_1 + \chi_2 + \chi_3 + \chi_4$$

according to the actions of U and P: $[P, P] \leq \mathrm{Ker}(\chi_1)$, all irreducible constituents of χ_2 are trivial on the unipotent radical U of P but non-linear, χ_3 lies over non-principal $\mathbf{O}^{p'}(L)$-invariant irreducible characters of U (where L is a Levi complement in P and p is the characteristic of G), and χ_4 accounts for the rest. (Here, if $N \lhd G$ and $\theta \in \mathrm{Irr}(N)$, then a character χ of G is said to *lie over* θ if θ is an irreducible constituent of the restriction $\chi|_N$.) Then one bounds $|\chi_i(g)|$, $1 \leq i \leq 4$, using various results including some on fixed point ratios of transitive permutation actions of finite groups of Lie type. $\square$

Gluck's bound, [21, Theorem 2.4], and [27, Lemma 2.23] leads to the following.

Corollary 9.2.2 *In Problem 9.1.2(A), if G is quasisimple then one can take $\gamma = 0.95$, unless $G = \mathsf{A}_n$ for some $n \geq 5$.*

In the exceptional case $G = \mathsf{A}_n$, it was shown in [21, Theorem 1.6] that

$$|\chi(g)/\chi(1)| \leq (1 + \mathsf{cycle}(g)/n)/2,$$

where $g \in \mathsf{S}_n$ is a disjoint product of $\mathsf{cycle}(g)$ cycles. Taking g to be a 3-cycle and χ the character of the deleted natural permutation module (of dimension $n-1$), one can see that the character ratio $|\chi(g)/\chi(1)|$ can be as close to 1 as one wishes when $n \to \infty$.

Together with [27, Corollary 2.14], Corollary 9.2.2 yields the following answer to Problem 9.1.2(A):

Corollary 9.2.3 *In general, in Problem 9.1.2(A) one can take $\gamma = 79/80$, unless $\mathsf{A}_n \leq G/\mathbf{Z}(G) \leq \mathsf{S}_n$ for some $n \geq 5$.*

As close to 1 this constant 79/80 is, one can't do much better than this. Indeed, the irreducible *Weil characters* of $\mathrm{Sp}_{2n}(3)$ (of degree $(3^n \pm 1)/2$), evaluated at transvections, yield character ratios $\approx 1/\sqrt{3}$ when n is large. Moreover, the reflection character of the Weyl group $W(E_8)$ (of degree 8), evaluated at a reflection, gives the character ratio 3/4.

Note that $\chi(g) = \chi(1) - 2$ in the example of $W(E_8)$ mentioned above. This brings up a natural question: *How small can $\chi(1) - |\chi(g)|$ be?* For instance, how close to n can the absolute value of the trace of non-scalar elements in a finite complex linear group G of degree n be? They can in fact be arbitrarily close, if G is imprimitive. In a sense, this turns out to be the only significant restriction:

Theorem 9.2.4 ([27, Theorem 1.3]) *Let $V = \mathbb{C}^d$ with $d \geq 5$, $G < \mathrm{GL}(V)$ a finite primitive, irreducible subgroup, and let $g \in G \smallsetminus \mathbf{Z}(G)$. If G is tensor induced, assume in addition that g acts nontrivially on the tensor factors of V. Then one of the following statements holds for*

$$\Delta(g) := \dim(V) - |\mathrm{Tr}(g)|.$$

(a) $\Delta(g) \geq 8 - 4\sqrt{2}$ (this is attained at a normalizer of a 2-group in $\mathrm{GL}_8(\mathbb{C})$).

(b) $\Delta(g) = 2$ and g is a multiple of a reflection.

(c) As a G-module, $V = A \otimes_{\mathbb{C}} B$ is tensor decomposable, $\dim(A) = 2$, $2 \leq \dim(B) \leq 6$, $g|_B$ is scalar, and $\Delta(g) \geq \dim(B) \cdot (3 - \sqrt{5})/2$.

The proof of Theorem 9.2.4 is based on the (79/80)-bound in Corollary 9.2.3, and a version of Aschbacher's theorem [1] on finite subgroups of $\mathrm{GL}(V)$, as given in [26, Proposition 2.8].

One should compare Theorem 9.2.4 to the classic theorem of Blichfeldt that the smallest arc of S^1 that can contain all eigenvalues of a non-central element in a finite primitive complex linear group has length $\geq \pi/3$.

Why should one be interested in $\Delta(g)$? Our motivation comes from algebraic geometry, in particular because of its connection to M. Reid's notion of the *age* of a unitary transformation.

Definition 9.2.5 [64] Let $g \in \mathrm{GU}_d(\mathbb{C})$ be any unitary transformation. Write it in diagonal form $g = \mathrm{diag}(e^{2\pi i r_1}, \ldots, e^{2\pi i r_d})$ with $0 \leq r_j < 1$. Then the *age* of g is defined to be $\mathsf{age}(g) = \sum_{j=1}^{d} r_j$. Furthermore, g is called *junior* if $0 < \mathsf{age}(g) \leq 1$.

Example 9.2.6 $g \sim \mathrm{diag}(-1, 1, \ldots, 1)$ a reflection $\Rightarrow \mathsf{age}(g) = 1/2$.
$g \sim \mathrm{diag}(z, 1, \ldots, 1)$ a complex reflection $\Rightarrow 0 < \mathsf{age}(g) < 1$.
$g \sim \mathrm{diag}(-1, -1, 1, \ldots, 1)$ a bireflection $\Rightarrow \mathsf{age}(g) = 1$.

Proposition 9.2.7 ([27]) *For any $g \in \mathrm{GU}(V) = \mathrm{GU}_d(\mathbb{C})$ we have*

$$\mathsf{age}(g) \geq \frac{\Delta(g)}{(1.45)\pi}.$$

Proof of Proposition 9.2.7 uses an L^2-version $d_2(g)$ of the Kollár-Larsen L^1-deviation [37]

$$d_2(g)^2 := \inf_{|\lambda|=1,\ B\in\mathcal{B}(V)} \left(\sum_{v\in B} ||g(v) - \lambda v||^2 \right)^{1/2}$$

where $\mathcal{B}(V)$ is the set of orthonormal bases of the unitary space V. A key fact is that $d_2(g)^2 = 2\Delta(g)$, see [27, Lemma 2.11]. □

The following problem and conjecture on finite complex linear groups $G < \mathrm{GL}(V)$, $V = \mathbb{C}^d$ with $d \geq 5$, were raised by Kollár and Larsen in [37]:

Problem 9.2.8 ([37]) Describe the finite irreducible subgroups $G < \mathrm{GL}(V)$ which are generated, up to scalars, by elements of $\mathsf{age} < 1$ (resp. by junior elements).

Conjecture 9.2.9 (Kollár-Larsen [37]) *Suppose $G < \mathrm{GL}(V)$ is a finite irreducible subgroup such that $G = \langle x^G \rangle$ for all non-central elements $x \in G$ with $\mathsf{age}(x) < 1$. Then, modulo scalars, G is a complex reflection group* (and so known by [67]).

The first motivation for Problem 9.2.8 and Conjecture 9.2.9 comes from work of Kollár and Larsen [37] on quotients X/G of Calabi-Yau varieties X by a finite group G. In particular, they show that the Kodaira dimension of X/G is controlled by whether the action of $\mathrm{Stab}_G(x)$ on the tangent space T_xX contains elements of $\mathsf{age} < 1$.

The second motivation is related to the notion of crepant resolutions. Given a singular variety Y and a resolution $f : X \to Y$, one can express the canonical class K_X as $K_X = f^*K_Y + \sum_i a_iE_i$, where $a_i \in \mathbb{Q}$ and the sum is over the irreducible exceptional divisors. Now, Y is called *terminal* if $a_i > 0$ for all i, and *canonical* if $a_i \geq 0$ for all i. Furthermore, the resolution f is called *crepant* if $a_i = 0$ for all i, i.e. if $K_X = f^*K_Y$.

Crepant resolutions play an important role in algebraic geometry, string theory, and quantum cohomology. For *quotient singularities*, the following criterion was obtained by Ito and Reid and explains the relevance of age and junior elements in complex linear groups:

Criterion 9.2.10 ([33]) Suppose $G < \mathrm{GL}_d(\mathbb{C})$ is finite and the quotient singularity $\mathbb{C}^d/G$ admits a crepant resolution. Then G contains junior elements.

Our main results on Problem 9.2.8 and Conjecture 9.2.9 can be summarized as follows:

Theorem 9.2.11 ([27]) *We have the following results.*

(i) A solution of Problem 9.2.8 is given in [27, Theorem 1.4].
(ii) The Kollár-Larsen Conjecture 9.2.9 holds.

As a consequence, we obtain:

Corollary 9.2.12 ([27, Corollary 1.6]) *Let $d \geq 11$ and let $G < \mathrm{GL}_d(\mathbb{C})$ be a finite irreducible, primitive, tensor indecomposable subgroup. Assume $\mathbb{C}^d/G$ is not terminal (e.g., it has a crepant resolution). Then one of the following statements holds.*

(a) $\mathbf{Z}(G) \times \mathsf{A}_{d+1} \leq G \leq (\mathbf{Z}(G) \times \mathsf{A}_{d+1}) \cdot 2$, with A_{d+1} acting on $\mathbb{C}^d$ as on its deleted natural permutation module.
(b) All junior elements of G are central, and $|\mathbf{Z}(G)| \geq d$.

Corollary 9.2.12 indicates that crepant resolutions seem to occur mostly in low dimensions.

Outline of the proof of Theorem 9.2.11. Let $G \leq \mathrm{GL}(V)$ and $\mathsf{age}(g) \leq 1$ for some $g \in G \smallsetminus \mathbf{Z}(G)$. By Proposition 9.2.7, we have $\Delta(g) \leq (1.45)\pi$. Next, one uses [26, Proposition 2.8] to reduce to the case where G is almost quasisimple, with the layer $E(G) \neq \mathsf{A}_n$. Applying Corollary 9.2.3, we obtain $\dim(V) \leq 364$. At this point, we can identify possible (V, G) using classification results on low-dimensional complex representations of finite groups of Lie type, see [31, 56, 75, 77]. To deal with these (V, G), as well as with the case $E(G) = \mathsf{A}_n$, ad hoc arguments are utilized. □

Next we describe the first result on Problem 9.1.2(A) after Gluck's bound, obtained in [44] and applied to finite classical groups $G = \mathrm{Cl}(V)$, $V = \mathbb{F}_q^n$. Set $\tilde{V} := V \otimes_{\mathbb{F}_q} \overline{\mathbb{F}_q}$. Empirical data (obtained using e.g. [6] or [15]) indicate that

$$\text{the further } g \in G \text{ is from } \mathbf{Z}(G), \text{ the smaller } \frac{|\chi(g)|}{\chi(1)} \text{ should be.} \quad (9.3)$$

A measure for this distance from $\mathbf{Z}(G)$ is given by the *support* of g:

$$\mathsf{supp}(g) = \inf_{\lambda \in \overline{\mathbb{F}_q}} \operatorname{codim} \operatorname{Ker}(g - \lambda \cdot 1_{\tilde{V}}). \quad (9.4)$$

Theorem 9.2.13 ([44, Theorem 1.2.1]) *For any quasisimple classical group $G = \mathrm{Cl}(V)$ on $V = \mathbb{F}_q^n$, any irreducible character $1_G \neq \chi \in \mathrm{Irr}(G)$, and any $g \in G$*

$$\frac{|\chi(g)|}{\chi(1)} \leq \frac{1}{q^{\sqrt{\mathsf{supp}(g)/481}}}.$$

Sketch of proof of Theorem 9.2.13. First we show that either

$$\frac{|\chi(g)|}{\chi(1)} \leq q^{1-n/4}$$

(which implies the desired bound if n is not too small), or g stabilizes a direct (orthogonal) decomposition

$$V = V_1 \oplus \ldots \oplus V_m \quad (9.5)$$

with $m > 1$. In the latter case, choose such a decomposition (9.5) with m as large as possible, and restrict χ to the stabilizer H of the decomposition, for which one has

$$[H, H] = \mathrm{Cl}(V_1) \times \mathrm{Cl}(V_2) \times \ldots \times \mathrm{Cl}(V_m).$$

Then one needs to control the dimension of the $[H, H]$-fixed point subspace in the G-module affording χ. Next, for any constituent θ of $\chi|_H$ which lies above an irreducible character $\theta_1 \otimes \ldots \otimes \theta_m \in \mathrm{Irr}([H, H])$ with $\theta_i(1) > 1$, the maximal choice of the decomposition (9.5) allows one to obtain good upper bounds on $|\theta_i(g)|/\theta_i(1)$. Extra care is needed in these arguments to go up from $[H, H]$ to H. □

On the one hand, the bound in Theorem 9.2.13 has proved to be useful in a number of applications, and it confirms the heuristic (9.3) explicitly. On the other hand, one believes that a *linear* upgrade of the function $\sqrt{\mathsf{supp}(g)}$ in Theorem 9.2.13 should be possible. For more about this see §9.6.

9.3 Problem 9.1.2(B) for symmetric groups and applications

The first breakthough on Problem 9.1.2(B) for symmetric groups was obtained by Fomin and Lulov [12], where they showed

Theorem 9.3.1 ([12]) *For any $m \in \mathbb{Z}_{\geq 1}$, if all disjoint cycles of a permutation $g \in \mathsf{S}_n$ have size m and $\chi \in \mathrm{Irr}(\mathsf{S}_n)$, then $|\chi(g)| \leq \chi(1)^{1/m+o(1)}$.*

This result has been generalized by Müller and Schlage-Puchta in [62] and by Larsen and Shalev in [43]:

Theorem 9.3.2 ([62]) *If the permutation $g \in \mathsf{S}_n$ has $f \geq 0$ fixed points, $\chi \in \mathrm{Irr}(\mathsf{S}_n)$, and $n \gg 0$, then $|\chi(g)| \leq \chi(1)^{1-1/\delta}$ with*

$$\delta := \frac{12 \log n/(\log(n/f))}{1 - 1/\log(n)} + 18 > 30.$$

Theorem 9.3.3 ([43]) *For any $g \in \mathsf{S}_n$, define the sequence $e_1, \ldots, e_n$ of non-negative real numbers such that, for any $1 \leq k \leq n$, $n^{e_1+\ldots+e_k} = \max(|\Sigma_k|, 1)$, where Σ_k is the union of all g-orbits on $\{1, 2, \ldots, n\}$ of length at most k. Then*

$$|\chi(g)| \leq \chi(1)^{\sum_{k=1}^{n}(e_k/k)+o(1)}$$

for any $\chi \in \mathrm{Irr}(\mathsf{S}_n)$.

In Theorems 9.3.1 and 9.3.3, $o(1)$ is a real number depending on n that tends to 0 as $n \to \infty$. We also note that the A_n-case of Problem 9.1.2(B) easily reduces to that of S_n, by branching.

The proof of Theorem 9.3.3 uses the *virtual degree* $D(\lambda)$ of a partition $\lambda \vdash n =: |\lambda|$. If the Young diagram $Y(\lambda)$ of λ has rows $\lambda_1 \geq \ldots \geq \lambda_m$ and columns $\mu_1 \geq \ldots \geq \mu_m$, then

$$D(\lambda) = \frac{(n-1)!}{\prod_{i=1}^{m}(\lambda_i - i)!(\mu_i - i)!}.$$

$D(\lambda)$ is a good replacement for the degree $\chi^\lambda(1)$ of the irreducible S_n-character labeled by λ, because

(i) $\lim_{|\lambda|\to\infty} \dfrac{\log D(\lambda)}{\log \chi^\lambda(1)} = 1$,

(ii) (Stability) $Y(\mu) \subseteq Y(\lambda)$ implies $\dfrac{\log D(\mu)}{\log|\mu|} \leq \dfrac{\log D(\lambda)}{\log|\lambda|}$,

(iii) $D(\lambda)$ works better with the Murnaghan–Nakayama rule.

□

Theorem 9.3.3 has many interesting consequences.

Corollary 9.3.4 ([43]) *The following statements hold.*

(i) For a fixed $m \geq 1$, if $g \in \mathsf{S}_n$ has at most $n^{o(1)}$ disjoint cycles of length $< m$ and $\chi \in \mathrm{Irr}(\mathsf{S}_n)$, then $|\chi(g)| \leq \chi(1)^{1/m+o(1)}$.

(ii) For a fixed $0 < \alpha < 1$, if $g \in \mathsf{S}_n$ has at most n^α disjoint cycles and $\chi \in \mathrm{Irr}(\mathsf{S}_n)$, then $|\chi(g)| \leq \chi(1)^{\alpha+o(1)}$.

The Larsen–Shalev bound in Theorem 9.3.3 led to significant results in a number of applications, see [43]. One of them is concerned with *word maps on simple groups.* Let F_d denote the free group on the (alphabet) $x_1, \ldots, x_d$, $1 \neq w = w(x_1, \ldots, x_d) \in F_d$ any nontrivial *word*, and let G be any group. The *word map* corresponding to w, also denoted $w : G^d \to G$, sends any $(g_1, \ldots, g_d) \in G^d$ to $w(g_1, \ldots, g_d)$, with image denoted

$$w(G) := \{w(g_1, \ldots, g_d) \mid g_i \in G\}.$$

Problem 9.3.5 (Non-commutative Waring Problem) Let $1 \neq w \in F_d$ and let G be a finite simple group.

(a) How large is $w(G)$?

(b) Assuming $w(G) \neq \{1\}$, what is the *width* $c(w)$ of w, where

$$c(w) := \min\{k > 0 \mid w(G)^k = G\} \;?$$

(c) How close is the distribution $\mathbf{P}_{w,G} : g \mapsto \dfrac{|w^{-1}(g)|}{|G|^d}$ to the uniform distribution $\mathbf{U}_G$?

To measure the closeness of a distribution $\mathbf{P}$ on G to $\mathbf{U}_G$, we will frequently use the L^1-norm:

$$||\mathbf{P} - \mathbf{U}_G||_{L^1} := \sum_{x\in G} \big|\mathbf{P}(x) - \mathbf{U}_G(x)\big|.$$

Example 9.3.6 (i) $c(w) = 1$ for $w(x,y) = xyx^{-1}y^{-1}$. This statement is known as *Ore's conjecture* [63], whose proof is completed in [47].

(ii) If $w(x) = x^2$ then $c(w) > 1$.

(iii) It was shown in [48] that the width $c(w)$ is well-defined for any $1 \neq w \in F_d$, and $c(w) < \infty$.

The Larsen–Shalev bound in Theorem 9.3.3 was instrumental to prove $c(w) \leq 2$ for $G = \mathsf{A}_n$. Likewise, Theorem 9.2.13, together with algebro-geometric tools, led to a similar result for simple groups of Lie type, culminating in the following solution of Problem 9.3.5:

Theorem 9.3.7 ([44, 45]) *Let $w = w_1w_2$ be a product of disjoint words $1 \neq w_1, w_2 \in F_d$ and let G be a finite simple group.*

(i) If $|G| \gg 0$, then $w(G) = G$.

(ii) If $|G| \gg 0$, then any nontrivial word w_1 has width 2.

(iii) When $|G| \to \infty$, w induces an almost uniform distribution $\mathbf{P}_{w,G}$ on G: $\lim_{|G|\to\infty} ||\mathbf{P}_{w,G} - \mathbf{U}_G||_{L^1} = 0$.

Sketch of proof of Theorem 9.3.7(i). Consider the case where G is a central quotient of the finite quasisimple group $\mathcal{G}^F$ for a suitable Frobenius endomorphism F on a simple simply connected algebraic group $\mathcal{G}$. The starting point is a result of Borel stating that the word map $w_i : \mathcal{G} \to \mathcal{G}$, $i = 1, 2$, is *dominant*, i.e. $w_i(\mathcal{G})$ contains an open dense subset. Hence $w_i(G)$ should hit a large, i.e. *regular semisimple*, conjugacy class $\mathcal{C}_i = s_i^G$. This implies that $|\chi(s_i)|$ is small, for $i = 1, 2$ and for all $\chi \in \mathrm{Irr}(G)$. Using Theorem 9.2.13, we show that

$$\left| \sum_{1_G \neq \chi \in \mathrm{Irr}(G)} \frac{\chi(s_1)\chi(s_2)\overline{\chi(g)}}{\chi(1)} \right| < 1.$$

By Lemma 9.1.1, $g \in \mathcal{C}_1\mathcal{C}_2 \subset w_1(G)w_2(G)$. □

For the next application, consider the following shuffle of a deck of 52 cards: pick two cards i and j uniformly at random from the deck and swap them. How many shuffles is needed to mix the deck?

More generally, let $G = \langle X \rangle$ be a finite group generated by a subset X. The (directed) *Cayley graph* $\Gamma = \Gamma(G, X)$ has G as its set of vertices, and its set of edges is $\{(g, gs) \mid g \in G, s \in X\}$. For a *random walk on* Γ, one starts from 1, and at each step one chooses the next edge at random uniformly with probability $1/|X|$. Then the *mixing time* $T = T(G, X)$ is defined to be the minimal number of steps required to reach an almost uniform distribution on G:

$$T := \min\Big\{k > 0 : ||\mathbf{P}^k - \mathbf{U}_G||_{L^1} < \frac{1}{e}\Big\}.$$

Here, $\mathbf{P}^k(x)$ is the probability that one is at the vertex $x \in G$ after k steps, and $\mathbf{U}_G$ is the uniform distribution on G.

The above card shuffle is the case $n = 52$ of the famous random walk on transpositions (and identity) studied by Diaconis and Shashahani in [9]. Here we have $G = \mathsf{S}_n$, $X = \{1, 2\text{-cycles}\}$, and it was shown in [9] that $T \approx (1/2)n \log n$.

A basic tool in the study of random walks on Cayley graphs defined by conjugacy classes in finite groups is provided by

Lemma 9.3.8 (Upper bound lemma of Diaconis–Shashahani) *Suppose $X = g^G$ is a single conjugacy class. Then*

$$||\mathbf{P}^k - \mathbf{U}||^2_{L^1} \leq \sum_{1_G \neq \chi \in \mathrm{Irr}(G)} \frac{|\chi(g)|^{2k}}{\chi(1)^{2k-2}}.$$

Another tool, which is particularly useful when one works with finite groups of Lie type, is the *Witten ζ-function* [79]:

$$\zeta_G(s) = \sum_{\chi \in \mathrm{Irr}(G)} \chi(1)^{-s}.$$

Theorem 9.3.9 (Liebeck–Shalev [49]) *Let $\mathcal{G}$ be a simple algebraic group with Coxeter number $h = (\dim \mathcal{G})/(\mathrm{rank}\, \mathcal{G}) - 1$, and $G = \mathcal{G}^F$ a quasisimple group of Lie type defined over $\mathbb{F}_q$. If $s > 2/h$, then*

$$\lim_{q \to \infty} \zeta_G(s) = 1.$$

If $s < 2/h$, then

$$\lim_{q \to \infty} \zeta_G(s) = \infty.$$

Now assume that the random walk on $\Gamma = \Gamma(G, X)$ is defined using $X = g^G$, with $G = \mathcal{G}^F$ and $|\chi(g)| \leq \chi(1)^{\boldsymbol{\alpha}}$ for all $\chi \in \mathrm{Irr}(G)$. Taking

$$t > (1 + 1/h)/(1 - \boldsymbol{\alpha}),$$

when $q \to \infty$ we have

$$||\mathbf{P}^t - \mathbf{U}_G||^2_{L^1} \leq \sum_{1_G \neq \chi \in \mathrm{Irr}(G)} \chi(1)^{2t\boldsymbol{\alpha} - 2t + 2} = \zeta_G\big(2t(1 - \alpha) - 2\big) - 1 \to 0.$$

Hence, when q is sufficiently large, $T(G, X) \leq (1 + 1/h)/(1 - \boldsymbol{\alpha})$.

9.4 Problem 9.1.2(B): the BLST-bounds

In this section, we will focus on Problem 9.1.2(B) for finite groups of Lie type. Let $\mathcal{G}$ be a connected reductive algebraic group in characteristic p and of semisimple rank r. Let $G = \mathcal{G}^F$ for a Steinberg endomorphism $F : \mathcal{G} \to \mathcal{G}$. For a proper F-stable Levi subgroup $\mathcal{L}$ of $\mathcal{G}$, define

$$\alpha(\mathcal{L}^F) := \begin{cases} 0, & \text{if } \mathcal{L} \text{ is a torus,} \\ \max\limits_{\substack{1 \neq u \in \mathcal{L}^F, \\ u \text{ unipotent}}} \dfrac{\dim u^{\mathcal{L}}}{\dim u^{\mathcal{G}}}, & \text{otherwise.} \end{cases}$$

Example 9.4.1 ([3]) (i) $\alpha(\mathcal{L}^F) \leq (n-2)/(n-1)$ if $\mathcal{G} = \mathrm{GL}_n$.
(ii) $\alpha(\mathcal{L}^F) \leq 17/29$ if $\mathcal{G} = E_8$ (attained at an E_7-Levi subgroup).
(iii) In general,

$$\alpha(\mathcal{L}^F) \leq \frac{1}{2}\left(1 + \frac{\dim \mathcal{L}}{\dim \mathcal{G}}\right)$$

if $\mathcal{G}$ is a simple classical group.

An F-stable Levi subgroup $\mathcal{L}$ of $\mathcal{G}$ is called *split*, if it is a Levi subgroup of an F-stable parabolic subgroup of $\mathcal{G}$.

Example 9.4.2 (i) Consider the case $\mathcal{G} = \mathrm{SL}_2$ and $F((x_{ij})) = (x_{ij}^q)$, so that $G = \mathrm{SL}_2(q)$. Then a proper split Levi subgroup $\mathcal{L}$ yields $\mathcal{L}^F \cong \mathbf{C}_{q-1}$, a diagonal torus, whereas a non-split Levi subgroup $\mathcal{L}$ yields $\mathcal{L}^F \cong \mathbf{C}_{q+1}$, a Coxeter torus.
(ii) Let $\mathcal{G} = \mathrm{GL}_n$, still with standard Frobenius map $F((x_{ij})) = (x_{ij}^q)$, so that $\mathcal{G}^F = \mathrm{GL}_n(q)$. Then a proper split Levi subgroup $\mathcal{L}$ is conjugate to a block-diagonal subgroup, and

$$\mathcal{L}^F \cong \mathrm{GL}_{n_1}(q) \times \ldots \times \mathrm{GL}_{n_k}(q),$$

where $n_1 \geq n_2 \geq \ldots \geq n_k \geq 1$. If $n_1 = 1$, then $\alpha(\mathcal{L}^F) = 0$. If $n_1 \geq 2$ and $n_1 = n_2 = \ldots = n_t > n_{t+1} \geq 1$, then it was shown in [3, Theorem 1.10] that

$$\frac{n_1 - 1}{n - t} \leq \alpha(\mathcal{L}^F) \leq \frac{n_1}{n}.$$

It remains an open question to find the exact value of $\alpha(\mathcal{L}^F)$.

In the case $\mathcal{G}$ is simple, recall that the characteristic p is said to be *bad* for $\mathcal{G}$, if $p = 2$ for types B, C, D, $p = 2, 3$ for types G_2, F_4, E_6, E_7, and $p = 2, 3, 5$ for type E_8. Furthermore, p is *good*, if it is not bad.

Theorem 9.4.3 ([3, Theorem 1.1]) *There exists an explicit function $f : \mathbb{N} \to \mathbb{N}$ such that the following statement holds. Let $\mathcal{G}$ be a connected reductive algebraic group in characteristic p, such that $[\mathcal{G}, \mathcal{G}]$ is simple of rank r and p is a good prime for $[\mathcal{G}, \mathcal{G}]$. Suppose $g \in \mathcal{G}^F$ is any element such that $\mathbf{C}_{\mathcal{G}^F}(g) \leq \mathcal{L}^F$ for some proper split F-stable Levi subgroup $\mathcal{L}$ of $\mathcal{G}$. Then*

$$|\chi(g)| \leq f(r) \cdot \chi(1)^{\boldsymbol{\alpha}(\mathcal{L}^F)}$$

for all $\chi \in \mathrm{Irr}(\mathcal{G}^F)$.

Remark 9.4.4 Suppose $\mathcal{G}^F = \mathrm{GL}_n(q)$ or $\mathrm{SL}_n(q)$. Then in fact Theorem 9.4.3 also holds for *Brauer characters* in any characteristic $\ell \nmid q$, evaluated at ℓ'-elements $g \in \mathcal{G}^F$, see [3, Theorem 1.4].

Remark 9.4.5 ([3]) How optimal is the character bound in Theorem 9.4.3, in particular the constant $f(r)$ and the exponent $\boldsymbol{\alpha}(\mathcal{L}^F)$?

(i) If r is large then $f(r)$, as constructed in the proof of Theorem 9.4.3, is of the magnitude of $((r+1)!)^2$. Indeed, its main term is $\max_W |W|^2$, where the maximum is taken over all Weyl groups W of simple algebraic groups of rank r. It may seem a bit inflated. However, there exist examples that show $f(r)$ must be at least of the magnitude of $\sqrt{(r+1)!}$.

(ii) The exponent $\alpha(\mathcal{L}^F)$ in Theorem 9.4.3 is sharp in many cases. For instance, it was shown in [3, Theorem 1.3] that in the case where $\mathcal{G} = \mathrm{GL}_n$, $G = \mathcal{G}^F = \mathrm{GL}_n(q)$ and $\mathcal{L}$ is a proper split Levi subgroup, then when q is large enough (in terms of n) there exist a semisimple element $g \in G$ with $\mathbf{C}_G(g) = \mathcal{L}^F$ and a unipotent character $\chi \in \mathrm{Irr}(G)$ such that

$$|\chi(g)| \geq (1/4) \cdot \chi(1)^{\boldsymbol{\alpha}(\mathcal{L}^F)}.$$

We now describe some of the main ideas of the proof of Theorem 9.4.3. Let $\mathcal{P}$ be an F-stable parabolic subgroup of $\mathcal{G}$, with unipotent radical $\mathcal{U}$ and $\mathcal{L}$ as an F-stable Levi subgroup. Write $P = \mathcal{P}^F$, $U = \mathcal{U}^F$, and $L = \mathcal{L}^F$. If a G-character χ is afforded by a $\mathbb{C}G$-module V, then the *Harish-Chandra restriction* ${}^*R^G_L(\chi)$ of χ is the L-character of the L-module V^U, where

$$V^U := \{x \in V \mid \forall u \in U, u(x) = x\}$$

is the U-fixed point subspace. The adjoint functor of *Harish-Chandra induction* R^G_L is defined as follows. If an L-character θ is afforded by a

$\mathbb{C}L$-module X, then we can inflate X to a P-module $\hat{X}$ (by letting U act trivially on it), and then induce to G to obtain a $\mathbb{C}G$-module, whose character is $R_L^G(\theta)$. (It would be more precise to write ${}^*R^{\mathcal{G}}_{\mathcal{L}\subset\mathcal{P}}$ and $R^{\mathcal{G}}_{\mathcal{L}\subset\mathcal{P}}$ instead of ${}^*R_L^G$ and R_L^G, see [7, Chapter 4].)

Lemma 9.4.6 *For any G-character χ and any element $g \in G$ with $\mathbf{C}_G(g) \leq L = \mathcal{L}^F$, we have $\chi(g) = {}^*R_L^G(\chi)(g)$.*

Proof Consider the map $f : U \to U$ given by $f(u) = g^{-1}ugu^{-1}$. For any $u, v \in U$, note that

$$f(u) = f(v) \Leftrightarrow v^{-1}u \in U \cap \mathbf{C}_G(g) \subseteq U \cap L = 1.$$

Thus f is injective, and hence bijective. Therefore, when u runs over U, ugu^{-1} runs over the coset gU, covering each element once.

Now we decompose $V|_P = V^U \oplus [V, U]$ and write $\Phi = \Phi_1 \oplus \Phi_2$ for the corresponding representations of P. Then no irreducible constituent of $(\Phi_2)|_U$ is trivial, so $\sum_{u\in U} \Phi_2(u) = 0$. It follows that

$$\begin{aligned}\sum_{u\in U} \Phi(ugu^{-1}) &= \sum_{u\in U} \Phi(gu) = \Phi(g) \sum_{u\in U} \Phi(u)\\ &= \mathrm{diag}\left(\Phi_1(g) \sum_{u\in U} \Phi_1(u), \Phi_2(g) \sum_{u\in U} \Phi_2(u)\right)\\ &= \mathrm{diag}(|U|\Phi_1(g), 0).\end{aligned}$$

Thus $\sum_{u\in U} \Phi(ugu^{-1}) = \mathrm{diag}(|U|\Phi_1(g), 0)$. Taking the trace of both sides, we obtain $|U|\chi(g) = |U|\psi(g)$ with $\psi := {}^*R_L^G(\chi)$. □

Note that Lemma 9.4.6 also works in the setting of ℓ-Brauer characters, if $\ell \nmid p|g|$.

Lemma 9.4.6 implies the bound $|\chi(g)| \leq {}^*R_L^G(\chi)(1)$. To bound the latter for any $\chi \in \mathrm{Irr}(G)$, we need to accomplish the following two tasks:

(A) Control the number of irreducible constituents of ${}^*R_L^G(\chi)$, and

(B) Control the degree of each of these constituents.

Task (A) is accomplished by the following statement, whose proof utilizes properties of ${}^*R_L^G$ and R_L^G.

Proposition 9.4.7 ([3, Proposition 2.2]) *There is an explicit function $A : \mathbb{N} \to \mathbb{N}$ such that the following statement holds. Let $\mathcal{G}$ be of semisimple rank r. Suppose that $\chi \in \mathrm{Irr}(G)$ is such that ${}^*R_L^G(\chi) \neq 0$ for some $L = \mathcal{L}^F$, where $\mathcal{L}$ is a proper split F-stable Levi subgroup of $\mathcal{G}$. Then the total number of irreducible constituents of ${}^*R_L^G(\chi)$ (counting multiplicities) is at most $A(r)$.*

To accomplish task (B), we rely Lusztig's notion of *unipotent support* $\mathcal{O}_\chi$ [60] and Kawanaka's dual notion of *wave front set* $\mathcal{O}^*_\chi$ for any irreducible character χ of $G = \mathcal{G}^F$.

Assume p is good for $\mathcal{G}$. Then to each unipotent class u^G one can associate a *generalized Gelfand–Graev character* (GGGR) Γ_u of G. Works of Lusztig [60], Geck–Malle [17], and Taylor [70], show that for each $\chi \in \mathrm{Irr}(G)$, there is a unique F-stable unipotent class $\mathcal{O}_\chi = u^{\mathcal{G}}$, called the *unipotent support* of χ, such that

(A1) $\sum_{g \in u^{\mathcal{G}} \cap G} \chi(g) \neq 0$, and

(A2) $\dim u^{\mathcal{G}}$ is largest possible subject to (A1).

Dually, there is a unique F-stable unipotent class $\mathcal{O}^*_\chi = v^{\mathcal{G}}$, called the *wave front set* of χ, such that

(B1) $[\sum_{i=1}^m \Gamma_{v_i}, \chi] \neq 0$, where $v^{\mathcal{G}} \cap G = \sqcup_{i=1}^m v_i^G$ (and $[\cdot,\cdot]$ denotes the inner product of class functions), and

(B2) $\dim v^{\mathcal{G}}$ is largest possible subject to (B1).

The duality between these two concepts is provided by the *Alvis-Curtis duality* $D_G : \mathrm{Irr}(G) \to \pm\mathrm{Irr}(G)$ [7, Chapter 8],

$$\mathcal{O}^*_\chi = \mathcal{O}_{D_G(\chi)},$$

see [70, §1.5]. (Note that $D_G(1_G) = \pm\,\mathsf{St}_G$, where St_G is the *Steinberg character* of G.)

A key property of wave front sets is that $\mathcal{O}^*_\chi = v^{\mathcal{G}}$ *controls* $\chi(1)$: If $G = \mathcal{G}^F$ is defined over $\mathbb{F}_q$, then

$$\chi(1) = \frac{1}{n_\chi}(q^{\frac{1}{2}\dim \mathcal{O}^*_\chi} + \text{lower powers of } q)$$

for some divisor n_χ of $[\mathbf{C}_{\mathcal{G}}(u) : \mathbf{C}_{\mathcal{G}}(u)^\circ]$. To make it useful for the aforementioned task (b), we need to show that wave front sets "grow" under Harish-Chandra induction:

Proposition 9.4.8 ([3, Proposition 2.6]) *Let p be a good prime for $\mathcal{G}$, $\mathbf{Z}(\mathcal{G})$ connected, and let $L = \mathcal{L}^F$ for a proper split F-stable Levi subgroup $\mathcal{L}$ of $\mathcal{G}$. Let $\eta \in \mathrm{Irr}(L)$ and let $\chi \in \mathrm{Irr}(G)$ be an irreducible constituent of $R^G_L(\eta)$. If $\mathcal{O}^*_\eta = u^{\mathcal{L}}$ and $\mathcal{O}^*_\chi = v^{\mathcal{G}}$ then $\dim u^{\mathcal{G}} \leq \dim v^{\mathcal{G}}$.*

Sketch of proof of Proposition 9.4.8. Let Γ^L_u denote the corresponding GGGR of L defined by $u^{\mathcal{L}}$. The hypotheses imply

$$\begin{aligned} 0 < [{}^*R^G_L(\chi), \eta]_L &\leq [{}^*R^G_L(\chi), \Gamma^L_u]_L = [\chi, R^G_L(\Gamma^L_u)]_G \\ &= [D_G(\chi), D_G(R^G_L(\Gamma^L_u))]_G = [D_G(\chi), R^G_L(D_L\Gamma^L_u)]_G. \end{aligned}$$

Hence one can find an element $w \in G$ such that $D_G(\chi)(w) \neq 0$ and

$R_L^G(D_L\Gamma_u^L)(w) \neq 0$. One shows that the latter condition implies that w is unipotent and $u^{\mathcal{G}} \subseteq \overline{w^{\mathcal{G}}}$. In turn, the former condition implies that $\dim w^{\mathcal{G}} \leq \dim v^{\mathcal{G}}$. Hence $\dim u^{\mathcal{G}} \leq \dim w^{\mathcal{G}} \leq \dim v^{\mathcal{G}}$. □

Sketch of proof of Theorem 9.4.3. First, we use *regular embeddings* to reduce to the case where $\mathbf{Z}(\mathcal{G})$ is connected. Recall that $|\chi(g)| \leq {}^*R_L^G(\chi)(1)$ by Lemma 9.4.6, and ${}^*R_L^G(\chi)$ has at most $A(r)$ irreducible constituents by Proposition 9.4.7. Let η be any of them, and write $u^{\mathcal{L}} = \mathcal{O}_\eta^*$, $v^{\mathcal{G}} = \mathcal{O}_\chi^*$. By Proposition 9.4.8 and the definition of $\boldsymbol{\alpha} := \boldsymbol{\alpha}(\mathcal{L}^F)$ we have

$$\dim u^{\mathcal{L}} \leq \boldsymbol{\alpha}(\dim u^{\mathcal{G}}) \leq \boldsymbol{\alpha}(\dim v^{\mathcal{G}}).$$

Also recall that $\chi(1) \approx q^{(\dim v^{\mathcal{G}})/2}$ and $\eta(1) \approx q^{(\dim u^{\mathcal{L}})/2}$. It follows that $\eta(1) \lessapprox \chi(1)^{\boldsymbol{\alpha}}$. □

Applying Theorem 9.4.3 to $G := \mathrm{GL}_n(q)$ and using Example 9.4.2(ii), we obtain

Corollary 9.4.9 *Let $G = \mathrm{GL}_n(q)$, $m < n$ a divisor of n, and let $L \leq G$ be a Levi subgroup of the form $L = \mathrm{GL}_{n/m}(q)^m$. If $\mathbf{C}_G(g) \leq L$ for $g \in G$, then*

$$|\chi(g)| \leq f(n-1)\chi(1)^{\frac{1}{m}}$$

for any $\chi \in \mathrm{Irr}(G)$, where f is as specified in Theorem 9.4.3.

It is illustrative to compare Corollary 9.4.9 to the Fomin–Lulov bound [12].

Certainly, Theorem 9.4.3 leaves out a number of cases:

• What if $\mathbf{C}_{\mathcal{G}}(g) \not\leq L$ for any proper Levi subgroup L? E.g. if g is unipotent?

• What about the elements $g \in G$ with $\mathbf{C}_{\mathcal{G}}(g) \leq \mathcal{L}$ for a non-split Levi subgroup $\mathcal{L}$?

Unipotent elements in split groups of type A are addressed in the following result:

Theorem 9.4.10 ([3, Theorem 3.3]) *There is a function $g : \mathbb{N} \to \mathbb{N}$ such that the following statement holds. For any integer $n \geq 2$, for $\ell = 0$ or any prime $\ell \nmid q$, any group $G := \mathrm{GL}_n(q)$ or $\mathrm{SL}_n(q)$, any irreducible ℓ-Brauer character χ of G, and any unipotent element $1 \neq u \in G$, we have*

$$|\chi(u)| \leq g(n) \cdot \chi(1)^{\frac{n-2}{n-1}}.$$

The proof of Theorem 9.4.10 relies on the property that any unipotent element $u \in G$ is *Richardson*, i.e. there is a parabolic subgroup $\mathcal{P} = \mathcal{U}\mathcal{L}$ of $\mathcal{G} = \mathrm{GL}_n$ such that $u^{\mathcal{G}} \cap \mathcal{U} = u^{\mathcal{P}}$ and is open dense in $\mathcal{U}$. It then proceeds by induction, using the partial order $\preceq$ on unipotent classes:

$$x^{\mathcal{G}} \preceq y^{\mathcal{G}} \text{ iff } x \in \overline{y^{\mathcal{G}}}.$$

□

Theorem 9.4.3 has been extended to the non-split case in [71]:

Theorem 9.4.11 ([71, Theorem 1.9]) *There exists an explicit function $f^* : \mathbb{N} \to \mathbb{N}$ such that the following statement holds. Let $\mathcal{G}$ be of semisimple rank r, p a good prime for $\mathcal{G}$, and let $\mathbf{Z}(\mathcal{G})$ be connected. Suppose $\mathcal{L}$ is a proper F-stable Levi subgroup of $\mathcal{G}$ and $g \in \mathcal{L}^F$ is such that $\mathbf{C}_{\mathcal{G}}(g)^\circ \leq \mathcal{L}$. Then*

$$|\chi(g)| \leq f^*(r) \cdot \chi(1)^{\alpha(\mathcal{L}^F)}$$

for all $\chi \in \mathrm{Irr}(\mathcal{G}^F)$.

Remark 9.4.12 Suppose $\mathcal{L}$ is an F-stable split Levi subgroup of $\mathcal{G}$. Then for any $g \in \mathcal{G}^F$ we have that

$$\mathbf{C}_G(g) \leq \mathcal{L}^F \Longrightarrow [g \in \mathcal{L}^F \text{ and } (\mathbf{C}_{\mathcal{G}}(g)^\circ)^F \leq \mathcal{L}^F] \Longrightarrow \mathbf{C}_{\mathcal{G}}(g)^\circ \leq \mathcal{L},$$

see [71, Lemma 13.4] for the proof of the (only non-obvious) second implication. Hence (the formulation of) Theorem 9.4.11 includes the $\mathbf{Z}(\mathcal{G})$ connected case of Theorem 9.4.3.

Sketch of proof of Theorem 9.4.11. First, we generalize Lemma 9.4.6 to the non-split case, which applies to *Lusztig restriction*, see [4, 7, 58], instead of Harish-Chandra restriction:

$$\text{If } g \in \mathcal{L}^F \text{ and } \mathbf{C}_{\mathcal{G}}(g)^\circ \leq \mathcal{L}, \text{ then } \chi(g) = {}^*R_L^G(\chi)(g).$$

Next, note that ${}^*R_L^G(\chi)$ is a *generalized* character, so one cannot bound $|\chi(g)|$ by its degree. Instead, for any $\eta \in \mathrm{Irr}(\mathcal{L}^F)$, we bound the scalar product $[R_{\mathcal{L}}^{\mathcal{G}}(\eta), R_{\mathcal{L}}^{\mathcal{G}}(\eta)]$ and hence the multiplicity of η in ${}^*R_L^G(\chi)$.

It remains to bound $\eta(1)$ for any irreducible constituent η of ${}^*R_{\mathcal{L}}^{\mathcal{G}}(\chi)$, via proving

$$\mathcal{O}_\eta^* \preceq \mathcal{O}_\chi^*.$$

Equivalently, one needs to prove

$$\mathcal{O}_\eta \preceq \mathcal{O}_\chi \text{ whenever } [\chi, R_{\mathcal{L}}^{\mathcal{G}}(\eta)] \neq 0.$$

Note that $\mathcal{O}_\chi = \mathcal{O}_{\mathsf{A}}$, if A is an F-stable character sheaf such that $[\chi, \chi_{\mathsf{A}}] \neq 0$. So we can work at geometric level, and prove the following statement for parabolic induction $\mathrm{Ind}_{\mathcal{L}}^{\mathcal{G}}$ of character sheaves:

$$\text{If } \mathsf{A} \text{ is a constituent of } \mathrm{Ind}_{\mathcal{L}}^{\mathcal{G}}(\mathsf{B}) \text{ then } \mathcal{O}_{\mathsf{B}} \preceq \mathcal{O}_{\mathsf{A}}.$$

To do so, we pass to F^d for a suitable $d \in \mathbb{Z}_{\geq 1}$ such that $\mathcal{L}$ is F^d-split, and apply the split case as in the proof of Theorem 9.4.3. □

The proof of Theorem 9.4.11 as given in [71] needs $\mathbf{Z}(\mathcal{G})$ to be connected so that one can apply results of Lusztig [59] and Shoji [68] on certain conjectures, which in particular ensure that, firstly, the χ_{A} with $\mathsf{A} \in \mathcal{F}$ span the vector space $\langle \chi \in \mathrm{Irr}(G), \chi \in \mathcal{F} \rangle_{\mathbb{C}}$ for any family $\mathcal{F}$, and, secondly, $\chi_{\mathrm{Ind}_{\mathcal{L}}^{\mathcal{G}}(\mathsf{B})} = R_{\mathcal{L}}^{\mathcal{G}}(\chi_{\mathsf{B}})$ for any character sheaf B on $\mathcal{L}$.

In certain cases one can relax the connectedness condition on $\mathbf{Z}(\mathcal{G})$:

Theorem 9.4.13 ([71, Corollary 1.16]) *The character bound in Theorem 9.4.11 holds for $\chi(g)$, if at least one of the following conditions holds.*

(i) There exists a regular embedding $(\mathcal{G}, F) \hookrightarrow (\tilde{\mathcal{G}}, F)$ such that either χ is $\tilde{\mathcal{G}}^F$-invariant, or $g^{\mathcal{G}^F} = g^{\tilde{\mathcal{G}}^F}$.
(ii) $\mathbf{C}_{\mathcal{G}}(g)$ is connected.
(iii) $[\mathcal{G}, \mathcal{G}]$ is simply connected and g is semisimple.
(iv) $\mathcal{G} = \mathrm{SL}_{r+1}$ or Sp_{2r}, and $q \gg r$.

Moreover, a solution to Problem 9.1.2(B) has been obtained in [71] for *arbitrary elements* in finite Lie-type groups of type A:

Theorem 9.4.14 ([71, Theorem 1.17]) *There exists an explicit function $h : \mathbb{N} \to \mathbb{N}$ such that the following statement holds. Let $G := \mathrm{GL}_n(q)$, $\mathrm{SL}_n(q)$, $\mathrm{GU}_n(q)$, or $\mathrm{SU}_n(q)$, with $n \geq 5$, and let $g \in G \smallsetminus \mathbf{Z}(G)$. Then*

$$|\chi(g)| \leq h(n) \cdot \chi(1)^{(n-2)/(n-1)}$$

for all $\chi \in \mathrm{Irr}(\mathcal{G}^F)$.

Note that the exponent $(n-2)/(n-1)$ is best possible, since it is the maximum value of $\boldsymbol{\alpha}(\mathcal{L}^F)$ among proper Levi subgroups of GL_n, cf. Remark 9.4.5 and Example 9.4.2.

Sketch of proof of Theorem 9.4.14. Reduce to the case of $\mathrm{GL}_n(q)$, respectively $\mathrm{GU}_n(q)$. These two groups share the nice property that for any $g \in G$, either $\mathbf{C}_{\mathcal{G}}(g)^\circ \leq \mathcal{L}$ or g is unipotent. If the former conclusion holds, we apply Theorem 9.4.11. In the latter case, if $G = \mathrm{GL}_n(q)$, then we are done by Theorem 9.4.10. To finish off the case $G = \mathrm{GU}_n(q)$, we use *Ennola duality* [11]. □

Before proceeding further, let us make some more comments about the BLST-bound 9.4.3 and its extension 9.4.11. First, if $\mathcal{G}$ is not of type A, then these character bounds cover a majority, but still not all elements in $\mathcal{G}^F = \mathcal{G}(q)$; for instance, they leave out all the nontrivial unipotent elements. Secondly, these results apply in *good* characteristics; also, they are meaningful only when $q \gg r := \mathrm{rank}([\mathcal{G}, \mathcal{G}])$, because of the constants $f(r)$, respectively $f^*(r)$. This certainly raises the questions: What can be done in the case of *bad* characteristics? What about finite groups of Lie type $\mathcal{G}(q)$ in the regime where r is not small compared to q, in particular when $\mathcal{G}$ is of *unbounded rank*?

For Lie-type groups of *bounded rank*, one can try to put the remaining elements of $\mathcal{G}^F$ in "good" subgroups H and then work with the restriction to H. This method works in the case of *exceptional* groups of Lie type, leading to strong exponential character bounds for these groups. See [54] for results in this direction for finite classical groups.

Theorem 9.4.15 ([55, Theorem 1]) *Let $G = \mathcal{G}(q)$ be a quasisimple exceptional group of Lie type in good characteristic. There exist explicit constants $C, a_1, a_2 > 0$ such that for all $g \in G \smallsetminus \mathbf{Z}(G)$ and for all non-linear $\chi \in \mathrm{Irr}(G)$,*

$$\frac{|\chi(g)|}{\chi(1)} \le \begin{cases} C/q^{a_1}, & g \text{ long-root element,} \\ C/q^{a_2}, & \text{otherwise.} \end{cases}$$

The constants a_1, a_2 are listed in [55, Table 1.1]; e.g. $(a_1, a_2) = (6, 10)$ when $G = E_8(q)$.

Sketch of proof of Theorem 9.4.15. For any element $g \in G$, one can show that at least one of the following conditions holds.

(a) $\mathbf{C}_G(g)$ is small, hence $|\chi(g)| \le \sqrt{|\mathbf{C}_G(g)|}$ is small.
(b) $g \in P = UL$ for some maximal parabolic subgroup P of G with radical U and Levi subgroup L. In this case, we bound

$$|{}^*R^G_L(\chi)(g)| \le {}^*R^G_L(\chi)(1)$$

using Theorem 9.4.3. Then we bound the values at g of the remaining constituents of $\chi|_P$ using detailed information (including fixed point ratios) on the action of L on $\mathrm{Irr}(U)$.

□

9.5 The GLT-bounds

In this section, we will change gear and shift the focus to the classical groups $G = \mathrm{Cl}_n(q)$.

In a number of applications, the most attention is needed to bound $|\chi(g)|$ for elements g in a finite almost quasisimple group G only when either $\chi(1)$ or $|\mathbf{C}_G(g)|$ is *not too large*. In these cases, the task of bounding character values can be solved by another approach, via the notion of *character level*. This notion is based on the empirical fact that character degrees of finite almost quasisimple groups seem to exist in "clusters", with characters in the same cluster displaying a similar behavior.

Let's first illustrate this on the example $G = \mathsf{S}_n$, whose complex irreducible characters are labeled by partitions $\lambda \vdash n$. For a given $\lambda \vdash n$, the degree of χ^λ can be read off from the hook formula. Based on this information, one can intuitively define the level of characters of S_n as demonstrated in Table 9.1.

Table 9.1 *Level of characters of* S_n

λ	$\chi^\lambda(1)$	"Level"
(n)	1	0
$(n-1,1)$	$\approx n$	1
$(n-2,2)$, $(n-2,1^2)$	$\approx n^2/2$	2
$(n-3,3)$, $(n-3,1^3)$ $(n-3,2,1)$	$\approx n^3/6$ $\approx n^3/3$	3
	$\ldots$	
$(n-k,\mu)$, $\mu \vdash k$ μ fixed, $n \to \infty$	$\approx \frac{n^k}{k!}\chi^\mu(1)$ [34]	k
(1^n)	1	?

In this example, the level of χ seems to be roughly $\log_n \chi(1)$. For a rigorous treatment of character level, which also works for Brauer characters, see [36].

In Tables 9.2 and 9.3, we collect some low-degree complex irreducible characters of $\mathrm{SL}_n(q)$ and $\mathrm{SU}_n(q)$, based on [77]. In these two examples, the level of χ is roughly $(\log_q \chi(1))/n$.

The question is: How to define the level in a rigorous, and more importantly, a useful way?

Table 9.2 *Characters of low degree of* $\mathrm{SL}_n(q)$

Number	Characters of degree	"Level"
1	1	0
$q-1$	$\approx q^{n-1}$	1
q^2-1	$\approx q^{2n-4},\ q^{2n-3}$	2

Table 9.3 *Characters of low degree of* $\mathrm{SU}_n(q)$

Number	Characters of degree	"Level"
1	1	0
$q+1$	$\approx q^{n-1}$	1
$(q+1)^2$	$\approx q^{2n-4},\ q^{2n-3}$	2

We now describe the approach of [22] to define the level of complex characters of $\mathrm{GL}_n(q) =: \mathrm{GL}_n^+(q)$ and $\mathrm{GU}_n(q) =: \mathrm{GL}_n^-(q)$.

First, for $\mathrm{GL}_n(q)$ we set

$$A := \mathbb{F}_q^n,\ G = \mathrm{GL}(A) = \mathrm{GL}_n(q) \text{ or } G = \mathrm{SL}(A) = \mathrm{SL}_n(q).$$

Let $\tau = \tau_A = \tau_{n,q}$ denote the permutation character of G acting on the vectors of A:

$$\tau(g) = q^{\dim_{\mathbb{F}_q} \mathrm{Ker}(g-1_A)}. \tag{9.6}$$

Then $\chi \in \mathrm{Irr}(G)$ is said to be of *true level* $\mathfrak{l}^*(\chi) = j$, if $j \in \mathbb{Z}_{\geq 0}$ is smallest such that $[\tau^j, \chi] \neq 0$. Furthermore, $\chi \in \mathrm{Irr}(G)$ has *level* $\mathfrak{l}(\chi) = j$, if $j \in \mathbb{Z}_{\geq 0}$ is smallest such that $[\tau^j, \chi\lambda] \neq 0$ for some character λ of G of degree 1.

We will see in Lemma 9.5.3 (below) that both of these notions are well-defined.

Example 9.5.1 (i) The principal character 1_G has level 0.

(ii) The Weil characters [78] of G have degree $(q^n - q)/(q-1)$ and $(q^n-1)/(q-1)$. They are irreducible constituents of $\mathrm{Ind}_P^G(1_P)\lambda$, for $P = \mathrm{Stab}_G(v)$, $0 \neq v \in A$, and for some $\lambda \in \mathrm{Irr}(G)$ of degree 1. Hence they are of level 1.

(iii) The Steinberg character St of G has degree $q^{n(n-1)/2}$. It turns out to have level $n-1$, the reason being that St is the unipotent character $\chi^{(1,1,\ldots,1)}$ of G.

Next, for $\mathrm{GU}_n(q)$ we set

$$A = \mathbb{F}_{q^2}^n,\ G = \mathrm{GU}(A) = \mathrm{GU}_n(q) \text{ or } G = \mathrm{SU}(A) = \mathrm{SU}_n(q).$$

It turns out that, to define the character level for G, one can no longer use the permutation character of G acting on A, but needs to use its "square root" instead. The key is to apply Ennola duality $q \mapsto -q$ [11] to $\tau_{n,q}$ and work with $\zeta = \zeta_A = \zeta_{n,q}$, where

$$\zeta(g) = (-1)^n(-q)^{\dim_{\mathbb{F}_{q^2}} \mathrm{Ker}(g-1_A)} \tag{9.7}$$

is a true character of G. Now, $\chi \in \mathrm{Irr}(G)$ is said to be of *true level* $\mathfrak{l}^*(\chi) = j$, if $j \in \mathbb{Z}_{\geq 0}$ is smallest such that $[\zeta^j, \chi] \neq 0$. Furthermore, $\chi \in \mathrm{Irr}(G)$ has *level* $\mathfrak{l}(\chi) = j$, if $j \in \mathbb{Z}_{\geq 0}$ is smallest such that $[\zeta^j, \chi\lambda] \neq 0$ for some character λ of degree 1.

Example 9.5.2 (i) The principal character 1_G has level 0.

(ii) The *Weil characters* [78] of G have degree $(q^n + (-1)^n q)/(q+1)$ and $(q^n - (-1)^n)/(q+1)$. They are irreducible constituents of $\zeta\lambda$ for some $\lambda \in \mathrm{Irr}(G)$ of degree 1. Hence they are of level 1.

(iii) The Steinberg character St of G has degree $q^{n(n-1)/2}$. It is the unipotent character $\chi^{(1,1,\ldots,1)}$ of G, and has level $n-1$.

The reason why the true level $\mathfrak{l}^*(\chi)$ is well-defined for any $\chi \in \mathrm{Irr}(G)$ is based on the following statement.

Lemma 9.5.3 (Burnside, Brauer) *Let Θ be a generalized character of a finite group G which takes exactly N different values $a_0 = \Theta(1)$, $a_1, \ldots, a_{N-1}$ on G. Suppose $\Theta(g) \neq \Theta(1)$ for all $1 \neq g \in G$. Then every $\chi \in \mathrm{Irr}(G)$ occurs in some Θ^k with $0 \leq k \leq N-1$.*

Proof Consider any $\chi \in \mathrm{Irr}(G)$. By assumption,

$$[\chi, \prod_{i=1}^{N-1}(\Theta - a_i \cdot 1_G)]_G = \frac{\chi(1)}{|G|}\prod_{i=1}^{N-1}(a_0 - a_i) \neq 0,$$

whence $[\chi, \Theta^k]_G \neq 0$ for some $0 \leq k \leq N-1$. □

In the case of $G = \mathrm{GL}_n(q)$, the character τ takes values q^k, $0 \leq k \leq n$. Hence $0 \leq \mathfrak{l}^*(\chi) \leq n$ for $\chi \in \mathrm{Irr}(G)$ by Lemma 9.5.3.

Theorem 9.5.4 ([22, Theorem 1.1]) *For each $\epsilon = \pm$ and prime power q, the following statements hold for $G_n := \mathrm{GL}_n^\epsilon(q)$.*

(i) *There is a canonical bijection* Θ *between*

$$\{\chi \in \mathrm{Irr}(\mathrm{GL}_n^\epsilon(q)) \mid \mathfrak{l}^*(\chi) = j\}$$

and

$$\{\alpha \in \mathrm{Irr}(\mathrm{GL}_j^\epsilon(q)) \mid \mathfrak{l}^*(\alpha) \geq 2j - n\}$$

given by

$$\chi \mapsto \left({}^*R^{G_n}_{G_j \times G_{n-j}}(\chi)\right)^{G_{n-j}}.$$

(ii) *If* $0 \leq j < n/2$, *restriction to* $\mathrm{SL}_n^\epsilon(q)$ *of* Θ *yields a canonical bijection*

$$\{\varphi \in \mathrm{Irr}(\mathrm{SL}_n^\epsilon(q)) \mid \mathfrak{l}(\varphi) = j\} \longleftrightarrow \mathrm{Irr}(\mathrm{GL}_j^\epsilon(q)).$$

Theorem 9.5.4 shows that characters of level j of G_n are controlled by G_j. We should mention that canonical bijections like the ones defined in Theorem 9.5.4 are good for many reasons: they are equivariant under outer automorphisms, Galois automorphisms, etc., and hence useful for many purposes; in particular, with regard to the *inductive McKay condition* of [32].

The proof of Theorem 9.5.4 relies on Deligne-Lusztig theory and Howe's *dual pair* philosophy. In the linear case, set

$$A = \mathbb{F}_q^n,\ G = \mathrm{GL}(A) \cong \mathrm{GL}_n(q),\ B = \mathbb{F}_q^j,\ S = \mathrm{GL}(B) \cong \mathrm{GL}_j(q).$$

Also consider

$$V = A \otimes_{\mathbb{F}_q} B = \mathbb{F}_q^{nj},\ \Gamma = \mathrm{GL}(V) \cong \mathrm{GL}_{nj}(q)$$

and the permutation character

$$\omega(g) = q^{\dim \mathrm{Ker}(g - 1_V)}$$

of Γ on V. Then we restrict ω to the image of $G \times S$ in Γ:

$$\omega|_{G \times S} = \sum_{\alpha \in \mathrm{Irr}(S)} D_\alpha \boxtimes \alpha$$

for suitable class functions D_α on G, and $D_\alpha \boxtimes \alpha : (g, s) \mapsto D_\alpha(g)\alpha(s)$ is the outer tensor product. It turns out that, for every α, the class function D_α is either 0, or a G-character of true level $\leq j$. Moreover, if $\mathfrak{l}^*(\alpha) < 2j - n$, then all irreducible constituents of D_α have true level $< j$. If $\mathfrak{l}^*(\alpha) \geq 2j - n$, then D_α has a unique irreducible constituent H_α

of true level j, and $\alpha \mapsto H_\alpha$ is the inverse of the desired map Θ. A key identity in the proof is

$$R^{G_n}_{G_j \times G_{n-j}}(\mathbf{reg}_{G_j} \times 1_{G_{n-j}}) = \prod_{i=0}^{j-1}(\tau - q^i \cdot 1_G)$$

(where $\mathbf{reg}_{G_j}$ denotes the regular character of G_j and τ as in (9.6)).

In the unitary case, key roles are played by the fact that the *Ennola product* of class functions [11] agrees with Lusztig's induction, and the identity

$$R^{G_n}_{G_j \times G_{n-j}}(\mathbf{reg}_{G_j} \times 1_{G_{n-j}}) = (-1)^j \prod_{i=0}^{j-1}((-1)^n\zeta - (-q)^i \cdot 1_G),$$

with ζ as in (9.7). □

The next result [22, Theorem 3.9] shows how to read off $\mathfrak{l}(\chi)$ from the Lusztig label of χ. Let us describe the result in the linear case. First, one can identify $G = \mathrm{GL}(A) \cong \mathrm{GL}_n(q)$ with its *dual*. For any semisimple element $s \in G$, we can decompose $A = A^0 \oplus A^1$ as a direct sum of s-invariant subspaces, where

$$A^0 = \bigoplus_{\delta \in \mathbb{F}_q^\times} A_\delta, \quad A_\delta = \mathrm{Ker}(s - \delta \cdot 1_A).$$

Correspondingly,

$$\mathbf{C}_G(s) = \Big(\prod_{\delta \in \mathbb{F}_q^\times} \mathrm{GL}(A_\delta)\Big) \times \mathbf{C}_{\mathrm{GL}(A^1)}(s).$$

Any unipotent character ψ of $\mathbf{C}_G(s)$ can be written as

$$\psi = \big(\boxtimes_{\delta \in \mathbb{F}_q^\times} \psi^{\gamma_\delta}\big) \boxtimes \psi_1,$$

where ψ^{γ_δ} is the unipotent character of $\mathrm{GL}(A_\delta)$ labeled by the partition γ_δ of $\dim(A_\delta)$ for $\delta \in \mathbb{F}_q^\times$. According to Lusztig's classification of irreducible characters of G [4, 7, 58], any given $\chi \in \mathrm{Irr}(G)$ is

$$\chi = \pm R^G_{\mathbf{C}_G(s)}(\hat{s}\psi)$$

for some such s (which is unique up to conjugacy), ψ, and a linear character $\hat{s}$ of $\mathbf{C}_G(s)$. Given this identification of χ, we have, see Theorems 3.9 and 1.2 of [22]

(i) $\mathfrak{l}^*(\chi) = j$, *if* $n - j$ *is the largest part of* γ_1.
(ii) $\mathfrak{l}(\chi) = j$, *if* $n - j$ *is the largest part of* γ_δ *among all* $\delta \in \mathbb{F}_q^\times$.

(iii) *If* $\mathfrak{l}(\chi) = j$, *then* $q^{j(n-j)} \leq \chi(1) \leq q^{nj}$.
(iv) *Moreover, if* $n \geq 18$ *and* $\chi(1) \leq q^{n\sqrt{n}/2}$, *then* $\mathfrak{l}(\chi) = \lceil (\log_q \chi(1))/n \rceil$ (which agrees with our intuitive definition of character level, cf. Table 9.2; note that $n \geq 18$ ensures that $\lceil \sqrt{n}/2 \rceil < \sqrt{n-1} - 1$ and hence [22, Theorem 1.2(iii)] applies).

Using level theory, we can prove an exponential bound on $|\chi(g)|$ for any irreducible character χ of $\mathrm{GL}_n^{\pm}(q)$ or $\mathrm{SL}_n^{\pm}(q)$ of *not too large* degree:

Theorem 9.5.5 ([22, Theorem 1.6], [46, Proposition 8.8]) *Let* $n \geq 21$ *and let* $G = \mathrm{GL}_n(q)$, $\mathrm{SL}_n(q)$, $\mathrm{GU}_n(q)$, *or* $\mathrm{SU}_n(q)$. *Suppose* $\chi \in \mathrm{Irr}(G)$ *with* $\chi(1) \leq q^{n\sqrt{n}/2}$. *Then*

$$|\chi(g)| \leq (1.93)\chi(1)^{1-1/n}$$

for all $g \in G \smallsetminus \mathbf{Z}(G)$.

Sketch of proof. The assumptions on n and $\chi(1)$ imply that

$$j := \mathfrak{l}(\chi) \leq \lceil \sqrt{n}/2 \rceil \leq \max(\sqrt{(8n-17)/12} - 1/2, \sqrt{n - 3/4} - 1/2).$$

By Theorem 9.5.4 and its proof, $\chi = \Theta^{-1}(\alpha) = H_\alpha$ for some $\alpha \in \mathrm{Irr}(\mathrm{GL}_j^{\epsilon}(q))$ if $G = \mathrm{GL}_n^{\epsilon}(q)$ or $\mathrm{SL}_n^{\epsilon}(q)$. Recall that

$$D_\alpha = H_\alpha + \text{ a sum of characters of level } < j.$$

So we bound

$$D_\alpha(g) = \frac{1}{|\mathrm{GL}_j^{\epsilon}(q)|} \sum_{x \in \mathrm{GL}_j^{\epsilon}(q)} \omega(gx)\overline{\alpha(x)}$$

in terms of $\mathsf{supp}(g)$. Then we bound the remainder $D_\alpha(g) - H_\alpha(g)$ in the range $j \leq \lceil \sqrt{n}/2 \rceil$. □

Next we prove an exponential bound on $|\chi(g)|$, again for groups of type A, in the case $|\mathbf{C}_G(g)|$ is not too large:

Theorem 9.5.6 ([22, Theorem 1.5]) *Let* $\epsilon = \pm$ *and* $G = \mathrm{GL}_n^{\epsilon}(q)$ *or* $\mathrm{SL}_n^{\epsilon}(q)$. *Suppose* $g \in G$ *satisfies* $|\mathbf{C}_{\mathrm{GL}_n^{\epsilon}(q)}(g)| \leq q^{n^2/12}$. *Then*

$$|\chi(g)| \leq \chi(1)^{8/9}$$

for all $\chi \in \mathrm{Irr}(G)$.

Sketch of proof. If $k := \mathfrak{l}(\chi) \leq 1$, then χ is a trivial or a Weil character, in which case the desired character bound holds by explicit character formulae for χ, see [78]. Also, if $1 \leq n \leq 4$, then the bound is vacuously true, since G contains no elements g with $|\mathbf{C}_{\mathrm{GL}_n^{\epsilon}(q)}(g)| \leq q^{n^2/12}$. So we will assume $n \geq 5$ and $k \geq 2$, whence $\chi(1) \geq q^{2n-5}$. Now, if $5 \leq n \leq 41$, then the centralizer bound $|\chi(g)| \leq \sqrt{|\mathbf{C}_G(g)|}$ implies the desired character bound. So we will assume $n \geq 42$.

If $k > n/15$, then $\chi(1) > q^{14n^2/225-1}$, whereas $|\chi(g)| \leq q^{n^2/24}$, and so we are done again. In the remaining, main case $k \leq n/15$, we will employ a divide-and-conquer strategy. For a suitable integer $1 \leq m \leq n/4k$, decompose

$$(\chi\bar{\chi})^m = \sum_i a_i\alpha_i + \sum_j b_j\beta_j,$$

where $\alpha_i, \beta_j \in \mathrm{Irr}(G)$, $\alpha_i(1) < q^{n^2/12}$ for all i and $\beta_j(1) \geq q^{n^2/12}$ for all j, $a_i, b_j \in \mathbb{Z}_{\geq 0}$. By the definition of $k = \mathfrak{l}(\chi)$, up to a linear character χ is an irreducible constituent of $\tau_{n,q}^k$ when $\epsilon = +$, see (9.6), and of $\zeta_{n,q}^k$ when $\epsilon = -$, see (9.7). This allows us to show

$$\sum_i a_i + \sum_j b_j \leq q^{19m^2k^2/4}.$$

For the values at g of individual constituents, we use the bounds

$$|\alpha_i(g)| \leq \alpha_i(1), \; |\beta_j(g)| \leq \sqrt{|\mathbf{C}_G(g)|} \leq q^{n^2/24}.$$

It follows that

$$|\chi(g)| \leq \left(\sum_i a_i\alpha_i(1)\right)^{1/2m} + \left(\sum_j b_j q^{n^2/24}\right)^{1/2m}.$$

Now, optimize the choice of m. □

Our next goal is to prove analogues of the above results for finite classical groups of types BCD.

Let $V = \mathbb{F}_q^n$ be endowed with a non-degenerate symplectic, respectively quadratic, form, and let $G = \mathrm{Cl}(V) = \mathrm{Cl}_n(q) = \mathrm{Sp}(V)$, $\mathrm{GO}(V)$, $\mathrm{SO}(V)$, $\Omega(V)$ denote one of the corresponding finite classical groups (which is almost quasisimple in general).

First we consider the case $G = \mathrm{Sp}_{2n}(q)$ where $2 \nmid q = p^f$. Then the natural faithful action of G on the *Heisenberg group* p_+^{1+2nf} (which is extraspecial, of exponent p) yields two Galois-conjugate *reducible Weil characters* ω and ω^*. Now, any $\chi \in \mathrm{Irr}(G)$ is said to be of *level* $\mathfrak{l}(\chi) = j$, if $j \in (1/2)\mathbb{Z}_{\geq 0}$ is smallest such that χ is an irreducible constituent of $(\omega + \omega^*)^{2j}$. In particular, the *irreducible Weil characters* of G (of degree $(q^n \pm 1)/2$) have level $1/2$.

For the other classical groups, let τ_V denote the permutation character of G acting on the point set of V, cf. (9.6). Next, we can also embed G in $\mathrm{GU}(n, \mathbb{F}_{q^2})$, and hence G inherits the reducible Weil character

$$\zeta_V(g) = (-1)^n(-q)^{\dim_{\mathbb{F}_q} \mathrm{Ker}(g-1_V)},$$

cf. (9.7). Then, any $\chi \in \mathrm{Irr}(G)$ is said to be of *level* $\mathfrak{l}(\chi) = j$, if $j \in \mathbb{Z}_{\geq 0}$ is smallest such that χ is an irreducible constituent of $(\tau_V + \zeta_V)^j$.

In all cases, Lemma 9.5.3 shows that $\mathfrak{l}(\chi)$ is well-defined, and in fact we have

$$\mathfrak{l}(\chi) \leq \lfloor (\dim V)/2 \rfloor + 1,$$

see [23, Lemma 3.4]. One of the main results of [23] relates the level $\mathfrak{l}(\chi)$ to the degree of any $\chi \in \mathrm{Irr}(G)$:

Theorem 9.5.7 ([23, Theorem 1.5]) *Let $G = \mathrm{Cl}_n(q)$ be a finite classical group of type BCD on $V = \mathbb{F}_q^n$. For any $\chi \in \mathrm{Irr}(G)$ of level $j = \mathfrak{l}(\chi)$, set $k := j/3$. Then*

$$q^{nk-2k(k+1)} \leq \chi(1) \leq q^{nj}.$$

The proof of Theorem 9.5.7 relies on similar results for type A, see [22, Theorem 1.2]: restrict χ to a *Siegel parabolic subgroup* of G (i.e. the stabilizer in G of a maximal totally singular subspace of V), and proceed by induction on k. □

The second main result of [23] gives an exponential character bound for finite, quasisimple, classical groups of types BCD.

Theorem 9.5.8 ([23, Theorems 1.3, 1.4]) *For any $0 < \varepsilon < 1$, there is an effective constant $\delta = \delta(\varepsilon) > 0$ such that the following statement holds. For any finite quasisimple classical group G, for any $g \in G$ with $|\mathbf{C}_G(g)| \leq |G|^\delta$, and for all $\chi \in \mathrm{Irr}(G)$,*

$$|\chi(g)| \leq \chi(1)^\varepsilon.$$

As stated in Theorem 9.5.6, in the case $G = \mathrm{GL}_n(q)$ or $\mathrm{GU}_n(q)$, if $\varepsilon = 8/9$ then one can take $\delta = 1/12$. In all cases, if $\varepsilon = 0.992$ then one can take $\delta = 0.001$. Looking forward, an immediate question arises: Can one "swap" ε and δ in Theorem 9.5.8?

The main steps of the proof of Theorem 9.5.8. A preliminary reduction allows us to assume $G = \mathrm{Sp}(V)$ or $\mathrm{SO}(V)$. Now we restrict to a Levi subgroup L of a Siegel parabolic of G, which is $\mathrm{GL}_k(q)$ for $k = \lfloor n/2 \rfloor$ or $k = \lfloor n/2 \rfloor - 1$. For a suitable $m \in \mathbb{Z}_{\geq 1}$, we decompose

$$(\chi\overline{\chi})^m = \sum a_i \chi_i,$$

where $\chi_i \in \mathrm{Irr}(G)$ and $a_i \in \mathbb{Z}_{\geq 0}$. We use level theory [22] for L to control $\sum_i a_i$ in terms of m and $\mathfrak{l}(\chi)$. We divide-and-conquer the constituents χ_i, and optimize the choice of m, as in the proof of Theorem 9.5.5. □

We will now take a moment and describe some applications of the obtained character bounds to mixing time $T = T(G, X)$ of random walks on Cayley graphs $\Gamma(G, X)$ of finite almost quasisimple groups. We already mentioned the celebrated 1981 result of Diaconis and Shashahani [9] in the case $G = \mathsf{S}_n$ and $X = \{1\} \cup \{2\text{-cycles}\}$ Results on general random walks on S_n are obtained in [62] and [43], relying on Theorems 9.3.2, 9.3.3 and similar character bounds.

In the case $G = \mathrm{SL}_n(q)$ and $X = \{\text{transvections}\}$, Hildebrand [30] showed that $T \approx n$. More generally,

Theorem 9.5.9 ([22, Theorem 1.7], [71, Corollary 1.19]) *Let $\epsilon = \pm$, $G = \mathrm{SL}_n^\epsilon(q)$, and let $x \in G \smallsetminus \mathbf{Z}(G)$.*

(i) Suppose $n \geq 10$, q is large enough compared to n, and

$$|\mathbf{C}_{\mathrm{GL}_n^\epsilon(q)}(x)| \leq q^{n^2/12}.$$

Then the mixing time of the random walk on $\Gamma(G, x^G)$ is at most 10. If $n \geq 19$ in addition, then $\Gamma(G, x^G)$ has diameter at most 19.

(ii) The mixing time of the random walk on $\Gamma(G, x^G)$ is at most n when q is large enough compared to n.

Theorem 9.5.10 ([23, Corollary 7.5]) *There is an absolute constant $0 < \gamma < 1$ such that the following statement holds for any finite simple group G of Lie type over $\mathbb{F}_q$. For any sufficiently large prime power q and for any $x \in G$ with $|\mathbf{C}_G(x)| \leq |G|^\gamma$,*

(i) The Cayley graph $\Gamma(G, x^G)$ has diameter ≤ 3, and

(ii) The random walk on $\Gamma(G, x^G)$ has mixing time 2.

9.6 LT-bounds

As mentioned above, the character bounds in Theorems 9.4.3, 9.4.11, 9.5.6, and 9.5.8 do not cover arbitrary elements and arbitrary irreducible characters of finite classical groups. Likewise, the character bound in Theorem 9.2.13, although applied to any element and any character, only yields the upper bound $q^{-\sqrt{\mathsf{supp}(g)}/481}$ for the character ratio (recall the definition (9.4) of the support $\mathsf{supp}(g)$). On the other hand, computational evidence suggests that one should be able to prove a *linear* bound $q^{-\gamma \cdot \mathsf{supp}(g)}$ (for some universal constant $\gamma > 0$), instead of the square root. Moreover, if $G = \mathrm{Cl}_n(q)$ and $\chi \in \mathrm{Irr}(G)$ is not linear, then $\chi(1) \geq q^{n/3}$ by the Landazuri–Seitz–Zalesski bound. This raises the question whether one can even upgrade the bound in Theorem 9.2.13 to

$$|\chi(g)/\chi(1)| \leq \chi(1)^{-\sigma \cdot \mathsf{supp}(g)/n} \tag{9.8}$$

for some universal constant $\sigma > 0$?

The following observation shows that, for any element g in any finite classical group $G = \mathrm{Cl}_n(q)$, $\log |g^G| / \log |G|$ and $\mathsf{supp}(g)/n$ are of the same magnitude.

Lemma 9.6.1 *If $G = \mathrm{Cl}_n(q)$ is quasisimple and $g \in G$ has support s, then*

$$|G|^{s/3n} \leq |g^G| \leq |G|^{5s/n}.$$

Proof Note that

$$(n-s)^2 \leq \dim \mathbf{C}_{\mathrm{GL}(V \otimes \overline{\mathbb{F}_q})}(g) \leq n(n-s).$$

(First, one proves the claim for unipotent elements. The general case then reduces to it.) See [46, Corollary 6.7] for more details. □

Next we show that, if it holds, the bound (9.8) would be best possible, up to the choice of σ.

Lemma 9.6.2 *Let $G = \mathrm{Cl}_n(q)$ with $n \geq 7$. If $|G|$ is sufficiently large and $g \in G$ has support $s = \mathsf{supp}(g) \leq n-2$, then there is a non-trivial $\chi \in \mathrm{Irr}(G)$ such that*

$$|\chi(g)/\chi(1)| \geq \chi(1)^{-6s/n}.$$

Proof Assume the contrary that $|\chi(g)| < \chi(1)^{1-6s/n}$ for all $\chi \in \mathrm{Irr}(G)$. Choosing $|G|$ sufficiently large, we have

$$\sum_{1_G \neq \chi \in \mathrm{Irr}(G)} \chi(1)^{-0.55} = \zeta_G(0.55) - 1 < 1$$

by Theorem 9.3.9. Taking $k := \lfloor (n-2)/s \rfloor$ we also have $6ks > 2.55n$. Hence, for any $x \in G$ we get

$$\sum_{1_G \neq \chi} \frac{|\chi(g)^k \bar{\chi}(x)|}{\chi(1)^{k-1}} \leq \sum_{1_G \neq \chi} \frac{1}{\chi(1)^{6ks/n-2}} \leq \sum_{1_G \neq \chi} \chi(1)^{-0.55} < 1.$$

By Lemma 9.1.1, this means that every $x \in G$ is a product of k conjugates of g, and so has $\mathsf{supp}(x) \le ks \le n-2$. But this is a contradiction, since G always contains elements of support $n-1$. □

The uniform exponential character bound (9.8) has been recently established in [46].

Theorem 9.6.3 ([46, Theorem A]) *There exists an absolute constant $c > 0$ such that for all finite quasisimple groups G of Lie type, irreducible characters χ of G, and elements $g \in G$, we have*

$$\frac{|\chi(g)|}{\chi(1)} \le \chi(1)^{-c\log|g^G|/\log|G|}.$$

As mentioned above, the exponent in the character bound in Theorem 9.6.3 is optimal for all g (up to a linear factor), see Lemma 9.6.2. Its proof, as given in [46], yields an explicit, but tiny constant $c \approx 10^{-29}$. The key idea of the proof is to reverse the usual order of things: we use *probability theory* to solve a representation-theoretic problem. This probabilistic approach is based on the following lemma, in which $\mathbb{E}(\mathsf{X})$ denotes the expected value of the random variable X.

Lemma 9.6.4 *Let G be finite group and $g \in G$. Choose $\mathsf{X}_1, \ldots, \mathsf{X}_b$ from G independently and uniformly at random. For any $\chi \in \mathrm{Irr}(G)$ and any $b \in \mathbb{Z}_{\ge 1}$, we have*

$$\mathbb{E}[\chi(g^{\mathsf{X}_1}\cdots g^{\mathsf{X}_b})] = \frac{\chi(g)^b}{\chi(1)^{b-1}}.$$

Proof By Lemma 9.1.1, for any $h \in G$ we have

$$\mathbf{P}[g^{X_1}\cdots g^{X_b} = h] = \frac{1}{|G|}\sum_{\varphi\in\mathrm{Irr}(G)}\frac{\varphi(g)^b\overline{\varphi}(h)}{\varphi(1)^{b-1}}.$$

Therefore,

$$\begin{aligned}
\mathbb{E}[\chi(g^{X_1}\cdots g^{X_b})] &= \sum_{h\in G}\chi(h)\frac{1}{|G|}\sum_{\varphi\in\mathrm{Irr}(G)}\frac{\varphi(g)^b\overline{\varphi}(h)}{\varphi(1)^{b-1}} \\
&= \sum_{\varphi\in\mathrm{Irr}(G)}\frac{\varphi(g)^b}{\varphi(1)^{b-1}}\frac{1}{|G|}\sum_{h\in G}\chi(h)\overline{\varphi}(h) \\
&= \frac{\chi(g)^b}{\chi(1)^{b-1}}.
\end{aligned}$$

□

The proof of Theorem 9.6.3 proceeds in subsequent bootstraps. The first bootstrap allows us to apply the GLT-bound in Theorem 9.5.8 to elements of large support.

Theorem 9.6.5 ([46, Theorem 4.4]) *There exist explicit constants $\gamma > 0$ and $C \geq 4$ such that the following statement holds for any positive integer n, any $0 < \beta < 1$, any $V = \mathbb{F}_q^n$ for any prime power q, any $G := \mathrm{SL}(V)$, $\mathrm{SU}(V)$, $\mathrm{Sp}(V)$, or $\Omega(V)$ (or $\mathrm{SO}(V)$ or $\mathrm{Spin}(V)$ if q is odd), any element $g \in G$, and any irreducible character $\chi \in \mathrm{Irr}(G)$. If $s := \mathsf{supp}(g) \geq \max(C, \beta n)$, then*

$$\frac{|\chi(g)|}{\chi(1)} \leq \chi(1)^{-\gamma s/(n \cdot \lceil 1/\beta \rceil)}.$$

Outline of the proof of Theorem 9.6.5. Take $\varepsilon = 0.992$, $\delta = 0.0011$. By Theorem 9.5.8, $|\chi(h)| \leq \chi(1)^{\varepsilon}$ whenever $h \in G$ satisfies $|\mathbf{C}_G(h)| \leq |G|^{\delta}$. Now choose

$$C = 2^{14} \cdot 10^{12}, \ \gamma = 2/C = 2^{-13} \cdot 10^{-12}.$$

Then we show that $s = \mathsf{supp}(g) \geq \beta n$ implies for $b := \lceil n/s \rceil$ that

$$\mathbf{P}[|\mathbf{C}_G(g^{X_1} \cdots g^{X_b})| \geq |G|^{\delta}] \leq \chi(1)^{-2\gamma s/n},$$

where $\mathbf{P}[\mathsf{Y}]$ denotes the probability of the event Y. Thus with high probability, the product of any b random conjugates of g satisfies the GLT-bound in Theorem 9.5.8. Now we can apply Lemma 9.6.4 to bound $|\chi(g)^b|$. □

The second bootstrap allows us to apply Theorem 9.6.5 to arbitrary elements.

Theorem 9.6.6 ([46, Theorem 5.5]) *There exists an explicit constant $\sigma > 0$ such that the following statement holds for any positive integer $n \geq 3$, any $V = \mathbb{F}_q^n$ for any prime power q, any $G := \mathrm{SL}(V)$, $\mathrm{SU}(V)$, $\mathrm{Sp}(V)$, or $\Omega(V)$ (or $\mathrm{SO}(V)$ or $\mathrm{Spin}(V)$ if q is odd), any $g \in G$, and any irreducible character $\chi \in \mathrm{Irr}(G)$:*

$$\frac{|\chi(g)|}{\chi(1)} \leq \chi(1)^{-\sigma \cdot \mathsf{supp}(g)/n}.$$

Outline of the proof of Theorem 9.6.6. Let $G = \mathrm{Cl}_n(q)$ and let C, γ be as in the proof of Theorem 9.6.5. First we apply Theorem 9.2.13 to handle the case $n < 9C$. Assuming $n \geq 9C$, we choose $\sigma = 7 \cdot 10^{-29}$.

If $s = \mathsf{supp}(g) \geq n/9$, we apply Theorem 9.6.5 with $\beta := 1/9$.

If $1 \leq s = \mathsf{supp}(g) \leq n/9$, we show that there exists an integer $1 \leq b \leq n/s$ such that

$$\mathbf{P}[\mathsf{supp}(g^{X_1} \cdots g^{X_b})| < n/9] < \chi(1)^{-1/10}.$$

Thus, with high probability, the product of any b random conjugates of g satisfies Theorem 9.6.5. Now we can apply Lemma 9.6.4 to bound $|\chi(g)^b|$. □

Proof of Theorem 9.6.3. By Lemma 9.6.1, $(\log |g^G|)/(\log |G|)$ and $\mathsf{supp}(g)/n$ are of the same magnitude. Hence Theorem 9.6.3 follows from Theorem 9.6.6, by choosing $c = 10^{-29}$. □

Theorem 9.6.5 allows us to "swap" ε and δ in Theorem 9.5.8:

Theorem 9.6.7 ([46, Theorem 4.5]) *For any $0 < \varepsilon < 1$, there exists a constant $0 < \delta < 1$ such that the following statement holds. For any quasisimple classical group $G = \mathrm{Cl}_n(q)$, any $g \in G$, and any $\chi \in \mathrm{Irr}(G)$, if $|\mathbf{C}_G(g)| \leq |G|^\varepsilon$ we have*

$$|\chi(g)| \leq \chi(1)^\delta.$$

As an immediate application of Theorem 9.6.6, we obtain a linear upgrade of Theorem 9.2.13:

Theorem 9.6.8 ([46, Corollary 5.6]) *There exists an absolute constant $\gamma > 0$ (which can be taken to be $2 \cdot 10^{-29}$) such that the following statement holds. For any $n \in \mathbb{Z}_{\geq 2}$, any prime power q, any quasisimple classical group*

$$G = \mathrm{SL}_n(q),\ \mathrm{SU}_n(q),\ \mathrm{Sp}_{2n}(q),\ \Omega_n^{\pm}(q),\ \mathrm{Spin}_n^{\pm}(q),$$

any $g \in G$, and any $\chi \in \mathrm{Irr}(G)$ of degree $\chi(1) > 1$, we have

$$\frac{|\chi(g)|}{\chi(1)} \leq q^{-\gamma \cdot \mathsf{supp}(g)}.$$

9.7 Applications

Let us discuss how character bounds have been used in [47] to complete the proof of Ore's conjecture [63].

Theorem 9.7.1 ([47]) *Every element g in any finite non-abelian simple group G is a commutator.*

Sketch of proof. By the main result of [10], we may assume that G is a simple group of Lie type over $\mathbb{F}_q$ with $q \le 7$. Say $G = \mathrm{Sp}_{2n}(2)$. We proceed by induction on $n \ge 3$, for which the induction base $3 \le n \le 5$ can be easily verified using Lemma 9.1.1.

For the induction step $n \ge 6$, if the element $g \in G$ is *breakable*, i.e. if $g \in \mathrm{Sp}_{2a}(2)$ with $3 \le a \le n-1$ or $g \in \mathrm{Sp}_{2a}(2) \times \mathrm{Sp}_{2b}(2)$ with $3 \le a \le b = n-a$, then we are done by using the induction hypothesis. Otherwise we show that $|\mathbf{C}_G(g)| \le 2^{2n+15}$.

By Lemma 9.1.1, it suffices to show that

$$\left| \sum_{1_G \neq \chi \in \mathrm{Irr}(G)} \chi(g)/\chi(1) \right| < 1.$$

According to [14, Corollary 1.2], $k(G) := |\mathrm{Irr}(G)| < 2^{n+4}$. Hence, with $D := 2^{4n-1}/15$, by Schwarz's inequality we have

$$\left| \sum_{\chi : \chi(1) \ge D} \chi(g)/\chi(1) \right| \le \frac{\sqrt{k(G) \cdot |\mathbf{C}_G(g)|}}{D} < 0.6.$$

On the other hand, by [25, Theorem 6.1], $\mathrm{Irr}(G)$ contains only 5 characters χ_i with $1 < \chi(1) < D$. So we bound $|\chi_i(g)/\chi_i(1)|$ to show that

$$\left| \sum_{\chi : 1 < \chi(1) < D} \chi(g)/\chi(1) \right| < 0.4.$$

□

We already mentioned that character bounds play an important role in the study of the non-commutative Waring problem 9.3.5. Let us formulate recent results on the *Waring problems for powers*, that is when $w(x) = x^n$.

Theorem 9.7.2 *For any fixed integer $n \ge 1$, the following statements hold for finite non-abelian simple groups G.*

(i) [44, 45] *If $|G|$ is sufficiently large, then every element $g \in G$ is a product of two n^{th} powers. Moreover, the induced distribution L^1-converges to the uniform distribution on G.*

(ii) [28] *If $|G| \ge n^{8n^2}$ or if $\exp(G) \nmid n$, then every element $g \in G$ is a product of two n^{th} powers.*

Theorem 9.7.3 ([24]) *The following statements hold for any finite non-abelian simple group G.*

(i) *If $n = p^a q^b$ is a product of (at most) two prime powers, then every $g \in G$ is a product of two n^{th} powers.*
(ii) *If $2 \nmid n$, then every $g \in G$ is a product of three n^{th} powers.*

Note that Theorem 9.7.3 generalizes the celebrated $p^a q^b$ Theorem of Burnside and Odd Order Theorem of Feit and Thompson.

Next we discuss the following conjecture, the second part of which was also raised by Shalev.

Conjecture 9.7.4 (Lubotzky [57, p.179]) *There exist absolute constants C_1, C_2, C_3, C_4 such that the following inequalities hold for all finite simple groups of Lie type G and all conjugacy classes $S = g^G \neq \{1\}$:*

(a) $C_1(\log|G|/\log|S|) \leq \mathsf{diam}(\Gamma(G,S)) < C_2(\log|G|/\log|S|)$.
(b) $C_3(\log|G|/\log|S|) \leq T(\Gamma(G,S)) < C_4(\log|G|/\log|S|)$.

Equivalently, the diameter of the Cayley graph $\Gamma(G,S)$ and the mixing time of the random walk on $\Gamma(G,S)$ are of the same magnitude as of $\log_{|S|}|G|$.

The diameter part 9.7.4(a) of Lubotzky's conjecture was proved in [48]. Note that the lower bound is clear: for any integer $N \geq 1$, if $\cup_{i=1}^{N} S^i = G$ then $|S|^N \gtrsim |G|$ and so $N \gtrsim \log_{|S|}|G|$. Hence the main bulk of work on it is to prove the upper bound. The mixing time part is the content of the following recent result:

Theorem 9.7.5 ([46, Theorem 8.1]) *Lubotzky's conjecture 9.7.4(b) holds for all simple groups of Lie type G.*

Proof We will show that the random walk on $\Gamma(G,S)$ mixes well after N steps if

$$N \geq (26/c+1)(\log|G|)/(\log|S|),$$

where c is the constant in Theorem 9.6.3. For such an N, Theorem 9.6.3 implies

$$|\chi(g)|^N \leq \chi(1)^{N-26}$$

for all $\chi \in \mathrm{Irr}(G)$. Let $\mathbf{U}_G$ denote the uniform distribution on G, and let $\mathbf{U}_S$ denote the uniform distribution concentrated on S, that is, $\mathbf{U}_S(x) = 1/|S|$ if $x \in S$ and 0 otherwise. By Lemma 9.1.1, the probability $\mathbf{P}^N(x)$ that the product of N i.i.d. random variables with distribution U_S yields a given element $x \in G$ is

$$\mathbf{P}^N(x) = \frac{1}{|G|} \sum_{\chi \in \mathrm{Irr}(G)} \frac{\chi(g)^N \bar{\chi}(x)}{\chi(1)^{N-1}}.$$

Also, if G is defined over $\mathbb{F}_q$, then $|\mathrm{Irr}(G)| \leq 27.2q^r$ by [14, Theorem 1.1]. Hence

$$\sum_{x\in G} |\mathbf{P}^N(x) - \mathbf{U}_G(x)| \leq \max_x \sum_{\chi\neq 1_G} \frac{|\chi(g)^N \bar{\chi}(x)|}{\chi(1)^{N-1}} \leq \sum_{\chi\neq 1_G} \chi(1)^{-24}$$
$$\leq \frac{|\mathrm{Irr}(G)|}{\min_{\chi\neq 1_G} \chi(1)^{24}} \leq \frac{27.2q^r}{q^{8r}} < 1/e.$$

□

The next application is concerned with the notion of (directed) *McKay graphs* $\mathcal{M}(G,\alpha)$, which in a sense is dual to the notion of Cayley graphs $\Gamma(G,S)$. If G a finite group and α a fixed, not necessarily irreducible, complex character of G then $\mathcal{M}(G,\alpha)$ has $\mathrm{Irr}(G)$ as its set of vertices, and its edges are (χ,φ) with φ any irreducible constituent of $\chi\alpha$. This kind of graph was first considered by McKay in [61] for any finite subgroup G of $\mathrm{SU}_2(\mathbb{C})$ and α the corresponding character on $\mathbb{C}^2$: he showed that $\mathcal{M}(G,\alpha)$ is then an affine Dynkin diagram of type ADE (see also [69]).

It follows from Lemma 9.5.3 that $\mathcal{M}(G,\alpha)$ is connected if and only if the character α is faithful. Hence, a natural question to study is $\mathsf{diam}\,\mathcal{M}(G,\alpha)$ when α is faithful. One can also study random walks on $\mathcal{M}(G,\alpha)$: start at 1_G, then at each vertex χ choose the next edge φ with probability $\dfrac{[\varphi,\chi\alpha]_G\cdot\varphi(1)}{\chi(1)\alpha(1)}$. In fact, random walks on complex representations have been studied by Steinberg [69] and Fulman [13]. Recently, random walks on modular representations have also been considered in [2].

For any character α of a finite group G, let $\alpha^\sharp$ denote the sum of distinct irreducible constituents of α.

Conjecture 9.7.6 ([52]) *There is an absolute constant $C > 0$ such that*

$$\mathsf{diam}\,\mathcal{M}(G,\alpha) \leq C(\log|G|/\log\alpha^\sharp(1))$$

for every finite non-abelian simple group G and for every faithful character α of G.

Note that, up to a constant, Conjecture 9.7.6 is optimal. Indeed, G admits an irreducible character $\chi\in\mathrm{Irr}(G)$ with $\log\chi(1)\approx(1/2)\log|G|$.

Assume now that α is irreducible. Then Conjecture 9.7.6 has been proved for $\mathrm{PSL}_n^\pm(q)$ when q is large enough in [52], using Theorem 9.4.3. In the case $G=\mathsf{A}_n$, it was shown in [80] that $\mathsf{diam}\,\mathcal{M}(G,\alpha)\leq n-\sqrt{n}$.

Moreover, Conjecture 9.7.6 holds for A_n by [53, Theorem 2] which relies on the recent result [66] that proves the conjecture for S_n. For simple classical groups $G = \mathrm{Cl}_n(q)$, the uniform bound $\mathsf{diam}\,\mathcal{M}(G,\alpha) \leq 32n$ was obtained in [53], relying on the main result of [29] that the tensor cube of the Steinberg character contains every irreducible character of G.

Theorem 9.7.7 ([46, Theorem 8.3]) *Conjecture 9.7.6 holds for all finite non-abelian simple groups.*

Proof Results of [14] and [50] allow one to reduce to the case where α is irreducible. One can certainly ignore sporadic simple groups. Moreover, using the aforementioned results of [52] and [53], we may assume that G is classical of unbounded rank r, defined over $\mathbb{F}_q$.

For any integer $N \geq 1$ and for any irreducible characters χ_1, χ_2 of G we have

$$[\alpha^N \chi_1, \chi_2] = \frac{1}{|G|} \sum_{S=g^G \subset G} |S|\alpha(g)^N \chi_1(g)\overline{\chi}_2(g).$$

So to show that $\mathsf{diam}\mathcal{M}(G,\alpha) \leq N$, it suffices to prove

$$\sum_{S=g^G \neq \{1\}} |S| \left(\frac{|\alpha(g)|}{\alpha(1)}\right)^N < 1.$$

Recall from [14] that $k(G) < 32q^r < |S|^6$ for any nontrivial conjugacy class S, which implies

$$\sum_S \frac{1}{|S|^6} < 1. \tag{9.9}$$

Now taking

$$N \geq (7/c)(\log|G|)/(\log\alpha(1))$$

and using Theorem 9.6.3, we get

$$\sum_{S \neq \{1\}} |S| \left(\frac{|\alpha(g)|}{\alpha(1)}\right)^N < \sum_{S \neq \{1\}} |S|\alpha(1)^{-7\log|S|/\log\alpha(1)} = \sum_{S \neq \{1\}} \frac{1}{|S|^6} < 1.$$

□

Motivated by results of Rodgers and Saxl [65] on products of conjugacy classes, the following conjecture is also concerned with products of irreducible characters:

Conjecture 9.7.8 (**Gill** [18]) *There exists an absolute constant $\delta > 0$ such that for all finite non-abelian simple groups G and all irreducible characters $\chi_1, \chi_2, \ldots, \chi_m$ of G, if $\chi_1(1)\chi_2(1)\ldots\chi_m(1) \geq |G|^\delta$, then $\chi_1\chi_2\cdots\chi_m$ contains every irreducible character of G as a constituent.*

Theorem 9.7.9 ([46, Theorem 8.5]) *Gill's conjecture 9.7.8 holds for all simple groups of Lie type.*

Proof Using [52, Theorem 3], we may assume that G is classical of unbounded rank r over $\mathbb{F}_q$. Take $\delta = 7/c$, where c is the constant in Theorem 9.6.3. For any $\alpha \in \mathrm{Irr}(G)$, we need to show

$$[\chi_1\chi_2\cdots\chi_m, \alpha]_G = \frac{1}{|G|}\sum_{S=g^G\subset G}|S|\chi_1(g)\ldots\chi_m(g)\bar{\alpha}(g) \neq 0.$$

It suffices to prove

$$\sum_{S=g^G\neq\{1\}}|S|\prod_{i=1}^{m}\frac{|\chi_i(g)|}{\chi_i(1)} < 1.$$

The choice of δ and Theorem 9.6.3 imply

$$\begin{aligned}\sum_{S\neq\{1\}}|S|\prod_i\frac{|\chi_i(g)|}{\chi_i(1)} &< \sum_{S\neq 1}|S|\prod_i\chi_i(1)^{-c\log_{|G|}|S|}\\ &< \sum_{S\neq 1}|S|^{1-\delta c} < \sum_{S\neq 1}\frac{1}{|S|^6} < 1\end{aligned}$$

using (9.9). □

The next application is concerned with *Fuchsian groups* which are finitely generated, discrete subgroups of $PSL_2(\mathbb{R})$, the isometry group of the hyperbolic plane. This important class of groups includes

- free groups,
- free products $C_2 * \mathbb{Z}$, $C_r * C_s$ of cyclic groups,
- the *modular group* $PSL_2(\mathbb{Z})$,
- the *surface groups* $\pi_1(\Sigma_g)$ of the closed, orientable genus g surface Σ_g:

$$\pi_1(\Sigma_g) = \langle a_1, b_1, \ldots, a_g, b_g \mid [a_1, b_1]\ldots[a_g, b_g] = 1\rangle,$$

- the *triangle groups*

$$\Delta(m, n, l) = \langle a, b, c \mid a^2 = b^2 = c^2 = (ab)^m = (bc)^n = (ca)^l = 1\rangle$$

and their index 2, *Von Dyck* subgroup

$$D(m,n,l) = \langle x,y,z \mid x^m = y^n = z^l = xyz = 1\rangle.$$

In particular, $D(2,3,7)$ is the *Hurwitz group.*

We will consider a co-compact Fuchsian group Γ of genus g having d *elliptic* generators of orders $m_1,\dots,m_d \geq 2$. Such a group has the following presentation:

$$\Gamma = \langle a_1,b_1,\dots,a_g,b_g,\ x_1,\dots,x_d \mid \\ x_1^{m_1} = \cdots = x_d^{m_d} = 1,\ x_1\cdots x_d \prod_{i=1}^{g}[a_i,b_i] = 1\rangle$$

if Γ is orientation-preserving, and

$$\Gamma = \langle a_1,\dots,a_g,\ x_1,\dots,x_d \mid \\ x_1^{m_1} = \cdots = x_d^{m_d} = 1,\ x_1\cdots x_d\, a_1^2\cdots a_g^2 = 1\rangle$$

if Γ is non-orientation-preserving. The *measure* of Γ is

$$\mu := \mu(\Gamma) = vg - 2 + \sum_{i=1}^{d}\left(1 - \frac{1}{m_i}\right) > 0,$$

where $v := 2$ if Γ is orientation-preserving and $v := 1$ otherwise. If Γ is a lattice in $PSL_2(\mathbb{R})$ then $-\mu(\Gamma) = \chi(\Gamma)$, the Euler characteristic of Γ.

We will also assume that Γ is not virtually abelian, so $vg + d \geq 3$.

We are interested in *representation varieties* of Γ, that is, $\mathrm{Hom}(\Gamma, G)$, where $G = \mathcal{G}$ is a simple algebraic group, or $G = \mathcal{G}^F = \mathcal{G}(q)$ a quasisimple group of Lie type. (If Γ is not co-compact, then Γ is a free product of cyclic subgroups, making it easy to find $|\,\mathrm{Hom}(\Gamma,G)|$. Hence the assumption that Γ is co-compact.)

Under the above assumptions, it was shown in [50] (for $g \geq 2$) and [51] (for $g \leq 1$) that

$$|\,\mathrm{Hom}(\Gamma,G)| \approx |G|^{\mu+1}$$

for $G = \mathcal{G}(q)$ (up to a multiplicative error, which is small when $r = \mathrm{rank}(\mathcal{G}) \gg 0$ and $q \gg r$). Using the Lang–Weil estimates [39], this implies

$$\dim \mathrm{Hom}(\Gamma,\mathcal{G}) \approx (\mu+1)\dim(\mathcal{G})$$

(up to a small error, which is bounded in terms of $d, m_1, \dots, m_d$). The proofs use the Frobenius character formula 9.1.1 and character bounds such as Theorem 9.4.3 to estimate the number of homomorphisms $\Gamma \to G$

with specified conjugacy classes of images of generators. (Here, using "good" classes one gets a lower bound, whereas upper bounds are obtained by considering general classes.)

Strong asymptotic results were also obtained in [40], [41], [42], using deformation theory.

Questions on representation varieties of Fuchsian groups are closely related to *random generation* of finite simple groups. Let G be a finite non-abelian simple group. The probability $\mathbf{P}'$ that $f \in \mathrm{Hom}(\Gamma, G)$ is *not onto* is

$$\mathbf{P}' \leq \sum_{\substack{M \text{ maximal} \\ \text{subgroup of } G}} |\mathrm{Hom}(\Gamma, M)|/|\mathrm{Hom}(\Gamma, G)|.$$

Note that most of M's are tiny in size. Hence, by bounding $|\mathrm{Hom}(\Gamma, M)|$, it was shown in [50] (for $g \geq 2$), and [51] (for $g \leq 1$, and under some conditions on q for $G = \mathcal{G}(q)$) that

the probability that a randomly chosen homomorphism in $\mathrm{Hom}(\Gamma, G)$ *is an epimorphism tends to* 1 *as* $|G| \to \infty$.

Let us formulate some celebrated instances of this statement.

(i) $\Gamma = F_2$: *Dixon's conjecture* on generation of finite simple groups by two elements, proved by Dixon, Kantor–Lubotzky, and Liebeck–Shalev.

(ii) $\Gamma = C_2 * \mathbb{Z}$: *Kantor-Lubotzky conjecture* on generation of finite simple groups by an involution and another element, proved by Liebeck–Shalev.

(iii) $\Gamma = \mathrm{PSL}_2(\mathbb{Z})$: $(2,3)$-*generation*, proved for all finite simple groups (but $\mathrm{PSp}_4(q)$ and ${}^2B_2(q)$) by work of Liebeck–Shalev, and Guralnick–Liebeck–Lübeck–Shalev.

(iv) $\Gamma = C_r * C_s$: (r,s)-*generation*, studied by Liebeck–Shalev.

In the next application, we study fibers of product morphisms on semisimple algebraic groups.

Theorem 9.7.10 ([46, Theorem 8.11]) *There exists a constant* $C > 0$ *such that the following statement holds. For any simple algebraic group* $\mathcal{G}$ *over* $K = \overline{K}$, *any conjugacy classes* $\mathcal{C}_1, \ldots, \mathcal{C}_m$ *in* $\mathcal{G}$, *if*

$$\sum_i \dim \mathcal{C}_i \geq C(\dim \mathcal{G}),$$

then the multiplication morphism

$$\mu : \mathcal{X} := \mathcal{C}_1 \times \ldots \times \mathcal{C}_m \to \mathcal{G}$$

is flat.

Sketch of proof. Since conjugacy classes are non-singular varieties, by *miracle flatness*, the flatness of μ is equivalent to all fibers $\mu^{-1}(g)$ having dimension $\dim \mathcal{X} - \dim \mathcal{G}$. First, one reduces to the case where $\mathcal{G}$ is simply connected. Next, we reduce to the case where $K = \overline{\mathbb{F}_p}$ for a prime p, and work inside $G = \mathcal{G}(\mathbb{F}_q)$. Using Lemma 9.1.1 and Theorem 9.6.3, we show that each fiber of $\mu|_X : X := \mathcal{X}(\mathbb{F}_q) \to G$ has size $O(q^{\dim \mathcal{X} - \dim \mathcal{G}})$. The claim then follows from the Lang–Weil estimates [39]. □

We will now discuss the following conjecture of John Thompson:

Conjecture 9.7.11 (Thompson) *Every finite non-abelian simple group G admits a conjugacy class C such that $G = C^2$.*

Note that Thompson's conjecture implies Ore's conjecture. (Indeed, $1 \in C^2$, so $g^G = C = C^{-1} = (g^{-1})^G$. Hence $G = g^G \cdot (g^{-1})^G$, i.e. every $x \in G$ is a commutator.)

Conjecture 9.7.11 has been established for many simple groups. It holds for $G = \mathsf{A}_n$ by Hsü (using a combinatorial proof). Neubüser, Pahlings, and Cleuvers verified it for all sporadic simple groups using Lemma 9.1.1. The case $G = \mathrm{PSL}_n(\mathbb{F})$ follows from work of Brenner, Sourour, and Lev, whereas the case $G = \mathrm{PSp}_{2n}(\mathbb{F})$, where $\mathrm{char}(\mathbb{F}) \neq 2$ and $-1 \in \mathbb{F}^{\times 2}$ was handled by Gow. The celebrated result of Ellers and Gordeev [10] proves Thompson's conjecture (and hence Ore's conjecture) for all finite simple Lie-type groups $G = \mathcal{G}(q)$ with $q > 8$. There has been no further progress on Thompson's conjecture, until very recently.

Theorem 9.7.12 ([46, Theorem 7.7]) *Let G be one of the simple groups $\mathrm{PSU}_n(q)$, $PSp_n(q)$, $\mathrm{P}\Omega^\epsilon_n(q)$. Suppose that $(q+1)|n$ in the* SU*-case, and that, if $2 \nmid q$ then $2|n$ and $\epsilon = (-1)^{n(q-1)/4}$ in the Ω-case. If n is sufficiently large, then Thompson's conjecture 9.7.11 holds for G.*

Strategy of proof. Work inside $G = \mathrm{SU}_n(q)$, $\mathrm{Sp}_n(q)$, or $\Omega^\epsilon_n(q)$ instead, and choose a class $C = g^G$ with small centralizer. Using the Frobenius character formula 9.1.1 and Theorem 9.6.3, we can show that C^2 contains every target element $x \in G$ with $\mathsf{supp}(x) \geq B$ for some B that depends on the choice of g. If $\mathsf{supp}(x) \leq B$, the assumptions on (n, q) allow us to assume that $x = \mathrm{diag}(y, I_{n-k})$ with k small. Our chosen element g is conjugate to $\mathrm{diag}(g_1, g_2)$, where g_2 is real. By choosing g carefully, we show that y can be written as a product of two conjugates of g_1. So x lies in $g^G \cdot g^G$. □

We will finish with another application, which is related to the α-*invariant* and yet another conjecture of John Thompson.

For a Kähler manifold X and a compact subgroup $G \leq \mathrm{Aut}(X)$, Tian [73], [74] defined an invariant $\alpha_G(X)$. He used it to prove the existence of a G-invariant Kähler-Einstein metric on X in some important cases, e.g. when

$$\alpha_G(X) > \frac{\dim(X)}{\dim(X)+1}.$$

Consider the case where a finite group $G < \mathrm{GL}_{n+1}(\mathbb{C})$ acts on the projective space $\mathbb{P}^n$. Then Tian's invariant $\alpha_G(\mathbb{P}^n)$ is the *log-canonical threshold*

$$\mathrm{lct}(\mathbb{P}^n, G) = \sup \left\{ \lambda \in \mathbb{Q} \;\middle|\; \begin{array}{c} (\mathbb{P}^n, \lambda D) \text{ has log-canonical singularity,} \\ \forall\, G\text{-invariant effective } \mathbb{Q}\text{-divisors } D \end{array} \right\},$$

see [8].

Theorem 9.7.13 (Thompson [72]) *Suppose that $G < \mathrm{GL}_{n+1}(\mathbb{C})$ is any finite group. Then $\alpha_G(\mathbb{P}^n) \leq 4(n+1)$.*

Thompson's theorem 9.7.13 was in fact proved in a different setup. Given a finite-dimensional complex space V and a subgroup $G < \mathrm{GL}(V)$, G is said to have a *semi-invariant* of degree k on V if $\mathrm{Sym}^k(V)$ contains a one-dimensional G-submodule. Let

$$d(G,V) := \min\{k \geq 1 \mid G \text{ has a semi-invariant of degree } k \text{ on } V\}.$$

It turns out that $d(G)$ gives an upper bound on the α-invariant:

$$\alpha_G(\mathbb{P}^n) \leq \frac{d(G, \mathbb{C}^{n+1})}{n+1},$$

see [5, §1], and Thompson actually proved that $d(G,V) \leq 4(\dim V)^2$.

Thompson conjectured that a much stronger bound should hold asymptotically:

Conjecture 9.7.14 (Thompson [72]) *There exists an absolute constant $C > 0$ such that*

(a) $\alpha_G(\mathbb{P}^n) \leq C$, *and*
(b) $d(G, \mathbb{C}^{n+1}) \leq C(n+1)$

for all finite subgroups $G < \mathrm{GL}_{n+1}(\mathbb{C})$.

Theorem 9.7.15 ([76]) *Thompson's conjecture 9.7.14 holds, with $C =$* 1184036.

A key inequality in the proof of Theorem 9.7.15 is the following. Suppose $|\chi(x)|/\chi(1) \leq \gamma$ for $\chi \in \mathrm{Irr}(G)$ of degree n and for all $x \in G \smallsetminus \mathbf{Z}(G)$. Then for any $g \in G \smallsetminus \mathbf{Z}(G)$ and any $k \in \mathbb{N}$ we have

$$|\mathrm{Sym}^k(\chi)(g)| \leq \frac{\prod_{j=0}^{k-1}(\gamma n + j/\gamma)}{k!}.$$

In particular, if k is such that $n \geq 3k/\gamma^{3/2}$, then

$$\frac{|\mathrm{Sym}^k(\chi)(g)|}{\mathrm{Sym}^k(\chi)(1)} \leq \gamma^{k/2}.$$

As a consequence of Theorem 9.7.15, we obtain

Corollary 9.7.16 ([76]) *Any finite subgroup $G < \mathrm{GL}_n(\mathbb{C})$ admits an invariant of degree $\leq Cn \cdot |G/G'|$, with $C = 1184036$.*

More generally, the question of occurrences of any irreducible representation of a finite subgroup $G < \mathrm{GL}(V)$ in symmetric powers of V, where V is any finite-dimensional vector space over any field, has been studied in [38].

Acknowledgements This survey paper is based on the lecture series given by the author at the '*Groups St Andrews 2022*' in Newcastle, UK. It is a pleasure to thank the organizers of the conference, Colin Campbell, Martyn Quick, Edmund Robertson, Colva Roney-Dougal and David Stewart, and the Clay Mathematics Institute for generous hospitality and support.

The author gratefully acknowledges the support of the NSF (grants DMS-1840702 and DMS-2200850), the Simons Foundation, and the Joshua Barlaz Chair in Mathematics.

Part of this paper was written while the author visited the Department of Mathematics, Princeton University, in Fall 2022. It is a pleasure to thank the Princeton University for hospitality and support.

The author is grateful to the referee for careful reading and insightful comments that helped greatly improve the exposition of the paper.

References

[1] M. Aschbacher, On the maximal subgroups of the finite classical groups, *Invent. Math.* **76** (1984), 469–514.

[2] G. Benkart, P. Diaconis, M.W. Liebeck, and Pham Huu Tiep, Tensor product Markov chains, *J. Algebra* **561** (2020), 17–83.

[3] R. Bezrukavnikov, M. W. Liebeck, A. Shalev, and Pham Huu Tiep, Character bounds for finite groups of Lie type, *Acta Math.* **221** (2018), 1–57.

[4] R.W. Carter, *Finite Groups of Lie type: Conjugacy Classes and Complex Characters*, Wiley, Chichester, 1985.

[5] I. Cheltsov and K. Shramov, On exceptional quotient singularities, *Geometry Topology*, **15** (2011), 1843–1882.

[6] J. H. Conway, R. T. Curtis, S. P. Norton, R. A. Parker, and R. A. Wilson, '*An ATLAS of Finite Groups*', Clarendon Press, Oxford, 1985.

[7] F. Digne and J. Michel, *Representations of Finite Groups of Lie Type*, London Math. Soc. Student Texts **21**, Cambridge Univ. Press, 1991.

[8] J.-P. Demailly and J. Kollár, Semi-continuity of complex singularity exponents and Kähler-Einstein metrics on Fano orbifolds, *Ann. Sci. École Norm. Sup.* **34** (2001), 525–556.

[9] P. Diaconis and M. Shahshahani, Generating a random permutation with random transpositions, *Z. Wahrsch. Verw. Gebiete* **57** (1981), 159–179.

[10] E.W. Ellers and N. Gordeev, On the conjectures of J. Thompson and O. Ore, *Trans. Amer. Math. Soc.* **350** (1998), 3657–3671.

[11] V. Ennola, On the characters of the finite unitary groups, *Ann. Acad. Scient. Fenn. A I*, no. 323 (1963).

[12] S. Fomin and N. Lulov, On the number of rim hook tableaux, *Zap. Nauchn. Semin. POMI* **223**(1995), 219–226.

[13] J. Fulman, Convergence rates of random walk on irreducible representations of finite groups, *J. Theoret. Probab.* **21** (2008), 193–211.

[14] J. Fulman and R. Guralnick, Bounds on the number and sizes of conjugacy classes in finite Chevalley groups with applications to derangements, *Trans. Amer. Math. Soc.* **364** (2012), 3023–3070.

[15] The GAP group, GAP - groups, algorithms, and programming, Version 4.4, 2004, `http://www.gap-system.org`.

[16] M. Geck, On the average values of the irreducible characters of finite groups of Lie type on geometric unipotent classes, *Doc. Math.* **1** (1996), 293–317.

[17] M. Geck and G. Malle, On the existence of a unipotent support for the irreducible characters of a finite group of Lie type, *Trans. Amer. Math. Soc.* **352** (1999), 429–456.

[18] N.P. Gill, blog post, `https://nickpgill.github.io/a-rodgers-saxl-conjecture-for-characters`.

[19] D. Gluck, Sharper character value estimates for groups of Lie type, *J. Algebra* **174** (1995), 229–266.

[20] D. Gluck, Characters and random walks on finite classical groups, *Adv. Math.* **129** (1997), 46–72.

[21] D. Gluck and K. Magaard, Base sizes and regular orbits for coprime affine permutation groups, *J. London Math. Soc.* **58** (1998), 603–618.

[22] R.M. Guralnick, M. Larsen, and Pham Huu Tiep, Character levels and character bounds, *Forum of Math. Pi* **8** (2020), e2, 81 pages.

[23] R.M. Guralnick, M. Larsen, and Pham Huu Tiep, Character levels and character bounds for finite classical groups, (submitted).

[24] R.M. Guralnick, M.W. Liebeck, E.A. O'Brien, A. Shalev, and Pham Huu Tiep, Surjective word maps and Burnside's $p^a q^b$ theorem, *Invent. Math.* **213** (2018), 589–695.

[25] R.M. Guralnick and Pham Huu Tiep, Cross characteristic representations of even characteristic symplectic groups, *Trans. Amer. Math. Soc.* **356** (2004), 4969–5023.

[26] R.M. Guralnick and Pham Huu Tiep, Symmetric powers and a problem of Kollár and Larsen, *Invent. Math.* **174** (2008), 505–554.

[27] R.M. Guralnick and Pham Huu Tiep, A problem of Kollár and Larsen on finite linear groups and crepant resolutions, *J. Europ. Math. Soc.* **14** (2012), 605–657.

[28] R.M. Guralnick and Pham Huu Tiep, Effective results on the Waring problem for finite simple groups, *Amer. J. Math.* **137** (2015), 1401–1430.

[29] G. Heide, J. Saxl, P.H. Tiep and A.E. Zalesski, Conjugacy action, induced representations and the Steinberg square for simple groups of Lie type, *Proc. Lond. Math. Soc.* **106** (2013), 908–930.

[30] M. Hildebrand, Generating random elements in $SL_n(\mathbb{F}_q)$ by random transvections, *J. Alg. Comb.* **1** (1992), 133–150.

[31] G. Hiss and G. Malle, Low-dimensional representations of quasi-simple groups, *LMS J. Comput. Math.* **4** (2001), 22–63; Corrigenda: Low-dimensional representations of quasi-simple groups, *LMS J. Comput. Math.* **5** (2002), 95–126.

[32] I.M. Isaacs, G. Malle, and G. Navarro, A reduction theorem for the McKay conjecture, *Invent. Math.* **170** (2007), 33–101.

[33] Y. Ito and M. Reid, The McKay correspondence for finite subgroups of SL(3, $\mathbf{C}$), in: *Higher-dimensional Complex Varieties* (Trento, 1994), pp. 221–240, de Gruyter, Berlin, 1996.

[34] G. D. James, On the minimal dimensions of irreducible representations of symmetric groups, *Math. Proc. Cam. Phil. Soc.* **94** 1983), 417–424.

[35] N. Kawanaka, Generalized Gelfand-Graev representations of exceptional simple algebraic groups over a finite field. I, *Invent. Math.* **84** (1986), 575–616.

[36] A.S. Kleshchev, M. Larsen, and Pham Huu Tiep, Level, rank, and tensor growth of representations of symmetric groups, arXiv:2212.06256.

[37] J. Kollár and M. Larsen, Quotients of Calabi-Yau varieties, in: *Algebra, Arithmetic, and Geometry: In Honor of Yu. I. Manin*, Vol. II, 179–211, Progr. Math., **270**, Birkhäuser, Boston, MA, 2009.

[38] J. Kollár and Pham Huu Tiep, Symmetric powers, (in preraration).

[39] S. Lang and A. Weil, Number of points of varieties over finite fields, *Amer. J. Math.* **76** (1954), 819–827.

[40] M. Larsen and A. Lubotzky, Representation varieties of Fuchsian groups, in: *From Fourier Analysis and Number Theory to Radon Transforms and Geometry* (eds. H. Farkas et al), *Dev. Math.* **28** (2013), Springer, New York, pp. 375–397.

[41] M. Larsen, A. Lubotzky and C. Marion, Deformation theory and finite simple quotients of triangle groups I, *J. Eur. Math. Soc.* **16** (2014), 1349–1375.

[42] M. Larsen, A. Lubotzky and C. Marion, Deformation theory and finite simple quotients of triangle groups II, *Groups Geom. Dyn.* **8** (2014), 811–836.

[43] M. Larsen and A. Shalev, Characters of symmetric groups: sharp bounds and applications, *Invent. Math.* **174** (2008), 645–687.

[44] M. Larsen, A. Shalev, and Pham Huu Tiep, The Waring problem for finite simple groups, *Annals of Math.* **174** (2011), 1885–1950.

[45] M. Larsen, A. Shalev, and Pham Huu Tiep, Probabilistic Waring problems for finite simple groups, *Annals of Math.* **190** (2019), 561–608.

[46] M. Larsen and Pham Huu Tiep, Uniform character bounds for finite classical groups, (submitted).

[47] M.W. Liebeck, E.A. O'Brien, A. Shalev and Pham Huu Tiep, The Ore conjecture, *J. Eur. Math. Soc.* **12** (2010), 939–1008.

[48] M.W. Liebeck and A. Shalev, Diameters of simple groups: sharp bounds and applications, *Annals of Math.* **154** (2001), 383–406.

[49] M.W. Liebeck and A. Shalev, Character degrees and random walks in finite groups of Lie type, *Proc. Lond. Math. Soc.* **90** (2005), 61–86.

[50] M.W. Liebeck and A. Shalev, Fuchsian groups, finite simple groups and representation varieties. *Invent. Math.* **159** (2005), 317–367.

[51] M.W. Liebeck, A. Shalev, and Pham Huu Tiep, Character ratios, representation varieties and random generation of finite groups of Lie type, *Adv. Math.* **374** (2020), 107386, 39 pp.

[52] M.W. Liebeck, A. Shalev, and Pham Huu Tiep, On the diameters of McKay graphs for finite simple groups, *Israel J. Math.* **241** (2021), 449–464.

[53] M.W. Liebeck, A. Shalev, and Pham Huu Tiep, McKay graphs for alternating groups and classical groups, *Trans. Amer. Math. Soc.* **374** (2021), 5651–5676.

[54] M.W. Liebeck, A. Shalev, and Pham Huu Tiep, Character ratios for finite classical groups, (in preparation).

[55] M.W. Liebeck and Pham Huu Tiep, Character ratios for exceptional groups of Lie type, *Int. Math. Res. Not. IMRN* 2021, no. 16, 12054–12076.

[56] F. Lübeck, Smallest degrees of representations of exceptional groups of Lie type, *Comm. Algebra* **29** (2001), 2147–2169.

[57] A. Lubotzky, Cayley graphs: eigenvalues, expanders and random walks, *Surveys in combinatorics*, 1995 (Stirling), 155–189, London Math. Soc. Lecture Note Ser. **218**, Cambridge Univ. Press, Cambridge, 1995.

[58] G. Lusztig, *Characters of Reductive Groups over a Finite Field*, Annals of Math. Studies **107**, Princeton Univ. Press, Princeton, 1984.

[59] G. Lusztig, Green functions and character sheaves, *Ann. of Math.* **131** (1990), 355–408.

[60] G. Lusztig, A unipotent support for irreducible representations, *Adv. Math.* **94** (1992), 139–179.

[61] J. McKay, Graphs, singularities and finite groups, *Proc. Symp. Pure Math* **37** (1980), 183–186.

[62] T.W. Müller and J.-C. Schlage-Puchta, Character theory of symmetric groups, subgroup growth of Fuchsian groups, and random walks, *Adv. Math.* **213** (2007), 919–982.

[63] O. Ore, Some remarks on commutators, *Proc. Amer. Math. Soc.* **2** (1951), 307–314.

[64] M. Reid, La correspondance de McKay, Séminaire Bourbaki, Vol. 1999/2000, *Astérisque* **276** (2002), 53–72.

[65] D.M. Rodgers and J. Saxl, Products of conjugacy classes in the special linear groups, *Comm. Algebra* **31** (2003), 4623–4638.

[66] M. Sellke, Covering Irrep(S_n) with tensor products and powers, *Math. Annalen*, 2022, `https://doi.org/10.1007/s00208-022-02532-3`.

[67] G.C. Shephard and J.A. Todd, Finite unitary reflection groups, *Can. J. Math.* **6** (1954), 274–304.

[68] T. Shoji, Character sheaves and almost characters of reductive groups. I, II, *Adv. Math.* **111** (1995), 244–313, 314–354.

[69] R. Steinberg, Finite subgroups of SU2, Dynkin diagrams and affine Coxeter elements, *Pacific J. Math.* **118** (1985), 587–598.

[70] J. Taylor, Generalised Gelfand–Graev representations in small characteristics, *Nagoya Math. J.* **224** (2016), 93–167.

[71] J. Taylor and Pham Huu Tiep, Lusztig induction, unipotent support, and character bounds, *Trans. Amer. Math. Soc.* **373** (2020), 8637–8776.

[72] J. G. Thompson, Invariants of finite groups, *J. Algebra* **69** (1981), 143–145.

[73] G. Tian, On Kähler-Einstein metrics on certain Kähler manifolds with $c_1(M) > 0$, *Invent. Math.* **89** (1987), 225–246.

[74] G. Tian and S.-T. Yau, Kähler-Einstein metrics metrics on complex surfaces with $C_1 > 0$, *Comm. Math. Phys.* **112** (1987), 175–203.

[75] Pham Huu Tiep, Finite groups admitting grassmannian 4-designs, *J. Algebra* **306** (2006), 227–243.

[76] Pham Huu Tiep, The alpha-invariant and Thompson's conjecture, *Forum of Math. Pi* **4** (2016), e5, 28 pages.

[77] Pham Huu Tiep and A.E. Zalesskii, Minimal characters of the finite classical groups, *Comm. Algebra* **24** (1996), 2093–2167.

[78] Pham Huu Tiep and A.E. Zalesskii, Some characterizations of the Weil representations of the symplectic and unitary groups, *J. Algebra* **192** (1997), 130–165.

[79] E. Witten, On quantum gauge theories in two dimensions, *Comm. Math. Phys.* **141** (1991) 153–209.

[80] I. Zisser, The character covering number for alternating groups, *J. Algebra* **153** (1992), 357–372.

10
Generalized Baumslag-Solitar groups: a topological approach

Mathew Timm[a]

Abstract

Generalized Baumslag-Solitar groups are a class of combinatorially interesting groups. Their group theory is also closely associated with the topology of a class of 2-dimensional spaces. These 2-dimensional spaces are Seifert fibred. We develop the basic topology of these fibrations and derive some of the most immediate group theoretic consequences of this topology.

MSC2020 primary 57M07, 57M15; secondary 20F34, 05C10, 05C25.

Key words generalized Baumslag-Solitar group, generalized Baumslag-Solitar graph, generalized Baumslag-Solitar complex, Schur multiplier, Seifert fibred.

A *generalized Baumslag-Solitar (GBS) group* is a group which has a graph of groups description in which all vertex and edge groups are infinite cyclic and all edge maps are monomorphisms. Each such group G can be described by a weighted directed graph (Γ, ω) where for each directed edge $e = [e^-, e^+]$ of Γ, the weight function $\omega(e) = (\omega^-(e), \omega^+(e))$ assigns an ordered pair of non-zero integers.

The weighted directed graphs (Γ, ω) as above are called *generalized Baumslag-Solitar (GBS) graphs.* Along with specifying a particular GBS group $\pi_1(\Gamma, \omega)$, the graph (Γ, ω) can be used to build a 2-dimensional

[a] Department of Mathematics, Bradley University, Peoria, IL 61604, USA.
Partially supported by U.S. Fulbright Scholars Grant PS00309131, the Croatian Agencija za Mobilnost i Programe EU, and the Department of Mathematics, Faculty of Science and Mathematics, University of Split.
mtimm@bradley.edu

space, a *generalized Baumslag-Solitar (GBS) complex*, $K(\Gamma, \omega)$. There are intimate connections between the group theory of $\pi_1(\Gamma, \omega)$, the topology of $K(\Gamma, \omega)$, and the combinatorics of (Γ, ω). Along with the relationships between the structures of the GBS groups and the GBS spaces imposed by the combinatorics of the graphs, the connections between the objects in these GBS classes are highly dependent upon the fact that GBS spaces are naturally Seifert fibered, that is, they are unions of collections of pairwise disjoint circles, and that all groups in the graph of groups description are copies of $\mathbb{Z}$, the fundamental group of the circle. In this paper we develop the basic topology of these Seifert fibrations. Much of this topology seems to be known but has not been collected together in one place. The intent of this paper is to do so and to indicate some of the more immediate group theoretic consequences of the topology. The straightforward applications of this topology to the group theory shows that this topological approach can augment the set of geometric and classical group theoretic tools already used to study GBS group. Thus, the topology provides another entry into the study of a class of groups which are of historical interest, e.g., [1], [8], [24], and of current interest, e.g., [3], [4], [5], [6], [7], [11], [14], [15], [12], [17], [18], [21], [23]. Derek Robinson gave a talk on the state of the GBS art at Groups St Andrews 2013, [19].

The paper is organized as follows. Section 10.1 develops the topology and includes simple, but instructive, examples. Section 10.2 contains a brief review of the generalized Bausmslag-Solitar groups, and in particular, a reminder of how to obtain a presentation for a GBS group from its graphical description. Section 10.3 shows that GBS complexes have fundamental groups which are GBS groups and that there is a natural system of generators which can be used to present these groups. Sections 10.4 and 10.5 include results on GBS groups which follow easily from the topology. Section 10.6 includes statements of some open problems.

10.1 Basic Constructions

Let Γ be a finite connected graph, loops and multiple edges allowed, with vertex set $V(\Gamma)$ and directed edge set $E(\Gamma)$. A directed edge e is a copy of an oriented closed interval. Assign an initial vertex e^- and terminal vertex e^+ indicating the direction on e. By analogy with the notation for a closed interval, write $e = [e^-, e^+]$. Set $\mathbb{Z}^* = \mathbb{Z} \setminus 0$. A *weight function* ω is a function $\omega : E(G) \to \mathbb{Z}^* \times \mathbb{Z}^*$ and we write

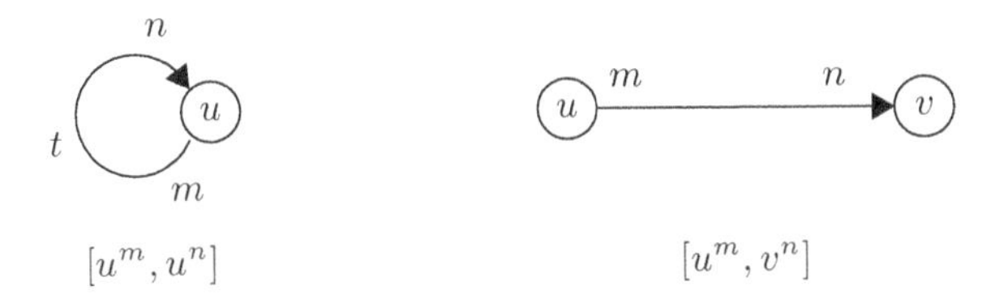

Figure 10.1 Simple GBS graphs.

$\omega(e) = (\omega^-(e), \omega^+(e))$. Given a graph G and weight function ω, the pair (Γ, ω) is called a *generalized Baumslag-Solitar (GBS) graph.*

If $e^- = u$, $e^+ = v$, and $w(e) = (m, n)$, we sometimes write the directed weighted edge $(e, w(e))$ as $(e, w(e)) = [u^m, v^n]$. When $e^- = e^+ = u$, i.e., when e is a loop, we may write $(e, w(e)) = [u^m, u^n]$. Pictures of $(\Gamma, \omega) = [u^m, u^n]$ and $(\Gamma, \omega) = [u^m, v^n]$ are in Figure 10.1. Implicit in this "weighted interval" notation for a directed edge is the fact that a graph Γ has the structure of a 1-dimensional CW or simplicial complex: the 0-cells/simplices are the vertices, the 1-cells/simplices are the edges. See Figures 10.2 and 10.4 for more complicated examples.

It is convenient to have a notation which allows us to combine GBS graphs. If (Γ, ω) and (Δ, χ) are GBS graphs with $\{v_1, \ldots, v_k\} \subset V(\Gamma)$ and $\{w_1, \ldots, w_k\} \subset V(\Delta)$, we write

$$(\Gamma \oplus \Delta, \omega \oplus \chi) = (\Gamma, \omega) \underset{v_i = w_i}{\oplus} (\Delta, \chi)$$

for the GBS graph obtained by identifying v_i with w_i. Note that when, for example, the vertices of an edge $[u_{i_1}, u_{i_2}]$ in Γ and $[w_{i_1}, w_{i_2}]$ in Δ are identified via $u_{i_1} = w_{i_1}$ and $u_{i_2} = w_{i_2}$, then $\Gamma \oplus \Delta$ has multiple edges between a pair of vertices. We typically denote multiple edges between the same pair u, v of vertices with subscripts $[u, v]_1, \ldots, [u, v]_k$.

Associated to each GBS graph (Γ, ω) is a 2-dimensional complex $K(\Gamma, \omega)$ called a *generalized Baumslag-Solitar (GBS) complex.* Define $K(\Gamma, \omega)$ as follows.

For each $v \in V(\Gamma)$, let

$$S_v^1 = \{\exp(i\theta) : 0 \leq \theta \leq 2\pi\}$$

be a copy of the oriented unit circle in the complex plane with the orientation induced by the parametrization. For each $e \in E(\Gamma)$, let

$$A_e = S_e^1 \times [0, 1]_e$$

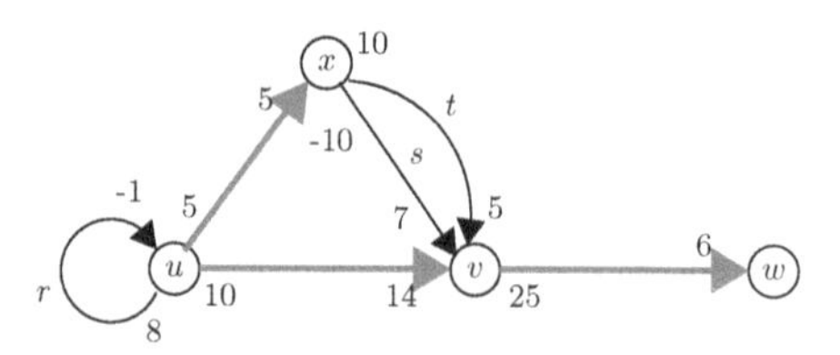

Figure 10.2 An example of a GBS graph (Γ, ω). The maximal subtree is in thick grey. The non-tree edges are r, s, t.

where S^1_e is again a copy of the oriented unit circle with the orientation induced by the parametrization above and $[0,1]_e$ is a copy of the unit interval parametrized in the usual way. Both boundary circles of A_e are oriented by the parametrizations they inherit from S^1_e. For each e define the oriented degree $\omega^{\mp}(e)$ attaching maps to be the $\omega^{\mp}(e)$-cyclic covers

$$\begin{aligned} p_{e^-} : S^1_e \times 0 \to S^1_{e-} \quad & \text{given by } (\exp(i\cdot\theta), 0) \mapsto \exp(i\cdot\omega^-(e)\theta), \text{ and} \\ p_{e^+} : S^1_e \times 1 \to S^1_{e+} \quad & \text{given by } (\exp(i\cdot\theta), 1) \mapsto \exp(i\cdot\omega^+(e)\theta). \end{aligned}$$

Let $K(\Gamma, \omega)$ be the resulting 2-complex. The collection of quintuples

$$\mathcal{A} = \mathcal{A}(\Gamma, \omega) = \left\{ \left(A_e, S^1_{e-}, S^1_{e+}, p_{e^-}, p_{e^+} \right) : e \in E(\Gamma) \right\}$$

forms the *annular decomposition* for $K(\Gamma, \omega)$.

We will not normally distinguish between the annulus A_e as an element of $\mathcal{A}$ and its image in $K(\Gamma, \omega)$. As a consequence, a subspace of $K(\Gamma, \omega)$ that might actually be a mapping cylinder or torus formed from one of the annuli in $\mathcal{A}$ may still be called an annulus. The context will make our intention clear. In a consistent abuse of the language, both the circles $S^1_e \times 0$ and $S^1_e \times 1$ and their images S^1_{e-} and S^1_{e+} in $K(\Gamma, \omega)$ are called *vertex circles*.

We emphasize that the circles, annuli, and attaching maps used in the construction of GBS complexes are always parametrized as indicated above. This introduces a certain "rigidity" which can be conveniently exploited. In particular, it leads to the definition of 2-dimensional Seifert fibred spaces presented in the next subsection and avoids technical difficulties discussed in two cases at its end.

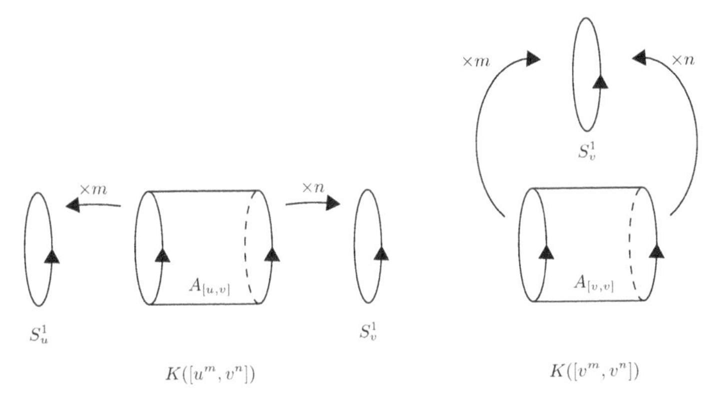

Figure 10.3 Pictures of $K([u^m, v^n])$ and $K([v^m, v^n])$.

10.1.1 Seifert fibered 2-complexes

Our approach to Seifert fibred 2-dimensional spaces is analogous to that of [9] for Seifert fibred 3-manifolds. In this context, the next example is fundamental.

Example 10.1.1 Let $m \in \mathbb{N}$. An *m-od* T_m is a space homeomorphic to

$$\left\{ r \cdot \exp\left(\frac{2\pi k \cdot i}{m}\right) : 0 \leq r \leq 1, 0 \leq k \leq m-1 \right\} \subset \mathbb{C}.$$

It can also be described as m copies of the unit interval $[0, 1]$ which have their origins identified to a common point. The second description emphasizes that an m-od has a graphical structure. Consequently, the product $S^1 \times T_m$ has the GBS graph description $S^1 \times T_m = K\left(\overset{m}{\underset{k=1}{\oplus}} [u^1, v_k^1]\right)$. Implicit in this GBS notation is the understanding that all m copies of the vertex u in the m edges $[u, v_k]$ are identified to a single point.

Definition 10.1.2 Let T_m be an m-od. The space $S^1 \times T_m$ is a *Seifert fibred space with base space* T_m and *projection* η, where the *fibers* of the Seifert fibration are the circles $S^1 \times t$, $t \in T_m$, and the projection is the map $\eta(\exp(i\theta, t)) = t \in T_m$. In general, a 2-dimensional space X is *Seifert fibred* with *projection* η and *base space* B if the following conditions are satisfied:

1. $X = \cup\{S_i^1 : i \in I\}$, I some index set, where $\{S_i^1 : i \in I\}$ is a collection

of pairwise disjoint circles. The circles S_i^1, are the *fibers* of the Seifert fibration.

2. For each fiber S_i^1 there is an $m_i \in \mathbb{N}$ and a neighborhood $N(S_i^1) \supset S_i$ such that $N(S_i^1)$ is a union of fibres in the Seifert fibration and there is a Seifert fibration preserving finite sheeted covering $\varphi_i : S^1 \times T_{m_i} \to N(S_i^1)$.
3. B has the quotient space topology determined by mapping each fiber in the Seifert fibration to a point and $\eta : X \to B$ is the quotient map.

The $N(S_i^1)$ and $S^1 \times T_{m_i}$ are the *model fibred neighborhoods* of the fibres S_i^1. When T_{m_i} is a single interval, there is a finite sheeted cover $\varphi_i : S^1 \times [0,1] \to N(S_i^1)$ which is a homeomorphism. In this case, S_i^1 is an *ordinary* fiber. Otherwise S_i^1 is an *exceptional* fibre. While a vertex circle S_u^1 may have a fibred neighborhood which is an annulus, it is also frequently convenient to think of every vertex circle S_u^1 in a $K(\Gamma, \omega)$ as an exceptional fiber since every vertex is in some sense special.

In Example 10.1.1, the entire space of $K([u^m, v^1])$ can be taken as the model fibred neighborhood of the the vertex circle S_u^1. $K([u^m, v^1])$ is finitely covered by $S^1 \times T_m$. The covering map is that induced by the m-fold cyclic self cover $(\exp(i \cdot \theta), t) \mapsto (\exp(i \cdot m\theta, t))$ defined on $S^1 \times T_m$. The minimum degree of a such a cover is $|m|$. Thus $S_u^1 \subset K([u^m, v^1])$ is an exceptional fiber when $|m| \geq 2$.

There are interesting spaces which are disjoint unions of circles which do not satisfy the second condition in the definition. One such is obtained by allowing annuli to be attached to circles via more complicated attaching maps. See Example 5.2 of [2]. A second, call it Y, can be obtained from Example 5.1 of [2] by sewing in one additional circle, S_0^1, which intersects each of the infinitely many 2-spheres in the construction in its "east" and "west" poles: one should think of S_0^1 as being the union of the diameters connecting these east and west cone-points of the 2-spheres in the picture. Then, consider $Y \setminus S_0^1$: it is the union of countable many open annuli $S^1 \times (0,1)$, each of which is Seifert fibered via the product structure. Thus, Y is the union of S_0^1 and the fibers in these annuli. By way of comparison, Y has many non-trivial self covers, as do all GBS complexes associated to finite GBS graphs, but GBS complexes are aspherical, while Y has infinitely generated π_2. I suspect that $\pi_2(Y)$ is uncountable generated, but have not made a serious attempt to verify this. Determining $H_2(Y, \mathbb{Z})$ is also of interest. Observe that one can also form a quotient space by sending each of the circles in the

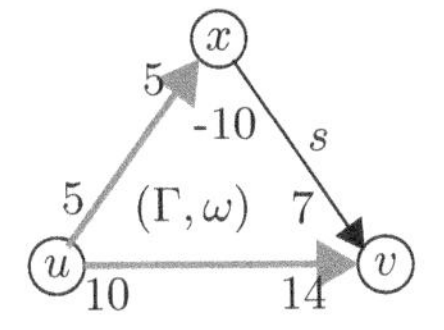

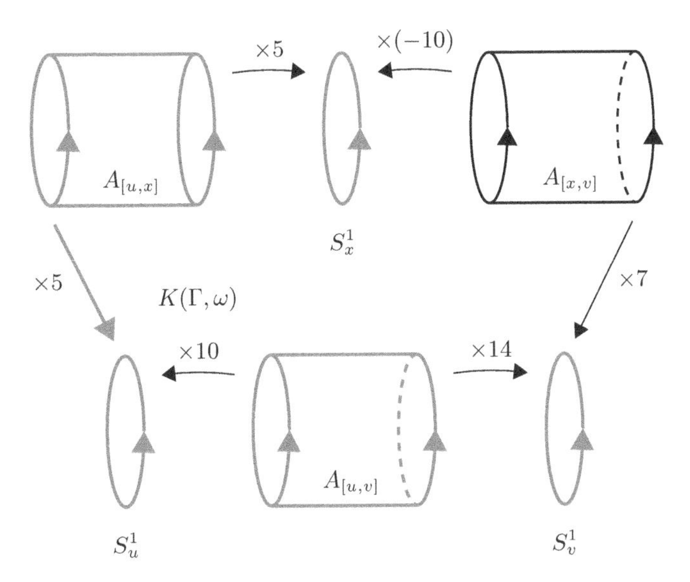

Figure 10.4 A GBS graph (Γ, ω) and the associated 2-complex $K(\Gamma, \omega)$.

above decomposition of Y to a point. The resulting space B is a union of infinitely many circles which share a common point. One wonders if its fundamental group is isomorphic to the fundamental group of the Hawaiian earring.

10.1.2 Basic Examples

Example 10.1.3 The GBS complex $K([u^m, v^1])$ is a locally trivial m-od bundle over the circle. Write $K([u^m, v^1]) = T_m \times [0, 1]/\sim$ where

T_m is the m-od and $\sim$ indicates the identification of $(exp(\frac{2\pi k \cdot i}{m}), 0)$ with $(exp(\frac{2\pi(k+1) \cdot i}{m}), 1)$. The addition in the numerator of the second fraction is $mod(m)$. $K([u^2, v^1])$ is the Mobius band, a 2-od (or interval) bundle over S^1. Figure 10.6 contains three different ways of illustrating $K([u^3, v^1])$.

Example 10.1.4 Pictorial representations of $K([u^m, v^n])$ and $K([v^m, v^n])$ are in Figure 10.3. Alternate views of these GBS complexes for the specific weights $(m, n) = (2, 3)$ are in Figure 10.5. When m and n are relatively prime, $K([u^m, v^n])$ is the 2-dimensional core of the (m, n)-torus knot. When m and n are not relatively prime the $K([u^m, v^n])$ are still interesting simple examples of GBS complexes. For example, as noted above, $K([u^2, v^2])$ is the Klein bottle.

Example 10.1.5 Let Γ be a graph and ω the constant weight function $\omega(e) = (1, 1)$ for every $e \in E(\Gamma)$. $K(\Gamma, \omega)$ is the topological product $K(\Gamma, \omega) = S^1 \times \Gamma$. The projection η of the Seifert fibration is $\eta(\exp(i\theta), t) = t$. That is, the base space of the Seifert fibration is a copy of Γ. Also note that the inclusion map $\iota : \Gamma \to K(\Gamma, \omega)$ given by $\iota(s) = (1, s) = (\exp(i \cdot 0), s)$, $s \in e = [e^-, e^+]$ for each edge e of Γ is such that $\eta \circ \iota = id_\Gamma$. Consequently, in this case, the Seifert fibration has a *section*. We will see that this is true in general.

The fact that $K([u^m, v^1])$ describes the locally trivial m-od bundle allows for another decomposition of the GBS complex $K(\Gamma, \omega)$. For each edge $e = [e^-, e^+] \in E(\Gamma)$, let μ_e be the midpoint of e. Then for each $e \in E(\Gamma)$ we can write its corresponding weighted edge as

$$[(e^-)^{\omega^-(e)}, (e^+)^{\omega^+(e)}] = [(e^-)^{\omega^-(e)}, \mu_e^1] \oplus [\mu_e^1, (e^+)^{\omega^+(e)}].$$

Thus, for each edge e of Γ the sub-GBS complex

$$K([(e^-)^{\omega^-(e)}, (e^+)^{\omega^+(e)}]) \subset K(\Gamma, \omega)$$

can be written as

$$\begin{aligned} K([(e^-)^{\omega^-(e)}, (e^+)^{\omega^+(e)}]) &= K([(e^-)^{\omega^-(e)}, \mu_e^1] \oplus [\mu_e^1, (e^+)^{\omega^+(e)}]) \\ &= K([(e^-)^{\omega^-(e)}, \mu_e^1]) \underset{S^1_{\mu_e}}{\cup} K([\mu_e^1, (e^+)^{\omega^+(e)}]). \end{aligned}$$

Since the above rewriting can be done for each edge, it follows that every GBS complex $K(\Gamma, \omega)$ can be written as a union of locally trivial $m_e^{\pm}$-od

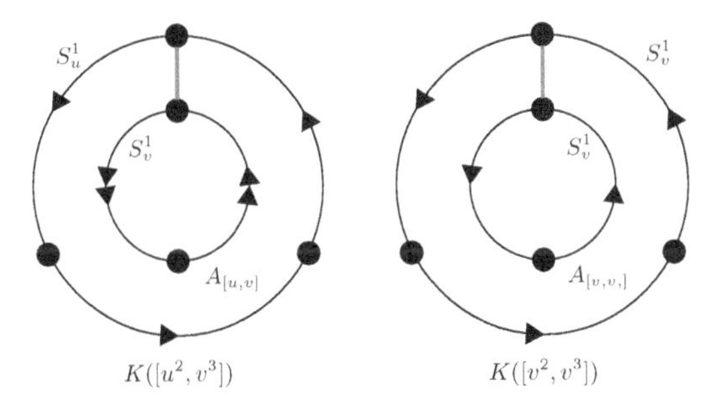

Figure 10.5 Another representation of $K([u^2, v^3])$ and $K([v^2, v^3])$. Directed arcs with the same number of arrowheads on them are glued together in the direction indicated by the arrowheads.

bundles, where $m_e^{\pm} = \omega^{\pm}(e)$ is determined by the weights on each edge. Thus, each $K(\Gamma, \omega)$ has a *locally trivial bundle decomposition.*

The fact that a GBS complex has a locally trivial bundle decomposition has interesting, and easily derived, consequences. For $k \geq 1$, each locally trivial bundle $K([u^m, v^1]$ has non-trivial k-fold self covers for each k relatively prime to m. It is induced by the k-fold self cover of the annulus of which $K([u^m, v^1])$ is a quotient. Consequently, when (Γ, ω) is a GBS tree, or more generally, when all weights on (Γ, ω) are positive, it is easy to see that $K(\Gamma, \omega)$ has a non-trivial k-fold self cover for each $k \geq 2$ which is relatively prime to all the weights on (Γ, ω): simply glue the self covers of the bundles of the indicated degree in the bundle decomposition together along vertex circles in the right way. Because of the correspondence between covering spaces of the GBS complexes and subgroups of their fundamental group, it then follows that $\pi_1(K(\Gamma, \omega), x_0)$ has index $k \geq 2$ subgroups which are isomorphic to itself, i.e, it is easy to see in these cases that $\pi_1(K(\Gamma, \omega), x_0)$ is *non-cohopfian.*

10.1.3 The standard system of loops in $K(\Gamma, \omega)$

Given a GBS graph (Γ, ω) with $K = K(\Gamma, \omega)$, there is a natural way to define a generating system for $\pi_1(K, x_0)$. We call this system a *standard system* for the GBS complex K. The process of determining a standard system for K follows.

Step 1. Choose a maximal subtree $T \subset K$, a vertex $v_0 \in V(\Gamma)$, and the associated circle $S_{v_0}^1$ in K. Let $x_0 = 1 = \exp(0) \in S_{v_0}^1$. The point x_0 is called the *base point* for the standard system. For each edge $e \in E(\Gamma)$. Let a_e be the parametrized arc $a_e = \{(1, s) : s \in [0, 1]_e\}$. For each e,

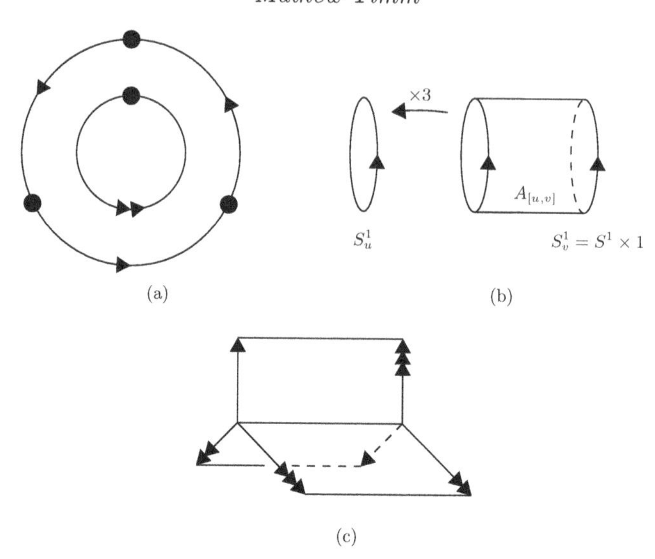

Figure 10.6 Three illustrations of $K([u^3, v^1])$. The drawings in (a) and (b) emphasis that $K([u^3, v^1])$ has an annular decompostion. Part (c) emphasises that it is a locally trivial 3-od bundle over S^1.

a_e is the *standard straight arc* in A_e. Note that in K, because of our rigidity conditions, the standard straight arc a_e is a loop if and only if both boundary circles of A_e are joined to the same vertex circle.

For each vertex $v \in V(\Gamma)$, let $\gamma(v_0, v)$ be the unique simple directed path in T from v_0 to v. Note that $\gamma(v_0, v_0) = v_0$. Let $\overline{a_e}$ denote a_e with the reverse orientation.

Step 2. Let $v \in V(\Gamma)$. The *vertex circuit* x_v *based at* x_0 is

$$x_v = \left(\bigcup_{f \in E(\gamma(v_0, v))} a_f \right) \cup \{\exp(i\theta) \in S^1_v : 0 \leq \theta \leq 2\pi\} \cup \left(\bigcup_{f \in E(\gamma(v_0, v))} \overline{a_f} \right).$$

Step 3. For each edge $e \notin T$, there are the simple paths $\gamma(v_0, e^-) \subset T$ and $\gamma(v_0, e^+) \subset T$. Thus, for each non-$T$ edge e, there is the *edge circuit*

t_e *based at* x_0 given by

$$t_e = \left(\bigcup_{f \in E(\gamma(v_0, e^-))} a_f \right) \cup a_e \cup \left(\bigcup_{f \in E(\gamma(v_0, e^+))} \overline{a_f} \right).$$

Step 4. The set $\mathcal{S}(\Gamma, \omega, T) = \{x_v, t_e : v \in V(\Gamma), e \in E(\Gamma \setminus T\}$ is the *standard system for* K *based at* x_0. By [2], the homotopy classes of the paths x_v and t_e in the standard system form a generating set for the fundamental group $\pi_1(K, x_0)$. The proof, very briefly, is an induction. Induct first on the number of edges in T, then on the number of non-T edges in Γ. Repeated applications of Van Kampen's Theorem gives a presentation for $\pi_1(K, x_0)$. The base case of the induction is illustrated in Figure 10.7. It decomposes the annuli in the pictures into the complements of the interiors (grey) of the small closed disks and the small disks themselves. Orient the boundary circles of the disks as indicated. There are strong deformation retracts of the complements of the interiors of the small disks onto the unions of the boundary circle(s) of each annulus together with the thick grey transverse arc. In Figure 10.7(a), the complement has the homotopy type of the wedge $S^1_u \vee S^1_v$ of the two vertex circles. Note that the "transverse" straight arc $[v, u]$, in thick grey the picture, is homotopically trivial. Ignoring some base point issues, and in the picture allowing the deformation retract to be along radial arcs outward from the center of the small grey circle, one sees that the (oriented) boundary of the small gray disk is homotopically equivalent to the product of homotopy classes $\left[S^1_v\right]^{-2} \star [[v, u]] \star \left[S^1_u\right]^3 \star [[v, u]]^{-1}$ which, after applying Van Kampen's Theorem produces the GBS relation $\left[S^1_v\right]^2 = \left[S^1_u\right]^3$. In a similar analysis of Figure 10.7(b), the deformation retract of the complement of the small gray disk shows it has the homotopy type of the wedge $S^1_v \vee [v, v]$ of the one vertex circle and the "transverse" loop, in thick grey, formed by the embedded copy of $[v, v]$ in $A_{[v,v]}$. In this case, when one applies Van Kampen's Theorem, one obtains the GBS relation $[[v, v]]^{-1} \left[S^1_v\right]^3 [[v, v]] = \left[S^1_v\right]^2$. In general, setting x_u and x_v equal to the homotopy classes of the vertex circle, t_e equal to the homotopy class of the loop $[v, v]$, and suppressing base points, from (a) one reads off the presentation $\pi_1(K([u^m, v^n])) = \langle x_u, x_v : x_u^m = x_v^n \rangle$ and from (b), one obtains $\pi_1(K([v^m, v^n]) = \langle x_v, t_e : t_e^{-1} x_v^m t_e = x_v^n \rangle$.

To complete the explanation of why a system of (homotopy classes of) circuits which the process described in Steps 1-4 produces should be

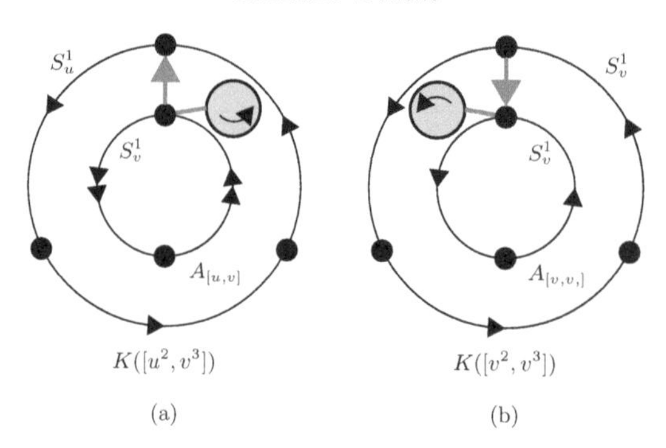

Figure 10.7 The base case of the induction for the proof of the Presentation Theorem.

called a *standard system* we need to look more closely at a method for obtaining a presentation for the fundamental group of a GBS group.

10.2 A very brief primer on GBS groups

As mentioned above, a *generalized Baumslag-Solitar (GBS) group* is the fundamental group of a graph of groups with infinite cyclic vertex and edge groups and for which all edge maps are monomorphisms. In our context, we are most interested in how this graph of groups description, i.e., a GBS graph, leads to a presentation for the associated GBS group.

As is the case for $K(\Gamma, \omega)$, Γ is a finite connected directed graph, loops and multiple edges allowed. The vertex set is $V(\Gamma)$ and the directed edge set is $E(\Gamma)$. For each directed edge e, assign an initial vertex e^- and terminal vertex e^+ indicating the direction on e. Associate to e, e^- and e^+ infinite cyclic groups $\langle u_e \rangle$, $\langle g_{e^-} \rangle$, and $\langle g_{e^+} \rangle$. Injective homomorphisms from $\langle u_e \rangle$, into $\langle g_{e^-} \rangle$, and $\langle g_{e^+} \rangle$ are determined by the assignments $u_e \mapsto g_{e^-}^{\omega^-(e)}$ and $u_e \mapsto g_{e^+}^{\omega^+(e)}$ where $\omega^\pm(e) \in \mathbb{Z}^* = \mathbb{Z} \setminus 0$. These data constitute a graph of groups (Γ, ω) which is completely determined by the finite connected graph Γ together with the *weight function* $\omega : E(\Gamma) \to \mathbb{Z}^* \times \mathbb{Z}^*$ whose values are written $\omega(e) = (\omega^-(e), \omega^+(e))$. The group associated to this pair (Γ, ω) is $G = \pi_1(\Gamma, \omega)$ and is called the *fundamental group* of (Γ, ω) or the *generalized Baumslag-Solitar (GBS) group associated to* (Γ, ω). Note, topologists need to be aware that $\pi_1(\Gamma, \omega)$ is not the fundamental group of the graph Γ – the fundamental group of a graph is always a free group and GBS groups are generally not free. Indeed, the only free group which is a GBS group is $\mathbb{Z}$, the free group on one generator.

To obtain a presentation for G, choose a maximal (spanning) subtree of Γ. The group G has a *vertex generator* g_v for each vertex v of Γ, an *edge generator* t_e for each edge $e \in E(\Gamma \setminus T)$, and two types of relations

1. *tree*, or *T-relations*, $g_{e^-}^{\omega^-(e)} = g_{e^+}^{\omega^+(e)}$ for $e \in E(T)$ and
2. *non-tree*, or *non-T relations*, $t_e^{-1} g_{e^-}^{\omega^-(e)} t = g_{e^+}^{\omega^+(e)}$ for $e \in E(\Gamma) \setminus E(T)$.

A presentation for a GBS group that has generators and relations as indicated above is called a *GBS presentation.* It is well known, see e.g., Serre [20], that $\pi_1(\Gamma, \omega)$ is independent of T. Our topological approach provides another proof of this.

To simplify the notation, when writing down presentations for the GBS group associated to the GBS graph (Γ, ω) we typically denote the vertex generator associated to the vertex u by the same symbol and the edge generator associated to a non-tree edge t by itself as well. With these conventions, setting $t = [v, v]$, the classical Baumslag-Solitar group $BS(m,n) = \pi_1([v^m, v^n]) \cong \langle v, t : t^{-1}v^m t = v^n \rangle$ and $\pi_1([u^m, v^n]) \equiv_{def} K(m,n) \cong \langle u, v : u^m = v^n \rangle$. When m and n are relatively prime, $K(m,n)$ is the fundamental group of the (m,n)-torus knot complement. We call $BS(m,n)$ and $K(m,n)$ *loop groups* and *edge groups* for obvious reasons. See Figure 10.1

A more complicated example appears in Figure 10.2. Call the GBS graph in the picture (Γ, ω). Let T be the spanning tree of Γ formed by the thick grey edges in the figure. The non-T edges of Γ are the black edges. Their labels are r, s, and t. We use the vertex labels themselves to represent the generators associated to each and the non-T edge labels to represent the corresponding edge generators. Then, $\pi_1(\Gamma, \omega)$ has a presentation with four vertex generators

$$x, u, v, w;$$

three non-T edge generators

$$r, s, t;$$

and six relations

$$u^5 = x^5, u^{10} = v^{14}, v^{25} = w^6,$$
$$r^{-1}u^8 r = u^{-1}, s^{-1}x^{-10}s = v^7, t^{-1}x^{10}t = v^5.$$

A directed weighted graph uniquely determines the presentation for its GBS group, however a GBS group does not uniquely determine the GBS graph. For instance, if $v \in V(\Gamma)$, then replacing the generator g_v by g_v^{-1} changes the signs of the weights adjacent to v on edges incident with v. If $e \in E(\Gamma \setminus T)$, then replacing t_e by t_e^{-1} reverses the direction on e. These actions define an equivalence relation on GBS graphs with groups defined by equivalent graphs being isomorphic. Note that every GBS graph is equivalent to one whose weights are positive on the edges of a spanning tree and which has at most one negative weight on each non-tree edge. Avoiding some of these technical issues are among the reasons for adopting the rigidity conditions for GBS complexes.

10.3 The Presentation Theorem

Combining the description of how to use a GBS graph to obtain a presentation for its associated GBS group and Step 4 in Section 10.1.3 gives an outline of the proof of our first theorem. For more details, see [2].

Theorem 10.3.1 (Presentation Theorem) *Let $K = K(\Gamma, \omega)$, T a maximal subtree of Γ and $\mathcal{S}(\Gamma, \omega, T)$ be a standard system for K. Then the homotopy classes of the vertex and edge circuits in the standard system (determined by T) give a GBS presentation for $\pi_1(K, x_0)$.*

The Presentation Theorem completes the justification for the designation of $\mathcal{S}(\Gamma, \omega)$ as a "standard system."

Example 10.3.2 For each of the following let the vertices be u or u, v. Let $x_0 = 1 \in S_u^1$ and, via another mild abuse of the notation, we let x_u, x_v denote both the circuits in the standard systems for the GBS complexes and their homotopy classes. Then the GBS edge graphs $[u^m, v^n]$ produce GBS presentations for fundamental groups

$$\pi_1(K([u^m, v^n]), x_0) = \langle x_u, x_v : x_u^m = x_m^n \rangle = K(m, n).$$

When the edge $e = [u, u]$ is a loop, the standard system for $K([u^m, u^n])$ gives the group presentation for the classical Baumslag-Solitar group

$$\pi_1(K([u^m, u^n]), x_0) = \langle x_u, t_e : t_e^{-1} x_u^m t_e = x_u^n \rangle.$$

See Figure 10.3 for picture of the $K([u^m, v^n])$ and $K([v^m, v^n])$ which have these GBS groups as fundamental groups. See Figures 10.5 and 10.7 for alternate ways of drawing these GBS complexes for the specific values

$m = 2$ and $n = 3$. One can read the given presentations for fundamental groups of $K([u^2, v^3])$ and $K([v^2, v^3])$ off the pictures in Figure 10.7.

In general, observe that for each edge $e \in E(\Gamma)$, there is the natural inclusion $\iota : [0,1]_e \to K = K(\Gamma, \omega)$ given by

$$s \mapsto 1 \times s \in A_e \subset K.$$

An induction, first on the number of edges in the maximal subtree T, then on the number of non-T edges, shows that ι embeds a copy of Γ onto $\cup\{a_e : e \in E(\Gamma)\}$. There is also a natural projection $\eta : K \to \Gamma$ given by $S^1_e \times s \mapsto s$. Consequently, by a similar induction, the GBS complex K is naturally Seifert fibred with each vertex circle an exceptional fibre and each circle $S^1_e \times s$, $e \in E(\Gamma)$, $s \in (0,1)_e$ an ordinary fibre.

Observe that each point on the exceptional fibre S^1_v, has a neighborhood homeomorphic to $(0,1) \times T_{n(v)}$ where $T_{n(v)}$ is an $n(v)$-od with $n(v)$ equal to the sum of the absolute values all weights adjacent to the vertex v on (Γ, ω). Thus, each vertex circle has a fibred neighborhood which is a locally trivial $n(v) - od$ bundle over S^1. Each such bundle is finitely-to-one, cyclically covered by the fibred product $S^1 \times T_{n(v)}$ where the Seifert fibration is that given by the product structure. An exercise in some elementary number theory gives that the minimum positive degree of such a covering is the least common multiple $\lambda(v)$ of the absolute values of all the weights adjacent to v in (Γ, ω). We collect some of these observations in the next proposition.

Proposition 10.3.3 *Let (Γ, ω) be given. Then $K = K(\Gamma, \omega)$ is a Seifert fibred 2-complex with base space a copy of Γ. In addition, if $\eta = \eta_K : K \to \Gamma$ is the projection of the Seifert fibration and $\iota : \Gamma \to K$ of Γ is the inclusion of Γ onto the standard copy of Γ in K, then $\eta \circ \iota = id_\Gamma$, i.e., ι is a section for the Seifert fibration.*

Remark 10.3.4 An immediate topological corollary of this proposition provides the connection between the GBS complex $K(\Gamma, \omega)$ and the standard 2-complex associated to a presentation for the GBS group $\pi_1(\Gamma, \omega)$, e.g, the one with a single vertex, one loop for each generator, and one 2-disk for each relation. Denote this second 2-complex by $K'(\Gamma, \omega)$. Then $K'(\Gamma, \omega)$ is obtained from the GBS complex $K(\Gamma, \omega)$ as follows. Assume that the GBS complex $K(\Gamma, \omega)$ and the presentation for the GBS group $\pi_1(\Gamma, \omega)$ are both built by choosing the same maximal subtree $T \subset \Gamma$. In $K(\Gamma, \omega)$, collapse this maximal subtree T to a point. The resulting 2-complex is the $K'(\Gamma, \omega)$ associated to the GBS presentation for $\pi_1(\Gamma, \omega)$ built using T. The proof of this is again an induction:

induct first on the number of edges in a spanning GBS tree, then on the number of non-tree edges in the GBS graph.

10.3.1 Simplicial structures on $K(\Gamma, \omega)$.

We will be using simplicial homology with $\mathbb{Z}$ coefficients in the homological calculations which follow. This section lays out the simplicial structure.

We henceforth denote the standard copy of Γ in K by the same symbol. The context should make clear what is intended. We assume that $K = K(\Gamma, \omega)$ is triangulated so that the following hold: (1) the set of 0-simplicies contains all vertices of Γ and each edge in Γ is a union of 1-simplicies; (2) for each edge e of Γ the annulus A_e in the annular decomposition is triangulated; (3) each standard straight arc a_e contained in the annulus A_e is a union of 0- and 1-simplices of the given triangulation and consequently, when T is a maximal subtree of Γ, v a vertex of Γ and e a non-T edge of Γ, it follows that the standard generators x_v and t_e is also a union of 0- and 1-simplices of the triangulation; (4) assume that the triangulation is fine enough that for each edge e there is a closed disk $D_e \subset int(A_e)$ that is a union of 2-simplices of the triangulation; (5) for each e choose some point $P_e \in int(D_e)$ and assume each P_e is a 0-simplex in the triangulation.

Let $\emptyset \neq S \subset E(\Gamma)$. Let $\mathcal{D}_S = \cup\{D_e : e \in S\}$ and let $\mathcal{C}_S$ be the closed complement of $\mathcal{D}_S$. That is, $\mathcal{C}_S \equiv_{def} K \setminus (\cup\{int(D_e) : e \in S\})$. Let Γ_S be the graph with vertex set $V(\Gamma_S) = V(\Gamma)$ and edge set $E(\Gamma_S) = E(\Gamma) \setminus E(S)$. We also write $\Gamma \setminus S$ for Γ_S. Let $K_S = K(\Gamma_S, \omega)$. Observe that $K_{E(\Gamma)} = K(\Gamma_{E(\Gamma)}, \omega) = \cup\{S^1_v : v \in V(\Gamma)\}$ is the disjoint union of the vertex circles.

10.4 Group theoretic consequences

The topology developed to this point has several immediate group theoretic consequence. We begin with a determination of the higher homology groups and the Euler characteristic $\chi(K(\Gamma, \omega))$ of the GBS complex. The third result gives a lower bound on the first Betti number of $K(\Gamma, \omega)$ and, consequently on the "complexity" of the fundamental group of $K(\Gamma, \omega)$. In fact, by [2], applying [10], $K(\Gamma, \omega)$ is an aspherical 2-complex. Consequently, by classical results, e.g., p. 344, [16], its homology groups are isomorphic to the homology groups of the GBS group $\pi_1(\Gamma, \omega)$. Thus, the results in this section are, in fact, results about the Euler characteristic and homology of the GBS group. Since $K(\Gamma, \omega)$ is 2-dimensional, the first lemma is immediate.

Lemma 10.4.1 *All the homology groups $H_i(K(\Gamma, \omega)) \cong H_i(\pi_1(\Gamma, \omega))$, $i \geq 3$, are trivial.*

Lemma 10.4.2 *The Euler characteristic $\chi(K(\Gamma, \omega)) = \chi(\pi_1(\Gamma, \omega))$ is 0.*

Proof By imposing a CW complex structure on $K(\Gamma, \omega)$, instead of a simplicial structure, there is an easy combinatorial proof that a GBS complex has Euler characteristic 0. The 0-cells are the vertices of Γ, the 1-cells are the edges of Γ together with one 1-cell for each vertex circle of $K(\Gamma, \omega)$, and there is one 2-cell in $K(\Gamma, \omega)$ for each each edge in Γ used to form the corresponding annulus in the annular decomposition for $K(\Gamma, \omega)$. Therefore

$$\begin{aligned}
\chi(K(\Gamma, \omega)) &= \sharp(\text{2-cells}) - \sharp(\text{1-cells}) + \sharp(\text{0-cells}) \\
&= |E(\Gamma)| - (|E(\Gamma)| + |V(\Gamma)|) + |V(\Gamma)| \\
&= 0.
\end{aligned}$$

□

Lemma 10.4.3 *If $\eta : K = K(\Gamma, \omega) \to \Gamma$ is the projection of the Seifert fibration for K to its base space Γ and $\iota : \Gamma \to K$ is the inclusion onto the standard copy of Γ in K, then the induced maps $\iota_* : \pi_1(\Gamma, x_0) \to \pi_1(K, x_0)$ and, for every i, $\iota_* : H_i(\Gamma, \mathbb{Z}) \to H_i(K, \mathbb{Z})$ are monomorphisms and the induced maps $\eta_* : \pi_1(K, x_0) \to \pi_1(\Gamma, x_0)$ and $\eta_* : H_i(\Gamma, \mathbb{Z}) \to H_i(K, \mathbb{Z})$ are epimorphisms. Consequently, the fundamental group of $K(\Gamma, \omega)$ is at least as complicated as is that of Γ and, in particular, $\beta_i(K) \geq \beta_i(\Gamma)$.*

Proof By Proposition 10.3.3, for $\iota : \Gamma \to K$ the inclusion of Γ onto the standard copy of Γ contained in K and $\eta : K \to \Gamma$ the projection of the Seifert fibration, $\eta \circ \iota$ is a homeomorphism. Hence the induced maps $(\eta \circ \iota)_* = \eta_* \circ \iota_* : \pi_1(\Gamma) \to \pi_1(\Gamma)$ and, for every i, $(\eta \circ \iota)_* = \eta_* \circ \iota_* : H_i(\Gamma) \to H_i(\Gamma)$ are isomorphisms. Consequently, each ι_* is a monomorphism and each η_* is an epimorphism of fundamental groups or of homology groups as needed. □

When Γ is a tree or contains a single loop, the above lemma does not have much content. However, as an immediate more interesting consequence, there is the following corollary.

Corollary 10.4.4 *If Γ is connected and contains $n \geq 2$ simple circuits, then $\pi_1(K(\Gamma,\omega),x_0)$ contains the free group F_n on $n \geq 2$ generators.*

Lemma 10.4.5 $\beta_2(K(\Gamma,\omega)) = \beta_1(K(\Gamma,\omega)) - 1.$

Proof By Lemma 10.4.2, the Euler characteristic of $K = K(\Gamma,\omega)$ is 0. It is also the alternating sum of the Betti numbers β_i of K. Since K is 2-dimensional, $\beta_i = 0$ for $i \geq 3$. Therefore, $\beta_0 - \beta_1 + \beta_2 = 0$. But K is connected. Therefore, $\beta_0 = 1$. It follows that $\beta_2 = \beta_1 - 1$. □

Corollary 10.4.6 *Let $K = K(\Gamma,\omega)$ and $\beta_i(K)$ be its ith Betti number. Let T be a maximal subtree of Γ. Then, $\beta_1(K) \geq |E(\Gamma \setminus T)|$. Consequently, $\beta_2(K) \geq |E(\Gamma \setminus T)| - 1$.*

Proof The graph Γ has the homotopy type of of a bouquet of $|E(\Gamma \setminus T)|$ loops. Therefore, its first homology is a direct sum of $|E(\Gamma \setminus T)|$ copies of $\mathbb{Z}$. Consequently, by Lemma 10.4.3,

$$\beta_1(K) \geq |E(\Gamma \setminus T)|.$$

Lemma 10.4.5 gives the indicated lower bound on $\beta_2(K)$. □

Additional, easy to obtain, results on the homology of a GBS complex can be deduced by looking at the upper end of the Mayer-Vietoris Sequence for the simplicial triple $(K, \mathcal{D}_S, \mathcal{C}_S)$. Observe that for each $e \in S$, the disk D_e strong deformation retracts to the point P_e. Therefore, $\mathcal{D}_S$ strong deformation retracts to the finite point-set $\{P_e : e \in S\}$. Also, for each $e \in S$, there is a strong deformation retract of $A_e \setminus int(D_e)$ onto the 1-complex $S^1_{e^-} \cup e \cup S^1_{e^+}$. This collection of strong deformation retracts induces a strong deformation retract of $\mathcal{C}_S$ onto the (connected) complex $K_S \cup \left(\cup\{S^1_{e^-} \cup e \cup S^1_{e^+} : e \in S\}\right) \subset K$.

Since all of K, $\mathcal{D}_S$, and $\mathcal{C}_S$ are 2-dimensional, all of the higher homology groups for all three are trivial. This produces the Mayer-Vietoris Sequence

$$\begin{aligned} 0 \to\ & H_2(\mathcal{C}_S \cap \mathcal{D}_S) \to H_2(\mathcal{C}_S) \oplus H_2(\mathcal{D}_S) \\ \to\ & H_2(K) \xrightarrow{\partial} H_1(\mathcal{C}_S \cap \mathcal{D}_S) \xrightarrow{\varphi} H_1(\mathcal{C}_S) \oplus H_1(\mathcal{D}_S) \to \cdots \end{aligned} \tag{10.1}$$

But, $\mathcal{D}_S \cap \mathcal{C}_S = \partial \mathcal{D}_S$ is a collection of $|S|$ circles and $\mathcal{D}_S$ has the homotopy type of a collection of $|S|$ points. Thus, all of $H_2(\mathcal{C}_S \cap \mathcal{D}_S)$, $H_2(\mathcal{D}_S)$, and

$H_1(\mathcal{D}_S)$ are trivial for dimensional reasons and $H_1(\mathcal{D}_S \cap \mathcal{C}_S) = \bigoplus_{i=1}^{|S|} \mathbb{Z}$. Therefore, equation (10.1) simplifies to

$$0 \to \quad H_2(\mathcal{C}_S) \oplus 0 \to H_2(K) \xrightarrow{\partial} \bigoplus_{i=1}^{|S|} \mathbb{Z} \xrightarrow{\varphi} H_1(\mathcal{C}_S) \oplus 0 \to \cdots \qquad (10.2)$$

Now set $S = E(\Gamma)$. In this case $K_S = K_{E(\Gamma)}$ is just the union of the vertex circles of K and $\mathcal{C}_S = \mathcal{C}_{E(\Gamma)}$ strong deformation retracts to the 1-complex

$$\cup\{S^1_{e^-} \cup e \cup S^1_{e^+} : e \in E(\Gamma)\} = \Gamma \cup \left(\cup\{S^1_v : v \in V(\Gamma)\}\right).$$

Therefore, again for dimensional reasons, equation (10.2), simplifies to

$$0 \to \quad H_2(K) \xrightarrow{\partial} \bigoplus_{i=1}^{|E(\Gamma)|} \mathbb{Z} \xrightarrow{\varphi} H_1(\mathcal{C}_{E(\Gamma)}) \oplus 0 \to \cdots \qquad (10.3)$$

From the exactness of equation (10.3), we easily obtain that $H_2(K)$ is free abelian. We have, in fact, the following:

Proposition 10.4.7 *If $K = K(\Gamma, \omega)$, then $H_2(\Gamma)$ is free abelian of rank $\beta_2(K) \leq |E(\Gamma)|$. Consequently, $|E(\Gamma \setminus T)| - 1 \leq \beta_2(K) \leq |E(\Gamma)|$ and $|E(\Gamma \setminus T)| \leq \beta_1(K) \leq |E(\Gamma)| + 1$.*

The proofs of the above results on the Euler characteristic and Betti numbers of the GBS complex and GBS group can be seen as steps towards understanding the structure of the first homology and Schur multiplier, i.e, the second homology group, of a GBS group. It is interesting to compare the proofs given here of these results to those given in Propostion 3.3 of [13] and [18]. What is most interesting about the above approach is how little effort is needed to determine that the Euler characteristic of the GBS group/space is 0.

The topological proofs of the full results on the homology require a detailed analysis of the Meyer-Vietoris sequence (10.1) for the triple $(K, \mathcal{C}_{E(\Gamma)}, \mathcal{D}_{E(\Gamma)})$. See [22]. The end results of this analysis give characterisations of the first homology and Schur multiplier (second homology) of the GBS group/space in terms of the graph Γ and a function $\epsilon(\Gamma, \omega)$ with values in $\{0, 1\}$ which encodes information about combinatorial properties of the weight function ω. The specific value of ϵ depends on the notion of *tree dependence*, [18], of the GBS graph where $\epsilon(\Gamma, \omega) = 1$ when (Γ, ω) is tree dependent and is 0 otherwise. Briefly, A GBS graph in which the graph is a tree is, by definition tree dependent. When (Γ, ω)

contains a circuit, one selects a maximal subtree T of Γ. Then, each non-tree edge e of Γ determines a circuit in Γ. Orient this circuit, then then compute the product of all the initial weights on the circuit and the product of all the terminal weights on the circuit. If these two products are equal, then the circuit determined by this non-tree edge e is tree dependent. If every non-tree edge in $E(\Gamma) \setminus E(T)$ is tree dependent, then (Γ, ω) is tree dependent. Note that a group theoretic version of tree dependence is that of *unimodularity* of a GBS group developed in [13].

The proofs deal with the tree and non-tree dependent cases and are the usual sorts of inductions: one inducts first on the number of edges in a spanning tree, then on the number of non-tree edges. There is an additional complication when considering the non-tree edges in that the tree and non-tree dependent cases must be dealt with separately. The final homological results are the following.

Theorem 10.4.8 (Robinson [18]) *Let (Γ, ω) be a GBS graph with associated GBS group G and $\epsilon = \epsilon(\Gamma, \omega)$. Then the Schur multiplier $H_2(G)$ is free abelian of rank $\beta_1(G) - 1 = |E(\Gamma)| - |V(\Gamma)| + \epsilon$.*

Theorem 10.4.9 (Levitt [14], Robinson [18]) *Let (Γ, ω) be a GBS graph with associated GBS group G and $\epsilon = \epsilon(\Gamma, \omega) \in \{0, 1\}$. Then*

$$\beta_1(G) = |E(\Gamma)| - |V(\Gamma)| + 1 + \epsilon.$$

10.5 Maps of GBS complexes

The pre-image of a circle in the range of a finite sheeted cover of a manifold or complex onto another is always a collection of circles in the domain space. However, when studying Seifert fibred spaces, it is reasonable to require that a stronger condition be satisfied: e.g, that each Seifert fibre in the domain be mapped onto a fiber in the range. Because the projection map of a GBS complex to its base space has a section, we impose a slightly stronger requirement in our definition of a map between GBS complexes. We will see that this stronger condition is satisfied in some specific cases, namely, for the examples of *geometric homomorphisms* introduced in [3].

Definition 10.5.1 Let (Γ_i, ω_i), $i = 1, 2$, be GBS graphs with spanning trees $T_i \subset \Gamma_i$. Let $j_i : \Gamma_i \subset K(\Gamma_i, \omega)$ be inclusions of Γ_i onto the standard copy of $\Gamma_i \subset K(\Gamma_i, \omega_i)$ and $\eta_i : K(\Gamma_i, \omega_i) \to \Gamma_i$ the projection of the Seifert fibration of $K(\Gamma_i, \omega_i)$ onto its base space Γ_i. A *map of GBS* *complexes* $\Phi : K(\Gamma_1, \omega_2) \to K(\Gamma_2, \omega_2)$ is a triple of continuous maps $(\varphi, \varphi', \varphi'')$ where

1. $\varphi : K(\Gamma_1, \omega_1) \to K(\Gamma_2, \omega_2)$ is a Seifert fibration preserving map of Seifert fibred spaces with the natural Seifert fibration on the GBS complexes;
2. φ' and φ'' map vertices of Γ_1 to vertices of Γ_2 and edges of Γ_1 to edges or vertices of Γ_2, in particular;
3. they map T_1-edges to T_2-edges or vertices in Γ_2 and non-T_1 edges to non-T_2 edges or vertices of Γ_2;
4. and which are such that the following diagram commutes.

$$\begin{array}{ccccc} \Gamma_1 & \xrightarrow{j_1} & K(\Gamma_1, \omega_1) & \xrightarrow{\eta_1} & \Gamma_1 \\ \downarrow{\varphi''} & & \downarrow{\varphi} & & \downarrow{\varphi'} \\ \Gamma_2 & \xrightarrow{j_2} & K(\Gamma_2, \omega_2) & \xrightarrow{\eta_2} & \Gamma_2 \end{array}$$

Typically, φ' and φ'' will map non-T_1 edges of Γ_1 to non-T_2 edges of Γ_2.

For GBS graphs (Γ_1, ω_1) and (Γ_2, ω_2) the geometric homomorphism $\psi : \pi_1(\Gamma_1, \omega_1) \to \pi_1(\Gamma_2, \omega_2)$ of [3] are induced by a map of the GBS graphs which define the groups. Our approach to these homomorphisms in the following is to use the topology to define a Seifert fibration respecting function φ on the domain GBS complex $K(\Gamma_1, \omega)$ and have its definition force the definition of the GBS graph (Γ_2, ω_2) which defines the range GBS complex. Because φ is continuous, it then follows, essentially for free, that there are induced group homomorphism $\varphi_* : \pi_1((\Gamma_1, \omega_1), x_0) \to \pi_1(K(\Gamma_2, \omega_2), y_0)$ and $\varphi_{*,i} : H_i(K(\Gamma_1, \omega_1)) \to H_i(K(\Gamma_2, \omega_2))$. We will see that the map φ_* of fundamental groups is a geometric map of fundamental groups of the associated GBS complexes. It is also the case that the rigidity conditions on the domain and range GBS complexes produce obvious definitions of the maps φ' and φ'' on the graphs: φ' turns out to be the restriction of φ to the standard copy of Γ_1 it contains. However, by analogy with what is done in [3], it is instructive to think of φ as being induced by the graph maps, φ' or φ''.

Recall that for each edge f of the GBS graph (Γ, ω), the annulus associated to f in $K(\Gamma, \omega)$ is $A_f = S^1_f \times [0, 1]_f$. We sometimes suppress the subscript on the circle and the interval in what follows. This should not cause any confusion.

Let (Γ, ω_1) be given. Let T_0 be a subtree of Γ contained in the spanning tree T of Γ. Note that T_0 is not required to be a spanning tree for Γ. Let $L(T_0)$ be the set of all leaves of T_0, that is, the set of all vertices of

T_0 incident with a single edge of T_0. A vertex $v \in V(T_0)$ is an *interior vertex of* T_0 if it is not a leaf of T_0. Note that when T_0 is a single vertex or a single edge, T_0 has no interior vertices. The set of all interior vertices of T_0 is denoted $In(T_0)$. A weight $\omega^{\pm}(e)$ is an *interior weight on* or *for* T_0 if e is an edge of T_0. A weight $\omega^{\pm}(e)$ is an *exterior weight for* T_0 if the respective vertex $e^{\pm}$ is a vertex of T_0, but $e \notin E(T_0)$.

10.5.1 The k-spin along a sub-tree T_0 of Γ

Assume that k is a non-zero integer which is relatively prime to all the weights on T_0. The *k-spin of* $K(\Gamma, \omega_1)$ *along* T_0 is a function φ defined on $K(\Gamma, \omega_1)$ which wraps each Seifert fiber $S^1 \times t$ of each annulus A_e for e an edge of T_0 to itself via a k-fold cyclic self cover, while for edges f of Γ which are not edges of T_0, each Seifert fiber $S^1 \times t$ of A_f which is not also a vertex circle associated to a vertex of T_0 is mapped to itself via the identity. Thus the domain and range graph used for a k-spin are the same graph. The above, together with the requirement that φ be continuous, then forces the weight function ω_2 to be defined as follows:

$$\omega_2(f) = \begin{cases} \omega_1(f), & \text{if } f \cap T_0 = \emptyset \text{ or } f \in E(T_0) \text{ and} \\ & \text{both } f^{\pm} \in V(T_0); \\ k \cdot \omega_1(f), & \text{if } f \notin E(T_0) \text{ and both } f^{\pm} \text{ are} \\ & \text{in } V(T_0); \\ (\omega_1^-(f), k \cdot \omega_1^+(f)), & \text{if } f^- \notin V(T_0) \text{ and } f^+ \in V(T_0); \\ (k \cdot \omega_1^-(f), \omega_1^+(f)), & \text{if } f^- \in V(T_0) \text{ and } f^+ \notin V(T_0). \end{cases}$$

Then, more formally, define $\varphi : K(\Gamma, \omega_1) \to K(\Gamma, \omega_2)$ via

$$\varphi(\exp(i \cdot \theta), t) = \begin{cases} (\exp(i \cdot k\theta), t), & \text{if } (\exp(i \cdot \theta), t) \in A_f \\ & \text{and } f \in E(T_0); \\ (\exp(i \cdot \theta), t), & \text{otherwise.} \end{cases}$$

Note that the first line in the above also says that for each vertex $v \in V(T_0)$, the vertex circle S_v^1 is wrapped k times around itself via $\exp(i \cdot \theta) \mapsto \exp(i \cdot k\theta)$. This follows because k is relatively prime to all the weights on T_0. The reader can confirm that φ is well defined and continuous.

The injections j_i, $i = 1, 2$, of Γ into the two GBS complexes and the projections η_i, $i = 1, 2$, of the Seifert fibrations of these GBS complexes onto the copies of Γ which are their base spaces are the ones defined earlier in this paper. This implies that the φ',φ'' forming the GBS map $\Phi = (\varphi, \varphi', \varphi'')$ are the obvious maps: in particular, φ' is the restriction of φ to the standard copy of Γ contained in $K(\Gamma, \omega_1)$ and, in fact, both φ' and φ'' are the identity. Since φ is continuous, the induced map $\varphi_* : \pi_1(K(\Gamma, \omega_1), x_0) \to \pi_1(K(\Gamma, \omega_2), x_0)$ is a well defined homomorphism of groups.

The definition of a geometric homomorphism of GBS groups requires specifying a collection of *parameters*, the so called *vertex-edge pair* (γ, δ) and *associated functions* (r, s). We illustrate this here, but leave it to the reader to determine them in the additional examples considered below. We introduce some notation to do so.

Let the copy of the homotopy class of the standard generator corresponding to vertex circle S^1_u in the domain GBS complex be denoted by g_u and that in the range be denoted by $\dot{g}_u$. Let t_e and $\dot{t}_e$, denote the (homotopy class of the) standard non-T edge generator corresponding to a non-T edge e in the domain and range, respectively. Then, it follows that

$$\begin{cases} \varphi_*(g_u) = \dot{g}_u^k, & \text{for each } u \in V(T_0); \\ \varphi_*(g_u) = \dot{g}_u, & \text{for each } u \in V(\Gamma) \setminus V(T_0); \\ \varphi_*(t_e) = \dot{t}_e, & \text{for each non-}T \text{ edge } e. \end{cases}$$

In the vertex-edge pair (γ, δ), where $\gamma : V(\Gamma) \to V(\Gamma)$ and $\delta : E(G\backslash T) \to E(\Gamma\backslash T)$, both γ and δ are the identity maps. For a vertex v of Γ, when v is not a vertex of T_0, the associated parameter is $r(v) = 1$, while $r(v) = k$ when v is a vertex of T_0. For each non-T edge, the associated parameter is $s(e) = 1$.

Definition 10.5.2 The above produces three functions; one, φ defined between GBS complexes; a second, φ_* defined between GBS groups; and an unnamed function mapping the GBS graph (Γ, ω_1) onto the GBS graph (Γ, ω_2). We expand the notion of k-spin and call each of theses maps of GBS objects a *k-spin* (of the appropriate GBS object) *along the sub-tree T_0 of Γ*.

Figure 10.8 contains an informative pictorial description of the k-spin.

It is easy to see that the topological k-spin $\varphi : K(\Gamma, \omega_1) \to K(\Gamma, \omega_2)$ is surjective. However, in general, the induced k-spin $\varphi_* : \pi_1(K(\Gamma, \omega_1), x_0) \to \pi_1(K(\Gamma, \omega), x_0)$ is an injective homomorphisms of fundamental groups. This is particularly easy to see in an important special case: when $\Gamma = T = T_0$, k is then relatively prime to all the weights on (Γ, ω). Consequently, the k-spin along T_0 is a k-fold cover of

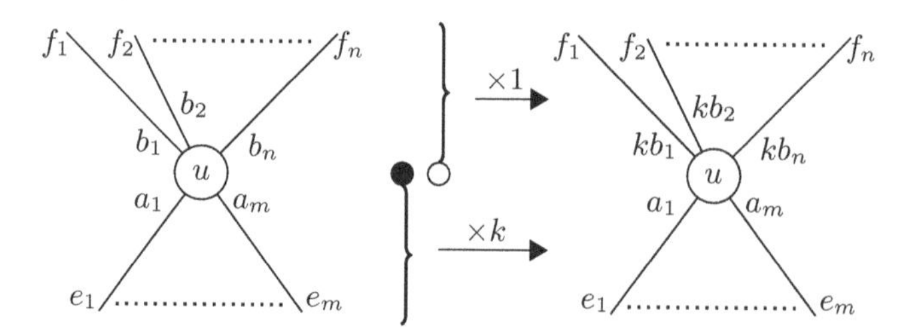

Figure 10.8 This figure shows the effect of a k-spin along the subtree T_0 of G at any vertex u of T_0. The edges e_i, $i = 1 \dots m$ are edges in T_0. By definition of k-spin, k is relatively prime to all the a_i. When $u = T_0$, then $m = 0$. When u is a leaf of T_0, then $m = 1$. Otherwise, $m \geq 2$. The edges f_j are edges of Γ which are not in T_0. There are $n \geq 0$ of them.

$K(\Gamma, \omega)$ by itself. But, when $k \geq 2$ and φ is a self cover, it follows that φ_* maps $\pi_1(K(\Gamma, \omega_1), x_0)$ isomorphically onto an index $k \geq 2$ subgroup of itself. This provides an easy topological proof that when Γ is a finite tree, $\pi_1(\Gamma, \omega) \cong \pi_1(K(\Gamma, \omega), x_0)$ is non-cohopfian.

10.5.2 Half pinch maps

Let $(e, \omega_1(e))$ be a directed weighted edge in (Γ_1, ω_1). Let d be a divisor of $\omega^-(e)$ and write $(e, \omega(e)) = [u^{dm}, v^n]$. It is possible that $u = v$, i.e., that e is a loop. See Figure 10.9 for pictures of the maps discussed in this section.

The maps under consideration in this section map each Seifert fiber $S^1 \times t$ contained in the interior of the annulus A_e in $K(\Gamma, \omega_1)$ to itself via a d-fold self cover and map each Seifert fiber contained in the complement $K(\Gamma, \omega_1) \setminus A_e$ to itself via the identity. These conditions, along with the condition that φ be continuous, impose restrictions on how φ is defined on the vertex circle(s) S^1_u (and S^1_v, when $u \neq v$) and on the weight function $\omega_2 : \Gamma \to \mathbb{Z}^* \times \mathbb{Z}^*$.

First, assume $(e, \omega(e)) = [u^{dm}, v^n]$, $e^- = u \neq v = e^+$. Define ω_2 and

φ as follows:

$$\omega_2(f) = \begin{cases} (m, \frac{n}{gcd(d,n)}), & \text{if } f = e = [u,v]; \\ \frac{n}{gcd(d,n)} \cdot \omega_1(f) & \text{if } f \neq e \text{ and } f^-, f^+ \in \{u,v\}; \\ (\omega_1^-(f), \frac{\omega^+(f)\cdot d}{gcd(d,n)}), & \text{if } f^- \notin \{u,v\} \text{ and } f^+ = v; \\ (\frac{\omega_1^-(f)\cdot d}{\gcd(d,n)}, \omega^+(f)), & \text{if } f^- = v \text{ and } f^+ \notin \{u,v\}. \\ \omega_1(f), & \text{if } f \cap e = \emptyset. \end{cases}$$

$$\varphi(\exp(i\theta), t) = \begin{cases} (\exp(i \cdot \theta d), t), & \text{if } (\exp(i\theta), t) \in A_e \\ & \text{and } t \neq 0, 1; \\ (\exp(i \cdot \theta \frac{d}{gcd(d,n)}), t), & \text{if } (\exp(i\theta), t) \in S^1_v, \text{ that is, if} \\ & t \in A_e \text{ and } t = 1; \\ (\exp(i\theta), t), & \text{if } (\exp(i\theta), t) \in S^1_u, \text{ that is, if} \\ & t \in A_e \text{ and } t = 0; \\ (\exp(i\theta), t), & \text{if } (\exp(i\theta), t) \in K(\Gamma, \omega_1) \setminus A_e. \end{cases}$$

The reader can verify that φ is continuous. We call the resulting GBS graph map, GBS complex map, and GBS group map a *d-half pinch along e in the positive direction.*

When $(e, \omega) = [u^{dm}, u^n]$, that is, when the edge e is a loop, one may arrive at definitions of ω_2 and φ in a manner analogous to the above. However, there is second way to do so. This method is useful in other contexts as well, so we illustrate it here.

Subdivide the directed weighted edge $(e, \omega_1(e)) = [u^{dm}, u^n]$ in (Γ, ω) by inserting the midpoint μ of $e = [u, u]$ and write $[u^{dm}, u^n] = [u^{dm}, \mu^1] \oplus [\mu^1, u^n]$. Call the resulting GBS graph $(\Gamma^{sub}, \omega_1^{sub})$. This divides A_e into two annuli: $A_e = A_{[u^{dm}, \mu^1]} \bigcup_{S^1_\mu} A_{[\mu^1, u^n]}$ and produces the GBS complex $K(\Gamma_1^{sub}, \omega_1^{sub})$ which is homeomorphic to $K(\Gamma, \omega_1)$ via a Seifert fibration preserving homeomorphism. Also, denote the subdivision operation on $K(\Gamma, \omega_1)$ by $sub : K(\Gamma, \omega_1) \to K(\Gamma^{sub}, \omega_1^{sub})$.

We are interested in the associated sequence of GBS graph maps along $[u^{dm}, u^n] = [u^{dm}, \mu^1] \oplus [\mu^1, u^n]$ shown below:

$$[u^{dm}, u^n] \xrightarrow{sub} [u^{dm}, \mu^1] \oplus [\mu^1, u^n] \xrightarrow{p_1} [u^m, \mu^1] \oplus [\mu^d, u^n] -$$
$$\xrightarrow{p_2} [u^{dm/\gcd(d,n)}, \mu^1] \oplus [\mu^1, u^{n/\gcd(d,n)}]$$
$$\xrightarrow{sub^{-1}} [u^{dm/\gcd(d,n)}, u^{n/\gcd(d,n)}].$$

where p_1 is the d-half pinch along $[u^{dm}, \mu^1]$ in the positive direction, p_2 is the d-half pinch along $[\mu^d, u^n]$ in the positive direction, and sub^{-1} is the "forgetful function" which removes the midpoint μ and changes the two edges $[u, \mu] \oplus [\mu, u]$ back to the single edge $[u, u]$. These changes along the edge(s) $[u^{dm}, u^n] = [u^{dm}, \mu^1] \oplus [\mu^1, u^n]$ produce a sequence of GBS graphs and maps between them. Using the same symbols to denote the maps, this sequence is

$$(\Gamma, \omega_1) \xrightarrow{sub} (\Gamma^{sub}, \omega_1^{sub}) \xrightarrow{p_1} (\Gamma^{sub}, (\omega_1^{sub})')$$
$$\xrightarrow{p_2} (\Gamma^{sub}, (\omega_2^{sub})') \xrightarrow{sub^{-1}} (\Gamma, \omega_2).$$

Thus the GBS graph d-half pinch around the loop $e = [u^d m, u^n]$ in the positive direction is

$$sub^{-1} \circ p_2 \circ p_1 \circ sub : (\Gamma_1, \omega_1) \to (\Gamma_2, \omega_2).$$

There are the associated d-half pinches of the associated the GBS complexes and groups.

Note in particular that when $\gcd(d, n) = 1$, this process does not change the weight $\omega^-(e) = dm$. Specializing to the case where the graph is a single loop, when $d \geq 2$ is relatively prime to n and $dm, n \geq 2$, a topological d-half pinch $\varphi : K([u^{dm}, u^n]) \to K([u^{dm}, u^n])$ along the edge $e = [u^{dm} u^n]$ produces a surjective (geometric) group homomorphism $\varphi_* : \pi_1([u^{dm}, u^n]) \to \pi_1([u^{dm}, u^n])$ of GBS groups with non-trivial kernel. See Theorem 2.5, [3]. This gives a topological proof of the classical result of [1] that when $\gcd(dm, n) = 1$ and $d, n \geq 2$ the Baumslag-Solitar groups $\pi_1([u^{dm}, u^n])$ are non-hopfian.

The above process of subdividing a weighted loop $[u^{dm}, u^n]$ in a GBS graph (Γ, ω) to obtain the d-half pinch along a loop by inserting its midpoint μ and making both weights adjacent to μ equal ± 1 is dependent upon the fact that this operation does not change the homeomorphism (isomorphism) class of the associated GBS complex (GBS group). There are other types of subdivision operations on an edge of a GBS graph which also do not change the homeomorphism (isomorphism) class of the associated GBS complex (group). One such,

$$[u^m, v^n] \mapsto [u^m, \mu^{-1}] \oplus [\mu^1, v^{-n}],$$

is referred to below.

The d-half pinches of GBS graphs, complexes, and groups along an edge e in the *negative* direction can be defined in the obvious analogous manner.

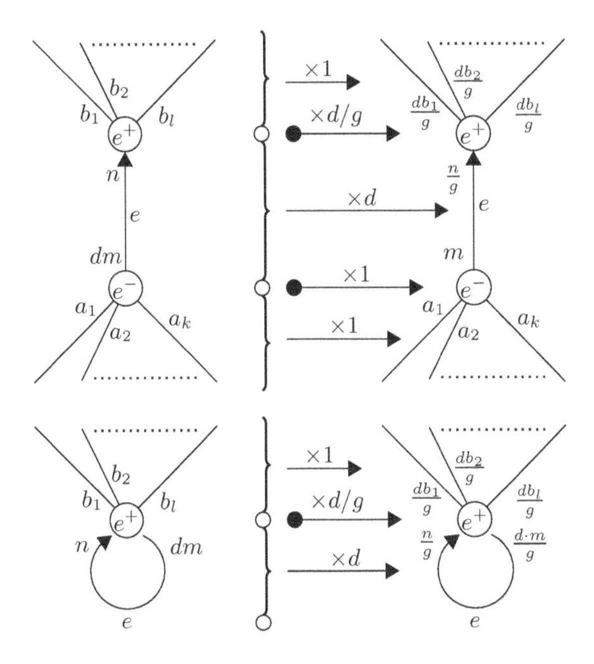

Figure 10.9 Let $g = gcd(d, n)$. The d-half pinch along the edge e in the positive direction for the non-loop and loop cases.

10.5.3 Pinch maps

When $(e, \omega_1(e)) = [u^{dm}, v^{dn}]$ ($u = v$ allowed), a d-half pinch along e in either direction produces the *d-pinch of GBS groups* defined in [3]. As is the case for the d-half pinch, we call the associated topological and group theoretic maps *d-pinches along the edge* e as well. The topological d-pinch $\varphi : K(\Gamma, \omega_1) \to K(\Gamma, \omega_2)$ wraps each Seifert fiber $S^1 \times t$ in the interior of the annulus A_e onto its image via a d-fold cover. All other Seifert fibers are mapped by φ to their image in the range GBS complex via the identity. The topological d-pinch along an edge is clearly surjective. The induced geometric homomorphisms φ_* between their fundamental groups is also an epimorphism. The reader is encouraged to write out the details for the topological and group theoretic GBS maps or see [3] for the group theoretic details.

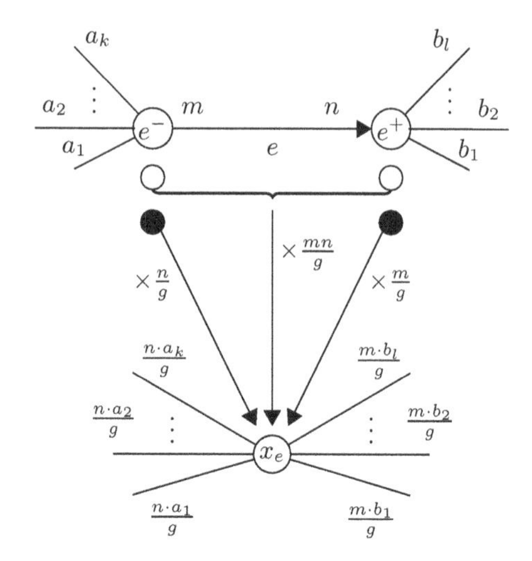

Figure 10.10 Let $g = gcd(m, n)$. The above shows the contraction map along the edge e.

10.5.4 Contraction maps

Let (Γ_1, ω_1) be a GBS graph. Let $e \in E(\Gamma_1)$ be an edge contained in a spanning tree T_1 of Γ_1 and write $(e, \omega(e)) = [u^m, v^n]$, $u \neq v$. We want a continuous map φ defined on $K(\Gamma_1, \omega_1)$ which satisfies the following conditions: (1) each Seifert fiber in the interior of the annulus A_e maps to the same circle, call is $S^1_{x_e}$, via an $lcm(m, n)$-fold covering map, and (2) each Seifert fiber in the complement $K(\Gamma_1, \omega_1) \setminus A_e$ maps to itself via the identity. These two conditions and the requirement that φ be continuous force φ to map the vertex circle S^1_u to $S^1_{x_e}$ via a $\frac{n}{gcd(m,n)}$-fold cover and S^1_v to map to $S^1_{x_e}$ via a $\frac{m}{gcd(m,n)}$-fold cover. Then the range graph (Γ_2, ω_2) is a GBS graph with vertex set $V(\Gamma_2) = (V(\Gamma_1) \setminus \{u, v\}) \cup \{x_e\}$ where $x_e = x_{[u,v]}$ is a vertex not appearing elsewhere in Γ_1, and the rest of Γ_1 is unchanged. The idea is that $e = [u, v]$ is contracted to the vertex x_e. For each weighted edge $(f, \omega_2(f)) \in (\Gamma_2, \omega_2)$, define the weight function

ω_2 as follows:

$$\omega(f) = \begin{cases} \omega_1(f), & \text{if } f \cap e = \emptyset; \\ (\omega_1^-(f), \frac{\omega_1^+(f)\cdot n}{gcd(m,n)}), & \text{if } f \cap e = f^+ = u; \\ (\frac{\omega_1^-(f)\cdot n}{gcd(m,n)}, \omega_1(f)), & \text{if } f \cap e = f^- = u; \\ (\omega_1^-(f), \frac{\omega_1^+(f)\cdot m}{gcd(m,n)}), & \text{if } f \cap e = f^+ = v; \\ (\frac{\omega_1^-(f)\cdot m}{gcd(m,n)}, \omega_1(f)), & \text{if } f \cap e = f^- = v. \end{cases}$$

Define the Seifert fibration preserving map $\varphi : K(\Gamma_1, \omega_1) \to K(\Gamma_2, \omega_2)$ as follows:

$$\varphi(\exp(i\cdot\theta), t) = \begin{cases} (\exp(i\cdot\theta), t), & \text{if } (\exp(i\cdot\theta), t) \notin A_e; \\ \exp(i\cdot lcm(m,n)\theta) \in S^1_{x_e}, & \\ \qquad \text{if } (\exp(i\cdot\theta), t) \in A_e \text{ and } t \neq 0, 1; & \\ \exp(i\cdot \frac{n\theta}{gcd(m,n)}) \in S^1_{x_e}, & \text{if } (\exp(i\cdot\theta), t) \in S^1_u; \\ \exp(i\cdot \frac{m\theta}{gcd(m,n)}) \in S^1_{x_e}, & \text{if } (\exp(i\cdot\theta), t) \in S^1_v. \end{cases}$$

The reader can again confirm that this definition produces a well defined continuous map of $K(\Gamma_1, \omega_1)$ onto $K(\Gamma_2, \omega_2)$ and that φ' and φ'' can be chosen to be the obvious maps. There is the induced homomorphism of fundamental groups $\varphi_* : \pi_1(K(\Gamma_1, \omega_1), x_0) \to \pi_1(K(\Gamma_2, \omega_2), y_0)$. Also, for the maximal subtree of T_1 of Γ_1 containing e, there is a maximal subtree T_2 of Γ_2 obtained by contracting e in T_1 to the vertex x_e. The standard systems of generators for the fundamental groups of the two GBS complexes relative to these two maximal subtrees can be used to produce presentations for the fundamental groups $\pi_1(K(\Gamma_i, \omega_i))$. The induced map φ_* on fundamental groups can be seen to be the type of geometric map relative to these maximal subtrees known as a *contraction map along the edge* e. The group theoretic proof in [3] shows it is an epimorphism of GBS groups. The map $\Phi = (\varphi, \varphi', \varphi'')$ and the map of GBS graphs $(\Gamma_1, \omega_1) \to (\Gamma_2, \omega_2)$ are, respectively, *contraction maps* of the appropriate objects.

The notion of a contraction map along an edge e which forms a loop is also defined in [3]. Depending on the direction on e, these maps along the loop e are either $\omega^+(e)$-half pinches in the negative direction along e or $\omega^-(e)$-half pinches in the positive direction along e. Thinking about these maps as half pinches in the appropriate direction along the looped edges better captures the topology of the situation.

10.5.5 Hopf maps

Let $(\Gamma, \omega) = \bigoplus_{i=1}^{k} [v^{m_i}, v^{n_i}]_i$ be a bouquet of loops with edges $e_i = [v, v]_i$ and $gcd(m_i, n_i) = 1$. The Hopf maps defined on the GBS objects (Γ, ω), $K(\Gamma, \omega)$, and $\pi_1(\Gamma, \omega) \cong \pi_1(K(\Gamma, \omega), x_0)$ are self maps of the respective objects. For the GBS groups, the spanning tree is $T = v$ and the desired effect of the geometric homomorphism on the GBS group, see [3], is to have each non-T edge generator mapped to itself and to have the single vertex generator g_v map to the $lcm\{\omega^-(e_i) : i = 1, \ldots, k\}$ power of itself.

This result can be obtained by defining a composition of a sequence of half pinches and pinches on various intermediary objects. The description of the topological Hopf map follows.

Assume $m_i \geq n_i \geq 1$ and that $gcd(m_i, n_i) = 1$. Let

$$m'_1 = m_1,\ m'_2 = \frac{m_2}{\gcd(m_1, m_2)},\ m'_3 = \frac{m_3}{\gcd(\text{lcm}(m_1, m_2), m_3)},$$
$$\ldots, m'_k = \frac{m_k}{\gcd(\text{lcm}(m_1, m_2, \ldots, m_{k-1}), m_k)}.$$

An easy induction shows that $lcm(m_1, m_2, \ldots, m_k) = m'_1 m'_2 \cdots m'_k$. Note that m'_i is relatively prime to n_i. Let (Γ_i, ω_i) be the graph Γ with weight function ω_i defined as follows:

$$\omega_i(e_j) = \begin{cases} (m_i, n_i), & \text{if } i = j; \\ m'_i \cdot \omega(e_j), & \text{if } i \neq j. \end{cases}$$

Let $h_i : K(\Gamma, \omega) \to K(\Gamma_i, \omega_i)$ be the m'_i-half pinch along $e_i = [v, v]_i$ in the positive direction. Then for each i, let $p_i : K(\Gamma_i, \omega_i) \to K(\Gamma, \omega)$ be the composition of the $(k-1)$ m'_i-pinches along the edges $e_j = [v, v]_j$, for $j \neq i$. The reader should provide the graph theoretic descriptions of the various m'_i-pinches and m'_i-half pinches. Note that p_i is independent of the order in which the m'_i-pinches are done. Define $\varphi : K(\Gamma, \omega) \to K(\Gamma, \omega)$ to be the composition

$$\varphi = p_k \circ h_k \circ p_{k-1} \circ h_{k-1} \circ \cdots \circ p_1 \circ h_1.$$

The effect of φ on the vertex circle S^1_v is to map it to itself via a $lcm(m_1, m_2, \ldots, m_k)$-fold self cover. The Seifert fibres $S^1_{e_i} \times t$, $t \in (0,1)_{e_i}$, also map to themselves via $m'_1 m'_2 \cdots m'_k = lcm(m_1 m_2 \cdots m_k)$-fold self covers. However, note that φ, in its entirety, is not a covering space projection. The proof in [3] shows that the induced map φ_* on $\pi_1(K(\Gamma, \omega), x_0)$ is a geometric epimorphism. It has non-trivial kernel when at least one m_i has $|m_i| \geq 2$ and (Γ, ω) is not the single loop $[v^{m_1}, v^{\pm 1}]$, that is, when $\pi_1(\Gamma, \omega)$ is not one of the soluble GBS groups.

10.5.6 Bouquet folding

The definition of the bouquet folding geometric homomorphism given in [3] requires that certain non-T edge generators map to the the trivial element in range GBS group. We see that this effect can be achieved topologically as follows.

Let (Γ, ω) be a bouquet of loops $(\Gamma, \omega) = \bigoplus_{i=1}^{k} [v^1, v^{\pm 1}]_i$ in which all weights are ± 1. When all weights are $+1$, $K(\Gamma, \omega)$ is a union of Seifert fibred tori $\tau_i = S_i^1 \times [0,1]_i/\{(t,0) \sim (t,1)\}$ with each of the k fibers $S_i^1 \times 1 \subset \tau_i$ attached to S_v^1 via the identity. This union can clearly be mapped onto a single torus $\tau = K([u^1, u^1] = S^1 \times [0,1]/\{(t,0) \sim (t,1)\}$ by mapping each τ_i to τ via the identity. Note that this maps Seifert fibres to Seifert fibres. The induced map of fundamental groups is clearly an epimorphism.

Thus, the more interesting case is the case where at least one edge of the bouquet (Γ, ω) is a $[v^1, v^{-1}]$. In this case, the bouquet folding map maps $K(\Gamma, \omega)$ to the Klein bottle $K([v^1, v^{-1}])$. Each Klein bottle $K([v^1, v^{-1}]_j)$ contained in $K(\Gamma, \omega)$ is mapped to $K([v^1, v^{-1}])$ via the identity. Consequently, the induced map on fundamental groups is again an epimorphism. However, for each weighted edge $[v^1, v^1]_i$ each fiber $S_i^1 \times t$, $0 \le t \le 1$ in the associated torus $K([v^1, v^1]_i)$ is mapped to the the vertex circle S_v^1 via $(\exp(i\theta), t) \mapsto \exp(i\theta)$. This maps the copy of the standard loop $[v, v]_i$ in the torus $K([v^1, v^1]_i) \subset K(\Gamma, \omega)$ to the point $1 \in S_u^1$. Consequently, at the fundamental group level, the non-T edge generator g_{e_i} is mapped to the trivial element in the range group.

Note when (Γ, ω) is a bouquet of $k \ge$ loops on which all weights are ± 1 and at least one edge has one weight $+1$ and the other a -1, there is an alternate natural map of GBS complex which produces an epimorphism of the $\pi_1(K(\Gamma, \omega), x_0)$ onto the fundamental group of the Klein bottle. Again, map each Klein bottle in $K(\Gamma, \omega)$ to the range Klein bottle $K([v^1, v^{-1}])$ via the identity. Then for each torus $K([u^1, u^1]_j) \subset K(\Gamma, \omega)$, apply the subdivision operation and write $K([u^1, u^1]_j) = K([u^1, \mu^{-1}]_j \oplus [\mu^1, u^{-1}]_j)$. Let $t_{[u,u]_j}$ be the homotopy class of the non-tree edge generator $[u^1, \mu^{-1}]_j \oplus [u^1, u^{-1}]_j)$ contained in $\pi_1(K([u^1, u^1]_j, x_0)) = \pi_1(K([u^1, \mu^{-1}]_j \oplus [\mu^1, u^{-1}]_j), x_0)$. Then, there is a Seifert fibration respecting double cover

$$K([u^1, u^1]_j) = K([u^1, \mu^{-1}]_j \oplus [\mu^1, u^{-1}]_j) \longrightarrow K([v^1, v^{-1}])$$

mapping $t_{[u,u]}$ onto $(t_{[v,v]})^2$ in $\pi_1(K([u^1, u^{-1}], y_0)$ where $t_{[v,v]}$ is the homotopy class of the non-tree edge generator corresponding to the loop

$[v, v]$ in the range Klein bottle. The resulting function defined on these GBS complexes is not a map of GBS complexes in the sense defined above and neither is the induced group map geometric in the sense of [3], but perhaps they should be. These are problems worth additional consideration.

10.6 Conclusion

The above suggests a number of questions. Two of the most interesting follow. First, to what degree can other results about the generalized Baumslag-Solitar groups be realized by these topological techniques. Second, let H be a non-cohopfian group with a proper finite index subgroup H_0 isomorphic to H. Let G be a group which has a graph of groups description in which all vertex and edge groups are copies of H and all edge maps are monomorphisms. Is G also non-cohopfian?

Acknowledgments

Thanks to Alberto Delegado and Derek Robinson for their many interesting conversations through the years about this mathematics and many other topics. Thanks to the members of the Topology Seminar, Department of Mathematics, Faculty of Science and Mathematics, University of Split for the opportunities they provided me for more recent interesting discussions of these topics. Thanks to the organizers of Groups St Andrews 2022 in Newcastle. Lastly, thanks to the referee and editors for their help preparing the final version of this paper.

References

[1] Baumslag, G., and Solitar, D. 1962. Some two-generator one-relator non-Hopfian groups. *Bull. Amer. Math. Soc.*, **68**, 199–201.

[2] Bedenikovic, T., Delgado, A., and Timm, M. 2006. A classification of 2-complexes with nontrivial self-covers. *Topology and Its Applications*, **153**, 2073–2091.

[3] Delgado, A., D.J.S., Robinson, and Timm, M. 2011. Generalized Baumslag-Solitar groups and geometric homomorphisms. *J. Pure and Applied Algebra*, **215**, 398–410.

[4] Delgado, A., Robinson, D.J.S, and Timm, M. 2014. Generalised Baumslag-Solitar Graphs with Soluble Fundamental Groups. *Algebra Colloquium*, **21**, 53–58.

[5] Delgado, A., Robinson, D.J.S., and Timm, M. 2017. Cyclic normal subgroups of generalized Baumslag-Solitar groups. *Communications in Algebra*, 1801–1818.

[6] Delgado, A., Robinson, D.J.S., and Timm, M. 2018. 3 manifolds and generalized Baumslag-Solitar groups. *Communications in Analysis and Geometry*, **26**, 571–584.

[7] Forester, M. 2003. On uniqueness of JSJ decompositions of finitely generated groups. *Comment. Math. Helv.*, **78**, 740–751.

[8] Heil, W.H. 1975. Some finetely presented non-3-manifold groups. *PAMS*, **53**, 497–500.

[9] Hempel, J. 1976. 3-Manifolds. Ann. of Math. Studies, vol. 86. Princeton, New Jersey: Princeton University Press.

[10] Howie, J. 1982. On locally indicable groups. *Math. Z.*, **180**, 445–461.

[11] Kropholler, P.H. 1990. Baumslag-Solitar groups and some other groups of cohomological dimension 2. *Commentarii Hatematicic Helvetici*, **65**, 445–461.

[12] Levit, G. 2015. Generalized Baumslag-Solitar groups: rank and finite index subgroups. *Ann. Inst. Fourier (Grenoble)*, **65**, 725–762.

[13] Levitt, G. 2005. Characteriszing rigid simplicial actions on trees. Pages 27–33 of: *Geometric Methods in Group Theory*. Contemporay Mathematics, vol. 372. Providence, Rhode Island: Amer. Math. Soc.

[14] Levitt, G. 2007. On the automorphism group of generalized Baumslag-Solitar groups. *Geom. Top.*, **11**, 473–515.

[15] Levitt, G. 2015. Quotients and subgroups of Baumslag-Solitar groups. *J. Group Theory*, **18**, 1–43.

[16] Mac Lane, S. 1963. *Homology*. Springer-Verlag.

[17] Margolis, A. 2022. *Model geometries of finitely generated groups.* arXiv:2207.10509v1

[18] Robinson, D.J.S. 2011. The Schur multiplier of a generalized Baumslag-Solitar group. *Rend. Se. Mat. Univ. Padova*, **125**, 207–215.

[19] Robinson, D.J.S. 2015. Generalized Baumslag-Solitar groups: a survey of recent progress. Page 457–468 of: Campbell, C. M., Quick, M. R., Robertson, E. F., and Roney-Dougal, C. M. (eds), *Groups St Andrews 2013*. London Mathematical Society Lecture Note Series. Cambridge University Press.

[20] Serre, J.P. 2003. *Trees*. Monographs on Mathematics. Berlin: Springer-Verlag. Translated from the French original by John Stillwell. Corrected 2nd printing of the 1980 English translation.

[21] Sokolov, E.V. 2021. Certain residual properties of the generalized Bausmslag-Solitar groups. *Journal of Algebra*, **582**(September), 1–25. https://doi.org/10.1016/j.jalgebra.2021.05.001.

[22] Timm, M. 2022. *A topologist's interactions with Derek J. S. Robinson and his mathematics*. pre-print.

[23] Whyte, K. 2001. The large scale geometry of the higher Baumslag-Solitar groups. *Geometric and Functional Analysis*, **11**, 1327–1343. doi 10.1007/s00039-001-8232-6.
[24] Yoshikawa, K. 1988. A ribbon knot group which has no free base. *Proceedings of the American Mathematical Society*, **102**, 1065–1070.